GEOMORPHOLOGY AND RIVER MANAGEMENT

To our families

"Every tool carries with it the spirit by which it has been created."
Werner Karl Heisenberg

Geomorphology and River Management

Applications of the River Styles Framework

Gary J. Brierley and Kirstie A. Fryirs

Blackwell Publishing

BLACKWELL PUBLISHING
350 Main Street, Malden, MA 02148-5020, USA
9600 Garsington Road, Oxford OX4 2DQ, UK
550 Swanston Street, Carlton, Victoria 3053, Australia

First published 2005 by Blackwell Science Ltd

Library of Congress Cataloging-in-Publication Data

Brierley, Gary J.
  Geomorphology and river management : applications of the river styles framework / Gary J. Brierley and Kirstie A. Fryirs.
     p. cm.
  Includes bibliographical references and index.
  ISBN 1-4051-1516-5 (pbk. : alk. paper) 1. Rivers. 2. Stream ecology. 3. Watershed management.
4. Geomorphology. I. Fryirs, Kirstie A. II. Title.

GB1203.2.B755 2005
551.48'3 – dc22

2004011686

ISBN-13: 978-1-4051-1516-2 (pbk. : alk. paper)

For further information on
Blackwell Publishing, visit our website:
www.blackwellpublishing.com

Contents

Preface: our personal, Australian, perspectives

Every country has its own landscape which deposits itself in layers on the consciousness of its citizens, thereby canceling the exclusive claims made by all other landscapes.

Murray Bail, 1998, p. 23

Any book reflects the personal histories and associated geographic and cultural values of its authors. In a number of ways it is increasingly difficult for us to separate our scientific perspective on rivers and their management from an emotional and aesthetic bond that has developed in our work. Working within a conservation ethos, we promote a positive sense of what can be achieved through effective implementation of rehabilitation practices.

Perspectives conveyed in this book undoubtedly reflect, to some degree, the distinctive nature of the Australian landscape and biota, the recent yet profound nature of disturbance associated with colonial settlement, and community involvement in river conservation and rehabilitation practices. The long and slow landscape evolution of the Australia landmass has resulted in rivers with a distinctive character and behavior, driven by factors such as the relative tectonic stability and topographic setting of the continent, pronounced discharge variability, and limited material availability. Remarkably few river systems comprise truly alluvial, self-adjusting streams. Many contemporary river forms and processes have been influenced by antecedent landscape controls, such as the nature of the bedrock or older alluvial materials over which they flow, and generally limited relief. Given the nature of the environmental setting, it is scarcely surprising that the Australian landscape is characterized by an array of river forms and processes that is seldom observed elsewhere. Across much of the continent, human disturbance has left a profound "recent" imprint on this largely ancient landscape, the consequences of which vary markedly from system to system (e.g. Rutherfurd, 2000).

Along with its unique environmental setting and history of human disturbance, a distinctive approach to natural resources management that is characterized by extensive on-the-ground involvement of community groups has developed in Australia. Rehabilitation strategies implemented through Catchment Management Committees (or Authorities/Trusts), Landcare Groups, Rivercare Groups, etc. have been complemented by core support through Federal and State Government programs. Adoption of participatory rather than regulatory approaches to river management has presented significant opportunity to incorporate research ideas into management practice.

Uptake of rehabilitation programs that strive to heal river systems in Australia has been driven by extensive involvement and leadership from the small group of professional geomorphologists in the country. A significant collection of tools and techniques for river rehabilitation has been provided, including the National Stream Rehabilitation Guide (Rutherfurd et al., 2000), the National Stream Restoration Framework (Koehn et al., 2001), and proceedings from various Stream Management Conferences (Rutherfurd and Walker, 1996; Rutherfurd and Bartley, 1999; Rutherfurd et al., 2001b). Our efforts in writing this book have been aided enormously by this invigorating set of research products, and the dedication of various river practitioners who have "made this happen."

In our quest to develop a logical set of principles with which to interpret the diversity and complexity of the real world, we have tried to communicate our understanding in as simple a way as possible. Duplications, inaccuracies, and inconsistencies may have arisen in cross-disciplinary use of terms, but hopefully we provide a useful platform that aids uptake and implementation of geomorphic principles in river rehabilitation practice.

Although this book has an unashamedly Australian flavor, we have endeavored to write it from a global perspective. We convey our apologies, in advance, to those readers to whom this book bears little semblance of reality in terms of the types of rivers you live and/or work with. However, we hope that the principles presented here bear relevance to the management issues that you face, and that the book provides useful guidance in the development of core understanding that is required if management activities are to yield sustainable outcomes.

Acknowledgments

The River Styles framework has its origins in river reach analysis of the Waiau River in New Zealand, in a project coordinated through Southland Regional Council, following a flash of inspiration generated by Glen Lauder. In 1994, Gary Brierley was invited to South Africa to participate in a river health workshop coordinated by Barry Hart (from the Australian half of the gathering). This built on initial contacts suggested by Brian Finlayson, who recommended an approach be made to a Federal Government body, the Land and Water Resources Research and Development Corporation (now Land and Water Australia; LWA) to seek support to continue this work. The award of a substantive grant effectively marked the birth of the River Styles framework. Phil Price provided invaluable guidance in these initial endeavors – his broadening of scope ensured that a generic, open-ended approach was developed, moving beyond a case study perspective. Further backing by Siwan Lovett and Nick Schofield in LWA aided the coordination of early work. Collaboration with Tim Cohen, Sharon Cunial, and Fiona Nagel fashioned initial endeavors, with willing sounding boards on hand at Macquarie University in discussions with Andrew Brooks, John Jansen, and Rob Ferguson.

Substantive external support through the State Government agency, then called the Department of Land and Water Conservation (DLWC), was generated at the outset of the project. Head Office leadership was guided by David Outhet, and on-the-ground support in the Bega Regional Office, initially by Justin Gouvernet and Don McPhee and substantially with Cliff Massey. The practical development and application of the River Styles work in Bega catchment was enormously enhanced by collaboration with the former Far South

Coast Catchment Management Committee, under leadership by Kerry Pfeiffer and funding generated through the Bega Valley Shire and the Natural Heritage Trust (NHT). Various workshops and reports promoted early findings of the work. At one of these meetings, Michael Pitt and various colleagues from the North Coast Office of DLWC envisaged potential applications of equivalent work in their catchments. Tony Broderick played a pivotal part in facilitating these applications. At this stage, Rob Ferguson, Ivars Reinfelds, and Guy Lampert extended the range of rivers to which the work was applied through characterizations of rivers in the Manning catchment. The primary role of differing forms of valley confinement, which formed a part of the PhD work completed by Rob Ferguson, advanced the framework.

Subsequent developments included research on stream power along longitudinal profiles in the Bellinger catchment, in work completed with Tim Cohen and Ivars Reinfelds. Insights into geological controls on patterns of River Styles was provided by Geoff Goldrick, in application of this work in the Richmond catchment. Eventually more than 10 catchment-based reports characterized the diversity of River Styles and their downstream patterns, in all North Coast catchments extending from the Hastings to the Tweed. Rob Ferguson coordinated this work, with field work completed by Guy Lampert. Paul Batten provided the initial algorithms to generate longitudinal profiles and stream power plots through use of Geographic Information Systems and Digital Elevation Models. Paula Crighton was invaluable in refining this procedure and processing the data for the North Coast catchments. Practical application of the work was enhanced through a subsequent contract in the Shoalhaven catchment where Rachel Nanson completed much of the field work.

A major advancement in the development of the River Styles framework occurred with extensions from assessment of river character and behavior to analyses of river condition and recovery potential. The PhD work of Kirstie Fryirs developed these procedures and applied them in the Bega catchment. These procedures now form Stages 2 and 3 of the framework. The development of these procedures was enhanced by a visit to Australia by Scott Babakaiff (funded by LWA) and development of the National River Restoration Framework (in a project with John Koehn and Belinda Cant funded by LWA).

The next phase of the River Styles work entailed fundamental research into ecological (habitat) associations with a geomorphic classification scheme. This work was completed by two Post-Docs (Mark Taylor and Jim Thomson), through collaborative funding provided by LWA and DLWC. Penny Knights and Glenda Orr supported this work. Collaboration with Bruce Chessman linked geomorphology and ecology in assessments of geoecological condition in Bega catchment.

Since its creation, promotion and adoption of the River Styles framework has occurred across the nation. Particular mention must be made of David Outhet, who promoted the adoption of the framework as a tool for management activities in New South Wales, and provided numerous insightful comments on its application. Sally Boon in Queensland and David Wright in Tasmania have also promoted the framework and have sourced funding for us to run courses and workshops in those states.

Numerous members of the Rivers Group at Macquarie University have provided many hours of enthusiastic and fruitful discussion about rivers. While most now roam further afield, they remain a large part of the "history" associated with this book. Particular mention must be made of Andrew Brooks, Tim Cohen, Rob Ferguson, John Jansen, Emily Cracknell, Paula Crighton, Mick Hillman, Pete Johnston, and Kahli McNab. Mick Hillman, in particular, provided the stimulus for greatly enhancing the 'extension science' component of our work.

Insightful and constructive review comments on this book were made by a range of academics and postgraduate students, including Ted Hickin, Malcolm Newson, Jonathan Phillips, Rob Ferguson, Jo Hoyle, Nick Preston, and John Spencer. These review comments substantially improved the clarity and communicability of the book.

Teaching River Styles Short Courses has occurred in parallel to development of the framework. We wish to thank the participants of these courses, who have spanned a wide range of professions and levels of experience from around the nation and overseas. Their contributions have improved the presentation of the River Styles framework and our ability to communicate and teach it. Each River Styles Short Course has been run through Macquarie Research Limited (MRL) with administrative support from Roslyn Green, Kerry Tilbrook, and Sophie Beauvais. Sophie Beauvais, Irina Zakoshanski, and Warren Bailey are thanked for their support in administering developments of the framework, promotion, trade-marking, and accreditation of the framework. The term River Styles® is a registered trademark held by Macquarie University and Land and Water Australia.

Most of the graphics in this book were designed by Kirstie Fryirs and drafted by Dean Oliver Graphics, Pty Ltd. We thank Dean for his commitment to this project. We also thank colleagues in the Department of Physical Geography, Macquarie University for their support.

Sincere thanks to Sue and Paul Gebauer who own Wonga Wildlife Retreat in Coffs Harbour. They provided us with a writer's paradise. Without Wonga the book would not be what it is today. We also extend our thanks to Chris and Rick Fryirs for use of their Woodford house during the postreview stage.

We extend our love to our families for their patience and support over the many years it has taken to write this book; Emmy, Zac, Whit, Chris, Rick, Steve, Sarah, Tim, Dee, Chris – thank you!

Self-evidently, many people have helped us along the way in a process that has provided many intellectual and personal challenges. Their insight and support have encouraged us to "maintain the rage," during countless phases when the project didn't quite want to come to fruition. Indeed, we hope the book is far from an endpoint. As in any book, ultimate responsibility in ideas presented lie with the authors. Our apologies, in advance, to anyone whose thoughts have been misrepresented.

CHAPTER 1

Introduction

Society's ability to maintain and restore the integrity of aquatic ecosystems requires that conservation and management actions be firmly grounded in scientific understanding.
LeRoy Poff, et al., 1997, p. 769

1.1 Concern for river health

Rivers are a much-cherished feature of the natural world. They perform countless vital functions in both societal and ecosystem terms, including personal water consumption, health and sanitation needs, agricultural, navigational, and industrial uses, and various aesthetic, cultural, spiritual, and recreational associations. In many parts of the world, human-induced degradation has profoundly altered the natural functioning of river systems. Sustained abuse has resulted in significant alarm for river health, defined as the ability of a river and its associated ecosystem to perform its natural functions. In a sense, river health is a measure of catchment health, which in turn provides an indication of environmental and societal health. It is increasingly recognized that ecosystem health is integral to human health and unless healthy rivers are maintained through ecologically sustainable practices, societal, cultural, and economic values are threatened and potentially compromised. Viewed in this way, our efforts to sustain healthy, living rivers provide a measure of societal health and our governance of the planet on which we live. It is scarcely surprising that concerns for river condition have been at the forefront of conservation and environmental movements across much of the planet.

In the past, the quest for security and stability to meet human needs largely overlooked the needs of aquatic ecosystems. In many instances, human activities brought about a suite of unintended and largely unconsidered impacts on river health, compromising the natural variability of rivers, their structural integrity and complexity, and the maintenance of functioning aquatic ecosys-

tems. Issues such as habitat loss, degradation, and fragmentation have resulted in significant concerns for ecological integrity, sustainability, and ecosystem health. As awareness and understanding of these issues has improved, society no longer has an excuse not to address concerns brought about by the impacts of human activities on river systems. Shifts in environmental attitudes and practice have transformed outlooks and actions towards revival of aquatic ecosystems. Increasingly, management activities work in harmony with natural processes in an emerging "age of repair," in which contemporary management strategies aim to enhance fluvial environments either by returning rivers, to some degree, to their former character, or by establishing a new, yet functional environment. Notable improvements to river health have been achieved across much of the industrialized world in recent decades. However, significant community and political concern remains over issues such as flow regulation, algal blooms, salinity, loss of habitat and species diversity, erosion and sedimentation problems, and water resource overallocation.

Rivers demonstrate a remarkable diversity of landform patterns, as shown in Figure 1.1. Each of the rivers shown has a distinct set of landforms and its own behavioral regime. Some rivers have significant capacity to adjust their form (e.g., the meandering, anastomosing, and braided river types), while others have a relatively simple geomorphic structure and limited capacity to adjust (e.g., the chain-of-ponds and gorge river types). This variability in geomorphic structure and capacity to adjust, which reflects the array of landscape settings in which these rivers are found, presents signifi-

Figure 1.1 The diversity of river morphology
Rivers are characterized by a continuum of morphological diversity, ranging from bedrock controlled variants such as
(a) gorges (with imposed sets of landforms), to fully alluvial, self-adjusting rivers such as (c) braided and (d) meandering
variants (with various midchannel, bank-attached and floodplain features). Other variants include multichanneled
anastomosing rivers that form in wide, low relief plains (e), and rivers with discontinuous floodplain pockets in
partly-confined valleys (b). In some settings, channels are discontinuous or absent, as exemplified by chain-of-ponds
(f). Each river type has a different capacity to adjust its position on the valley floor. (a) Upper Shoalhaven catchment,
New South Wales, Australia, (b) Clarence River, New South Wales, Australia, (c) Rakaia River, New Zealand, (d)
British Columbia, Canada, (e) Cooper Creek, central Australia, and (f) Murrumbateman Creek, New South Wales,
Australia.

cant diversity in the physical template atop which ecological associations have evolved.

Developing a meaningful framework to recognize, understand, document, and maintain this geodiversity is a core theme of this book. Working within a conservation ethos, emphasis is placed on the need to maintain the inherent diversity of riverscapes and their associated ecological values. Adhering to the precautionary principle, the highest priority in management efforts is placed on looking after good condition remnants of river courses, and seeking to sustain rare or unique reaches of river regardless of their condition.

Just as there is remarkable diversity of river forms and processes in the natural world, so human-induced disturbance to rivers is equally variable (see Figure 1.2). Many of these actions have been intentional, such as dam construction, channelization, urbanization, and gravel or sand extraction. Far more pervasive, however, have

Figure 1.2 Human modifications to river courses
Human modifications to rivers include (a) dams (Itaipu Dam, Brazil), (b) channelization (Ishikari River, Japan), (c) urban stream (Cessnock, New South Wales, Australia), (d) native and exotic vegetation removal (Busby's Creek Tasmania, Australia), (e) gravel and sand extraction (Nambucca River, New South Wales, Australia), and (f) mine effluent (King River, Tasmania, Australia)

been inadvertent changes brought about through adjustments to flow and sediment transfer regimes associated with land-use changes, clearance of riparian vegetation, etc. Across much of the planet, remarkably few river systems even approximate their pristine condition. Most rivers now operate as part of highly modified landscapes in which human activities are dominant.

The innate diversity of river courses is a source of inspiration, but it presents many perplexing challenges in the design and implementation of sustainable management practices. Unless management programs respect the inherent diversity of the natural world, are based on an understanding of controls on the nature and rate of landscape change, and consider how alterations to one part of an ecosystem affect other parts of that system, efforts to improve environmental condition are likely to be compromised. River management programs that work with natural processes will likely yield the most effective outcomes, in environmental, societal, and economic terms. Striving to meet these challenges, truly multifunctional, holistic, catchment-scale river management programs have emerged in recent decades (e.g., Gardiner, 1988; Newson, 1992a; Hillman and Brierley, in press). Procedures outlined in this book can be used to determine realistic goals for river restoration and rehabilitation programs, recognizing the constraints imposed by the nature and condition of river systems and the cultural, institutional, and legal frameworks within which these practices must be applied.

1.2 Geomorphic perspectives on ecosystem approaches to river management

Rivers are continuously changing ecosystems that interact with the surrounding atmosphere (climatic and hydrological factors), biosphere (biotic factors), and earth (terrestrial or geological factors). Increasing recognition that ecosystems are open, nondeterministic, heterogeneous, and often in nonequilibrium states, is prompting a shift in management away from maintaining stable systems for particular species to a whole-of-system approach which emphasizes diversity and flux across temporal and spatial scales (Rogers, 2002). Working within an ecosystem approach to natural

resources management, river rehabilitation programs apply multidisciplinary thinking to address concerns for biodiversity and ecosystem integrity (Sparks, 1995). Inevitably, the ultimate goals of these applications are guided by attempts to balance social, economic, and environmental needs, and they are constrained by the existing hydrological, water quality, and sediment transport regimes of any given system (Petts, 1996). Ultimately, however, biophysical considerations constrain what can be achieved in river management. If river structure and function are undermined, such that the ecological integrity of a river is compromised, what is left? River rehabilitation programs framed in terms of ecological integrity must build on principles of landscape ecology. The landscape context, manifest through the geomorphic structure and function of river systems, provides a coherent template upon which these aspirations must be grounded. The challenge presented to geomorphologists is to construct a framework with which to meaningfully describe, explain, and predict the character and behavior of aquatic ecosystems.

Biological integrity refers to a system's wholeness, including presence of all appropriate biotic elements and occurrence of all processes and interactions at appropriate scales and rates (Angermeier and Karr, 1994). This records a system's ability to generate and maintain adaptive biotic elements through natural evolutionary processes. Ecosystem integrity requires the maintenance of both physico-chemical and biological integrity, maintaining an appropriate level of connectivity between hydrological, geomorphic, and biotic processes. While loss of biological diversity is tragic, loss of biological integrity includes loss of diversity and breakdown in the processes necessary to generate future diversity (Angermeier and Karr, 1994). Endeavors to protect ecological integrity require increased reliance on preventive rather than reactive management, and a focus on landscapes rather than populations.

In riparian landscapes, aquatic, amphibious, and terrestrial species have adapted to a shifting mosaic of habitats, exploiting the heterogeneity that results from natural disturbance regimes (Junk et al., 1989; Petts and Amoros, 1996; Naiman and Decamps, 1997; Ward et al., 2002). This mosaic includes surface waters, alluvial aquifers, riparian vegetation associations, and geomorphic features

(Tockner et al., 2002). Because different organisms have different movement capacities and different habitat ranges, their responses to landscape heterogencity differ (Wiens, 2002). Fish diversity, for example, may peak in highly connected habitats, whereas amphibian diversity tends to be highest in habitats with low connectivity (Tockner et al., 1998). Other groups attain maximum species richness between these two extremes. The resulting pattern is a series of overlapping species diversity peaks along the connectivity gradient (Ward et al., 2002). Given the mutual interactions among species at differing levels in the food chain, ecosystem functioning reflects the range of habitats in any one setting and their connectivity.

Landscape ecology examines the influence of spatial pattern on ecological processes, considering the ecological consequences of where things are located in space, where they are relative to other things, and how these relationships and their consequences are contingent on the characteristics of the surrounding landscape mosaic. The pattern of a landscape is derived from its composition (the kinds of elements it contains), its structure (how they are arranged in space), and its behavior (how it adjusts over time; Wiens, 2002). A landscape approach to analysis of aquatic ecosystems offers an appropriate framework to elucidate the links between pattern and process across scales, to integrate spatial and temporal phenomena, to quantify fluxes of matter and energy across environmental gradients, to study complex phenomena such as succession, connectivity, biodiversity, and disturbance, and to link research with management (Townsend, 1996; Tockner et al., 2002; Ward et al., 2002; Wiens, 2002).

Principles from fluvial geomorphology provide a physical template with which to ground landscape perspectives that underpin the ecological integrity of river systems. Although landscape forms and processes, in themselves, cannot address all concerns for ecological sustainability and biodiversity management, these concerns cannot be meaningfully managed independent from geomorphological considerations. Working from the premise that concerns for ecological integrity are the cornerstone of river management practice, and that landscape considerations underpin these endeavors, interpretation of the diversity, patterns, and changing nature of river character and behavior

across a catchment is integral to proactive river management. This book outlines a generic set of procedures by which this understanding can be achieved.

Rehabilitation activities must be realistically achievable. Most riverscapes have deviated some way from their pristine, predisturbance condition. Hence, practical management must appraise what is the best that can be achieved to improve the health of a system, given the prevailing boundary conditions under which it operates. In instances where human changes to river ecosystems are irreversible or only partially reversible, a pragmatic definition of ecological integrity refers to the maintenance of a best achievable condition that contains the community structure and function that is characteristic of a particular locale, or a condition that is deemed satisfactory to society (Cairns, 1995). Specification of the goals of river management, in general, and river restoration, in particular, has provoked considerable discussion, as highlighted in the following section.

1.3 What is river restoration?

The nature and extent of river responses to human disturbance, and the future trajectory of change, constrain what can realistically be achieved in river management (Figure 1.3; Boon, 1992). At one extreme, *conservation* goals reflect the desire to preserve remnants of natural or near-intact systems. Far more common, however, are endeavors to rectify and repair some (or all) of the damage to river ecosystems brought about by human activities. Various terms used to describe these goals and activities can be summarized using the umbrella term "restoration."

Restoration means different things to different people, the specific details of which may promote considerable debate and frustration (Hobbs and Norton, 1996). Although the term has been applied to a wide range of management processes/activities, its precise meaning entails the uptake of measures to return the structure and function of a system to a previous state (an unimpaired, pristine, or healthy condition), such that previous attributes and/or values are regained (Bradshaw, 1996; Higgs, 2003). In general, reference is made to predisturbance functions and related physical,

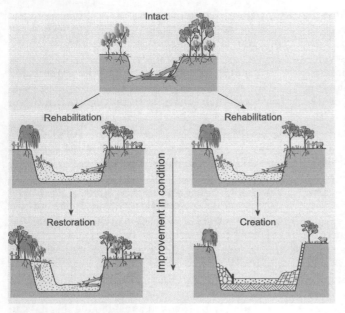

Figure 1.3 Framing realistic management options – what can be realistically achieved? Determination of river rehabilitation goals is constrained primarily by what it is realistically possible to achieve. This reflects system responses to human disturbance, the prevailing set of boundary conditions, and the likely future trajectory of change (as determined by limiting factors and pressures operating within the catchment and societal goals). Maintenance of an intact condition is a conservation goal. If a return to a predisturbance state is possible and desirable, rehabilitation activities can apply recovery principles to work towards a restoration goal. In many instances, adoption of a creation goal, which refers to a new condition that previously did not exist at the site, is the only realistic option.

chemical, and biological characteristics (e.g., Cairns, 1991; Jackson et al., 1995; Middleton, 1999).

The few studies that have documented geomorphic attributes of relatively intact or notionally pristine rivers (e.g., Collins and Montgomery, 2001; Brooks and Brierley, 2002), and countless studies that have provided detailed reconstructions of river evolution over timescales of decades, centuries, or longer, indicate just how profound human-induced changes to river forms and processes have been across most of the planet. It is important to remember the nonrepresentative nature of the quirks of history that have avoided the profound imprint of human disturbance. Intact reaches typically lie in relatively inaccessible areas. They are seldom representative of the areas in which management efforts aim to improve river health. However, it is in these reaches, and adjacent good condition reaches, that efforts at restoration can meaningfully endeavor to attain something akin to the pure definition of the word.

Viewed in a more general sense, restoration refers to a management process that provides a means to communicate notions of ecosystem recovery (Higgs, 2003). For example, the Society for Ecological Restoration (SERI, 2002) state that restoration refers to the process of assisting the recovery of an ecosystem that has been degraded, damaged, or destroyed. The notion of recovery describes the process of bringing something back.

Endeavors that assist a system to adjust towards a less stressed state, such that there is an improvement in condition, are more accurately referred to as river *rehabilitation*. Rehabilitation can mean the process of returning to a previous condition or status along a restoration pathway, or *creation* of a new ecosystem that previously did not exist (Fryirs and Brierley, 2000; Figure 1.3). In landscapes subjected to profound human disturbance, such as urban, industrial, or intensively irrigated areas, management activities inevitably work towards creation goals. Both restoration and creation goals require rehabilitation strategies that strive to improve river condition, applying recovery notions to work towards the best attainable ecosystem values given the prevailing boundary conditions. The essential difference between restoration and creation goals lies in the perspective of regenerating the "old" or creating a "new" system (Higgs, 2003).

Various other terms have been used to characterize practices where the goals are not necessarily framed in ecosystem terms. For example, *reclamation* refers to returning a river to a useful or proper state, such that it is rescued from an undesirable condition (Higgs, 2003). In its original sense, reclamation referred to making land fit for cultivation, turning marginal land into productive acreage. Alternatively, *remediation* refers to the process of repairing ecological damage in a manner that does not focus on ecological integrity and is typically applied without reference to historical conditions (Higgs, 2003). Reclamation and remediation are quick-fix solutions to environmental problems that address concerns for human values, viewed separately from their ecosystem context.

The purpose and motivation behind any rehabilitation activity are integral to the goal sought. Specification of conservation, restoration, or creation goals provides an indication of the level and type of intervention that is required to improve riverine environments.

1.4 Determination of realistic goals in river rehabilitation practice

The process of river rehabilitation begins with a judgment that an ecosystem damaged by human activities will not regain its former characteristic properties in the near term, and that continued degradation may occur (Jackson et al., 1995). Approaches to repair river systems may focus on rehabilitating "products" (species or ecosystems) directly, or on "processes" which generate the desired products (Neimi et al., 1990; Richards et al., 2002). However, unless activities emphasize concerns for the rehabilitation of fundamental processes by which ecosystems work, notions of ecosystem integrity and related measures of biodiversity may be compromised (Cairns, 1988).

The goal of increasing heterogeneity across the spectrum of river diversity represents a flawed perception of ecological diversity and integrity. In some cases, the "natural" range of habitat structure may be very simple. Hence, heterogeneity or geomorphic complexity does not provide an appropriate measure of river health (see Fairweather, 1999). Simplistic goals framed in expressions such as "more is better" should be avoided (Richards et al., 2002). Use of integrity as a primary management goal avoids the pitfalls associated with assumptions that greater diversity or productivity is preferred.

Unlike many biotic characteristics, physical habitat is directly amenable to management through implementation of rehabilitation programs (Jacobson et al., 2001). Hence, many management initiatives focus on physical habitat creation and maintenance, recognizing that geomorphic river structure and function and vegetation associations must be appropriately reconstructed before sympathetic rehabilitation of riverine ecology will occur (Newbury and Gaboury, 1993; Barinaga, 1996). Getting the geomorphological structure of rivers right maximizes the ecological potential of a reach, in the hope that improvements in biological integrity will follow (i.e., the "field of dreams" hypothesis; Palmer et al., 1997). The simplest procedure with which to determine a suitable geomorphic structure and function is to replicate the natural character of "healthy" rivers of the same "type," analyzed in equivalent landscape settings.

In any management endeavor, it is imperative to identify, justify, and communicate underlying goals, ensuring that the tasks and plan of action are visionary yet attainable. Although setting goals for rehabilitation is one of the most important steps in designing and implementing a project, it is often either overlooked entirely or not done very well (Hobbs, 2003). Success can only be measured if a definitive sense is provided of what it will look like. Unfortunately, however, there is a tendency to jump straight to the "doing" part of a project without clearly articulating the reasons why things are being done and what the outcome should be (Hobbs, 1994, 2003).

While sophisticated methodologies and techniques have arisen in the rapidly growing field of rehabilitation management, the conceptual foundations of much of this work remain vague (Ebersole et al., 1997). The pressure of timeframes, tangible results, and political objectives has lead to a preponderance of short-term, transitory rehabilitation projects that ignore the underlying capacities and developmental histories of the systems under consideration, and seldom place the study/treatment reach in its catchment context (Ebersole et al., 1997; Lake, 2001a, b). Unfortunately, many of these small-scale aquatic habitat enhancement projects have failed, or have proven to be ineffective (e.g., Frissell and Nawa, 1992).

Ensuring that goals are both explicit enough to be meaningful and realistic enough to be achievable is a key to the development of successful projects. Ideally, goals are decided inclusively, so that everyone with an interest in the outcomes of the project agrees with them (Hobbs, 2003). Scoping the future and generating a realistic vision of the desired river system are critical components of the planning process. The vision should be set over a 50 year timeframe (i.e., 1–2 generations; Jackson et al., 1995), such that ownership of outcomes can be achieved. A vision must be based on the best available information on the character, behavior, and evolution of the system, providing a basis to interpret the condition and trajectory of change from which desired future conditions can be established (Brierley and Fryirs, 2001). These concepts must be tied to analysis of biophysical linkages across a range of scales, enabling off-site impacts and lagged responses to disturbance events and/or rehabilitation treatments to be appraised (Boon, 1998).

To maximize effectiveness, rehabilitation efforts should incorporate spatiotemporal scales that are large enough to maintain the full range of habitats and biophysical linkages necessary for the biota to persist under the expected disturbance regime or prevailing boundary conditions. Although emphasis may be placed on a particular component or attribute, ultimate aims of long-term projects should focus on the whole system at the catchment scale (Bradshaw, 1996). Desired conditions for each reach should be specified as conservation, restoration, or creation goals, indicating how they fit within the overall catchment vision. Appropriate reference conditions should be specified for each reach.

Defining what is "natural" for a given type of river that operates under a certain set of prevailing boundary conditions provides an important step in identification of appropriate reference conditions against which to measure the geoecological integrity of a system and to identify target conditions for river rehabilitation. A "natural" river is defined here as "a dynamically adjusting system that behaves within a given range of variability that is appropriate for the river type and the boundary conditions under which it operates." Within this definition, two points of clarification are worth noting. First, a "natural" condition displays the full range of expected or appropriate structures and processes for that type of river under prevailing catchment boundary conditions. This does not necessarily equate to a predisturbance state, as human impacts may have altered the nature, rate, and extent of river adjustments (Cairns, 1989). Second, a dynamically adjusted reach does not necessarily equate to an equilibrium state. Rather, the river adjusts to disturbance via flow, sediment, and vegetation interactions that fall within the natural range of variability that is deemed appropriate for the type of river under investigation.

Determination of appropriate reference conditions, whether a fixed historical point in time or a suite of geoecological conditions, represents a critical challenge in rehabilitation practice (Higgs, 2003). Systems in pristine condition serve as a point of reference rather than a prospective goal for river rehabilitation projects, although attributes of this ideal condition may be helpful in rehabilitation design. Identification of reference conditions aids interpretation of the rehabilitation potential of sites, thereby providing a basis to measure the success of rehabilitation activities.

Reference conditions can be determined on the basis of historical data (paleo-references), data derived from actual situations elsewhere, knowledge about system structure and functioning in general (theoretical insights), or a combination of these sources (Petts and Amoros, 1996; Jungwirth et al., 2002; Leuven and Poudevigne, 2002). The morphological configuration and functional attributes of a reference reach must be compatible with prevailing biophysical fluxes, such that they closely equate to a "natural" condition for the river type. Ideally, reference reaches are located in a similar position in the catchment and have near equivalent channel gradient, hydraulic, and hydrologic conditions (Kondolf and Downs, 1996). Unfortunately, it is often difficult to find appropriate reference conditions for many types of river, as "natural" or minimally impacted reaches no longer exist (Henry and Amoros, 1995; Ward et al., 2001). In the absence of good condition remnants, reference conditions can be constructed from historical inferences drawn from evolutionary sequences that indicate how a river has adjusted over an interval of time during which boundary conditions have remained relatively uniform. Selection of the most appropriate reference condition is situated within this sequence. Alternatively, a suite of desirability criteria derived for each type of river can be used to define a natural reference condition against which to compare other reaches (Fryirs, 2003). These criteria must encapsulate the forms and processes that are "expected" or "appropriate" for the river type. They draw on system-specific and process-based knowledge, along with findings from analysis of river history and assessment of available analogs. This approach provides a guiding image, or *Leitbild*, of the channel form that would naturally occur at the site, adjusted to account for irreversible changes to controlling factors (such as runoff regime) and for considerations based on cultural ecology (such as preservation of existing land uses or creation of habitat for endangered species; Kern, 1992; Jungwirth et al., 2002; Kondolf et al., 2003). *Leitbilds* can be used to provide a reference network of sites of high ecological status for each river type, as required by the European Union Water Framework Directive.

1.5 Managing river recovery processes in river rehabilitation practice

Exactly what is required in any rehabilitation initiative will depend on what is wrong. Options may range from limited intervention and a leave-alone policy, to mitigation or significant intervention, depending on how far desired outcomes are from the present condition. In some instances, sensitive, critical, or refuge habitats, and the stressors or constraints that limit desirable habitat, must be identified, and efforts made to relieve these stressors or constraints (Ebersole et al., 1997). Controlling factors that will not ameliorate naturally must be identified, and addressed first. Elsewhere, rehabilitation may involve the reduction, if not elimination, of biota such as successful invaders, in the hope of favoring native biota (Bradshaw, 1996). For a multitude of reasons, ranging from notions of naturalness that strive to preserve "wilderness," to abject frustration at the inordinate cost and limited likelihood of success in adopted measures (sometimes referred to as basket cases, or "raising the *Titanic*"; Rutherfurd et al., 1999), it is sometimes advisable to pursue a passive approach to rehabilitation. This strategy, often referred to as the "do nothing option," allows

the river to self-adjust (cf., Hooke, 1999; Fryirs and Brierley, 2000; Parsons and Gilvear, 2002; Simon and Darby, 2002). Although these measures entail minimal intervention and cost, managers have negligible control over the characteristics and functioning of habitats (Jacobson et al., 2001).

In general terms, however, most contemporary approaches to river rehabilitation endeavor to "heal" river systems by enhancing natural recovery processes (Gore, 1985). Assessment of geomorphic river recovery is a predictive process that is based on the trajectory of change of a system in response to disturbance events. Recovery enhancement involves directing reach development along a desired trajectory to improve its geomorphic condition over a 50–100 year timeframe (Hobbs and Norton, 1996; Fryirs and Brierley 2000; Brierley et al., 2002). To achieve this goal, river rehabilitation activities must build on an understanding of the stage and direction of river degradation and/or recovery, determining whether the geomorphic condition of the river is improving, or continuing to deteriorate.

Assessment of geomorphic river condition measures whether the processes that shape river morphology are appropriate for the given setting, such that deviations from an expected set of attributes can be appraised (Figure 1.4; Kondolf and Larson, 1995; Maddock, 1999). Key consideration must be given to whether changes to the boundary conditions under which the river operates have brought about irreversible changes to river structure and function (Fryirs, 2003). Identification of good condition reaches provides a basis for their conservation. Elsewhere, critical forms and processes may be missing, accelerated, or anomalous, impacting on measures of geoecological functioning.

Understanding of geomorphic processes and their direction of change underpins rehabilitation strategies that embrace a philosophy of recovery enhancement (Gore, 1985; Heede and Rinne, 1990; Milner, 1994). Helping a river to help itself presents an appealing strategy for river rehabilitation activities because they cost nothing in themselves (although they may cost something to initiate), they are likely to be self-sustaining because they originate from within nature (although they may need nurturing in some situations), and they can be applied on a large scale (Bradshaw, 1996). Design

Increasing heterogeneity ⟶ = improved condition

Poor   Moderate   Good

a

b

Good   Moderate   Poor

Increasing heterogeneity ⟶ does not = improved condition

Figure 1.4 Habitat diversity for good, moderate, and poor condition variants of the same river type
Natural or expected character and behavior varies for differing types of river. Some may be relatively complex, others are relatively simple. Natural species adaptations have adapted to these conditions. Assessments of geomorphic river condition must take this into account, determining whether rehabilitation activities should increase (a) or decrease (b) the geomorphic heterogeneity of the type of river under investigation. Increasing geomorphic heterogeneity is not an appropriate goal for all types of river, and may have undesirable ecological outcomes. More appropriate strategies *work with* natural diversity and river change.

and implementation of appropriate monitoring procedures are integral in gauging the success of these strategies.

The process of river rehabilitation is a learning experience that requires ongoing and effective monitoring in order to evaluate and respond to findings. Measuring success must include the possibility of measuring failure, enabling midcourse corrections, or even complete changes in direction (Hobbs, 2003). If effectively documented, each project can be considered as an experiment, so that failure can be just as valuable to science as success, provided lessons are learnt. Goals or performance targets must be related to ecological outcomes and be measurable in terms such as increases in health indicators (e.g., increasing similarity of species or

structure with the reference community), or decreases in indicators of degradation (e.g., active erosion, salinity extent or impact, nonnative plant cover). The choice of parameters to be monitored must go hand in hand with the setting of goals, ensuring that they are relevant to the type of river under consideration, so that changes in condition can be meaningfully captured. Baseline data are required to evaluate changes induced by the project, including a detailed historical study (Downs and Kondolf, 2002). Monitoring should be applied over an extensive period, at least a decade, with surveys conducted after each flood above a predetermined threshold (Kondolf and Micheli, 1995). These various components are integral parts of effective river rehabilitation practice.

1.6 Overview of the River Styles framework

Best practice in natural resources management requires appropriate understanding of the resource that is being managed, and effective use of the best available information. In river management terms, catchment-scale information on the character, behavior, distribution, and condition of different river types is required if management strategies are to "work with nature." Given that rivers demonstrate remarkably different character, behavior, and evolutionary traits, both between- and within-catchments, individual catchments need to be managed in a flexible manner, recognizing what forms and processes occur where, why, how often, and how these processes have changed over time. The key challenge is to understand why rivers are the way they are, how they have changed, and how they are likely to look and behave in the future. Such insights are fundamental to our efforts at rehabilitation, guiding what can be achieved and the best way to get there.

This book presents a coherent set of procedural guidelines, termed the River Styles framework, with which to document the geomorphic structure and function of rivers, and appraise patterns of river types and their biophysical linkages in a catchment context. Meaningful and effective *description* of river character and behavior are tied to *explanation* of controls on why rivers are the way they are, how they have evolved, and the causes of change. These insights are used to *predict* likely river futures, framed in terms of the contemporary condition, evolution, and recovery potential of any given reach, and understanding of its trajectory of change (Figure 1.5).

The River Styles framework is a rigorous yet flexible scheme with which to structure observations and interpretations of geomorphic forms and processes. A structured basis of enquiry is applied to develop a catchment-wide package of physical information with which to frame management activities (Figure 1.6). This package guides insights into the type of river character and behavior that is expected for any given field setting and the type of adjustments that may be experienced by that type of river. A catchment-framed nested hierarchical arrangement is used to analyze landscapes in terms of their constituent parts. Reach-scale forms and processes are viewed in context of catchment-

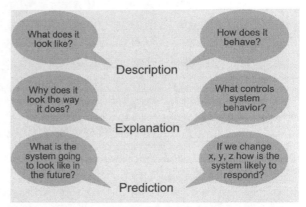

Figure 1.5 Routes to description, explanation, and prediction

scale patterns and rates of biophysical fluxes. Separate layers of information are derived to appraise river character and behavior, geomorphic condition, and recovery. Definition of ongoing adjustments around a characteristic state(s) enables differentiation of the behavioral regime of a given river type from river change. Analysis of system evolution is undertaken to appraise geomorphic river condition in context of "expected attributes" of river character and behavior given the reach setting. Interpretation of catchment-specific linkages of biophysical processes provides a basis with which to assess likely future patterns of adjustment and the geomorphic recovery potential of each reach. The capacity, type, and rate of recovery response of any given type of river are dependent on the nature and extent of disturbance, the inherent sensitivity of the river type, and the operation of biophysical fluxes (both now and into the future) at any given point in the landscape. When these notions are combined with interpretations of limiting factors to recovery and appraisal of ongoing and likely future pressures that will shape river forms and processes, a basis is provided to assess likely future river condition, identify sensitive reaches and associated off-site impacts, and determine the degree/rate of propagating impacts throughout a catchment.

The strategy outlined in this book emphasizes the need to understand individual systems, their idiosyncrasies of forms and processes, and evolutionary traits and biophysical linkages, as a basis to

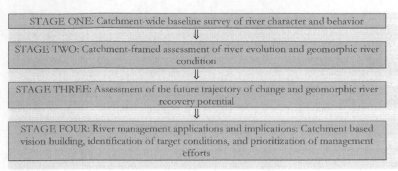

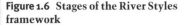
Figure 1.6 Stages of the River Styles framework

determine options for management – in planning, policy, and design terms. System configuration and history ensure that each catchment is unique. In making inferences from system-specific information, cross-reference is made to theoretical and empirical relationships to explain system behavior and predict likely future conditions. Principles outlined in this book provide a conceptual tool with which to read and interpret landscapes, rather than a quantitative approach to analysis of river forms and processes. Application of these procedures provides the groundwork for more detailed site- or reach-specific investigations.

However, application of geomorphic principles in the determination of sustainable river management practices is far from a simple task. The need for system-specific knowledge and appropriate skills with which to interpret river evolution and the changing nature of biophysical linkages (and their consequences) ensure that such exercises cannot be meaningfully undertaken using a pre-scriptive cook-book approach. The cautious, data intensive measures applied in this book are considered to present a far better perspective than managing rivers to some norm! Hopefully, lessons have been learnt from the homogenization of river courses under former management regimes.

Management applications of the River Styles framework focus on the derivation of a catchment-scale vision for conservation and rehabilitation, identification of reach-specific target conditions that fit into the bigger-picture vision, and application of a geomorphologically based prioritization procedure which outlines the sequencing of actions that best underpins the likelihood of management success. The framework does not provide direct guidance into river rehabilitation design and selection of the most appropriate technique. Rather, emphasis is placed on the need to appraise each field situation separately, viewed within its catchment context and evolutionary history. The underlying catchcry in applications of the River Styles framework is "KNOW YOUR CATCHMENT."

1.7 Layout and structure of the book

This book comprises four parts (Figure 1.7). Part A outlines the geoecological basis for river management. Chapter 2 documents the use of geomorphology as a *physical template* for integrating biophysical processes, *working with linkages* of biophysical processes within a catchment framework, and the need to *respect diversity* (work with nature). Chapter 3 outlines how geomorphic principles provide a basis for river management programs to *work with change* through understanding of controls on river character and behavior and prediction of likely future adjustments.

Geomorphic principles that underpin applications of the River Styles framework are documented in Part B. Pertinent literature is reviewed to assess river character (Chapter 4), interpret river behavior (Chapter 5), analyze river evolution and change (Chapter 6), and appraise river responses to human disturbance (Chapter 7).

The River Styles framework is presented in Part C. An overview of the framework in Chapter 8 is followed by a brief summary of practical and logistical issues that should be resolved prior to its application. Chapter 9 presents the step-by-step procedure used to classify and interpret river character and behavior in Stage One of the framework.

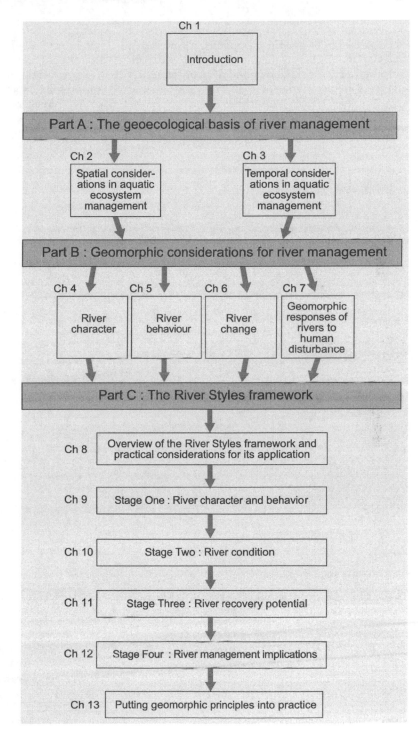

Figure 1.7 Structure of the book

Procedures used to assess geomorphic condition of rivers in Stage Two of the framework are presented in Chapter 10. Evolutionary insights are used to interpret the future trajectory and recovery potential of rivers in Stage Three of the framework (Chapter 11). Finally, Chapter 12 outlines Stage Four of the River Styles framework, which can be used to develop catchment-framed visions for management, identify target conditions for river rehabilitation, and prioritize where conservation and rehabilitation should take place.

The concluding chapter, in Part D, outlines an optimistic (aspirational) perspective on future river management practices and outcomes (Chapter 13).

PART A

The geoecological basis of river management

(A)n understanding of the nature of the building blocks that compose a particular landscape is fundamental to understanding how geomorphological processes function as ecological disturbance processes at the watershed or landscape scale.

Dave Montgomery, 2001, p. 249

Overview of Part A

This part demonstrates how principles from fluvial geomorphology can be used to develop an ecosystem approach to river analysis and management. In Chapter 2, spatial considerations in geomorphology and management practice are framed in terms of a nested hierarchical approach to catchment characterization. Principles from fluvial geomorphology are shown to provide an *integrative physical template* with which to assess habitat associations and *linkages of biophysical processes* in landscapes. Finally, the concept of *respecting diversity* is introduced, indicating why management strategies should strive to maintain unique or distinctive attributes of river courses.

Chapter 3 outlines how theoretical and field-based insights must be combined to meaningfully describe and explain river systems. These insights provide a critical platform for our efforts at prediction. Themes discussed in this chapter include the need for management programs to *work with change*, moving beyond notions of equilibrium and stability used in engineering applications. Timeframes of river adjustment, assessment of controls on river character and behavior, and approaches to prediction are also outlined.

CHAPTER 2

Spatial considerations in aquatic ecosystem management

The average textbook of fluvial geomorphology devotes equal space to channel-scale structure, process, and dynamics and to basin-scale structure, process, and dynamics. Stream ecology has focussed almost exclusively on the former. Abundant tools exist for a fruitful, creative incursion into the realm of the latter.

Stuart Fisher, 1997, p. 313

2.1 Introduction and chapter structure

The diversity and complexity of biophysical interactions that fashion the structure and function of aquatic ecosystems present an intriguing and demanding challenge for river managers. In this chapter, spatial considerations in the management of aquatic ecosystems are addressed. Emphasis is placed on geomorphic principles that underpin ecological considerations in river rehabilitation practice.

Application of a nested hierarchical framework aids the differentiation of scalar components of river systems (Section 2.2). Research and management applications of geoecological insights are framed in terms of their landscape context in Section 2.3, where habitat, flow, sediment transfer, and vegetation management considerations are examined atop a geomorphic template. Longitudinal, lateral, and vertical dimensions of biophysical fluxes are outlined in relation to these considerations in Section 2.4. Finally, the catchment-specific nature of these relationships prompts the need to respect diversity in the management of aquatic ecosystems (Section 2.5).

2.2 Spatial scales of analysis in aquatic geoecology: A nested hierarchical approach

The scale at which observations of the natural world are made constrains what is seen. A head

bowed over a gravel bar measuring the sizes of pebbles provides a very different perspective to that derived by viewing the landscape from the highest local point, or an aerial overview. Physical scale imposes various limitations on system structure and function. For example, bed material texture in flume studies does not scale in a linear manner to channel size, presenting considerable problems with dimensionality in extrapolating laboratory findings to field situations. In general, landscape complexity increases as size increases (Schumm, 1991). For example, while a small subcatchment may lie within one climatic region, form on one lithologic unit, and be subjected to one type of land use, a larger catchment may span climatic, lithologic, and land-use boundaries, and is thus more complex. Scalar considerations can also be appraised in relational terms. For example, a point bar scales in size relative to its channel, whether measured for a third-order stream with a catchment area of $50 \, km^2$, or the trunk stream of the Amazon.

Relationships between scales, and their significance in determining measures of system functioning, vary for differing branches of enquiry. For example, predation and species–species interactions operate at differing physical scales to geomorphic interactions that shape river morphology. The challenge for riverine ecologists is to match the scales of their observations and experiments to the characteristic scales of the phenomena that they investigate (Cooper et al., 1998). A coherent analytical framework is required to meaningfully interlink these scales.

Recognition of the controls imposed on small-scale (and short-term) physical features and processes in rivers by larger-scale (and longer-term) factors has led to the development of nested hierarchical models of physical organization (Table 2.1, Figure 2.1; Frissell et al., 1986; Naiman et al., 1992; Poole, 2002). Characteristics that vary over small spatial and temporal scales are constrained by, or nested within, boundaries set by characteristics that vary over large scales. In general terms, the larger the scale of analysis, the greater the level of generality of forms and processes involved. Large-scale attributes are delineated using large-scale characteristics such as relief and valley slope, and necessarily include a great deal of variation in small-scale characteristics such as flow type and substrate. Different scalar units in the nested hierarchy are commonly not discrete physical entities. Rather, they are part of an inordinately complex continuum in which the dimensions of units at each scale overlap significantly. Interaction between units, at each scale and between scales, determines system character and behavior (Ward, 1989; Naiman et al., 1992; Parsons et al., 2003, 2004).

A given parameter may exert a different influence on system structure and function at different spatial and temporal intervals (Schumm and

Table 2.1 A nested hierarchy of geoecological associations (modified from Frissell et al., 1986 and Poole, 2002).

Scale	Timeframe of evolutionary adjustment Frequency of disturbance event	Geomorphic influence on aquatic ecology
Ecoregion	10^5–10^6 years 10^3 months	Tectonic influences on relief, slope, and valley width are combined with lithologic and climatic controls on substrate, flow, and vegetation cover (among other factors) to determine the boundary conditions within which aquatic ecosystems function.
Catchment	10^5–10^6 years 10^3 months	The nature, rate, and pattern of biophysical fluxes are influenced by catchment geology, shape, drainage density, tributary–trunk stream interactions, etc. Vegetation cover indirectly influences river character through its impacts on flow and sediment delivery.
Landscape unit	10^3–10^4 years 10^2 months	Landscape units are readily identifiable topographic features with a characteristic pattern of landforms. The nature, rate, and pattern of biophysical fluxes are influenced by landscape configuration (i.e., the pattern of landscape units and how they fit together in any given catchment), and the connectivity, linkage, and coupling of ecological processes. At this scale, the channel, riparian zone, floodplain, and alluvial aquifer represent an integrated fluvial corridor that is distinct from, but interacts with, the remaining catchment. Inundation frequency and duration determine surface elements and their boundaries. Sediment source and water residence time in the aquifer determine aquifer element boundaries.
Reach	10^1–10^2 years 10^1 months	Geomorphic river structure and function are relatively uniform at the reach scale, as characterized by particular patterns of channel and/or floodplain landforms and their linkages. The presence and assemblage of landforms such as backswamps (wetlands), billabongs, pools, riffles, etc. define differing river reaches. At this scale, the lotic ecosystem within the fluvial corridor is divided into its distinct components (channel, floodplain, vegetation, and alluvial aquifer), which are measured and studied as separate, but interconnected systems. The character, pattern, and assemblage of these features exert a major influence on habitat diversity along a river course. Instream patterns of water, sediment, and vegetation interactions at this scale shape habitat availability and viability at differing flow stages, including patterns of water flux in the alluvial aquifer. Reach-scale dynamics determine channel geometry and planform attributes.

Table 2.1 *Continued*

Scale	Timeframe of evolutionary adjustment Frequency of disturbance event	Geomorphic influence on aquatic ecology
Geomorphic unit	10^0–10^1 years 10^0 months	These landform-scale features reflect formative processes that determine river structure and function. Sediment transport processes at this scale create and rework bars, bedforms, and particle motions, sustaining ecological dynamics at equivalent scales. Channel and floodplain features such as pools, riffles, cascades, bar platforms, benches, levees, and cutoff channels form relatively discrete functional habitat units which represent individual, but interactive features of the landscape. For example, different geomorphic units may act as feeding (runs), resting (backwater), and spawning (gravel bars) sites for fish, such that the reach-scale assemblage of geomorphic units may influence the composition of fish assemblages. Alternatively, cutoff channels may support a number of breeding, feeding, and nesting habitats, while floodplain wetlands may be important for waterfowl, amphibians, reptiles and some mammals. The pattern of geomorphic units along a river course also provides a basis to analyze edge effects, and their associated ecotones.
Hydraulic unit	10^{-1}–10^0 years 10^{-1} months	This scale of feature is determined by (and shapes) flow–sediment interactions that reflect the energy distribution along a river course. Ecohydraulic interactions at this mesohabitat or biotope scale provide the physical context within which patch dynamics are appraised. These relationships vary markedly with flow stage, acting as a key determinant of the presence and pattern of refugia (e.g., in pools and secondary channels). Hydraulic interactions with particle clusters of differing caliber and structure provide an array of environments for a variety of benthic organisms. Substrate size, heterogeneity, frequency of turnover of sediment, and the rates of erosional and depositional processes are determinants of invertebrate diversity and abundance. Larger substrate provides insects with a firm surface to hold onto and provides some protection from the force of the current. At low discharges a variety of conditions are provided for feeding, breeding and cover, ranging from slow, deep flow in pools to fast, shallow flow on riffles. Alternatively, pools may be all that remains in flow terms, acting as refugia for a myriad of aquatic species. At higher discharges, sheltered areas such as overhanging concave banks adjacent to pools provide protection from high water velocities, providing cover for fish.
Microhabitat	10^{-1}–10^0 years 10^{-1} months	This scale of feature encompasses individual clasts or elements (e.g., logs, rocks, gravel patches) within a stream. The boundaries between these features are determined by changes in substratum type, character, or position. The diversity within functional habitats is examined by measuring internal structural gradients and patchiness. This local scale variability in surface roughness, flow hydraulics, or sediment availability and movement provides the conditions within which certain types of species assemblage develop. Surface–subsurface flow linkages through differing substrates fashion hyporheic zone processes and associated biotic interactions (such as nutrient spiraling). While highly sensitive to disturbance, this scale of feature and its associated biotic assemblage may recover quickly after disturbance.

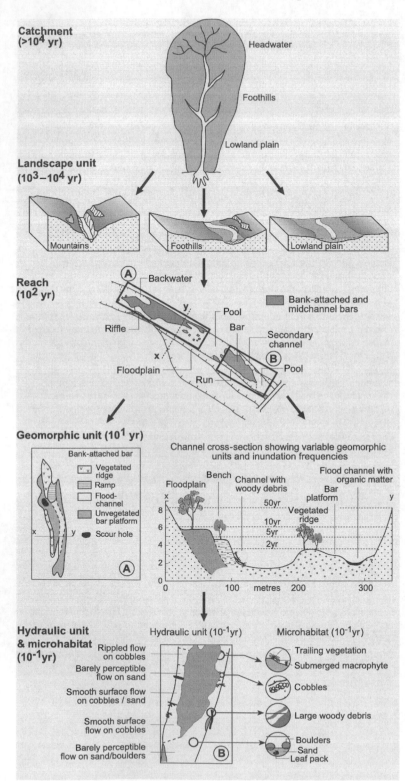

Figure 2.1 A catchment-framed nested hierarchy of geoecological interactions
Geoecological interactions can be considered within a hierarchy whereby smaller-scale landforms are nested within physical features at larger scales. Ecological interactions operate second-by-second at the microhabitat scale, as hydraulic interactions over geomorphic surfaces shape habitat availability. The geomorphic unit assemblage reflects reach scale controls that determine the distribution of energy and associated erosional and depositional forms. These factors, in turn, are controlled by valley confinement at the landscape unit scale. The spatial configuration of the catchment, represented by the pattern of landscape units, determines the distribution of geomorphic process zones. These larger-scale considerations set the boundary conditions within which rivers operate. The timeframe over which these interactions occur varies at differing scales in the hierarchy. Parts modified from Thomson et al. (2001) © John Wiley and Sons Limited. 2004. Reproduced with permission.

Lichty, 1965). For example, catchment-scale vegetation cover indirectly influences river character through its impacts on flow and sediment delivery. However, at the reach scale, vegetation may have a major role in determining stream boundary conditions and hydraulic resistance, as riparian corridors act as a buffer for flow, sediment, and nutrient throughput from slopes and adjacent floodplains. Finally, at the scale of geomorphic units, vegetation directly influences flow–sediment interactions at differing flow stages.

When used effectively, nested hierarchical frameworks provide an elegant tool with which to organize information, thereby presenting a coherent platform for management applications. In the sections that follow, various aspects of these scalar components are outlined.

2.2.1 Catchment-scale considerations: The boundary conditions within which rivers operate

Catchments, also referred to as drainage basins or watersheds, are clearly defined topographic and hydrological entities that have been described as the fundamental spatial unit of landscapes (Chorley, 1969) (Figure 2.1; Table 2.1). Catchment-scale considerations frame the boundary conditions within which rivers operate, constraining the range of river behavior and associated morphological attributes. For instance, regional geology and climate, among other factors, determine topography, sediment transport regime, and the discharge regime. These factors, in turn, influence patterns and rates of flow–sediment interaction through controls on the distribution of available energy.

In this book, catchment-scale boundary conditions are differentiated into two forms. First, *imposed boundary conditions* are considered to determine the relief, slope, and valley morphology (width and shape) within which rivers adjust. In a sense, these factors influence the potential energy of a landscape. They also constrain the way that energy is used, through their control on valley width and hence the concentration (or dissipation) of flow energy. Imposed boundary conditions effectively dictate the pattern of landscape units, thereby determining the valley setting within which a river adjusts.

Second, catchment-scale controls influence river character and behavior through the operation of *flux boundary conditions*, in particular the flow and sediment transfer regimes. Catchment-scale controls on the flow regime are determined largely by the climate setting. Stark contrasts in discharge regime are evident in arid, humid-temperate, tropical, Mediterranean, monsoonal and other climate settings, marking the differentiation of perennial and ephemeral systems, among many things. Climate also imposes critical constraints on magnitude–frequency relationships of flood events, and the effectiveness of extreme events (e.g., Wolman and Gerson, 1978). Secondary controls exerted by climatic influences at the catchment scale are manifest through effects on vegetation cover and associated rates of runoff and sediment yield in different environmental settings. Given their core influence on sediment production and fluxes, geological and climatic imprints are key considerations in determination of geomorphological provinces and ecoregions (Table 2.1).

Imposed and flux boundary conditions are appraised at the catchment scale. This entails analysis of factors such as landscape configuration, geology, catchment shape, drainage network, drainage density, tributary–trunk stream relationships, geographic location (connectivity and upstream–downstream relations), and landscape history. For example, catchment shape may exert a major influence on the pattern and rate of water and sediment fluxes. Factors that influence catchment shape include the history of uplift, the degree of landscape dissection, and the distribution of differing lithologies in a region. These boundary conditions, tied to long-term geological history, influence the shape of drainage networks and drainage density, thereby influencing within-system connectivity and the operation of biophysical fluxes (Figure 2.2). For example, catchment shape determines the relative size and frequency with which tributary streams join the trunk stream (among many factors). In elongate catchments, lower-order streams systematically and recurrently join the trunk stream (see Figure 2.2a). Progressive downstream increases in flow and relatively uniform increases in sediment loading (other things being equal) enable the trunk stream to maintain its capacity to transport its load. In

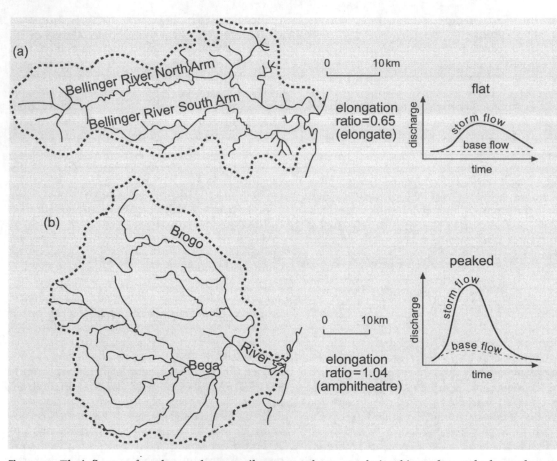

Figure 2.2 The influence of catchment shape on tributary–trunk stream relationships and storm hydrographs
Catchment shape and regional geology influence the pattern of tributary–trunk stream relationships. This exerts a secondary influence on biophysical fluxes such as the peakedness of flow during flood events. An elongate catchment (with a low elongation ratio) (Figure 2.2a) has a relatively suppressed flow duration curve (Bellinger catchment, New South Wales, Australia). In contrast, an amphitheatre, almost circular catchment (Figure 2.2b) is characterized by the convergence of several tributaries within a short distance along the trunk stream resulting in more peaked flood events (Bega catchment, New South Wales, Australia).

contrast, amphitheatre-shaped catchments may be characterized by dramatic increases in catchment area over relatively short distances along their long profiles, as several higher-order tributaries join the trunk stream (Figure 2.2b). This may lead to abrupt increases in water and sediment loadings, influencing the distribution of sediment stores along the trunk stream.

Within any catchment, individual subcatchments may have quite different physical attributes, with differing types and proportions of landscape units and associated variability in geo-morphic process zones. As such, interpretation of controls on river character and behavior is best framed in terms of subcatchment-specific attributes such as shape of the longitudinal profile, lithology, etc.

2.2.2 Landscape units: Topographic controls on river character and behavior

Just as drainage basins comprise a series of subcatchments, so each subcatchment can be differentiated into physiographic compartments based

Figure 2.3 The catchment-scale sediment conveyor belt
In general terms, rivers convey slope-derived sediments from source zones in headwaters, through transfer zones, to alluvial valleys in accumulation zones. The efficiency of this process depends upon the connectivity of differing landscape compartments. Rates of sediment input and the capacity of flow events to transport materials determine how jerky the operation of the conveyor belt is at any given time, and associated patterns of geomorphic response at differing positions along river courses. Modified from Kondolf (1994) and reprinted with permission from Elsevier, 2003.

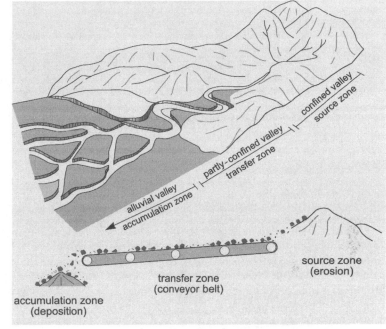

on relief variability (i.e., landscape morphology, assessed in terms of elevation, slope, and degree of dissection) and landscape position. In this book, areas of similar topography that comprise a characteristic pattern of landforms are referred to as *landscape units* (Figure 2.1; Table 2.1). Key factors used to identify landscape units include measures of relief, elevation, topography, geology, and position (e.g., upland versus lowland settings). As landscape units are a function of slope, valley confinement, and lithology, they not only determine the caliber and volume of sediment made available to a reach, they also impose major constraints on the distribution of flow energy that mobilizes sediments and shapes river morphology. Catchment to catchment variability in river character and behavior and the operation of biophysical fluxes are largely determined by the type and configuration of landscape units.

Downstream changes in slope and valley confinement result in widely differing settings in which rivers are able to adjust their morphology to varying degrees. These catchment-scale controls influence the nature and rate of erosional and depositional processes in differing landscape settings, determining the pattern of sediment source, transfer and accumulation zones (Schumm, 1977; see Figure 2.3; Table 2.2). Although sediments are eroded, transported, and deposited in each zone, the dominance of these processes varies spatially and temporally in each landscape compartment. Connectivity between geomorphic process zones in any given catchment influences the pattern and rate of flow and sediment transfer and other biophysical fluxes.

Relief variability, manifest primarily through the slope and confinement of the valley floor, is a key determinant of the *valley setting* in which a river is formed. In this book, three broad classes of valley setting are differentiated, namely confined, partly-confined and laterally-unconfined (see Chapter 4). The rate and extent of bedrock incision relative to valley widening determines valley width and shape. Tectonic setting is a primary control on this relationship, influencing the distribution of degrading and aggrading settings and resulting combinations of erosional and depositional landforms. Different sets of river character and behavior are found in zones that are dominated by erosional processes (where bedrock-confined rivers dominate), transfer zones (in partly-confined valley-settings where floodplains are

Table 2.2 Relationship between geomorphic process zones and landscape units.

	Examples of landscape units	Dominant fluvial process	Valley setting
Source	Mountain ranges, escarpment zones	Erosion via vertical-cutting; minimal sediment storage	Confined or partly-confined
Transfer	Tablelands, hills, erosional piedmonts	Erosion via lateral-cutting; fluctuating sediment storage	Partly-confined of laterally-unconfined with bedrock base
Accumulation	Coastal plains, alluvial plains, depositional piedmonts, playas	Deposition and net sediment accumulation	Laterally-unconfined with fully alluvial channel boundaries

discontinuous), and in accumulation zones (where alluvial streams are dominant).

Elevated areas such as mountain ranges, tablelands, and escarpments primarily comprise erosional landforms cut into bedrock. In mountain zones, vertical downcutting is the dominant fluvial process, producing steep and narrow (i.e., confined) valleys. These areas are dominated by processes of denudation (degradational zones) and act as sediment *source* zones (Figure 2.3). Sediments supplied from slopes are fed directly into the channel (i.e., coupling/connectivity is high). The bedrock valley condition is sustained as the transport capacity of rivers exceeds the rate of sediment supply from tributaries and slopes. Bedload transport mechanisms are dominant. A different set of processes is observed in tablelands, exemplified by low-relief plateau settings above escarpments. In these settings, underfit streams flow atop thin veneers of alluvial material in relatively wide valleys (*sensu* Dury, 1964). Slopes are largely disconnected from valley floors. In the main, these areas act as sediment storage zones made up primarily of suspended load deposits.

Downstream of upland sediment source zones, materials are typically conveyed through *transfer* reaches in landscape units such as rounded foothills or piedmont zones (Figure 2.3). In these areas, available energy remains sufficiently high to sustain the dominance of bedload transport along rivers, but a balance between net input and output is approached. Accumulation of debris exported from the headwaters, albeit in temporary stores, generally reflects downstream reduction in valley slope and increase in valley width. Considerable energy is expended eroding the base of confining hillslopes. These processes, combined with verti-

cal incision, create the space in which floodplain pockets form in partly-confined valley-settings. Floodplain pockets locally disconnect hillslopes from the channel. The character of the valley trough, in combination with slope and bed/bank material, exerts considerable control on channel planform and geometry. As these reaches have sufficient stream power to rework materials, any deviation in the flow–sediment balance may prompt changes to river morphology. For example, rates of bank erosion tend to be at a maximum in transfer reaches in midcatchment (Lawler, 1992).

Materials eroded and transported from upstream parts of catchments are deposited in flanking sedimentary basins (aggradational zones), such as lowland plains or broad alluvial plains in endorheic basins (Figure 2.3). The *accumulation* zone or sediment sink is marked by alluviation, aggradation, and long-term sediment storage. Alluvial channels develop, with continuous floodplains along each bank. Flow energy is dissipated across broad alluvial surfaces. In these lower slope settings, long-term prevalence of lateral-cutting has produced a broad valley trough in which the channel infrequently abuts the bedrock valley margin (i.e., slopes and channels are decoupled). As a consequence, sediments are delivered to the channel almost entirely from upstream sources. Relatively low stream power conditions reflect low slopes as base level is approached. The decline in stream power is marked by a decrease in bed material texture. Indeed, these tend to be suspended load systems.

Structural and lithological controls influence the degree of landscape dissection and resulting drainage patterns and density of river networks, thereby affecting the rate of water and sediment

Figure 2.4 Drainage patterns and examples from coastal New South Wales, Australia
(a) A wide range of drainage patterns has been described, influenced primarily by the regional geology (see text). Reprinted from Howard (1967) with permission of the AAPG © 2004 whose permission is required for further use. In the examples shown in (b), a parallel drainage pattern in the upper part of the Shoalhaven catchment, New South Wales, Australia, reflects the dissected plateau country of the Lachlan Fold Belt. This is transitional to a contorted pattern in the escarpment-dominated lower catchment, which is characterized by gorge retreat in the sandstone country of the Sydney Basin. The Goulburn catchment, a subcatchment of the Hunter River, New South Wales, Australia, has a rectangular drainage pattern reflecting the joint controlled sandstone landscape.

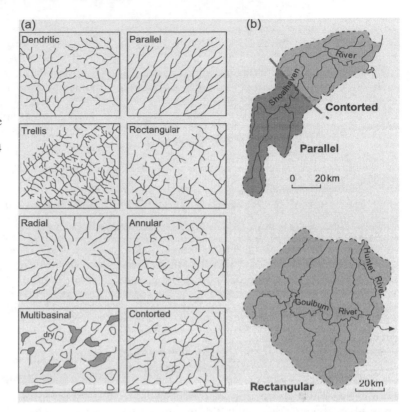

fluxes (Figure 2.4; Howard, 1967). Drainage patterns describe the ways tributaries are connected to each other and the trunk stream. Distinct patterns are commonly observed at the landscape unit scale (Thorne, 1997). The simplest form of drainage pattern, *dendritic*, develops in areas of homogeneous terrain in which there is no distinctive geological control. This configuration promotes relatively smooth downstream conveyance of sediment, because of a lack of structurally-controlled impediments (Ikeya, 1981; Takahashi et al., 1981). In many other settings, however, geological structure exerts a dominant influence on drainage pattern. For example, a *trellis* pattern is indicative of both a strong regional dip and the presence of folded sedimentary strata. Tight angle tributary junctions in these settings may induce short runout zones for debris flows (Johnson et al., 2000). A *parallel* pattern is found in terrains with a steep regional dip and marked lithological contacts that impose a preferred drainage direction. In areas of right-angled jointing and faulting, a rec-

tangular pattern is commonly observed. *Radial* and *annular* drainage patterns reflect differential erosion of volcanoes and eroded structural domes respectively. *Multibasinal* networks are typically observed in limestone terrains or in areas of glacially-derived materials. Finally, *contorted* drainage networks are generally associated with landscapes subjected to neotectonic and volcanic activity.

In general terms, more readily erodible rocks tend to have higher drainage densities. Maximum efficiency of flow and sediment transfer is achieved in basins with short slopes with complex bifurcating networks of small channels (i.e., badland settings). These conditions promote rapid geomorphic responses to disturbance events, and resulting transfer of flow and sediment. Drainage density also tends to be high in steep headwater areas where a multitude of lower order drainage lines occur, and in semiarid areas where a lack of vegetation cover facilitates landscape dissection (Knighton, 1990). Other climatic influences on geomorphic process activity and landscape forms in differing landscape units reflect variability in temperature and precipitation regimes and associated controls on runoff relationships (i.e., the discharge regime). For example, arid plains present a stark contrast to tropical steepland settings.

2.2.3 River reaches

Topographic constraints on river forms and processes result in differing ranges of river character and behavior in differing valley settings. In the nested hierarchical framework presented here, reaches are differentiated within each landscape unit (Figure 2.1; Table 2.1). Reaches are defined as "sections of river along which boundary conditions are sufficiently uniform (i.e., there is no change in the imposed flow or sediment load) such that the river maintains a near consistent structure" (Kellerhals et al., 1976). Alternating patterns of reach-scale river behavior may be termed segments (Frissell et al., 1986).

The critical issue in identification of reaches is determination of the attributes that are used to classify the river (see Chapter 4). Ultimately, reach boundaries must reflect discernible changes to river character and behavior. Reach boundaries may be distinct or gradual. Transitions in river type are generally coincident with a downstream change in one or more of the catchment boundary conditions within which the river operates. For example, a change in valley width may be coincident with a lithological break and differential resistance to erosion. Alternatively, major changes to flow and sediment discharge downstream of a tributary confluence may induce an abrupt change in river morphology. Any given landscape unit may contain multiple reaches. However, reach boundaries are not always coincident with landscape unit boundaries. In some instances, river morphology may be imposed by historical influences, such as former sediment supply conditions or climatic/tectonic fluvial regimes.

River character and behavior are influenced to a considerable degree by the space within which the river is able to move (see Chapter 4). In confined valleys the channel has limited capacity to adjust. In partly-confined valleys the channel adjusts around floodplain pockets. Given relatively high energy conditions, these sediment stores

are prone to reworking. Finally, a wide range of alluvial river forms may be evident in laterally-unconfined settings, where continuous floodplains line both banks of the channel. The assemblage of erosional and depositional landforms observed along rivers in these differing settings may vary markedly, as recorded by the assemblage of geomorphic units.

2.2.4 Geomorphic units

Rivers comprise reach-scale arrays of erosional and depositional landforms, referred to as geomorphic units (Figure 2.1; Table 2.1). The availability of material and the potential for it to be reworked in any given reach determines the distribution of geomorphic units, and hence river structure. Some rivers comprise forms sculpted into bedrock (e.g., cascades, falls, pools), while others comprise channel and floodplain forms that reflect sediment accumulation in short- or long-term depositional environments (e.g., midchannel bars versus a backswamp).

Geomorphic units are the building blocks of river systems (Brierley, 1996). Each landform has a distinct form–process association. Analysis of its morphology, bounding surface, and sedimentological associations, along with interpretation of its distribution and genetic associations with adjacent features, provides a basis to interpret formative processes (Miall, 1985). Given the specific set of flow (energy) and sediment conditions under which each type of geomorphic unit is formed and reworked, they are often found in characteristic locations along river courses. For example, point bars are found on the convex banks of bends, backswamps occur along distal floodplains, ridges and swales are located along the convex floodplains of some meandering rivers, and cascades are observed in steep headwater settings. Adjacent geomorphic units are commonly genetically linked, such as pool–riffle sequences, point bars with chute channels, and levee–floodchannel assemblages. Analysis of these features, and their assemblages, guides assessments of how rivers work, and are reworked, at differing flow stages. This enables interpretation of the range of formative events that sculpt the river, deposit materials, and rework and remold materials, providing insight into river behavior (see Chapter 5). In some instances, geomorphic units may reflect former conditions, such as extreme flood events.

Instream geomorphic units include a variety of bedrock and alluvial forms along a continuum of available energy (as determined by flow and slope) and sediment considerations (primarily the texture and volume of material) (see Chapter 4). They range from features sculpted in bedrock, such as falls, steps, and plunge pools, through to depositional features such as boulder dominated cascades, gravel riffles, and various sand and gravel bar types. Floodplain geomorphic units include a variety of laterally and vertically accreted features such as ridge and swale topography, billabongs, and backswamps. Interpretation of the character and juxtaposition of floodplain geomorphic units provides an insight into river history (e.g., formation of cutoffs, palaeochannels, etc.). Channel-marginal features, such as levees, influence the connectivity between channel and floodplain processes.

Various geomorphic units are shown for reaches in different process zones in Figure 2.5. In general terms, sculpted erosional forms and high energy depositional features such as boulder bars characterize confined (bedrock) rivers (Figure 2.5a). Partly-confined valleys have an array of instream features with a mix of erosional and depositional forms on floodplain pockets (Figure 2.5b). An amazing diversity of geomorphic units is evident along alluvial rivers such as anastamosing, braided, meandering, straight and wandering gravel bed rivers (Figure 2.5c). Although individual geomorphic units may be observed along reaches in a range of river types (e.g., pools are common along many variants of river morphology), the range and assemblage of geomorphic units provide a basis to differentiate among river types. At finer levels of resolution, hydraulic units may be characterized for instream geomorphic units, as indicated in the following section.

2.2.5 Hydraulic units

Hydraulic units are spatially distinct patches of relatively homogeneous surface flow and substrate character (Table 2.3; Kemp et al., 2000; Newson and Newson, 2000; Thomson et al., 2001). These can range from fast flowing variants over a range of coarse substrates to standing water environments on fine substrates. Flow–substrate interactions vary at differing flow stages. Several hydraulic units may comprise a single geomorphic unit. For example, distinct zones or patches may be evident within individual riffles, characterized by differing substrate, the height and spacing of roughness elements, flow depth, flow velocity, and hydraulic parameters such as Froude and Reynolds Number.

Table 2.3 Classification of surface flow types. From Thomson et al. (2001) © John Wiley and Sons Limited. 2004. Reproduced with permission.

Flow type	Description
Free fall	Water falling vertically without obstruction. Often associated with a bedrock or boulder step.
Chute	Fast, smooth boundary turbulent flow over boulders or bedrock. Flow is in contact with the substrate and exhibits upstream convergence and divergence.
Broken standing waves	White-water tumbling waves with crest facing in an upstream direction.
Unbroken standing waves	Undular standing waves in which the crest faces upstream without breaking.
Rippled	Surface turbulence does not produce waves, but symmetrical ripples that move in a general downstream direction.
Upwelling	Secondary flow cells visible at the surface by vertical "boils" or circular horizontal eddies.
Smooth surface flow	Relative roughness is sufficiently low that very little surface turbulence occurs. Very small turbulent flow cells are visible, reflections are distorted and surface "foam" moves in a downstream direction.
Scarcely perceptible flow	Surface foam appears stationary, little or no measurable velocity, reflections are not distorted.
Standing water/swamp	Abandoned channel zone or backswamp with no flow except at flood stage.

(a) Source zone

Figure 2.5 Geomorphic units in source, transfer, and accumulation zones
These photographs provide examples of geomorphic units in source (a), transfer (b), and accumulation (c) zones (see Chapter 4 for further details).

(c) Accumulation zone

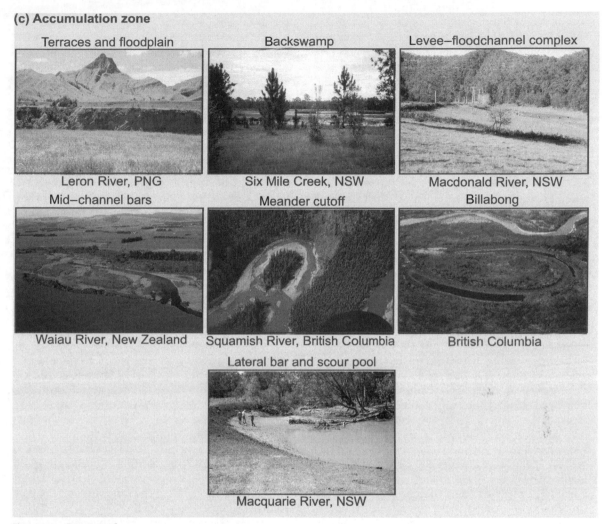

Terraces and floodplain — Leron River, PNG

Backswamp — Six Mile Creek, NSW

Levee–floodchannel complex — Macdonald River, NSW

Mid–channel bars — Waiau River, New Zealand

Meander cutoff — Squamish River, British Columbia

Billabong — British Columbia

Lateral bar and scour pool — Macquarie River, NSW

Figure 2.5 *Continued*

Aquatic habitat zonation along river courses can be related to the character and distribution of hydraulic units (Harper and Everard, 1998; Wadeson and Rowntree, 1998; Table 2.1). Indeed, ecohydrologic or ecohydraulic relationships form an integral component of many flow management strategies. In many settings, this instream focus must be linked with concerns for channel–floodplain connectivity, as a major component of the ecological diversity of a river system may be evident in wetlands on the floodplain.

2.2.6 Summary of geoecological considerations at different scales of the nested hierarchy

Primary linkages between geomorphological and ecological forms and processes at differing scales in the nested hierarchy are summarized in Table 2.1. Geoecological considerations at the catchment scale determine the boundary conditions within which aquatic ecosystems operate. Comparisons among stream networks aid interpretation of habitat/ecosystem diversity, biophysical linkages (connectivity), network efficiency, species distributions/migrations, sediment sources, and transport, etc. The operation of biophysical fluxes at the landscape unit scale is a critical determinant of river character and behavior, shaping the pattern and rate of flow–sediment transfer and the longitudinal, lateral, and vertical connectivity of the system. Assemblages of geomorphic units at the reach scale represent a primary guide to habitat diversity and community structure along a river course. For example, geo-ecological differentiation of an upland swamp and a gorge, or a braided river with no floodplain and an anastamosing channel network with multiple backchannels and a network of wetlands on its floodplain is self-evident. Richards et al. (2002) note that dominant fluvial and ecological processes seem to be fundamentally congruent at the reach scale, wherein dynamic changes to channel geometry and planform can be related to patch dynamics of ecosystem turnover and maintenance. Geomorphic units are the building blocks of river systems, in geomorphological and ecological terms, presenting particular constraints on the availability and viability of habitat at differing flow stages. As such, spatial and temporal adjustments to geomorphic units affect patterns and rates of resource utilization, competition, and habitat selection along river courses. Finally, flow–sediment interactions at the hydraulic unit scale are critical determinants of instream habitat viability, fashioning the operation of various biophysical processes that determine ecosystem functionality. Comparisons among microhabitats might include nutrient availability, substratum preference, feeding habits, etc. Collectively, these various interactions at differing scales indicate the vital part played by geomorphic (landscape) considerations in providing a physical template to examine aquatic ecosystems and develop river rehabilitation initiatives.

2.3 Use of geomorphology as an integrative physical template for river management activities

Principles from fluvial geomorphology provide an ideal starting point from which to evaluate the interaction of biophysical processes within a catchment, as geomorphological processes determine

the structure, or physical template, of a river system. In this section, geomorphic interactions with biophysical attributes of river courses are analyzed in terms of habitat availability (Section 2.3.1), flow considerations (Section 2.3.2), substrate conditions (Section 2.3.3), and vegetation associations (Section 2.3.4).

2.3.1 The geomorphic basis for management of habitat availability along river courses

Habitats refer to the places or environments in which species live. The physical structure of the river (i.e., substrate conditions, channel shape and size, assemblage of geomorphic units, floodplain attributes, etc.) provides a template for biotic interactions and associations (Southwood, 1977; Newson et al., 1998a; Newson and Newson, 2000; Thomson et al., 2001, 2004). Geomorphic diversity of river structure determines the amount and diversity of physical habitat along river courses (see Figure 2.6). As noted by Jacobson et al. (2001, p. 201), "It is generally accepted that physical habitat determines a template for aquatic ecosystem functions, but realization of the potential is highly dependent on other ecological processes." Similarly, Montgomery (2001, p. 247) noted that "ecologists have come to increasingly recognize the importance of the 'geomorphic template' that can structure ecological processes, habitat characteristics, and their dynamic interactions." The geomorphic diversity of rivers, and adjustments to river structure and function, affect the range of biotic opportunities along a river course.

Geomorphic processes, fashioned by the flow regime, vegetation associations, and sediment availability, induce direct controls on the distribution of energy within a system. These interactions determine the character and distribution of channel and floodplain forms, the local-scale pattern of erosion and deposition (and hence substrate character), and associated hydraulic diversity at differing flow stages. Collectively, these components induce different ranges of habitat availability in different settings at differing flow stages, exerting a strong influence on relationships between habitat and community structure and species richness. When integrated with measures such as elevation and riparian vegetation cover, and their effects on temperature and nutrient availability, these

Figure 2.6 Geomorphic controls on habitat availability for three contrasting river types

The pattern of geomorphic units along a reach affects the structural heterogeneity and hydraulic diversity of the river, along with channel–floodplain connectivity, vegetation associations (shading and temperature), etc. Variability in these biophysical interactions influences the availability and viability of habitat along these different river types, along with other measures of aquatic ecosystem functioning, such as nutrient fluxes, food web processes, predation, etc., and hyporheic zone processes (surface–subsurface flow interactions). The intact valley fill example (a) has a simple geomorphic structure, with occasional ponds and a discontinuous channel on an otherwise undifferentiated valley floor. Swampy deposits and vegetation associations contrast with surrounding open forest. Habitat diversity is limited by the suspended load nature of the system. In example (b), discontinuous floodplain pockets alternate from side to side of the valley floor. Significant instream and floodplain diversity in geomorphic structure, habitat availability, and vegetation associations is evident. In example (c), heterogeneity of geomorphic and hydraulic diversity is most pronounced, presenting the widest range of habitat availability in the three examples shown. (a, b) reproduced from Brierley et al. (2002) with permission from Elsevier, 2003.

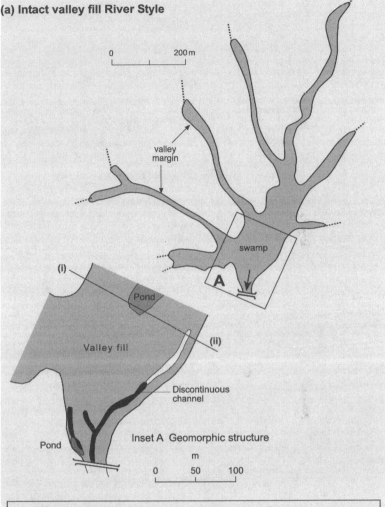

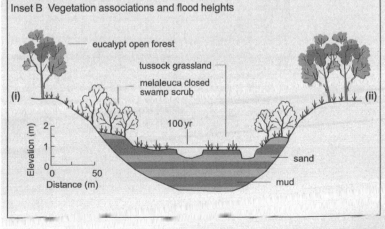

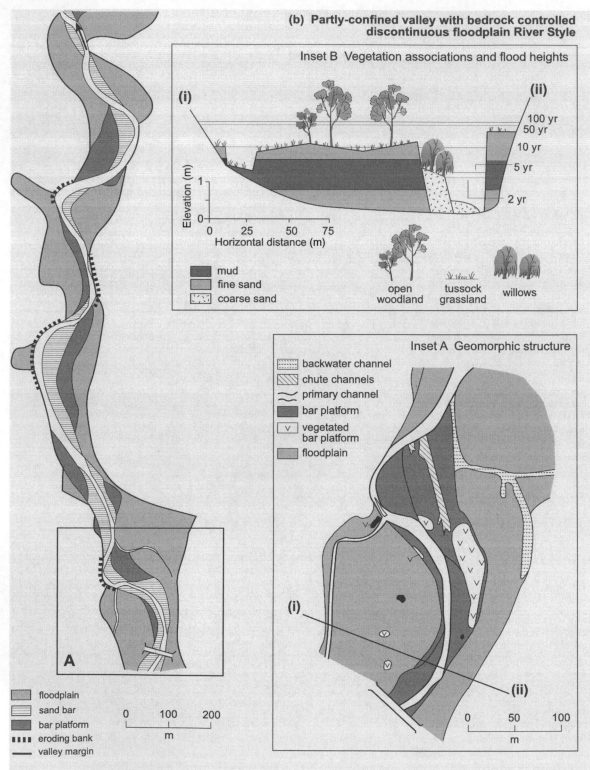

Figure 2.6 *Continued*

(c) Meandering gravel bed River Style

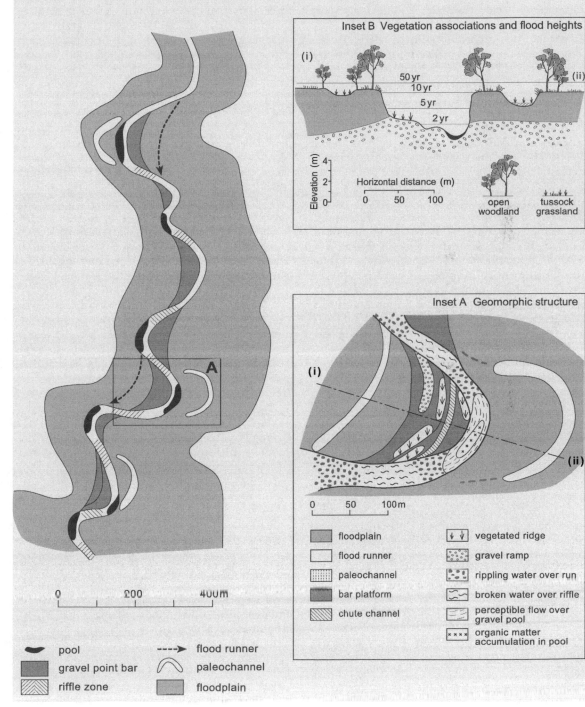

Inset B Vegetation associations and flood heights

Elevation (m)

Horizontal distance (m)

open woodland   tussock grassland

Inset A Geomorphic structure

floodplain	vegetated ridge
flood runner	gravel ramp
paleochannel	rippling water over run
bar platform	broken water over riffle
chute channel	perceptible flow over gravel pool
	organic matter accumulation in pool

pool   flood runner

gravel point bar   paleochannel

riffle zone   floodplain

0   200   400m

Figure 2.6 *Continued*

principles provide a coherent physical template for geoecological research and management applications.

Because geomorphic units form relatively discrete habitat units, and different types of river are characterized by their assemblage of geomorphic units, geomorphological classification procedures provide a reasonable basis to characterize habitat and biotic assemblages along river courses. For example, billabongs and backswamps with significant wetland habitats may be prominent (Table 2.1) along different types of meandering or anabranching river, while an array of geomorphic units with their own assemblage of aquatic species may occur along steep headwater streams. Similarly, differing assemblages of landforms present a range of habitats that may be important at different life stages for different organisms. For example, different geomorphic units act as feeding (runs), resting or holding/refuge (pools, backwater areas), and spawning (gravel bars) sites, such that the reach-scale assemblage of geomorphic units influences the composition of fish assemblages (e.g., Wesche, 1985; Aadland, 1993). Over the course of a day, fish may use pools, riffles, and adjacent units for different activities (e.g., Hawkins et al., 1993; Montgomery et al., 1995a; Bisson and Montgomery, 1996).

At a finer scale of resolution, hydraulic units also exert a direct influence on local habitat availability (Table 2.1). Several hydraulic biotopes may be nested within a single geomorphic unit. In general terms, macroinvertebrates respond to these local habitat variables rather than to larger scale factors (e.g., Thomson et al., 2004). Interactions at differing flow stages shape patch dynamics and edge effects along river courses. These interactions are especially important at channel margins, marking the interface between aquatic and terrestrial ecosystems of the riparian zone. Ultimately, these local habitat differences are driven by, and possibly predictable from, larger-scale geomorphic processes at reach, landscape unit, and catchment scales.

Assessments of aquatic habitat must also consider the floodplain compartment, ensuring that the integrity of hydrologically contiguous areas is maintained. Dependent on the setting, a range of geomorphic features and associated habitats may be evident in different types of floodplain, including wetlands, billabongs (oxbow lakes), secondary

channels, cutoff channels, backwater channels, floodplain ponds, backswamps, etc. For example, habitat heterogeneity in complex, laterally active reaches may be essential in the maintenance of aquatic communities (Power et al., 1995; Beechie et al., 2001). Similarly, many species have adapted their life cycles to the periodicity of flooding and the reliability of floodplain inundation. For example, many tropical wetlands are characterized by areas of remarkable nutrient productivity and comprise extremely diverse ecosystems (e.g., Amoros et al., 1987). In some instances, wetlands may provide a wide range of habitat for short intervals of time, providing a rich food source for various species (e.g., frogs, fish), impacting profoundly on bird migratory pathways.

Over time, ecological interactions have been influenced to a greater and greater degree by human activities. In many instances, the consequences of these changes are effectively intractable. Habitat loss associated with modification of river character and behavior has instigated loss of native flora and fauna, marked changes in spatial ranges and interactions, and incursion of exotic species. In many instances, simplification of river courses has reduced the geomorphic complexity of channels, and altered channel–floodplain connectivity, reducing the diversity of habitat (see Chapter 7).

Biological, chemical, and hydrological interactions along river courses are fashioned, to varying degrees, by the geomorphic template upon which biophysical interactions occur. Although principles from geomorphology provide a critical physical template for analysis of *habitat availability* along river courses, many other considerations must be appraised to determine *habitat viability* and unravel controls on ecological functioning. Various ecological processes are influenced by factors that are largely independent of river morphology, such as climate, elevation, aspect, and biogeographic factors, biotic interactions (e.g., food web processes, predator–prey relationships, nutrient availability, etc.), and aquatic geochemistry. Indeed, any limiting factor that affects an aquatic system can affect its viability, potentially compromising critical linkages in the chain of life. Disruptions to ecosystem functioning may be imposed by loss of habitat or refugia (at any stage in the life cycle), chemical or thermal limitations (e.g., nutrient deficiencies, contaminants, loss of shade), or any factor that breaks the continuity of the food web and/or predator–prey relationships. Spatial and temporal variability in species mobility must also be considered in evaluations of geomorphic and ecological interactions (e.g., Frothingham et al., 2001). Hence, evaluations of habitat should emphasize the diversity of available habitat, rather than the extent of a single type of habitat, ensuring that viable habitat is available for all stages in the life cycles of species (McKenney, 2001). Given these considerations, two reaches of the same geomorphic type within the same catchment will not necessarily have equivalent viable habitat and biotic assemblages, despite physical similarities (Kershner et al., 1992; Thomson et al., 2004). The potential range of habitats within comparable reaches, along with the dynamic nature of stream habitats and the biotic and abiotic factors that influence them, may ensure that identification of truly representative geoecological reaches is unachievable.

2.3.2 The geomorphic basis for management of river flows

The magnitude, frequency, timing, and duration of overbank flows, formative (bankfull stage) flows, low flows, and periods of no flow, shape, and in turn are shaped by, the geomorphic structure of a river. Many channel features, such as bars and riffle–pool sequences, are formed and maintained by dominant (or bankfull) discharge events, while overbank events form and reshape floodplains. Linkages between geomorphic river structure and function and biotic attributes of aquatic ecosystems are largely mediated through biophysical interactions at different flow stages. Indeed, species have evolved to exploit the habitat mosaic that is created and maintained by hydrologic variability (Poff et al., 1997). For many riverine species, completion of their life cycle requires an array of habitat types, the availability of which may be determined by the flow regime (e.g., Sparks, 1995) Hence, adaptation to a range of conditions is an integral component in the evolution of aquatic ecosystems, as disturbance events destroy and recreate habitat elements.

Natural variability in the timing and duration of flows varies over differing timescales in different environmental settings, exerting a profound influence on river structure and function, and associated ecological considerations (Poff and Ward, 1989;

Bailey and Li, 1992; Puckridge et al., 1998). Patterns and rates of hydrological connectivity determine biogeochemical exchanges between landscape patches (Petts and Amoros, 1996; Amoros and Bornette, 2002; Ward et al., 2002). These interactions, which are flow stage dependent, determine patterns and rates of biophysical fluxes in longitudinal (upstream–downstream), lateral (channel–floodplain), and vertical (surface–subsurface) dimensions.

Hydrological connectivity varies markedly for different types and scales of rivers at different positions in landscapes (and catchments), and in differing environmental settings. For example, patterns and rates of biophysical fluxes operate quite differently in small, headwater streams that are largely bedrock-based and have no floodplain, relative to expansive lowland plains with differing substrate conditions and an array of floodplain features such as billabongs and wetlands. Alternatively, flash floods in arid environments are irregular occurrences, while seasonal flooding in monsoonal settings is a relatively predictable phenomenon. In the former setting, the periodicity of inundation and ecosystem responses are particularly important in the maintenance of aquatic ecosystems, as extreme yet irregular flood events may trigger breeding cycles that have been dormant for many years.

A river's flow is derived from some combination of surface water, soil water, and groundwater. Climate, geology, topography, soils, and vegetation affect the supply of water and the pathways (and timing) by which it reaches the channel (Poff et al., 1997). Collectively, overland and shallow subsurface flow generates flood peaks following storm events. In contrast, deeper groundwater stores maintain base flow during periods of little rainfall, providing a steady source of generally high quality water at a relatively constant temperature (Potter and Gaffield, 2001). Moving downstream, discharge reflects the sum of flow derived from multiple subcatchments. The time taken for flow to be transferred down the system, combined with nonsynchronous tributary inputs and larger downstream channel and floodplain storage capacities, attenuate and dampen flood peaks in downstream reaches (Poff et al., 1997).

Flow variability determines the frequency and duration of inundation of differing geomorphic surfaces. Geomorphic units are formed and reworked by flow events of differing magnitude and recurrence. River structure is also shaped by the history of flow events. These relationships influence the sedimentary makeup of geomorphic units, the vegetation that grows on them, and the formation, maintenance, and disruption to habitat that occurs under differing flow conditions (Poole, 2002; Richards et al., 2002). Habitat diversity also changes as a function of flood magnitude (e.g., McKenney, 2001).

Flow regulation tends to reduce river complexity and habitat heterogeneity. Changes to the natural pattern of hydrologic variation and disturbance alter habitat dynamics, creating new conditions to which native biota may be poorly adapted (Poff et al., 1997). Many studies have demonstrated that altering hydrologic variability in rivers is ecologically harmful (e.g., Sparks, 1995; Stanford et al., 1996; Richter et al., 1997). Indeed, manipulation of flow probably represents the greatest single human impact on river ecosystems (Postel and Richter, 2003). Bunn and Arthington (2002) suggest four major reasons why flow modifications have been so devastating to river species and ecosystems. First, because river flows – and particularly floods – shape the physical habitats of rivers and their floodplains, changes to flow affect the distribution and abundance of plants and animals. In some instances, loss of a particular component of the flow regime may completely eliminate species that are dependent upon habitats that are no longer available after the flow alteration (Poff et al., 1997). Second, survival and reproductive strategies of various aquatic species have evolved such that they are keyed to natural flow conditions. If the flow conditions for a species to successfully complete its life cycle no longer exist, the species will quickly decline or disappear. Third, many species require adequate water depth at critical times of the year to facilitate their movements upstream and downstream and from the channel to the floodplain (and vice versa). Flow alterations that inhibit these movements may prevent them from reaching feeding and breeding sites that are critical to their growth and reproduction. Fourth, altered flow conditions often favor nonnative species that have been introduced into river systems, placing greater survival pressures on native species. For example, a regulated flow regime with less variability may give exotic fish a competitive

advantage over native species that are better adapted to flooding, impacting on patterns of fish migration, body size, and species richness (Poff and Allan, 1995). As the flow regime becomes more predictable and less variable, biotic interactions such as competition and predation may dominate over abiotic (physical and chemical) processes (Poff and Ward, 1989).

Impacts of flow regulation on aquatic ecosystems vary greatly, depending on the hydrogeomorphic setting of the river, and the operation of reservoirs relative to the prevailing sediment transport regime. The ecological significance of changes in flow may be accentuated or mitigated by other factors, including presence/absence of additional dams, contributions from unregulated tributaries, or by regional differences in the natural hydrologic regime and/or sediment availability (Pitlick and Wilcock, 2001). In general terms, reservoirs operated for flood control, water storage, and power generation typically alter the natural hydrograph by reducing peak flows and increasing base flows. The net effect is a flow regime with less variability that lowers the frequency of disturbance and reduces the diversity of riverine habitat. This creates conditions that are preferentially used by species that are better adapted to more stable flows (Stanford et al., 1996). Resulting shifts in ecosystem structure and function are likely to be most severe in arid and semiarid regions where natural flow variability is high and the amount of water impounded by reservoirs is large in proportion to the annual flow (Graf, 1999).

At the local scale, ecohydrological effects of dams and water diversions include adjustments to substrate and imposition of barriers to fish migration (Pitlick and Wilcock, 2001). Broader-scale impacts may include alterations to processes that affect nutrient cycles (Newbold, 1992), channel–floodplain interactions (Stanford and Ward, 1993), riparian vegetation (Friedman et al., 1998), food webs (Wooten et al., 1996), sediment loads (Andrews, 1996), and channel geomorphology (Ligon et al., 1995; Van Streeter and Pitlick, 1998).

Profound human disturbance to flow regimes has necessitated the implementation of environmental flows to maintain or reinvigorate ecosystem functioning (Dollar, 2000; Whiting, 2002; see Table 2.4). A range of flow disturbance regimes is required. Maintenance of the viability of habitat

Table 2.4 Forms of environmental flow management strategies. Modified from Dollar (2000) and reprinted with permission from Hodder Arnold, 2004.

Name	Definition
Fish maintenance flows	Minimum flows that are necessary for short-term maintenance of fish populations.
Channel maintenance flows	Minimum flows that maintain the erosional and depositional processes in the channels, and the immediately adjacent riparian zone.
Riparian maintenance flows	Overbank flows that are needed for the establishment of off-channel riparian vegetation and off-channel aquatic habitat (e.g., in backswamps).
Valley forming flows	Large, infrequent flows that are overbank and are needed to maintain channel patterns, riparian areas and buffer hillslope infringement processes.
Sediment maintenance flows	Designed to maintain the sediment transport and storage capacities of a reach.
Low flows	Flows that have the longest duration and provide seasonal habitat for individual species. These flows also maintain base flow conditions, providing habitat refugia in dry times.
Freshes	Small, short-lived increases in flow. This flow variability initiates scour and cleanses the river bed, dilutes poor water quality, and triggers the spawning of fish.
Floods	Substantial flow increases that cause significant bed scour, bank erosion, and sediment transport. Overtopping of banks provides the hydraulic link between the channel and the floodplain.

throughout species' life cycles requires that refugia are maintained across the range of flow conditions, from low flow stage to inundation of floodplain and wetland features, as necessary. Low flow strategies address concerns for the flow velocity and depth characteristics of hydraulic units or maintenance of low flow stage refugia (e.g., sustaining connectivity between pools and riffles). Designated flows termed freshes may be allocated to flush silt from the interstices of gravel bars, aiding hyporheic zone functioning. Turning the bed over creates fresh surfaces along a reach. For a fresh to be effective, fine material must be fully suspended and removed from the reach (Kondolf and Wilcock, 1996). However, the coarse material cannot become so mobile that it causes high mortality among aquatic organisms, or sediments are depleted. Alternatively, short-term evacuation of fine sediment from pools or deposition of coarse sediment in bars may increase the quality of habitat used for spawning, juvenile cover, and invertebrate food production (Pitlick and Wilcock, 2001).

At flow stages up to bankfull, sediment maintenance flows of differing frequency and magnitude may be applied to change substrate characteristics, potentially altering channel morphology and associated habitat. Channel maintenance flows may be applied to scour sand from pools, or maintain pool

riffle morphology and associated active channel width and topographic diversity. In striving to maintain or improve channel structure, heterogeneity of bed material, and associated diversity of habitat, it is important to determine whether the various sizes of sediment will move as bedload or suspended load, and whether transport rates will significantly alter the sediment balance of any given reach (Pitlick and Wilcock, 2001). Unless the balance of water and sediment transfer is maintained, accentuated rates of deposition or removal may result in loss or simplification of habitat. In settings where significant habitat value must be protected in the riparian zone and on the adjacent floodplain, riparian maintenance flows may address concerns for floodplain connectivity (including nutrient cycling, sediment transfer, etc.). Finally, valley-forming flows may be required to instigate overbank floods that modify floodplain morphology and activate wetlands, affecting flora and fauna life cycles.

Maintenance of the inherent variability and diversity of natural flow regimes represents a vital component in river management plans that address concerns for ecosystem integrity and native biodiversity. Ideally, flow management plans mimic the natural duration, timing, magnitude, and sequencing of flows to which a range of organisms have adapted and evolved (Poff et al., 1997). Quantitatively achievable objectives must be defined for flows of a given magnitude, duration, and timing, emphasizing that these goals are specific to the type of river under investigation and its position in the catchment.

2.3.3 The geomorphic basis for management of substrate conditions along rivers

Substrate conditions, whether characterized by rooted vegetation, dead wood, or varying sizes of inorganic particles (silt, sand, gravel, and cobble), are primary determinants of the abundance and diversity of many aquatic organisms, especially macroinvertebrates, macrophytes, and algae. Differing substrate conditions are also preferentially utilized at different stages in the life cycles of aquatic fauna such as fish. The frequency and efficiency with which flow reworks bed material exerts a major influence on geomorphic river structure (i.e., the type and distribution of geomorphic units) and associated hydraulic interactions (e.g., extent of inundation, flow type). These relationships vary markedly in bedload, mixed load, and suspended load rivers (see Chapter 3). Along suspended load rivers, bed heterogeneity tends to be limited and the geomorphic structure of rivers is relatively simple. Turbidity levels tend to be naturally high. Along gravel or coarser bedload rivers, bed material organization and the degree of armoring and packing determine the mobility of materials and the prevailing river structure. Once the surface armor is broken, large volumes of readily movable material may be released. A wide range of textures may be evident in these systems. In contrast, sand bed channels often have a remarkably uniform texture, and are characterized by nonturbid flows (i.e., the concentration of suspended load materials is very low under most flow conditions).

The relationship between sediment availability and the capacity of a reach to transport it determines the type and distribution of geomorphic units along a reach. This reflects, among many considerations, the balance of erosional and depositional processes along the reach, as determined by the nature and extent of impelling and resisting forces (see Chapter 3). Forcing elements such as bedrock constrictions, bedrock steps, riparian vegetation, and woody debris may exert a profound impact upon these relationships. At the extremes of river behavior, channels are scoured to bedrock or smothered by slugs of bedload caliber materials (supply- and transport-limited situations respectively). Most reaches fall between these extremes. Changes to the flow–sediment balance may alter the proportion of bedrock and bed material caliber along a reach, or modify the degree of armoring. Alternatively, an influx of suspended load materials may clog the interstices between gravels. These adjustments may change river structure, altering the ability of the river to convey sediments, thereby exerting a critical impact on the availability and viability of aquatic habitat (e.g., Pitlick and Van Streeter, 1998).

The importance of substrate heterogeneity and stream bed disturbance varies widely within and between streams (Pitlick and Wilcock, 2001). Benthic invertebrates, and fish that rely on coarse substrates for spawning, benefit from periodic flushing flows or floods (Kondolf and Wilcock, 1996). Although the movement of coarse bed material during floods may reduce invertebrate populations and species abundance, periodic, low level disturbance marked by bedload transport events of low to intermediate intensity and frequency are often necessary to maintain habitat quality and species diversity. Such events are very important for maintaining aquatic habitats, especially for migratory fish species or generalist fish that utilize a wide range of habitats (Pitlick and Wilcock, 2001). With increasing discharge and sediment transport intensity, refuge opportunities for these mobile organisms decrease (McKenney, 2001). Reworking of surface clasts is needed to remove fine sediment from the bed in order to maintain spawning habitat or interstitial void spaces for invertebrates and juvenile fish. If the supply of inorganic sediments increases, habitats may be smothered, potentially eliminating reproductive habitats, transforming riverine productivity, and altering food web processes, thereby bringing about biotic adjustments. For example, the filtering capacity of gravel bars and the bed influence hyporheic zone processes, impacting on ecological functioning and nutrient cycling.

Oversupply of materials may generate a sediment slug (see Figure 2.7). Downstream transmission of the slug is marked by a cycle of aggradation

Figure 2.7 The importance of sediment transfer in river management applications
Sediment flux is a key component in effective river management planning. Understanding where sediments are derived from, how often and where they are transferred to, and where they are stored along river courses (and for how long) influences determination of what is realistically achievable in river rehabilitation. Identification of sediment-starved and sediment-choked reaches is a key consideration. Assessment of river recovery potential is influenced by reach position in the catchment relative to reaches that are sediment-starved or sediment-choked and the off-site impacts that upstream sediment availability or exhaustion exert upon the recovery processes. Sediment-choked rivers often contain sediment slugs: (a) is a piggy-back bridge on the Macdonald River north of Sydney, Australia built in response to the movement of a sediment slug through the system, (b) is on the Waiapu River in New Zealand where tens of meters of gravel have infilled valleys, burying floodplains and houses, and reducing geomorphic heterogeneity. Sediment-starved rivers often have low recovery potential as sediments required to narrow channels and build geomorphic features are limited. The Australian examples shown are from the upper Latrobe catchment in Victoria (c) and Greendale Creek in Bega catchment, New South Wales (d).

and degradation, with accompanying changes to channel planform and cross-sectional geometry. Initially, aggradation promotes the development of a multichannel configuration and channel widening, decreasing channel heterogeneity and smothering habitat. The subsequent degradation phase is characterized by reversion to a single, narrower channel. Typical migration rates range from 0.1–0.5 km yr^{-1} for smaller slugs to 1–5 km yr^{-1} for slugs generated from mining waste (Nicholas et al., 1995). Hyporheic zone functioning and various other measures of aquatic ecosystem functioning vary markedly during these different phases of geomorphic adjustment. In contrast, limited sediment availability in sediment-starved systems may promote "hungry water" (Kondolf, 1997), potentially obliterating aquatic habitat (Figure 2.7).

Substrate composition is one of the most easily manipulated habitat characteristic in rehabilitation projects. Consideration must be given to the type of substrate, the degree of embeddedness, size of particles, contour of the substrate, and heterogeneity of substrate types in the source and recipient areas (Gore, 1985, 2001). These analyses must be placed in context of the broader sediment flux (see Chapter 3). Understanding the spatial and temporal distribution of sediment source, transfer and accumulation zones, and the connectivity between different landscape compartments, provides an initial basis to predict how a river will respond to changes in the sediment regime. Projects that fail to consider current trends in sediment delivery are likely to require costly maintenance, or fail to achieve their intended goal (Simon, 1995; Sear et al., 1995; Sear, 1996).

2.3.4 The geomorphic basis for riparian vegetation management

Morphodynamic relationships between geomorphic river structure and riparian vegetation associations lie at the heart of the physical template that records the diversity of river forms and processes. Mutual adjustments among these components exert a dramatic impact upon the availability and viability of aquatic and terrestrial habitat along valley floors. Indeed, riparian zones and floodplains link aquatic and terrestrial ecosystems, regulating fluxes of water, nutrients, and organic matter along river corridors (Gregory et al., 1991).

The affect of riparian vegetation and woody debris on the geomorphic structure of a river is manifest at various scales. At finer scales, patterns of erosion and deposition are influenced, resulting in different substrate types and sediment mixes in the channel zone. Significant local-scale heterogeneity in the assemblage of geomorphic units is evident along rivers with intact riparian forests and high loadings of woody debris (e.g., Collins and Montgomery, 2001; Brooks and Brierley, 2002). At coarser scales, the distribution of resisting elements along valley floors influences channel capacity, channel planform, and the character and rate of river adjustment (e.g., Millar, 2000). These considerations are largely influenced by, and in turn exert an influence upon, the distribution of energy within a river system. The nature of these interactions primarily reflects the environmental

setting (climate, topography, ecoregion, etc.), the periodicity of inundation (frequency, recurrence, and extent of flow), substrate conditions (including soil characteristics, access to water table, etc.), and system history (e.g., the sequence of disturbance events and the interval since the last major flood or fire).

Flow stage relationships (as influenced by channel morphology and capacity), determine the frequency and periodicity of inundation of differing geomorphic surfaces, affecting the capacity for sediment reworking, patterns of deposition, soil characteristics, depth to water table, and availability of nutritional resources (Malanson, 1993; Hupp and Osterkamp, 1996). These factors influence the potential for germination and growth of vegetation on different surfaces. Once established, vegetation patterns induce resistance along river courses, thereby affecting the geomorphic effectiveness of flow events and the evolution of geomorphic surfaces (Richards et al., 2002). Resulting vegetation patterns may range from systematic successional associations induced by lateral migration of channels, to patches characterized by abrupt transitions, such as those associated with channel abandonment along wandering gravel bed or anastomosing rivers.

Riparian vegetation associations vary markedly for different types of river in different environmental settings. For example, the geomorphic role of riparian vegetation and woody debris is quite different in arid and semiarid climates (whether hot or cold desert conditions) relative to humid-temperate or tropical settings. Markedly different patterns of riparian vegetation are evident along, say, upland swamps, gorges, meandering rivers in rainforests, and alluvial plains in arid zones (Figure 2.8). In some settings, nutrient availability may be a limiting factor to vegetation growth; elsewhere, excess nutrients may promote weed infestation and channel choking. Changes to the physical template of a river may alter the presence, character, and growth strategies of terrestrial and aquatic vegetation and invasive weeds (see Table 2.5). Secondary impacts may include alterations to organic matter and nutrient input, water temperature, and habitat availability.

Given the genetic link between riparian vegetation and recruitment of wood along rivers, loadings of woody debris vary markedly in different

Figure 2.8 Vegetation associations along different types of river in different environmental settings
A range of vegetation associations is shown for different types of river in Australia. River red gums characterize the banks of arid zone rivers of central New South Wales such as the Macquarie River (a), and the Darling River at Bourke (b). Swampland vegetation and reeds grow along discontinuous water courses such as those in tributary systems of the Bellinger River, New South Wales (c), and the Macquarie Marshes of central New South Wales (d). Temperate rainforest associations are found along rivers such as the Donaldson River, Tarkine Wilderness, Tasmania (e) and the Thurra River, East Gippsland, Victoria (f). Along many rivers, exotic willows are a significant weed infestation problem, as shown along the Lower Bega River, New South Wales (g).

Table 2.5 Response of aquatic vegetation to channel adjustment (modified from Brookes, 1987).

Channel adjustment	Vegetation response
Width increase	Increased biomass as available bed area for colonization is increased; species composition unchanged.
Depth increase	Reduced biomass as light required for plant growth is attenuated with depth; species composition unchanged.
Coarsening of substrate	Reduced biomass as availability of mixed substrate suitable for rooting of plants is decreased. Plant cover may be eliminated from exposed bedrock areas. Species composition is altered.
Planform adjustment	Relocated channel colonized by plants; biomass and species composition remains the same as substrate and channel morphology are likely to be similar in the relocated channel.

environmental settings. In general, the type of woody debris structures reflects vegetation type (e.g., size, root networks, wood density) and river morphology. For example, widely spreading or multiple-stemmed hardwoods are more prone to forming snags than accumulating as racked members of large log jams because they extend laterally as well as beyond their bole diameter. In contrast, coniferous woody debris tends to produce cylindrical pieces that are more readily transported through river systems, resulting in local concentrations of log jams (Montgomery and Piégay, 2003). Channel size determines whether spanning or instream woody debris structures are formed. The preservation potential of wood may also differ in different climatic settings. For example, the decay rate of wood in tropical rivers may be rapid, while temperate streams may retain the same pieces of wood for extended periods of time (Nanson et al., 1995).

Changes to riparian vegetation and the distribution of woody debris modify the geomorphic structure of rivers, and vice versa. In some settings, clearance of riparian vegetation and removal of woody debris may result in near-instantaneous geomorphic adjustments via incision and channel expansion (e.g., Brooks et al., 2003; Brierley et al., in press). Resulting changes to the depth, area, and frequency of pools may induce significant loss of habitat (e.g., Montgomery et al., 1995b; Abbe and Montgomery, 1996; Beechie et al., 2001; Collins and Montgomery, 2001). Given the enlarged channel capacity, the role of remaining woody debris differs from its original function. Indeed, the lack of large trees along riparian corridors inhibits natural recruitment of woody debris.

Maintenance of a riparian buffer strip in forestry management reduces impacts from increased suspended loads, protects banks, controls direct deposition of pollutants, and preserves stream habitat (through shade and food production). Even a narrow riparian strip may fulfill important functions in the filtering of water, nutrients, and sediment, and regulation of water temperature and light.

Given the importance of riparian vegetation as a determinant of geomorphic and biotic interactions along rivers, it is scarcely surprising that management of riparian vegetation and woody debris is a core component of many river rehabilitation programs. However, it must always be recognized that many types of river in different environmental settings naturally have sparse vegetation associations and little in the way of woody debris along their channels. Each river plan must therefore consider what is appropriate or "natural" for that type of river in that environmental setting. Vegetation that is endemic to the region should be planted on suitable geomorphic surfaces. This should be tied to weed eradication or suppression strategies. Similarly, wood placement can enhance habitat formation. In many cases, however, lessons learnt from near-pristine remnants may provide little guidance in endeavors to rehabilitate degraded rivers.

2.3.5 Use of geomorphology as an integrative physical template for river management

A geomorphic template aids the design and implementation of river rehabilitation strategies that aim to enhance the functioning of aquatic ecosys-

Figure 2.9 Secondary geomorphic controls on various biophysical attributes of rivers
Water quality is a critical factor in assessments of river health. Many rivers may naturally be turbid or contain high levels of tannin. For example, Frankland River in Tasmania (a) has a high level of tannins because of slow water retention/release from buttongrass swamps and riparian rainforest cover. The storage and residence time of organic matter is controlled by the geomorphic structure and hydraulic conditions at different positions across and down the channel. Backwaters and pools, shown along Upper Tantawangalo River, New South Wales (b), tend to be areas of high organic matter storage. Analysis of water chemistry (quality and pH) must be placed in context of the catchment geology and soils, and the sediment transport regime of the river. For example, fine grained rivers such as the Darling River at Wilcannia (c) in central New South Wales, tend to be more turbid than gravel bed rivers such as the Wilson River on the North Coast of New South Wales (d).

tems. Changes to the geomorphic character and behavior of rivers may induce secondary adjustments to biotic and chemical interactions, impacting upon the thermal regime of a river, the production, processing, and retention of nutrients and organic matter, and their role in food web processes. Other measures of aquatic ecosystem functioning may also be affected, such as water quality, pH, etc. (see Figure 2.9). For example, flow–sediment interactions are primary determinants of the distribution and retention of coarse, particulate, organic matter along rivers. Altered hydraulic conditions may change the stream's species assemblage and associated predator–prey relationships. An ecosystem approach that appraises the interaction of biophysical processes is required if river health is to be conserved or improved. Long-term management success will not be achieved by considering individual reaches in isolation from their catchment context. The operation of biophysical fluxes requires that due regard must be given to appraisal of natural patterns and rates of biophysical linkages (or connectivity) in any given setting.

2.4 Working with linkages of biophysical processes

Ecosystem approaches to river management are framed in terms of the spatial organization of landscapes and associated linkages of biophysical processes (see Figure 2.10). Indeed, various core concepts in aquatic ecology analyze patterns, processes, and linkages in space and time.

Examples include the River Continuum Concept (Vannote et al., 1980; Sedell et al., 1989), the Serial Discontinuity Concept (Ward and Stanford, 1983; 1995a, b; Stanford et al., 1988), the Nutrient Spiralling Model (Newbold, 1992), the Flood Pulse Concept (Junk et al., 1989; Tockner et al., 2000), and the Hyporheic Corridor Concept (Stanford and Ward, 1993). These various concepts view ecologi-

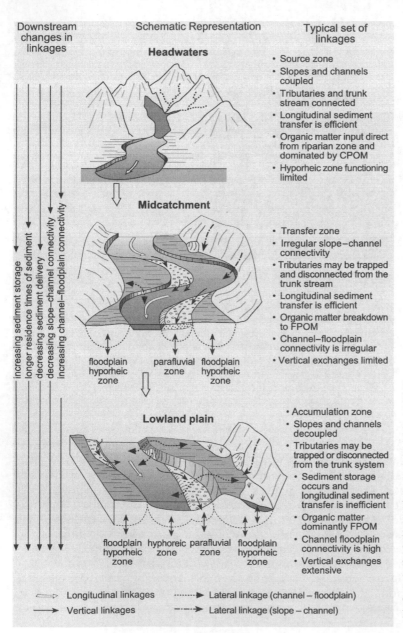

Figure 2.10 Spatial dimensions of landscape connectivity
Patterns of longitudinal, lateral, and vertical connectivity vary in headwater, midcatchment, and lowland sections of rivers, affecting the function of sediment source, transfer, and accumulation zones. In this idealized catchment, an array of biophysical linkages is portrayed in each zone (see text for detail). In river rehabilitation planning, these considerations must be appraised on a catchment-by-catchment basis. Modified from Sear and Newson (1993). CPOM = coarse particulate organic matter; FPOM = fine particulate organic matter.

cal connectivity and biotic response as a function of physical stream structure at different spatial and temporal scales (Poole, 2002).

Within these frameworks, linkages are variously appraised in terms of longitudinal (upstream–downstream, tributary–trunk stream), lateral (slope–channel, channel–floodplain), vertical (surface–subsurface and inundation levels), and temporal dimensions (Ward, 1989; Stanford and Ward, 1996). The River Continuum Concept and Serial Discontinuity Concept address community structure as a function of longitudinal connectivity in water movement at the same hierarchical scales (catchment and segment scales). In both concepts, longitudinal connectivity is influenced by transitions in stream structure between stream segments. While the River Continuum Concept assumes that the stream segment metastructure contained within the network forms a continuum, the Serial Discontinuity Concept focuses on abrupt transitions between adjacent segments with dissimilar physical structure (e.g., canyon to floodplain, lake to stream; Poole, 2002). In contrast, the Flood Pulse Concept and the Hyporheic Corridor Concept address lotic ecosystem function at finer spatial scales (segment to habitat unit) and focus on lateral and vertical connectivity as drivers of community structure and dynamics. Changes in flow energy associated with flood pulses modify patterns, processes, and rates of erosion and sedimentation, impacting on the diversity, distribution, and function of habitat patches (Hughes, 1997; Tockner et al., 2000). For example, channel dynamics create new surfaces for colonization and regeneration (e.g., Richards et al., 2002).

In river management applications, these linkages are most meaningfully analyzed and integrated at the catchment scale. At this scale, segments are ecologically connected in the longitudinal dimension. However, the longitudinal arrangement of segments within each catchment is unique and dynamic, such that patterns and/or phases of discontinuity may be evident over time (e.g., Townsend, 1996; Rice et al., 2001; Poole, 2002). Changes to the arrangement of patches along a river's course may alter the ecological dynamics of the system, even if the relative proportion of patch types remains the same (Fisher et al., 1998; Poole, 2002).

The spatial configuration of a catchment reflects the character and distribution of landscape units, framed in context of their broader tectonic, climatic, and ecoregion setting (Table 2.1). Longitudinal linkages include upstream–downstream and tributary–trunk stream fluxes of water, sediment, and nutrients (e.g., Lane and Richards, 1997; Figure 2.10). The cascading nature of these interactions is influenced by the pattern and extent of coupling in each subcatchment. Appraisal of these linkages is required to assess how off-site impacts such as sediment release, decreased water supply, or nutrient influxes affect reaches elsewhere in the catchment. The distribution of river types in each subcatchment, and how they fit together in the catchment as a whole, provides a physical basis to interpret these linkages.

Lateral linkages include interactions between slopes and channels, and channels and floodplains (Figures 2.10–2.12). In coupled systems, water, sediment, and nutrients are transferred directly from hillslopes to the channel network. Conversely, in decoupled systems, materials are stored for differing intervals of time in various features between the hillslope and the channel. Channel–floodplain connectivity reflects the two-way transfer of water, sediment, and nutrients between channel and floodplain compartments. These relationships exert a critical influence on the ecological health of some aquatic ecosystems. For example, the Flood Pulse Concept theorizes that various characteristics of the flood regime act as the principal driving force behind the existence, productivity, and interaction of diverse biota at different positions within catchments (Junk et al., 1989; Naiman and Bilby, 1998).

Vertical linkages record elevation-induced variations in biophysical fluxes (Figure 2.10). Examples include inundation levels of differing geomorphic units and the connectivity of surface and subsurface flow pathways. Hyporheic zone processes maintain hydrologic exchange and nutrient transformation between surface waters and alluvial groundwaters of the phreatic zone. This zone may extend a considerable distance beyond the channel margin beneath the floodplain (Stanford and Ward, 1988).

Building on the River Continuum Concept proposed by Vannote et al. (1980), various forms of biophysical linkage can be inferred for an idealized

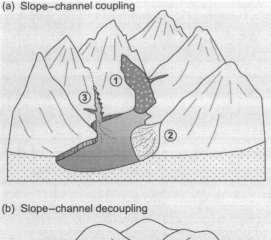

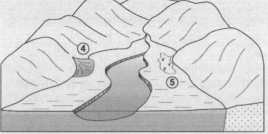

Figure 2.11 Slope–channel connectivity in different landscape settings
Slope–channel connectivity varies depending on landscape setting and the position of the channel on the valley floor. In settings where slopes and channels are coupled (a), sediment is directly delivered from slopes to the channel network. This may take a number of forms including (1) landslides, (2) alluvial fans, or (3) gully/badlands. In settings where slopes and channels are decoupled (b), sediments derived from hillslopes are not delivered directly to the channel network. The hillslopes and channels are often buffered by floodplains, and the system is disconnected. Sediments are either restored on-slope, at the base of slopes as fans (4), or in trapped tributary fills behind levees (5).

catchment, as shown in Figure 2.10. In confined headwater reaches, slopes and channels are coupled, such that water, sediment, and nutrients are readily conveyed from the surrounding catchment. These zones also act as vegetation seed sources. Coarse particulate organic matter, such as leaves and woody debris, is contributed directly from adjacent hillslopes and riparian vegetation to the channel network, where it is processed by macroinvertebrate "shredders." Floodplain pockets are virtually nonexistent, limiting channel–floodplain connectivity. Similarly, hyporheic zone functioning is limited by the imposed bedrock nature of these settings.

Midcatchment locations are characterized by transfer reaches with strong longitudinal connectivity. Discontinuous pockets of floodplain produce irregularities in slope–channel and channel–floodplain connectivity. Tributaries may be locally trapped behind floodplain pockets, disconnecting some lower order drainage lines from the trunk stream. Coarse particulate organic matter is broken down to produce finer particles that are readily transferred downstream. As bedrock is prominent on the channel bed, water and nutrient exchange in the hyporheic zone is restricted to areas of gravel bars.

Lowland plains act as accumulation zones with extensive sediment storage both instream and on the floodplain. Channel–floodplain connectivity is high, with ongoing exchange of water (surface or subsurface), sediment, and nutrients. Flood pulses may maintain wetlands and their related habitats. Sediments and attached nutrients can reside in these floodplains for considerable periods of time. Slopes and channels tend to be decoupled, as materials supplied from low-slope hillslopes are stored for extended periods at valley margins (Figures 2.11 and 2.12). These sediments only reach the channel network if floodplain reworking occurs. Some lower order tributaries may be disconnected from the trunk stream, effectively trapped behind levees. As channels tend to be wider in these locations, inputs of fine, particulate, organic matter from upstream tend to outweigh direct input of coarse, particulate, organic matter from the adjacent riparian zone. The macroinvertebrate assemblage has shifted towards "collectors" that feed on fine, particulate, organic matter. Vertical linkages of biophysical processes are more pronounced in these zones, as surface–subsurface exchange of water and nutrients is promoted by permeable alluvial materials stored along the valley floor.

Recognizing that the effectiveness of biophysical linkages may vary markedly in both space and time, an emerging perspective in ecology and geomorphology examines controls on the disconnectivity and the discontinuum of various processes.

Figure 2.12 Different forms of (de)coupling
Forms of coupling include (1) landslides, (2) alluvial fans, or (3) gully/badlands. (1) The Tarndale landslide in New Zealand has delivered massive volumes of sediment to the Waiapoa River, North Island, New Zealand. (2) The Fortaleza landslide in Brazil led to the formation of an alluvial fan that delivered sediment directly to the channel network. (3) Badlands in the Rhone Basin, France, deliver sediment directly from hillslopes to the channel. Forms of decoupling include alluvial fans formed at the base of slopes (4) and trapped tributary fills behind levees (5). The Mararoa fan in New Zealand (4), is shown spilling onto the floodplain. A tributary system is trapped behind a low lying levee along the Macdonald River, New South Wales (5). The levee is located in the background of the photograph.

For example, a discontinuum view of river systems can be applied to analyze the ecological importance of each stream's individual pattern of habitat transitions along longitudinal, lateral, or vertical vectors at any scale (Ward, 1989; Ward and Stanford, 1995a, b; Naiman and Bilby, 1998; Poole, 2002). In geomorphology, these notions are framed primarily in terms of the coupling or connectivity between landscape compartments (e.g., Fryirs and Brierley, 1999; Harvey, 2001, 2002; Hooke, 2003). Various buffers, barriers, and blankets may disrupt sediment transfer processes and the operation of associated biophysical fluxes (Figure 2.13). Buffers disrupt longitudinal and lateral linkages within catchments, preventing sediment from entering the channel network. Barriers impede downstream conveyance of sediment once it has reached the channel network. Finally, blankets refer to features that disrupt vertical linkages in landscapes, by smothering other landforms. The distribution and effectiveness of buffers, barriers, and blankets reflect catchment spatial configura-

tion, landscape history, and system responses to disturbance events. Any change to the pattern and operation of these features impacts significantly on sediment conveyance and other biophysical fluxes.

Given that catchment configuration influences the pattern and rate of process linkages in any given system, changes to geomorphic river structure and function in any compartment may result in changes to the longitudinal, lateral, and vertical linkages of biophysical processes within the catchment as a whole. Responses to (dis)connectivity may be manifest over differing time periods, with varying lag times. An example of how human disturbance has altered a catchment-wide pattern and operation of biophysical fluxes is presented in Figure 2.14.

Given the biophysical feedbacks inherent to healthy aquatic ecosystems, proactive river rehabilitation programs will not be derived unless reach-based understanding of biophysical processes is framed within a catchment context. The

catchment specific nature of linkages such as sediment movement, water transfer, and seed dispersion are critical factors in determining what can realistically be achieved in each reach. Concern for big-picture linkages in landscapes, among many factors, has lead to the conclusion that large-scale projects, although not always economically or socially feasible, may offer the greatest potential for effective river rehabilitation (Shields et al., 2003). It is only through bigger-picture appraisals of landscape forms and processes, and their ecological associations, that appropriate understanding of the diversity of river systems can be gained, enabling due regard to be placed on representative or unique sites.

2.5 Respect diversity

Unless river management frameworks respect diversity, appreciate controls on the nature and rate of landscape change, and recognize how alterations to one part of ecosystems may affect other parts of that system, efforts at environmental management are likely to be compromised. Ecosystem approaches to river rehabilitation embrace the inherent complexity of the natural world, promoting "naturalness" as a goal. Such endeavors strive to attain the best achievable river structure and function under prevailing catchment conditions. If management programs aim to maintain or establish a truly "natural" river character, with naturally adapted flora and fauna, target conditions must replicate the inherent variability and evolutionary tendencies of river forms and processes in any given environmental setting.

Any perception of how the real world looks or behaves requires the derivation of intellectual constructs with which to break down reality into meaningful parts, such that understanding can be communicated in an effective way. Categorization is the main way of making sense of experience. Attachment of labels to designated categories that possess discrete qualities provides a common basis for communication. As noted by Bail (1998, p. 70), naming and classifying things lies at the heart of human understanding of the world – at least it offers that illusion. Ideally, ordering of information provides a clear, systematic, and organized method with which to view reality. Classification refers to the determination of groups with a common yet discrete set of attributes, enabling interpretation of the complexity and variability that exists within a large group of objects. If successfully applied, classification schemes enable items to be described quickly, easily, and accurately, providing a basis for effective comparison and communication.

Intellectual constructs are required with which to interpret, record, and communicate discrete variants of river morphology. Ideally, such frameworks convey an understanding of formative processes, enhancing their explanatory and predictive capability. Adopted procedures should be open-ended, yet applied in a consistent and non-prescriptive manner. By allowing each field situation to "speak for itself," due regard is given to the uniqueness and/or rarity of differing river types. Indeed, exceptions to generalized observations may be the things that should be targeted in management programs, as they may represent critical areas for maintenance of biodiversity. Distinctive or unique attributes of river courses must first be identified and characterized if they are to be maintained, protected, or enhanced. Identification of "unique" reaches or remnants of "rare" reaches forms an integral starting point in conservation programs. Seemingly, exercises that document variability in river character and behavior in any given catchment continue to reveal further

---

Figure 2.13 Buffers, barriers, and blankets
Buffers are landforms that prevent sediment from entering the channel network by storing it in (a) swamps, Wingccaribee Swamp, New South Wales, (b) piedmonts, Flinders Ranges, South Australia, (c) trapped tributary fills, Wolumla Creek, NSW, (d) alluvial plains or floodplains, Aberfoyle River, New South Wales. Once sediment is supplied to the channel network, barriers impede downstream conveyance through their control on base level. Barriers include (e) bedrock steps, Far North Queensland and (f) dams, Gordon Dam, Tasmania. Blankets are features that smother other landforms and include (g) floodplain sand sheets, Lower King River, Tasmania, and (h) fine material that clogs the interstices of gravels, King River, Tasmania.

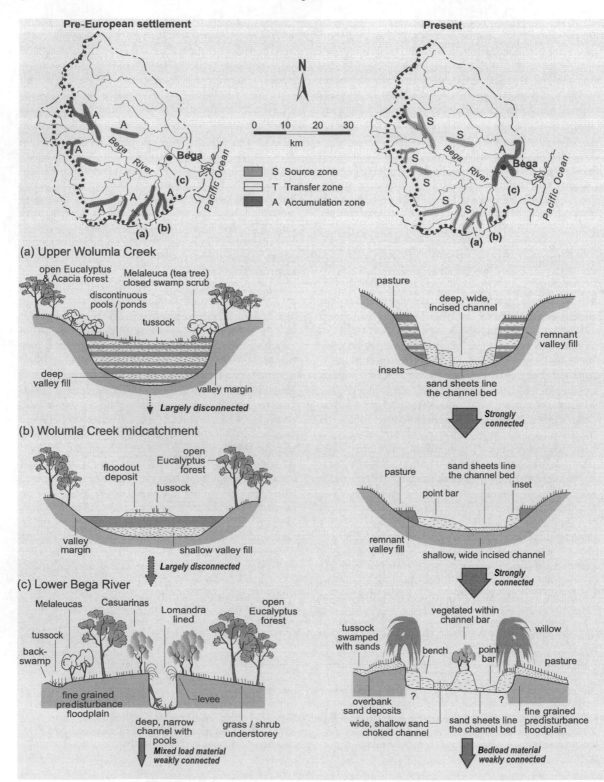

Pre-European settlement

Present

N

0	10	20	30

km

S　Source zone
T　Transfer zone
A　Accumulation zone

Bega River

Bega

Pacific Ocean

(a) (b) (c)

(a) Upper Wolumla Creek

open Eucalyptus & Acacia forest

Melaleuca (tea tree) closed swamp scrub

discontinuous pools / ponds

tussock

deep valley fill

valley margin

Largely disconnected

pasture

deep, wide, incised channel

remnant valley fill

insets

sand sheets line the channel bed

Strongly connected

(b) Wolumla Creek midcatchment

open Eucalyptus forest

floodout deposit

tussock

valley margin

shallow valley fill

Largely disconnected

pasture

sand sheets line the channel bed

point bar

inset

remnant valley fill

shallow, wide incised channel

Strongly connected

(c) Lower Bega River

Melaleucas

Casuarinas

Lomandra lined

open Eucalyptus forest

tussock

back-swamp

fine grained predisturbance floodplain

levee

deep, narrow channel with pools

grass / shrub understorey

Mixed load material weakly connected

vegetated within channel bar

tussock swamped with sands

bench

point bar

willow

pasture

overbank sand deposits

wide, shallow sand choked channel

sand sheets line the channel bed

fine grained predisturbance floodplain

Bedload material weakly connected

variants of previously undocumented types of river. For example, in the Richmond catchment of northern New South Wales, Australia, previously uncharacterized discontinuous sand bed and multichannel sand belt rivers were identified (Goldrick et al., 1999). Sadly, systematic baseline data on the diversity of river forms and processes are not available across much of the planet. Ultimately, however, no magic number of river types can meaningfully represent the spectrum of morphological complexity.

Information on the diversity and abundance of river types, their condition, and associated ecological values provides a critical starting point for management activities in any given catchment/region. Such information enables reaches with simi-lar/dissimilar characteristics to be differentiated, providing a basis to compare like-with-like and extrapolate from one area to another. This enables lessons learnt to be transferred in a meaningful manner from reach to reach. Identification of river type aids analysis of what types of management problems are to be expected where. Meaningful river classification procedures also provide a basis to assess biodiversity value.

The reductionist nature of modern science has resulted in numerous discipline-based approaches to river classification (see reviews in Mosley, 1987; Naiman et al., 1992; Downs 1995). Unfortunately, most schemes merely describe components of system structure, they are neither integrative nor functional, and they cannot be applied reliably

◄───

Figure 2.14 Human-induced changes to catchment-scale biophysical linkages in Bega catchment, Australia: Implications for river management

Human-induced changes to the geomorphic structure and function of river courses in Bega catchment in the period since European settlement have significantly altered the sedimentary cascade and longitudinal, lateral, and vertical connectivity throughout the catchment. The catchment maps note the distribution of source, transfer, and accumulation zones in Bega catchment prior to European settlement and today. (a), (b), and (c) show pre- and postdisturbance cross-sectional geomorphic structure of rivers in upper, middle, and lowland parts of the catchment. The thickness of the arrows between each section indicates dramatic adjustments in the longitudinal connectivity of the sediment flux.

Prior to European settlement, most base of escarpment settings in Bega catchment contained discontinuous watercourses or intact valley fills with no channels. These valley fills had accumulated sediments since around 6,000 years BP (Fryirs and Brierley, 1998), acting as buffers that disconnected much of the upper catchment from the sediment cascade (Brierley et al., 1999). The unincised nature of the valley floor buffered many lower order tributaries from gullying processes (Brierley and Fryirs, 1999). Many upland hillslopes were disconnected from the sedimentary cascade (Fryirs and Brierley, 1999). Although bedrock-controlled valleys dominate midcatchment locations, low rates of sediment supply are inferred prior to European settlement. Hence, rates of sediment accumulation on the lowland plain were relatively low, and the channel capacity decreased downstream (Brooks and Brierley, 1997). While channel–floodplain connectivity was high (i.e., sediments were continually being added to the floodplain), the floodplain acted as a significant buffer to sediment evacuation from the catchment.

Following European settlement, incision has transformed most intact valley fills into continuous watercourses, increasing the longitudinal connection of sediment movement through the upper parts of the catchment (Brierley et al., 1999). As these channels widened, they locally removed the fill that previously buffered tributaries from the trunk stream, prompting the formation of extensive gully networks (Brierley and Fryirs, 1999). Tributary–trunk stream connectivity has increased across much of the upper catchment. To accommodate the resulting sediment slug, downstream channels that previously contained floodouts became incised and enlarged, greatly increasing the flushing potential of the system. Bedrock controlled, high energy, midcatchment reaches efficiently conveyed large volumes of material downstream in a relatively short period of time (i.e., within decades). In response, and in conjunction with the removal of riparian vegetation, the capacity of the channel along the lowland plain increased by over 300% (Brooks and Brierley, 1997). Large sand sheets blanketed the floodplain, maintaining high channel–floodplain connectivity. However, the channel became transport-limited, as it was unable to transport the large volume of sediment supplied from upstream (Fryirs and Brierley, 2001). This forms a barrier to sediment movement, restraining the transfer of the sediment slug which now acts as a plug along the lowland plain. Although pulses of sediment move episodically to the estuary, sediment delivery to the coast remains low. This understanding has provided a critical platform for river rehabilitation initiatives in the catchment (see Brierley et al., 1999, 2002).

outside the disciplines from which they were derived (Mosley, 1987; O'Keeffe et al., 1994; Goodwin, 1999). River management frameworks need a clear, systematic, and organized method with which to view reality. Despite many concerted efforts, a cross-disciplinary river classification framework of general applicability has yet to be developed (Naiman et al., 1992; Uys and O'Keefe, 1997). Many procedures are unduly rigid and categorical in their structure and application. Simplification of real world complexity may mean that significant components of the natural world are overlooked, misrepresented, or misunderstood. Unexplored situations may be unique. If the integrity of such a site is compromised, society may lose something that it didn't even know existed! An open-mind must be maintained. While individual reaches may have similar attributes and behavioral traits, their evolutionary trajectories may be quite different. A precautionary approach to river management ensures that minimally invasive strategies are applied whenever uncertainty prevails.

It is increasingly clear, however, that a classification scheme based on geomorphology alone will not provide a means to order ecological variability (Newson et al., 1998b; Thomson et al., 2004). Complex interactions of geomorphic, hydrologic, and biotic feedbacks that operate over multiple spatial and temporal scales ensure that considerable natural variation may be evident in the range of biophysical processes observed in any particular geomorphic river type. Moves towards the development and application of more integrative biophysical perspectives in river management should not be derailed, in any way, by the challenges faced in the quest to develop a unifying river classification framework. Ultimately, intellectual constructs must have practical and meaningful application "on-the-ground." Prescriptive approaches to analysis based on black box procedures will not provide sustainable answers in the long term. The reemergence of landscape ecology, with its emphasis on biophysical patterns, processes, scale, and evolution, framed in a landscape context, provides fertile ground for development of new approaches to enquiry and management ap-

plications. Appreciation of the geomorphic template of any given system provides a platform upon which to ground these cross-disciplinary developments.

2.6 Summary

Landscapes comprise mosaics of genetically-linked components. In different process zones, landforms of differing sizes and longevity are formed and reworked by different sets of processes over a range of timescales. This controls the nature and strength of biophysical linkages and fluxes along a river course. Analysis of the structure, function, and distribution of river types enables reach-specific attributes to be set within their catchment context, assessing catchment–scale linkages of geomorphic processes, and propagatory and cumulative effects of disturbance responses that drive geomorphic processes and change. Geomorphic responses to changes in biophysical fluxes are likely to be landscape and timeframe specific. Classification of river type at the reach scale does not, in its own right, provide an appropriate basis for river management activities.

Proactive management programs that strive to preempt change and improve river condition by enhancing "natural" recovery must "work with nature," respect the inherent diversity of river forms and processes, and consider catchment-scale linkages of biophysical processes. A *geomorphic template* provides a physical basis to link aquatic and terrestrial ecosystems in a coherent approach to landscape ecology. *Catchment-scale* perspectives integrate our understanding of longitudinal, lateral, and vertical process linkages. Such insights are required to explain the present-day condition of a river, to interpret off-site and/or lagged responses to disturbances, to predict likely future river adjustments, and to ensure that river rehabilitation strategies focus on the underlying causes rather than the symptoms of change. Hence, spatial considerations in river management applications must be linked directly to temporal patterns and rates of adjustment, as emphasized in the following chapter.

CHAPTER 3

Temporal considerations in aquatic ecosystem management

All classifications based on existing channel morphology . . . fail to account for dynamic adjustment or evolution of the fluvial system. Increasing recognition of the fact that rivers are seldom in dynamic equilibrium has driven a desire on the part of engineers and managers to be able to predict channel changes in the short and medium term. In response, geomorphologists have begun to develop new schemes of river classification based on adjustment processes and trends of channel change rather than existing channel morphology and sediment features.

Colin Thorne, 1997, p. 213

3.1 Chapter structure

In this chapter, variable patterns and rates of river response to disturbance events are framed in their evolutionary context. Rivers are viewed to operate in a state of perpetual adjustment, rather than oscillating around an equilibrium form, presenting an intriguing contrast between engineering and ecosystem perspectives on river change (Section 3.2). In Section 3.3, timeframes of river adjustment are assessed. Spatial and temporal themes are combined in Section 3.4 to examine the balance of impelling and resisting forces along river courses, and how they change over time. These considerations are used to develop a framework with which to explain controls on river character and behavior. Longitudinal profiles are used as a basic tool to perform this synthesis. Finally, approaches to prediction in geomorphology are outlined in Section 3.5.

3.2 Working with river change

Rivers are never static entities. Indeed, change is a natural and a vital component of aquatic ecosystem functioning. A stable, nonadjusting ecosystem is an unhealthy one – it is functionally dead! Increasing efforts to promote the values of vibrant, living systems mark a transition beyond engineer-

ing perspectives on river health which subsume "natural" values within a perspective of human dominance of the physical world.

The character, rate, and permanence of river changes have enormous implications for river management. River management programs that *work with change* must build on knowledge of how and why rivers adjust their morphology, and what fashions the magnitude and rate of change. In order to address the causes rather than the symptoms of river change, practitioners must integrate assessments of what rivers look like with appraisals of *how they work*. Whether adjustments occur through barely perceptible changes over millennia, systematic patterns of change over hundreds of years, or catastrophic responses to floods over hours or days, understanding of *evolution* underpins applications of principles from fluvial geomorphology in river management practice. System dynamic, sensitivity to change, and proximity to threshold conditions are critical considerations in interpretations of river condition and trajectories of change, and related assessment of the potential for geoecological recovery.

The nature and rate of river responses to human disturbance vary markedly from reach to reach and from catchment to catchment, reflecting the sensitivity of the landscape to change and the type and extent of disturbance. In many instances, human actions have altered "natural" patterns and rates of change. In some cases, these changes are irreversible, and river systems now operate within a different set of boundary conditions.

In this book, river behavior is defined as *adjustments to river morphology induced by a range of erosional and depositional mechanisms by which water moulds, reworks and reshapes fluvial landforms, producing characteristic assemblages of landforms at the reach scale* (see Chapter 5). Some systems are inherently more prone to adjustment and are more sensitive to physical and biological disturbance than others. For example, the natural proportion of eroding banks varies markedly for differing types of river. Bank erosion is unexpected along discontinuous watercourses such as a chain-of-ponds, but natural patterns of bend migration along gravel-bed meandering rivers may result in active erosion along up to 50% of banks.

Some disturbance events may fundamentally alter geomorphic structure and function, potentially inducing a different suite of biophysical interactions. This is referred to as river change, and is defined as *adjustments to the assemblage of geomorphic units along a reach that record a marked shift in river character and behavior* (see Chapter 6). In general terms, river change occurs when an intrinsic or extrinsic threshold is breached, transforming channel geometry, planform, and the assemblage of geomorphic units along a reach. These wholesale changes severely impact on the structural integrity and ecological functioning of a river.

In recent decades there has been a shift in the way river change is perceived and tackled in river rehabilitation practice. The river engineering paradigm that emerged during the industrial revolution viewed nature to provide boundless resources to be conquered and utilized by human endeavor. Wherever rivers could be readily exploited by society through measures to control, divert, channelize, or dam them, they generally were. Management practices reflected human desires for simple, efficient, and predictable systems that enhanced prospects for economic development. Rivers were viewed as conduits with which to maximize the conveyance of water, sediments, and environmental "waste products" through uniform, stable, hydraulically smooth channels. Despite, or maybe because of, the profound changes to river character that have taken place

following human disturbance, management efforts emphasized the need for "stable" rivers. Principles of regime theory, originally devised by hydraulic engineers as part of canal design specifications, were applied to create uniform channels with a prescribed hydraulic regime (the classic trapezoidal channel). Roughness elements such as riparian vegetation and woody debris were considered to produce messy, complex, and irregular channels, creating uncertainty and reducing predictability in what was ostensibly a "controlled" environment (Williams, 2001).

When ecological impacts became so pronounced that societal alarm was raised, the engineering mind-set engendered a sense that problems could be rectified using the next variant of technological development (technofix). Reactive management practices were applied to maintain and protect infrastructure, navigation, and flood protection/mitigation networks. However, in endeavors to stabilize channels, many engineering practices accentuated their instability (e.g., Leeks et al., 1988; Bravard et al., 1997). Imposing a stable channel through "training," "improvement," or "stabilization" techniques, or "normalized" flow regimes, will not result in sustainable or healthy river systems. Indeed, planning for "mean" conditions is unsympathetic to the natural range of biotic and geomorphic process activity. In many ways, the inherent variability of river behavior and responses to disturbance drives the functioning of aquatic ecosystems. Instability is a key attribute of many systems. Suppression of the natural tendency of rivers to adjust limits the capacity of systems to self-heal. Anything that compromises the ecological integrity of river systems leaves society worse off in financial, cultural, social, and environmental terms.

Rivers that were physically transformed and ecologically ruined to facilitate industrial and agricultural developments are now receiving increasing societal demands for rehabilitation. Contemporary management activities endeavor to rectify human impacts on aquatic ecosystems through more harmonious approaches to environmental management, aiming to minimize the prospects of further damage, and repair the damage that has already occurred. A paradigm shift in river management practice is underway, marking the transition from an engineering-dominated

Table 3.1 Attributes of engineering- and ecosystem-based approaches to river rehabilitation (based on Williams, 2001; Carr, 2002; Hillman and Brierley, in press).

Engineering-based approach	Ecosystem-based approach
Top-down approach driven by Government agencies (i.e., politically driven).	Bottom-up (or middle ground, consultative) approach expressed within a participatory framework, reflecting interdisciplinary understanding within the broader community.
Single purpose, deterministic, emphasizing the desire for certainty in outcomes.	Multiobjective, probabilistic, with explicit recognition of uncertainty and complexity.
Outcome-driven, goal-oriented.	Emphasis placed on processes and outcomes, means and ends.
Perceives technical problems requiring technical solutions.	Perceives problems as symptomatic of wider socioeconomic, cultural, and biophysical considerations.
Site-specific or reach-scale applications are typically framed in the quest for stability over decadal timeframes with a construction focus.	Catchment-framed rehabilitation programs recognize natural variability over centuries or millennia using a continuum of interventions.
Limited accountability. If present, monitoring is externalized, with maintenance divorced from design.	Long-term commitment. Monitoring is internalized and maintenance is a core management (community) activity.
Extension science engaged to educate people about the environment.	Action research used for mutual learning by all practitioners in river management.

emphasis that endeavors to *control* nature, towards a more inclusive ecosystem-framed approach that strives to *work with* nature (Williams, 2001; Hillman and Brierley, in press; see Table 3.1). Such endeavors emphasize the physical and ecological integrity of living, variable, dynamic, and evolving aquatic ecosystems, endeavoring to let the rivers "run free" (Everard and Powell, 2002).

Whereas river engineering activities emphasize empirical "solutions" to site-specific or reach-scale issues, fluvial geomorphologists tend to adopt a broader, catchment-scale perspective that emphasizes the physical integrity of the drainage basin and the close links between catchment and river dynamics (Knighton, 1998). In the ecosystem-based approach to river management, skills and insights from geomorphologists and engineers are considered to be complementary. Regime assumptions are reframed in light of evolutionary insights, providing a clearer understanding of the variable sensitivity of rivers, interpretation of threshold conditions, and the circumstances that lead to instability. River managers now have at their disposal a range of strategies that extend from minimally invasive maintenance programs to large-scale structural engineering works (Figure 3.1). Rather than placing undue emphasis on attempts to restore the population of a single species

or individual forms, the emerging river management paradigm aims to restore ecological processes and habitats by enhancing natural recovery processes using environmentally sympathetic (soft or sensitive) engineering practices (e.g., Hey, 1994; Gilvear, 1999; Hooke, 1999).

The mind-set with which river rehabilitation practices are applied bears critical relation to perceptions of their success. Unless project designers recognize the diversity, variability, and uncertainty of river systems, their activities continue to work within an ethos of "control." Although many management programs now refer to goals expressed in terms of environmental restoration rather than flood control or navigation, the simplification of the riverine system and the faith in human modifications is strikingly similar (Kondolf, 1995). Piecemeal, reactive strategies do not provide efficient and cost-effective ways to achieve rehabilitation success (e.g., Bravard et al., 1997). Costs for repair and maintenance are excessive. Catchment-scale rehabilitation programs integrate spatial and temporal dimensions of change, recognizing explicitly system-specific linkages of biophysical processes and unique landscape histories (Brierley et al., 2002). Proactive plans are designed with a future-focus, framing target conditions for individual reaches within their

Figure 3.1 Examples of "hard" and "soft" engineering practices
Hard engineering structures are applied to protect infrastructure. Examples include (a) rockwalls, Singleton, New South Wales, Australia, (b) Sabo dams, Hokkaido, Japan, and (c) rock groynes, Wilson River, New South Wales, Australia. Soft engineering practices are ecologically based practices that aim to enhance natural recovery processes and include (d) engineered log jams, Williams River, New South Wales, Australia, (e) T-jacks, Gloucester River, New South Wales, Australia, and (f) fencing off and planting of riparian zones, Dorrigo, New South Wales, Australia.

catchment context, recognizing the changing character, pattern, and linkages of landscape forms and processes. As systems adjust to management treatments, ongoing maintenance is required to ensure that rehabilitation and conservation objectives are attained.

Evolutionary thinking that emphasizes system dynamics and the capacity for change is a core component of ecosystem approaches to river management. Reconstruction of the past provides a means to test models of change which, if verified, can be used to predict likely future river behavior. Timeframes of river adjustment that underpin this thinking are appraised in the following section.

3.3 Timescales of river adjustment

Adjustments to biophysical processes in river systems occur over a wide range of timescales (Table 3.2). Interactions among flow, sediment, and biotic considerations are shaped in differing ways in different environmental settings, reflecting catchment-specific sets of imposed boundary conditions, the operation of flux boundary conditions, and system history. In simple terms, short-term adjustments to the balance of flow and sediment along a reach trigger different biotic responses over daily, seasonal, or annual timeframes. Despite ongoing sediment movement, landscapes may appear to be relatively static over short (hourly, daily, weekly) intervals. Although erosion, sediment transport, and deposition modify landscape forms over annual or decadal timeframes, adjustments from a characteristic state may be barely discernible. For example, bedforms move downstream, bars adjust their morphology and position, and bends migrate, without altering the broad configuration of a reach. Channel adjustments are commonly observed over decadal intervals, while floodplain adjustments typically occur over longer (centennial) timeframes. When considered in context of environmental changes over thousands of years, the boundary conditions under which rivers operate may themselves be modified, bringing about interconnected sets of river adjustments.

Table 3.2 Timescales of some major phenomena that structure patterns and processes in riverine landscapes (modified from Ward et al., 2002 and Richards et al., 2002).

Timescale	Phenomenon
Seasonal	Spates, flow pulses, expansion/contraction of the active channel zone, bedform adjustments, seed dispersal
Annual	Flood pulse, seedling establishment, animal migration, reproduction, shallow groundwater exchange, bed material reorganization
Decadal	Drought cycles, episodic events (extreme floods, debris flows), bar formation and reworking, lateral channel migration, channel avulsion, island formation, channel abandonment, local adjustments to vegetation associations
Centennial	Floodplain formation, migration of sediment waves/slugs, hydrosere and riparian succession, deep ground water exchange
Millennial	Terrace formation, glaciation, climate change, sea-level fluctuation, orogeny, base level adjustments, channel incision, aggradation

Climatic, tectonic, and base level influences drive landscape evolution, resulting in sustained erosion or deposition (or a balance thereof) in differing landscape compartments. Confounding such simplistic notions, however, is the fact that dramatic change may occur almost instantaneously, as systems change state (i.e., evolve) in response to breaching of threshold conditions. This may occur in response to a catastrophic event, or in response to incremental adjustments when a system is particularly vulnerable, such that a small trigger drives dramatic responses (the so-called butterfly effect of chaos theory).

Some systems evolve via progressive incremental adjustments; others are characterized by long periods of relative inactivity followed by short intervals of dramatic change. For example, geomorphic responses to floods of varying magnitude, frequency, and duration may vary markedly from system to system (e.g., Wolman and Miller, 1960; Wolman and Gerson, 1978). Analysis of the evolutionary pathway of a reach provides a basis to interpret whether contemporary adjustments form part of an "expected" range of behavior, or whether the reach is experiencing anomalous or accelerated behavior. Such insights also determine whether the system is adjusting around a characteristic state or is evolving towards a different state. Understanding the present in the light of past adjustments underpins the effectiveness of foresighting scenarios that scope the future.

Hickin (1983) refers to intervals of long-term landscape evolution, intermediate scale equilib-

rium, and short-term stasis (*sensu* Schumm and Lichty, 1965) as geologic, geomorphic, and engineering timescales that operate over intervals of roughly 10^{4+}, 10^2, and 10^0 years respectively. *Geologic time* refers to long-term, tectonic uplift and progressive downwearing brought about by denudational processes, the nature and effectiveness of which are influenced by climatic conditions. Processes operative at this timescale determine the nature and distribution of landscape units, thereby shaping relief, slope, and valley confinement, and associated patterns of aggradation and degradation along a river.

Geomorphic time refers to periods over which a near equivalent form is maintained in any given reach, as channels adjust to average water and sediment fluxes. Much of the notional "theory" that has been developed in fluvial geomorphology is based on the premise that under these conditions, channels are able to self-regulate via negative feedback mechanisms, thereby retaining an equilibrium morphology (Hickin, 1983). Impacts of disturbance events are damped out or self-corrected via internal adjustments such that the suite of geomorphic features that make up an equilibrium reach remains uniform over time (Renwick, 1992).

Finally, *engineering time* refers to near instantaneous passages of time in which the landscape is viewed to be effectively static. From this perspective, engineering "solutions" to river problems assume that the governing conditions at the geomorphic timescale are constant (Hickin, 1983).

Based on this assumption, principles of fluid mechanics provide quantitative insights into flow fields, the capacity/rate of sediment transport, and the associated nature and rate of channel bed and channel form adjustments.

Regardless of the timeframe of investigations, patterns and rates of river adjustment or change reflect the nature of disturbance events. Useful differentiation can be made between *pulse* and *press* events based on the intensity and duration of the disturbance (Brunsden and Thornes, 1979; Schumm, 1979). *Pulsed* disturbance events are episodic events of low frequency, high magnitude, and limited duration whose effects tend to be localized. Off-site impacts are minimal and short-lived, such as the reorganization of bed materials within a reach. Floods are the primary form of pulsed natural disturbance along rivers. Although most reaches readily adjust once the disturbance event has passed, extreme events can produce a lasting effect, especially if a threshold is breached.

During a *press* type of disturbance, controlling variables are sustained at a new level as a result of more permanent shifts in input conditions. Such changes are likely to apply over much larger areas than pulsed events, although responses are not spatially uniform and tend to be more permanent. These disturbances alter the evolutionary pathway of a reach. Knock-on effects can induce geomorphic changes along reaches that were not directly impacted by the initial disturbance, often a considerable period after the initial disturbance.

System responses to press and pulse disturbance events vary markedly. Among many factors, this reflects the condition of the landscape at the time of any given event (i.e., how close to a threshold the system sits), and the connectivity of the system. In strongly coupled catchments, disturbance effects are often conveyed efficiently through the landscape. In contrast, responses to disturbance are inefficiently propagated through decoupled or disconnected landscapes, as barriers or buffers inhibit conveyance of water and sediment, absorbing or damping the impacts of disturbance.

Interpretation of trajectories of change must build on understanding of controls on river character and behavior, as highlighted in the following section.

3.4 Interpreting controls on river character and behavior

Explanation of controls on river character and behavior provides the foundations for predicting likely future river adjustments. Spatial and temporal considerations must be integrated in these assessments. In this book, analysis of ongoing system responses to disturbance is framed in terms of the imposed boundary conditions within which rivers operate (i.e., the catchment and landscape unit scale controls) and the flux boundary conditions that reflect the flow–sediment balance along a reach (and related vegetation associations). Various geomorphological considerations that underpin assessment of controls on river character and behavior are outlined in this section. First, spatial and temporal variability in the balance of impelling and resisting forces along river courses is appraised (Section 3.4.1). Second, patterns of river forms and processes are examined along longitudinal profiles, emphasizing downstream variation in controls and linkages of physical processes (Section 3.4.2).

3.4.1 The balance of impelling and resisting forces as a determinant of river character and behavior

Contemporary river morphology is fashioned by the operation and balance of various processes by which channels transport, rework, and deposit sediments. The sediment transport regime of a river reflects the volume and caliber of sediment that is delivered to the channel, the capacity of the channel to transport it, the distribution of excess energy that drives erosion, and the type of sediment that is deposited instream and on the floodplain. Over time, these adjustments fashion bed and bank composition and the capacity of the river to adjust its form. A simple summary of these concepts is encapsulated by the sediment balance diagram proposed by Lane (1954), as reproduced in Figure 3.2.

Impelling or driving forces that promote geomorphic work along rivers are expressed by a given volume of water acting on a given slope. Available energy is able to erode, entrain, transport, and deposit varying amounts of material, dependent upon the caliber and volume of available sediment.

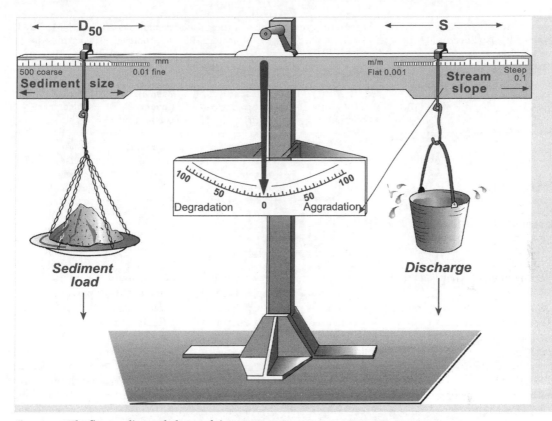

Figure 3.2 The flow–sediment balance of river courses
Flow–sediment interactions determine the aggradational–degradational balance of river courses. If excess flow occurs over steep slopes, the balance is tilted towards degradation. Alternatively, if excess sediment of a sufficiently coarse nature is made available to the river, the balance is tilted towards aggradation. D_{50} refers to the median bed material grain size (in mm), while s refers to slope (m/m). From Chorley et al. (1984) p. 290. Reproduced with permission from Thomson Publishing Services, 2004.

The relationship between the caliber and volume of sediment stored along a river, and the competence and capacity of flow to transport it, results in channels with differing ranges of bed material caliber, with associated sets of bedforms, flow resistance, and roughness coefficients (e.g., Darcy–Weisbach resistance coefficient and Manning's n; see Table 3.3).

The behavior of any given reach is subject to adjustment if the scales shown in Figure 3.2 are affected. A tip in the balance in favor of discharge, such that the system becomes sediment supply-limited, promotes degradation. Alternatively, a tip in the balance in favor of sediment load, such that the system becomes sediment transport-limited, leads to aggradation. Ultimately, resisting forces along the valley floor, whether induced by valley shape/alignment, vegetation cover, bedrock outcrops, or morphologic adjustments of the river itself, determine the distribution and use of flow energy on the valley floor, and the resulting pattern of erosional and depositional forms. These interactions vary markedly in different valley settings and at different positions along longitudinal profiles.

Impelling forces at any given location are fashioned by potential energy, as determined by elevation and slope, and kinetic energy, as determined

Table 3.3 Characteristics of different channel types and typical values of their flow resistance coefficients. Modified from Bathurst (1993) in Beven and Kirkby (1993) © John Wiley and Sons Limited. 2003. Reproduced with permission.

Type of channel	Approximate range of				
	Channel slope (%)	Bed material size (D_{50} in mm)	Darcy–Weisbach resistance coefficient (ff)	Manning's n (roughness coefficient)	Relative submergence *
Sand-bed	≤0.1	≤2	0.01–0.25	0.01–0.04	May be >1000
Gravel/cobble-bed	0.05–0.5	10–100	0.01–1	0.02–0.07	Usually 5–100
Boulder-bed	0.5–5	≥100	0.05–5	0.03–0.2	Often <1
Bedrock (steep-pool/fall)	≥5	variable	0.1–100	0.1–5	Generally <1

Note

* The ratio of mean flow depth to median sediment size

by the volume of water that is available to perform geomorphic work. Work performed by a given body of water acting on a given slope reflects the total stream power that is generated. These considerations are shaped largely by upstream catchment area, which acts as a surrogate for discharge, and the valley floor slope. In theory, channel capacity and geometry adjust to the available energy. These responses influence the distribution of unit stream power (i.e., the energy acting at any given point of the channel bed or floodplain surface). Rivers adopt an array of mechanisms to expend energy, some of which are imposed (e.g., bedrock steps), while others are shaped by the boundary conditions within which each reach operates (e.g., lithological controls on the availability and caliber of bed materials). The volume and caliber of available sediments influence their capacity to act as abrasive tools. A critical balance is maintained here. If too much sediment is stored on the bed, it protects the bed from erosion during all flow stages other than events that mobilize the entire body of sediments. If excess energy is available, sediments may simply be flushed through the reach, resulting in a bedrock channel (Montgomery et al., 1996).

Differences in bed material texture and the energy conditions under which rivers operate can be used to differentiate among three primary variants of sediment transport regime, namely bed material load, suspended load, and mixed load rivers (Schumm, 1960; Figure 3.3). *Bed material load* comprises particles that are transported in a shallow zone only a few grain diameters thick via rolling, saltation (in which grains hop over the bed in a series of low trajectories), and sliding. A critical entrainment threshold must be exceeded before movement commences. As grain size increases, the intermittent nature of this jerky conveyor belt becomes more pronounced. In boulder bed streams, only extreme flows may be able to mobilize the larger clasts. In gravel bed rivers, rolling is the primary mode of bedload transport, whereas saltation is largely restricted to sands and small gravels. The primary source of bedload material is the channel bed itself. In general terms, transport velocities of entrained particles are so low relative to flow velocities, that travel distances are short (Knighton, 1998). The intermittency of bed material transport, and the possibility of prolonged storage, results in long residence times for these coarser sediments.

The bed material load is much coarser than materials carried in suspension, typically comprising particles coarser than 62 μm. In terms of critical flow velocity, medium sand (0.25–0.5 mm) is the most readily eroded fraction and, other things being equal, is the first fraction of the sediment mix to be entrained. Sand-sized particles tend to move as migrating bedforms such as ripples, dunes, and antidunes (see Chapter 4). When particles smaller than about 0.2 mm are submerged in the laminar sublayer, they are no longer subjected to the stresses associated with turbulent flow, and greater threshold stresses are required to entrain

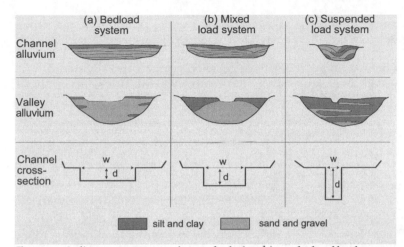

Figure 3.3 Sediment transport regime and relationships to bed and bank texture and channel size (from Schumm, 1968)
Bedload-dominated rivers are characterized by high width : depth ratio channels that adjust readily atop the valley floor. Bank materials are relatively noncohesive and channels tend to be relatively straight. In contrast, suspended load rivers have low width : depth ratios, sinuous planforms, and cohesive banks. They adjust relatively slowly atop the valley floor and accumulate materials primarily through vertical accretion mechanisms. Mixed load rivers are an intermediary variant characterized by moderate sinuosity channels with moderate width : depth ratios. Composite banks are common. Floodplains are formed by a mix of lateral and vertical accretion mechanisms. Note that the channel cross-sections in this diagram are not drawn to scale.

them (Knighton, 1998). Compared to other components of load, the movement of bed material load contributes least to the total. However, the bedload fraction is the primary determinant of river morphology. Schumm (1977) characterizes bedload rivers as systems that carry more than 11% of their sediment as the bedload fraction. In general terms, *bedload systems* tend to have wide, shallow channels with width : depth ratios <40 (see Figure 3.3a). Given the unconsolidated nature of the sediments, and the low silt : clay ratio in bank materials, bedload channels tend to be relatively straight, with a sinuosity (S) less than 1.3 (measured as the ratio of channel length to valley length).

The *suspended load* comprises particles that are maintained in suspension by turbulent eddies. These materials are typically > 62 μm, and are finer than particles usually found in the bed. Once entrained, suspended load materials remain in transport for an extended period of time. Most material is supplied by surface erosion in the catchment, by processes such as overland flow and gullying, and by erosion of cohesive banks (Knighton, 1998). The suspended load fraction tends to be carried at approximately the same speed as the flow. Particles

only settle out when flow velocities are much reduced. Suspended sediment contributes around 70% of the total load delivered to the world's oceans each year. In general terms, suspended load rivers transport less than 3% of their total load as bedload (Schumm, 1977). Banks are cohesive, given their high silt : clay ratio, promoting the development of highly sinuous (S<2.0), narrow and deep channels, with width : depth ratios >10 (see Figure 3.3c).

Mixed load channels are intermediary between bedload and suspended load variants (Figure 3.3b). Between 3–11% of their total load is conveyed as bedload (Schumm, 1977). Unconsolidated sediments on the bed contrast markedly with the silt : clay deposits of the floodplain (and hence the channel banks). These conditions promote the development of sinuous channels (1.3 > S > 2.0) with a moderate width : depth ratio (typically between 10–40).

In general terms, the key control on the movement of fine-grained material is the amount of material *supplied* to the river, rather than the transport capacity of the flow. In contrast, movement of coarser material (<62 μm) is *transport-*

limited, as it is determined by the energy of flow. Differentiation of supply- and transport-limited rivers provides a useful basis to characterize various attributes of river types. Supply-limited rivers have a capacity to transport materials that is greater than the volume of materials made available to them. As a consequence, they commonly have degradational tendencies. In contrast, transport-limited rivers are relatively overloaded with sediments, such that they have aggradational tendencies. Rates of sediment reworking are unable to keep up with rates of sediment supply. This commonly promotes the development of a multichanneled river, as the *capacity limit* of the system is breached and within-channel deposition occurs. This is typically associated with sediment-charged sand-bed rivers. Conversely, development of multichanneled configurations in gravel-bed rivers tends to be associated with the limited frequency with which bedload materials are transported, such that the coarsest fraction forms the core of midchannel bars in a *competence-limited* situation. In many instances, transport-limited rivers are characterized by discrete sediment inputs that generate larger scale and longer duration pulses of sediment that translate downstream as waves or slugs (Nicholas et al., 1995).

Inevitably, the use of energy by rivers is dependent upon how much energy is available. Dissipation of energy is constrained by various forms of resistance (Figure 3.4). In many instances, a significant proportion of available energy is consumed by resisting elements imposed by changes in valley alignment or constrictions (Figure 3.4a). Total flow resistance within the channel comprises several components, including boundary resistance (resulting from the frictional effect of the channel bed expressed through form and grain roughness), channel resistance (associated with bank irregularities and changes in channel alignment), and free surface resistance (stemming from the distortion of the water surface by waves and hydraulic jumps; Bathurst, 1993). While the latter component reflects hydraulic adjustments within the channel, changes to boundary resistance and channel resistance represent geomorphic adjustments that fashion, and are in turn reshaped by, river responses to flow events.

Channel resistance is determined, in part, by morphological configuration (e.g., curvature) and adjustments to channel position on the valley floor. These factors include bank and planform roughness elements associated with channel alignment, channel geometry, and the role of vegetation and woody debris (Figure 3.4b). Through controls on increased bank shear strength, and/or reduced boundary layer shear stress, riparian vegetation influences channel size and shape. Empirical studies have shown that channels with dense bank vegetation (i.e., trees and shrubs) are on average 50–70% times the width of an equivalent channel vegetated only by grass (e.g., Charlton et al., 1978; Hey and Thorne, 1983, 1986; Andrews, 1984). Instream vegetation and the loading of woody debris can comprise a significant proportion of channel roughness and hence total hydraulic resistance. Whether as individual pieces or through its influence as a determinant of the size, type, and evolution of instream features such as pools, bars, and steps, woody debris can impart significant hydraulic resistance, dramatically reducing bedload transport rates in rivers (Gippel, 1995). Variations in the height, density, and flexibility of aquatic vegetation influence reach-scale flow resistance. The role of instream vegetation and woody debris as agents of hydraulic resistance depends upon the size of the obstruction relative to the scale of the channel (Zimmerman et al., 1967; Montgomery and Piégay, 2003).

At a finer scale of resolution, boundary resistance may be differentiated into grain roughness and form roughness (Figure 3.4c). In general terms, *grain roughness* refers to the relationship between grain size and flow depth. In gravel or coarser textured streams, grain roughness can be the dominant form of roughness, exerting a considerable drag on the flow. As depth increases with discharge at a cross-section, the effect of grain roughness is drowned out and flow resistance decreases (see Table 3.3; Church, 1992; Bathurst, 1993, 1997). *Form roughness* is derived from features developed in the bed material. In sand-bed streams this commonly exceeds grain roughness in importance, and in streams with coarser bed material, where grain friction might be expected to dominate, it can still be a major contributor to flow resistance (Knighton, 1998). Bed configuration may vary with flow stage, altering form roughness (see Chapter 4). Types of form roughness in gravel-bed rivers include pebble clusters or bar forms that induce

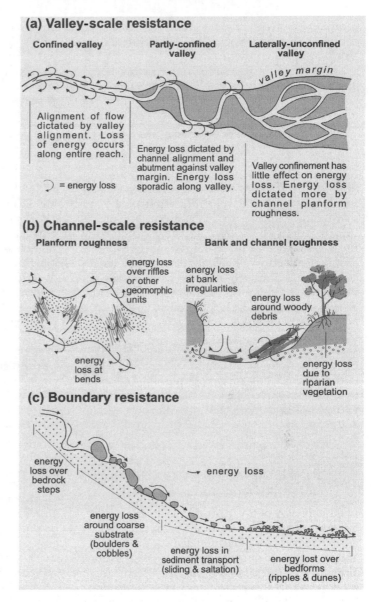

Figure 3.4 Forms of flow resistance
(a) Valley-scale resistance imposed by valley confinement and alignment. (b) Channel-scale resistance induced by planform and bank roughness (including vegetation and woody debris). (c) Boundary resistance imposed by bedrock outcrops, bedforms, and grain roughness.

resistance because of ponding upstream of steps, riffles, or bars. Although this effect is most pronounced at low flow stage, bar resistance can still account for 50–60% of total resistance at higher bankfull stages. In bedrock rivers, step-pool sequences and coarse substrate (i.e., boulders) are the key resisting bed elements that result in dissipation of energy through hydraulic jumps and ponding.

The nature, extent, and distribution of resisting elements along a valley floor are major determinants of the ease with which rivers are able to adjust their form. Alluvial rivers on sandy substrates with minimal vegetative protection are perhaps the most sensitive variant of river morphology to change. If banks are cohesive, or an armor layer is present, or dense vegetation cover lines the banks, or there is a high loading of woody debris, or bedrock/cohesive terrace materials line the bed/bank, the ease with which adjustments are able to take place is reduced.

The balance of impelling and resisting forces along a reach is a critical determinant of a river's behavioral regime, indicating whether there is a relative dominance of erosional or depositional tendencies. Interpreting controls on this balance represents a key step in understanding why a river looks and behaves in the way that it does. Adjustments to the magnitude or distribution of impelling or resisting forces are primary agents of river change (see Chapter 6). Adjustments to river morphology, in turn, drive changes to the operation of various physical fluxes, thereby modifying the way that energy is distributed and used within a system, affecting the balance of impelling and resisting forces in any given reach. These mutual adjustments are a natural part of system evolution. Across most of the planet, human disturbance has severely modified these relationships (see Chapter 7).

Analysis of the balance of impelling and resisting forces along longitudinal profiles provides a useful tool that aids interpretation of the distribution of geomorphic process zones and resulting patterns of river types and their linkages within any given catchment, as discussed in the following section.

3.4.2 Use of longitudinal profiles to interpret controls on river character and behavior

Landscape position is a vital consideration in determining why a river looks and behaves the way it does. Rivers have differing balances of erosional and depositional processes at different positions along their longitudinal profiles. Analysis of process linkages and landscape connectivity along longitudinal profiles provides a critical basis for interpretation and prediction, guiding insights into lagged and off-site responses to disturbance. Noting the position of each reach along its longitudinal profile, and considering the slope and upstream catchment area as determinants of flow–sediment fluxes and stream power, provides an initial guide into controls on river forms and processes. Determination of parallel patterns of river forms and processes along longitudinal profiles provides a meaningful basis to compare and

interpret these relationships. Assessment of trends in regions of similar geology, topography, and climate may provide insights into broader-level controls on patterns of river forms and processes, enabling elimination of some factors as underlying controls, and isolation of limiting factors that influence the presence/absence of certain forms and processes. However, local considerations, or unique combinations thereof, may fashion catchment-specific forms and processes. In performing such analyses, it must always be remembered that similar-looking forms can be explained by different sets of processes (equifinality).

In general terms, boundaries of landscape units are demarcated by significant breaks in slope on longitudinal profiles. When lengths of river between breaks in slope are related to valley confinement, a first order guide to the distribution of imposed boundary conditions along the river is provided. Various other parameters can be added to interpret controls on river types. For example, lithology and geological structure exert controls on the distribution of slope, valley width, stream power, and the availability/caliber of material supplied to the stream. In some settings, resistant bands of bedrock may induce local base-level control, impeding the rate and progression of downcutting in the drainage network upstream, while maintaining shallow, bedrock-based alluvial fills (e.g., Tooth et al., 2002). Variability in valley confinement concentrates or dissipates flow energy, altering the capacity of flow to erode, transport, and deposit sediments, in turn affecting river morphology. Differences in catchment shape, and associated patterns of tributary–trunk stream relations, result in catchment-specific patterns of discharge variability along longitudinal profiles.

A further factor that induces pronounced variability in river morphology along longitudinal profiles, and related variability in the capacity of these rivers to adjust their form, is sediment availability. Significant differences are observed, for example, in glaciated and nonglaciated landscapes. Alternatively, systems may have been subjected to massive sediment inputs via geologic (e.g., tectonic uplift, volcanic) or climatic (e.g., cyclonic) conditions. In some settings, the influence of antecedent controls preconditions river responses to disturbance events.

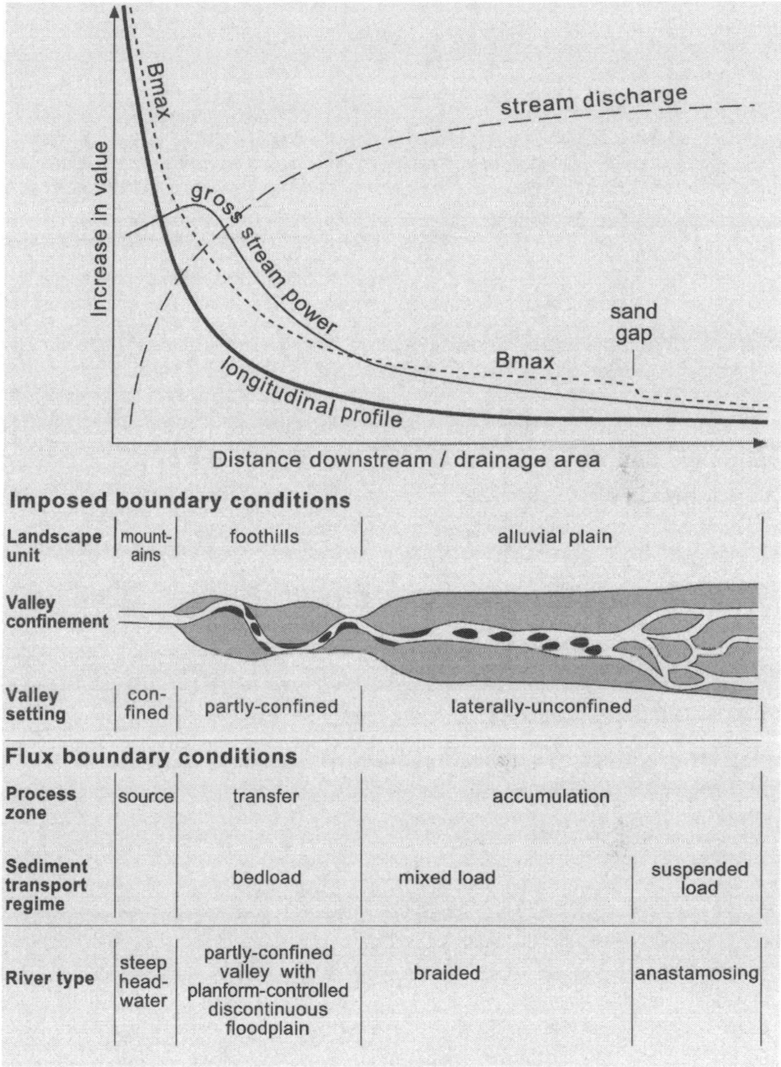

Figure 3.5 Typical longitudinal profile in a tectonically active setting

In general, longitudinal profiles in tectonically active terrains have a concave upwards (graded) form. Assuming that discharge increases progressively with catchment area, this results in a nonuniform downstream trend in gross stream power, with a peak immediately downstream of the headwaters when sufficient flow acts on sufficiently steep slopes. Bed material size (Bmax) closely follows the trend of longitudinal profile, other than a sand gap that commonly occurs towards the coastal interface. These characteristic trends result in discernable downstream patterns in the types of valleys that are found, the geomorphic process zone activity and associated patterns of river types (with a transition from bedload to mixed load to suspended load). Lighter shading highlights floodplain, dark shading highlights instream bars.

Interactions among a suite of imposed and flux boundary conditions that result in characteristic patterns of river types along longitudinal profiles are shown for differing landscape settings in Figures 3.5–3.7. Homogeneity in geological and climatic regimes is inferred in these idealized examples. Combinations of slope and discharge result in differing downstream patterns of total stream power, fashioning the capacity of the river to transport materials of varying size. These process relationships result in different river mor-

phologies at different positions along the longitudinal profile.

Figure 3.5 shows a classic concave upwards longitudinal profile in a tectonically active terrain (Figure 3.7a). In this example, no bedrock steps exert local base level control. The relationship between available energy and bed material size shapes the pattern of geomorphic work along the longitudinal profile, and the resulting distribution of erosional and depositional processes. In a general sense, these relationships maintain the smooth,

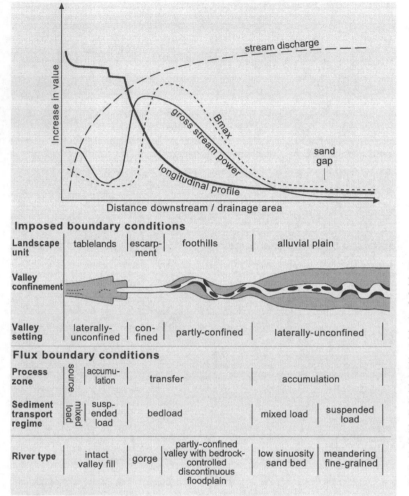

Imposed boundary conditions

Landscape unit	tablelands	escarp-ment	foothills	alluvial plain
Valley confinement				
Valley setting	laterally-unconfined	con-fined	partly-confined	laterally-unconfined

Flux boundary conditions

Process zone	source	accumu-lation	transfer	accumulation
Sediment transport regime	mixed load	susp-ended load	bedload	mixed load / suspended load
River type	intact valley fill	gorge	partly-confined valley with bedrock-controlled discontinuous floodplain	low sinuosity sand bed / meandering fine-grained

Figure 3.6 Typical longitudinal profile in a tectonically stable, escarpment-dominated setting
The "classic" sequence of landscape units, river morphology, and river processes characterized for tectonically-active terrains are not necessary repeated in different landscape settings. In the example presented here, denuded, upland settings drain into escarpment country, resulting in differing patterns of geomorphic process zones and related river morphologies (see text). Lighter shading highlights floodplain, dark shading highlights instream bars.

concave-upward shape of the longitudinal profile (Mackin, 1948; Hack, 1960; Leopold and Bull, 1979).

Erosional processes are dominant in upstream reaches, producing incised rivers that are sculpted into bedrock. In contrast, downstream reaches have aggradational tendencies, and materials accumulate on the valley floor. In the intervening transfer reaches, water and sediment budgets maintain a balance of erosional and depositional tendencies over timescales of 10^1–10^3 years. A progressive downstream increase in catchment area is accompanied by a reduction in slope and bed material texture, and an increase in valley width, dis-

charge, and channel capacity. The downstream trend for discharge (a surrogate for catchment area) is virtually opposite to that shown by slope. In relative terms, catchment area increases rapidly in headwater settings, but the rate of increase decreases with distance downstream. As stream power is the product of discharge and slope, the stream power maxima is attained in lower order streams where sufficient flow acts on relatively steep slopes (Fonstad, 2003).

In general terms, sediment availability increases downstream, such that there is a transition from supply-limited to transport-limited conditions. However, the capacity of the river to perform

Figure 3.7 Landscape units in differing tectonic settings

Pronounced variability in landscape units and patterns of landscape connectivity are evident in tectonically active and tectonically-stable, escarpment-dominated settings. These factors influence patterns of sediment storage and the rate and caliber of sediment input into river systems and the ease with which materials are conveyed through landscapes. As a consequence, marked differences in types of river may be observed in different settings. In the examples shown, the highly dissected mountainous terrains of tectonically active landscapes (a) supply large amounts of material to valley floors. High energy rivers rapidly convey materials to lowland plains. These are transport-limited landscapes. In example (b), which depicts a tectonically stable, escarpment-dominated landscape, relatively subdued relief in plateau areas delivers relatively small amounts of sediment to gorges and partly-confined valleys downstream of the escarpment. Sediments stores are limited. The upper parts of catchment are characterized by denuded and rounded hills. These landscapes are more disconnected than tectonically active landscapes. In these supply-limited settings, rates of sediment delivery to the lowland plain are limited. As a consequence, low energy rivers in the latter setting may be sediment starved.

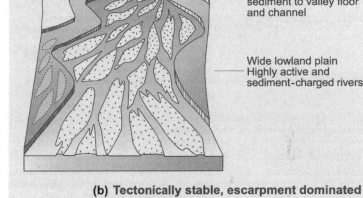

(a) Tectonically active landscape

Rugged mountains Highly disected

Piedmont zone where alluvial fans deliver sediment to valley floor and channel

Wide lowland plain Highly active and sediment-charged rivers

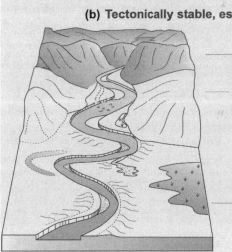

(b) Tectonically stable, escarpment dominated

Low relief plateau above a steep escarpment May contain alluvial fills

Denuded and rounded hills

Narrow lowland plain with low energy, sediment-starved rivers

geomorphic work is restricted by a decrease in available energy (i.e., stream power). Assuming a near-uniform lithology, the downstream trend in bed material size (represented by Bmax) is related in a nonlinear manner to flow energy. In lower order streams, systematic trends in texture or sorting are not evident because of more or less continuous recruitment from adjacent hillsides (Rice and Church, 1996) and the inability to rework material, other than on an irregular basis. As such, bed material size is commonly much coarser than in the zone of stream power maxima. Beyond this

point, materials are more easily reworked by abrasion and selective transportation mechanisms, resulting in a relatively smooth downstream gradation in bed material size. In general, the largest size that can be moved is proportional to the force that can be exerted on the stream-bed by the flow. As competence declines downstream, characteristic sediment sizes become finer. The downstream transition to a laterally-unconfined (alluvial) valley setting marks the onset of persistent and long-term accumulation of sediments (Church, 2002).

Quite different sets of process relationships and river morphologies are evident along the schematic longitudinal profile presented in Figure 3.6. In some tectonically stable settings, stepped longitudinal profiles reflect the influence of escarpment retreat on long-term landscape evolution (Figure 3.7b). Catchment area increases systematically along the longitudinal profile, in a similar manner to that described for the previous example, with a significant increase along the first few kilometers of river length and a gradual increase towards the river mouth. In contrast to Figure 3.5, however, the downstream variation in slope produces notable differences in plots for stream power and bed material size (represented by Bmax), as bedrock steps exert a major influence on base level. Resulting patterns of river morphology are related directly to landscape setting.

In the tablelands, low slopes and catchment areas produce low stream powers. Given limited relief, low energy floodplains develop in these upland settings, which function as sediment accumulation zones. Flow often has insufficient energy to maintain continuous channels, and discontinuous watercourses are common. Hillslopes are decoupled from watercourses. Suspended load materials accumulate on the valley floor, often in swamps. Stream power and bed material size maxima occur in the escarpment zone, where sufficient flow and steep slopes combine. The high flushing capacity of the channel ensures that only the coarsest component of the bedload fraction lines the bed. Incision produces narrow valleys with high hillslope–channel connectivity. As slope, stream power and bed material size decrease downstream of the escarpment zone, there is a similar set of relationships to those described in Figure 3.5. The transition from the foothills to the alluvial plain is characterized by a shift from bedload to mixed and suspended load transport. The lowland plain acts as an accumulation zone, with fully alluvial rivers.

The differing configuration of the landscape settings portrayed in Figures 3.5–3.7 ensures that disturbance responses in headwater areas are conveyed through the systems in differing ways. Longitudinal connectivity is likely to be much greater in the system portrayed in Figure 3.5 relative to that in Figure 3.6, assuming there are no artificial barriers such as dams along the longitudi-

nal profile. Catchment-specific patterns of geomorphic process zones and their downstream linkages are key determinants of river morphology and the propensity for river adjustments. Longitudinal profiles provide a valuable tool to examine these linkages, aiding interpretations of the balance of impelling and resisting forces along a river. Understanding of process linkages, and their sensitivity to change, is a critical consideration in endeavors to predict likely future river character and behavior.

3.5 Predicting the future in fluvial geomorphology

Unlike experimental studies undertaken in controlled settings, geomorphic enquiry is confronted by a perplexing array of factors that induce a bewildering array of landscape forms. Some individuals gain immense satisfaction in trying to unravel the complexities and uncertainties of real world landscapes, and commit themselves to a career in geomorphology. Others see training in geomorphology as an inordinately frustrating exercise in futility, in which exceptions are found for each principle that was thought to be well-grounded. Exposure to new experiences requires a perpetual process of unlearning and relearning. While some practitioners have a genuine flair for "reading the landscape" and interpreting controls on river character and behavior, others seem destined to never quite "get it"!

Geomorphologists are concerned with large-scale temporal and spatial analysis of complex natural experiments that are neither reversible nor repeatable, precluding exact reproduction (Schumm, 1991). As landscapes are emergent phenomena, characterized by nonlinear dynamics and contingent behavior, their analysis is not amenable to reductionist explanations (Harrison, 2001; Phillips, 2003). Emergent behavior of landscapes cannot be predicted using small-scale physics; rather, it must be analyzed using relations defined at the scale of the emergent behavior (Murray, 2003; Werner, 2003). Geomorphic models share with ecology a strong dependence on landscape history, because present processes and conditions are strongly contingent on past events and because processes operating over widely different time scales can be of coequal importance (Wilcock and Iverson, 2003).

While river rehabilitation activities are largely concerned with rectifying problems that have arisen from cumulative river responses to past disturbance events, proactive and strategic planning must relate to concerns for future river condition. Realistic expression of prospective future states must build on sound physical information on the system of concern. Strategies must be subjected to recurrent reappraisal in light of ongoing developments. Insights into system responses to disturbance events guide appraisals of the type and priority of differing intervention strategies, providing a basis to evaluate their likely success. For example, useful distinction can be made between predictions of present landscape behavior and predictions of landscape change (Wilcock and Iverson, 2003). This section appraises approaches to predict landscape futures used by geomorphologists in endeavors to provide a coherent landscape platform for management activities.

Predictions of likely future scenarios must be based on solid understanding of controls on contemporary landscape forms and processes. Such analyses must integrate the cumulative effects of system responses to multiple forms of disturbance. These responses are influenced by within-system connectivity of landscape compartments, and associated implications for the operation of biophysical fluxes. Human agency has added a further layer of complexity to the inherent diversity of the natural world. Predictability is complicated by the fact that differing components of rivers may respond at different rates or in different ways. Past events set the evolutionary pathway upon which the contemporary system sits, guiding interpretations of what the system is adjusting towards, assuming that boundary conditions remain roughly constant (Schumm, 1991). Without knowledge of the trajectory of change of the system, future changes due to changed conditions, or lags and off-site responses already instigated within the system, cannot be predicted. In many instances, these considerations introduce a level of complexity into extrapolations that defies realistic quantification. While some responses to disturbance are predictable, others are not. Without an appreciation of underlying causality, predictions are speculative at best; perhaps fanciful and dangerous at worst.

Three primary approaches to prediction are outlined in this section. First, comparative frameworks are used to relate different states of evolutionary adjustment in different areas that have a similar landscape configuration, enabling the application of space for time substitution (also known as ergodic reasoning) (Section 3.5.1). Second, equilibrium-based theoretical insights are merged with empirical relationships to derive models that predict how components of systems work (Section 3.5.2). Finally, Section 3.5.3 presents a real-world perspective on unraveling causality and predicting river change based on detailed analyses of system-specific evolutionary traits. Ultimately, these various approaches must be merged to provide practical guidance for the prediction process, recognizing explicitly that information bases and associated knowledge are incomplete and imperfect.

3.5.1 Comparative frameworks with which to predict river changes

Explanation of past events and prediction of future events frequently requires reasoning by analogy, which is the recognition of similarity among different things. Ultimately, these analogies must be based on meaningful insights into both the structure and function of phenomena. Geomorphic enquiry seeks repeatable patterns with which to guide our interpretations of likely future changes. Although case studies facilitate our understanding of changes, it is risky to generalize findings to derive conceptual frameworks that represent reproducible models of how systems work and their sensitivity to change. Inferences gained from other systems must be used in a precautionary manner. All too often, inappropriate inferences have been drawn from insufficient data or poorly framed comparisons. Little can be gained, and much may be lost or compromised, in approaches to enquiry that argue by correlation without underlying explanation, in what can cynically be viewed as "painting by number" exercises. The small number of detailed case studies that are available imposes critical limits on the opportunity for repeatability, negating the reliability of predictions (Schumm, 1991).

Geomorphologists use data collected from multiple locations to construct evolutionary

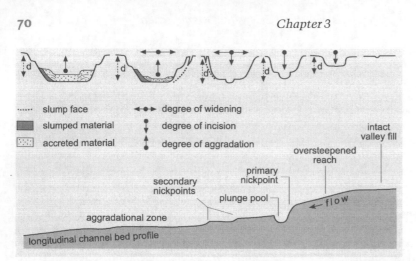

Figure 3.8 Phases of incised channel development using location for time substitution
Location for time substitution, also known as ergodic reasoning entails determination of different phases of landscape evolution, either within the same system or through analysis of different stages in different systems. In the example shown, different phases of incised channel development are indicated along a schematic channel bed profile. Initially the entire profile comprised an intact valley fill as indicated on the right of the diagram. Moving towards the left, the differing cross-sections indicated progressive phases of channel adjustment, marked by incision, channel expansion, phases of secondary headcuts, and finally by phases of refilling of the incised trench in the aggradational zone. Modified from Schumm et al. (1984). Reproduced with permission from Water Resources Publications, LLC, 2004.

sequences of events or phases of landscape activity. This is referred to as location for time substitution (or ergodic reasoning). A related approach can be applied to determine the sensitivity of landforms. This is termed location for condition evaluation by Piégay and Schumm (2003). However, to avoid confusion with notions of geomorphic condition in the River Styles framework, this is referred to here as proximity to threshold analysis. These procedures provide practical approaches to prediction and environmental reconstruction. Both approaches require that the investigator back away from a single site and look at many sites, thereby providing a "big picture" appraisal as a basis for generalization.

Location for time substitution provides an important basis with which to compare features produced by the same set of processes that operate under an equivalent set of conditions (Schumm, 1991; Piégay and Schumm, 2003). Samples of landforms at differing stages of evolutionary adjustment are arranged in a sequence to outline a model that differentiates among phases of landscape evolution. Hence, interpretation of a range of field sites is used to derive a schematic framework with which to analyze landscape evolution as a series of

timeslices. In fluvial terms, each timeslice records a differing stage of adjustment for a particular type of river. From this, it is possible to interpret the pathway of adjustment that is likely to be experienced for any given reach of that river type under a certain set of conditions. This tool has both academic and practical value. For example, determination of phases of incised channel evolution can be used to estimate agricultural land loss, sediment production, identify reaches that require treatments, and determine appropriate treatments to alleviate problems (Figure 3.8; Schumm et al., 1984).

In applications of location for time substitution, study sites must have equivalent initial conditions, such that differences in site behavior may be attributed to differences in treatment. The reliability of predictions is dependent on the similarity of the places to be compared (Paine, 1985; Schumm, 1991). These procedures have limited spatial scope and will usually only provide an indication of the range and rate of processes that are likely to occur within individual landscape elements after a particular type of disturbance or change in boundary conditions. At larger scales, process linkages may produce system-specific responses that override or damp-out the trajectories of change of individual reaches. In general terms, the longer the timespan and the larger the area, the less accurate will be predictions that are based upon observations of contemporary forms and processes. It is also important to note that similar results may arise from different processes and causes (the principle of convergence or equifinality). A common origin and equivalent causality are prerequisites for effective comparison.

Proximity to threshold analysis appraises the state of the landscape (or components thereof) at various sites to interpret the relative sensitivity of any given setting to change. From this, characteristics of relatively sensitive and insensitive landforms of the same generic type are appraised to identify threshold conditions under which change is inferred to occur. Data are collected from a range of localities in which the same set of processes operate (i.e., landscape setting is equivalent), such that a relationship can be established to differentiate among states of system behavior (i.e., condition). Rather than presenting an evolutionary basis with which to interpret landscapes, proximity to threshold analysis interprets the sensitivity of landforms (or landscapes) to change. Proximity to threshold analysis has been used to characterize the sensitivity of valley floors to gullying (Figure 3.9; Patton and Schumm, 1975). Critical relationships between valley floor slope and catchment area are used to differentiate among settings that have developed gullies and settings that retain an intact valley floor. An equivalent drainage area-slope relationship has been established to describe the susceptibility of alluvial fans to fan-head incision (Schumm et al., 1987). This approach has also been utilized to identify the susceptibility of a reach to a change in channel planform based on a relationship between sinuosity and valley slope (Schumm and Khan, 1972; Schumm, 1991). Piégay et al. (2000) used proximity to threshold analysis to analyze the dynamics of sediment infilling of cutoff channels.

Models that describe pathways of geomorphic response to disturbance events provide a basis to predict future changes to river morphology and hence guide mitigation measures. Determination of the sensitivity to change of a given reach to events of differing magnitude, or differing sets of external (driving) factors, provides critical guid-

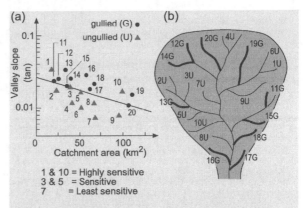

Figure 3.9 **Using proximity to threshold analysis as a basis for assessing the sensitivity of rivers to change (modified from Patton and Schumm, 1975)**
A long-standing relationship, developed initially by Patton and Schumm (1975), has indicated that the distribution of gullied and ungullied tributary stream lines can be predicted using discriminant analysis based on valley slope for a given catchment area (for drainage lines greater than 10 km^2). In the schematic example presented here (a), gullied tributaries plot above the discriminating function, while ungullied tributaries plot beneath it. However, this is not an entirely consistent relationship, as some systems are yet to become incised (b). Hence, while tributaries 1 and 10 are highly sensitive, as they sit above the threshold of gullying, tributaries 3 and 5 also lie very close to this threshold condition, while tributary 7 is most distant from the threshold and therefore considered to be least sensitive to change. From this perspective, if a major storm was to impact upon this catchment, the tributaries which are most likely to be subjected to dramatic change are tributaries 1 and 10.

ance in assessments of the direction and magnitude of river response to disturbance. Such tools can aid analyses of the likely potential for, and rate of, recovery after disturbance. Although response times are highly variable and poorly predictable phenomena, an exponential decay factor has been shown to have a reasonable degree of application (e.g., Graf, 1977; Kasai et al., 2004). Rate of response is controlled by many factors, including the inherent resistance, available energy and scale of river systems, and the severity of the disturbance event.

Location for time substitution and proximity to threshold analysis provide insights into the trajectory of change, isolate threshold conditions under which change is likely to occur, and determine how likely it is that change will occur. However, they provide little quantitative information. For example, they cannot be used to interpret the timeframe over which change is likely to take place. Although quantitative prediction of relaxation paths following disruption represents the ultimate objective, conceptual models of river evolution may provide a useful guide to interpret past and present changes and anticipate future trends associated with natural or human disturbance. For example, Simon (1995) used such an approach to identify dominant adjustment processes, to differentiate between local and more widespread instability, and to suggest the type of mitigation measures that are required to reduce future damage associated with impacts of channelization. Changes in channel form were quantified for different stages of evolution.

3.5.2 Theoretical predictions of river adjustments based on equilibrium channel morphology

Fluvial geomorphologists have tended to look to fluid mechanics and engineering research to provide explanations and deeper (predictive) meaning to their largely qualitative, case-study observations of rivers (Hickin, 1983). These theoretical principles generally build on continuity equations for water and sediment and the flow momentum equation. A notional equilibrium condition implies that stable width, depth, slope, and planform can be expressed as functions of the controlling variables: discharge, sediment supply, and channel bed and bank caliber. Under circumstances in which the boundary conditions within which a reach operates remain relatively constant, such that there is no change in slope, discharge, or sediment load, any small-scale disturbance to equilibrium channel morphology will set in motion processes that will return the channel to its stable form and pattern. Hydraulically-based models used to predict stable channel forms are based on the assumption that an equilibrium channel morphology is quickly reestablished (Hickin, 1983). Although this assumption may be valid for short periods of time (say a few years), it becomes less

defensible as the time period is lengthened. In reality, the timeframe over which a reach is adjusting is seldom such that truly steady state conditions prevail. Although equilibrium-based approaches provide an initial guide to predict short-term, reach-specific responses to flow events, these procedures are unreliable in assessment of longer-term trends, as assumptions of uniformity and stationarity that are applied in derivation of equilibrium-based probabilistic relationships are invalidated (e.g., Phillips, 1992; Richards, 1999).

Despite these limitations, qualitative expressions derived to describe the manner of channel adjustments associated with, say, changes to discharge or sediment load (e.g., Schumm, 1969), continue to form a basis for river rehabilitation practice. If these expressions are to be of any use in prediction, they must be related to the type of river under investigation, and the associated ease with which it is able to adjust. Relationships derived for meandering rivers by Schumm (1969) describe adjustments to various channel attributes in response to changes in discharge and sediment yield. In general terms, wider, less sinuous channels are formed when both discharge and bed material load increase. Predicted responses are reversed when both discharge and bed material load decrease. In many instances, changes in discharge and sediment load have an opposite tendency. For example, increasing discharge and decreasing bed material load tend to result in narrower, deeper channels of greater sinuosity and lower gradient. Many changes are indeterminate because the magnitude of opposed responses is unspecified (Knighton, 1998).

Empirical approaches to the study of rivers entail the collection and analysis of data to establish relationships between form variables, or between a form variable and factors that summarizes some aspect of process (Knighton, 1998). If sample sizes are sufficiently large and representative (of a certain type of river in a particular region), insights can be used to calibrate models based on physical laws, thereby providing a basis to validate their applicability. Comparative analyses and flume studies can be applied to test and extend these analyses. For example, empirically based power relationships have been used to describe the hydraulic geometry of channels, relating adjustment to channel geometry to changes in discharge (e.g., Leopold and Maddock, 1953; Hey and Thorne, 1986).

Inherent uncertainties are faced in making generalizations from empirical results founded on limited or unrepresentative databases. Strong correlations in statistical tests do not necessarily convey a genuine understanding of underlying processes. Indeed, it is naïve to think that simplistic cause-and-effect relationships, whether linear or otherwise, can be expected to apply from system to system independent of other causative factors. Significant challenges are faced in extrapolating findings to other systems in a meaningful manner, ensuring that analyses truly compare like with like. Hence, determinations of stable channel size and form should relate directly to the type of river under investigation, reflecting data collected in geologically and hydrologically homogenous regions. However, there continues to be a remarkable lack of such primary data from most parts of the planet. Indeed, the mind-set that underpins the development of theory in fluvial geomorphology is undoubtedly value-laden, and field-based data that researchers work from are far from representative. Until recently, for example, remarkably little research was carried out in fully vegetated river systems, limiting insights into naturalness and the predisturbance condition from which rivers have subsequently adjusted (see Montgomery and Piégay, 2003). Such oversights effectively mean that contemporary understanding is largely restricted to a subset of modified rivers, rather than the breadth of diversity of the natural world *per se*. Hence, extreme caution must be used when applying these relationships outside the area for which they were derived. Regionally specific empirical relationships must specify the type of river under investigation and the boundary conditions under which the river operates (or has operated), separating local-scale controls on channel geometry from broader scale controls. In addition, the geomorphic condition of the reach must be consistent from site to site, ensuring that the database is derived from equivalent sites. If these criteria are not fulfilled, derived empirical expressions represent a variant of proximity to threshold analysis (or location for condition evaluation), rather than representing broader predictive trends that have general application.

Significant errors can be made when (mis)applying these notions in a practical context. For example, mistakes are made when an apparently stable reach is used as a template for the restored channel, or when general empirical relations between channel geometry and flow frequency are used as the basis for design (Wilcock, 1997). This approach assumes that the channel has adjusted to the amount of water and sediment supplied to it. By definition, a river requiring rehabilitation is unlikely to be in such a state, precisely because the quantity and timing of water and sediment supply have changed and, in many cases, will continue to do so. Although empirical relations for channel geometry provide an initial guide in predictions of river behavior, they provide an inappropriate basis for river rehabilitation. A more useful paradigm views rivers as adjusting to the water and sediment supplied to them (Wilcock, 1997).

Approaches to geomorphology that consider landscapes over timeframes of operation under conditions of idealized thermodynamic equilibria must be viewed as ideals with limited practical relevance outside relatively "static" modeling applications. Landscapes are perpetually adjusting to disturbance of one form or another, and have system-specific imprints of geomorphic memory (Trofimov and Phillips, 1992). The extent to which contemporary river morphology is adjusted to former conditions constrains the use of empirical relationships to characterize the nature and rate of morphometric adjustments. More recent thinking in geomorphology no longer views stability as an endpoint, or even necessarily a "normal" state, as landscapes are viewed to be in phases of relative stability for varying intervals of time as they progressively adjust to ongoing perturbations (e.g., Phillips, 2003). Landscape evolution does not follow deterministic principles, moving through predictable stages towards a stable endpoint. Rather, disturbance events at differing spatial and temporal scales result in variable forms of instability. Ultimately, the natural world is so complex that a complete deterministic explanation can never be achieved. As noted by Richards and Lane (1997, p. 289), whilst simulation models may be used for predictive purposes, their ultimate value may be little more than their role as tools for probing the depths of our uncertainty. If empirically derived relationships are to be of a lasting and predictive nature, inductive reasoning needs to be embodied within a theoretical structure, even if only qualitatively. In practical terms, predictions must relate specifically to the catchment of concern, with its own configuration and history of adjustment, as indicated in the following section.

3.5.3 A real-world perspective on unraveling causality and predicting river changes: Know your catchment!

Rivers are complex, interactive systems that are subjected to an unceasing barrage of perturbations or disturbance events. Their unique configuration and history fashion catchment-specific patterns and rates of physical fluxes. These considerations ensure that rivers seldom operate as simple, linear, cause-and-effect systems. Adjacent systems in a relatively homogeneous region are unlikely to respond to external stimuli in a directly equivalent manner, making it difficult to interpret the historical record and to attach underlying causes to specific responses. In many cases it is difficult to isolate cause–effect relationships that directly link changes in river morphology to discrete underlying factors. Among many considerations, this may reflect the cumulative nature of disturbance impacts, nonsynchroneity in the character of forcing events (e.g., the sequence of large floods), or spatial variability in the extent to which individual reaches may be primed for change (i.e., reach sensitivity to threshold conditions that shape the magnitude and direction of change in any given system). For example, local instability that reflects an immediate response to some form of environmental change can easily be confused with upstream feedback effects that date from some past environmental perturbation, which is only now reaching the site in question (Macklin and Lewin, 1997). These considerations, along with appraisal of pressures and limiting factors, and lagged and off-site implications of disturbance response, must be integrated into catchment-specific predictions of likely future river character and behavior. In many instances, the whole is very different to the sum of the parts.

Nonsynchroneity in the timing, pattern, and rates of responses to differing forms of disturbance, and associated river changes, can be related to the character and configuration of individual systems. Given the catchment-specific patterns of river types and their linkages, predictions of reach and catchment-scale responses to disturbance events must be considered as part of a nondeterministic science. Appraisals of likely future river character must also be made in context of river history and understanding of spatial linkages of physical processes in the catchment of concern.

Unraveling system-specific evolutionary histories is a powerful means for assessing how a system has changed in the past, how it adjusts today, and what future trajectories of change are possible. Interpretations of the past provide no guarantees as to what the future will be like. However, they provide critical guidance in the derivation of predictions. Indeed, such exercises are surely compromised without this information. For example, catchment-specific knowledge of upstream sediment availability is required to infer potential downstream changes in river character or recovery potential. Historical (evolutionary) insights are required to develop an understanding of threshold conditions and lag effects that may shape behavioral responses in each system. Elements of randomness, local threshold conditions, and chaotic responses to perturbations ensure that interpretations of river change are best considered as a probabilistic science, in which differing degrees of likelihood are ascribed to differing forms and rates of adjustment.

Under this premise, the evolutionary pathway of each reach must be viewed in its catchment context. This provides a critical context with which to interpret landscape responses to natural events, appraise the impacts of human disturbance, and assess likely trajectories of future changes. If evolutionary pathways can be shown to be near-equivalent for various systems, ergodic reasoning can be usefully applied to appraise differing stages of evolution.

Foresighting (scenario-building) approaches that "scope the future" must be framed in terms of changes to physical boundary conditions. This involves assessment of the ways in which systems adjust to a range of limiting factors and pressures, recognizing that multiple trajectories may eventuate. These trajectories then aid the identification of target conditions for river rehabilitation. Limiting factors may include such factors as changes to sediment availability (e.g., passage of sediment slugs or sediment starvation), changes to runoff relations, and changes to vegetation cover. These factors are internal to the system. Pressures refer to factors that are external to the system, such as climate variability, human-induced changes to landscape forms and processes, and a myriad of socioeconomic and cultural changes (e.g., population, land use, direct/indirect adjustment to rivers, etc). These various considerations underpin the development of the River Styles framework documented in this book, prompting the management catch cry "know your catchment."

3.6 Summary and implications

This chapter has focused on temporal principles for effective river management. To be meaningful, temporal analysis and assessment of river evolution must extend over timeframes that capture the range of behavior and change in response to natural and human-induced disturbance events or changes in physical fluxes. Analyses of the nature, rate, and consequence of river change must be applied in a catchment-specific manner. These analyses then provide a basis to interpret controls on river character and behavior, placing reaches in their spatial and temporal context, from which predictions of future trajectories of change can be made. The mind-set with which approaches to prediction are undertaken is a major factor in determining the viability and reliability of the outcomes. Whether proponents favor a "best case" or "worst case" scenario, foresighting exercises must construct realistic future scenarios based on the present character, behavior, and condition of the landscape/ecosystem, cognizant of ongoing patterns and rates of adjustment (i.e., the evolutionary trajectory).

PART B

Geomorphic considerations for river management

*We have to abandon the arrogant belief that the world is merely a puzzle to be solved,
a machine with instructions for use waiting to be discovered, a body of information to be fed
into some computer in the hope that, sooner or later, it will spit out a universal solution.*
Vaclav Havel, quoted in Fletcher 2001, p. 194

Overview of Part B

In this part, the geoecological foundations of river rehabilitation practice developed in Part A of this book are placed in context of geomorphological understanding of river character, behavior, evolution, and human-induced change. To address these issues, various geomorphological classification procedures have gained significant application in the river management arena (e.g., Rosgen, 1996; Heritage et al., 1997, 2001). The approach adopted in this book builds on the underpinnings of these schemes, endeavoring to establish a set of procedures that can be applied across the spectrum of morphological complexity demonstrated by rivers.

In Chapter 4, a set of guidelines with which to meaningfully deconstruct river forms and processes at a range of scales is documented. These components can be readily reintegrated as a platform for coherent analysis of river character. River behavior is documented in Chapter 5. A conceptual framework with which to assess river dynamics, termed the river evolution diagram, is introduced. Chapter 6 examines river change, focusing on river responses to disturbance, and spatial and temporal controls on pathways of change (i.e., evolutionary tendencies). In Chapter 7 the role of human disturbance as an agent of river change is appraised. This contextual information provides the underpinnings of the River Styles framework that is presented in Part C.

CHAPTER 4

River character

The present trend in fluvial geomorphology is towards increasingly detailed understanding of smaller and smaller features – more and more about less and less. The variety of fluvial forms (in the natural world) . . . illustrate(s) the very limited way in which detailed fluvial research has been able to contribute to a broad understanding of rivers and channelways. It is time to draw conclusions from all that has been learned in the past 50 years, and apply them, in a suitably simplified way and at relevant scales, to entire fluvial systems, and to studying how such systems have evolved over time.

Mike Kirkby, 1999, p. 514

4.1 Introduction: Geomorphic approaches to river characterization

Despite the seeming simplicity of flow interactions with sediments on valley floors of differing slope, there is remarkable diversity in river morphology. Various laboratory, empirical, and theoretical studies have demonstrated that there is a continuum of river character and behavior, in which individual variants of river morphology fit along a gradation of stream power and grain size trends (Bridge, 1985). Increasing recognition and documentation of the diversity of river morphology is emphasized by the book edited by Miller and Gupta (1999) entitled *Varieties of Fluvial Form*. This chapter appraises approaches to the characterization of river morphology.

In simple terms, rivers comprise an array of bedrock and alluvial (free-forming) variants. Differentiation of bedrock and alluvial systems reflects the balance between sediment supply and channel transport capacity (Montgomery et al., 1996). Along bedrock rivers, transport capacity exceeds the rate of sediment supply (i.e., they are sediment supply limited). As such, their morphology is a function of the physical characteristics of the bedrock rather than the hydraulic and sediment transport characteristics of the river. Banks are imposed, so bed morphology drives adjustment in channel shape and the assemblage of geomorphic units. In contrast, alluvial rivers either attain a balance between rates of sediment supply and

sediment transfer, or there is excess sediment supply such that aggradation is induced (i.e., they are transport limited). Channel morphologies adjust to prevailing flow and sediment conditions, enabling them to convey a wide range of discharge and sediment loads. In most situations, the assemblage of geomorphic units is maintained by small-scale, short-term adjustments and the range of flow conditions that occur within the channel zone. Abrupt thresholds exist between gravel-bed, sand-bed and fine-grained channels (Howard, 1980, 1987). Compared to bedrock channels, alluvial rivers tend to have lower gradients, higher sinuosity with more pools, and greater capacity for morphological adjustment (Montgomery et al., 1996; Montgomery and Buffington, 1998; Tinkler and Wohl, 1998; Wohl, 1998).

The inordinate diversity of river forms and processes presents significant challenges in the development of a flexible and generic approach to river classification. However, these challenges must be met head-on by river scientists; otherwise, managers will adopt strategies that conform to their own criteria, often divorced from, or misapplying, geomorphic principles. The challenge of embracing morphological continua in derivation of river classification procedures is compounded by temporal variability. Many alluvial rivers adopt differing planform configurations at differing flow stages (e.g., ephemeral and monsoonal rivers). Inevitably, as rivers change over time, so must their classification category. Indeed, capacity to change and system dynamic may, in themselves, provide a basis for river classification!

Practical approaches to river classification must move beyond description of the visual character of a reach to include interpretation of river behavior, explaining why that particular morphology has been adopted. Ideally, this understanding can be related to the landscape setting, framing insights in terms of reach position in the catchment, upstream and downstream controls, the balance of impelling and resisting forces, sediment and flow regimes, and river evolution (see Chapter 3). Principles of geomorphic convergence or equifinality may ensure that any given river type may reflect a range of controlling variables and processes.

The approach to river classification adopted in this book endeavors to allow each field situation to "speak for itself." Attributes of river character are assessed at a range of scales. Channel morphology is differentiated into two components: the bed and the banks. Bed morphology is appraised at two scales: transient bedforms (Section 4.2.1) and form-process associations of instream geomorphic units (whether erosional (Section 4.2.2), midchannel (Section 4.2.3), or bank-attached features (Section 4.2.4)). Bank morphology and a summary of bank erosion processes are presented in Section 4.3. Bed and bank morphology are then combined to appraise channel shape (Section 4.4) and channel size (Section 4.5). Floodplain formation and reworking processes, and related geomorphic units, are discussed in Section 4.6. Variants of channel planform in laterally-unconfined settings, and their controls, are outlined in Section 4.7. Finally, the role of valley confinement as a determinant of river morphology and associated bedrock river variants is discussed in Section 4.8.

4.2 Channel bed morphology

Bed material can be molded into coherent structures that may be broadly classed as "hydraulic" features (microscale and mesoscale), in that development is related to local flow conditions over the bed, or "sediment storage" features (macroscale and megascale) which represent larger scale instream landforms. These various features affect flow resistance, the dynamics of sediment

transport, and the form of the channel bed (see Chapter 3). In general terms, gravel- and sand-bed rivers adjust their morphology around a range of morphodynamic features over an array of scales, while bedrock and boulder streams tend to be located in high-energy, erosional settings in which flows either flush materials through the reach or coarse bedload materials impose an irregular bed morphology.

The size and shape characteristics of bed material at any point along a river are determined by the volume and caliber of materials supplied and the capacity of flow to rework it. Suites of bedforms reflect local sorting under differing flow energy conditions, controlled primarily by relationships between velocity, flow depth, and bed material size. Broader, within-reach, and downstream changes in bed material caliber exert a dominant influence on the geomorphic unit structure of a reach, and hence river morphology.

The nature and pattern of instream geomorphic units are fashioned by flow energy within a reach, and the capacity of flow to mould available materials. Among many considerations, this is influenced by the volume, caliber, and mobility (packing) of bed materials. If a reach has excess energy relative to available sediment of sufficient size, flushing is likely to occur. Alternatively, with excess sediment availability or insufficient flow energy, continuous instream sedimentation is likely to occur, commonly in the form of near-homogenous sheets. In some instances, the array of observed features may record past events, possibly extending back over hundreds or even thousands of years. Elsewhere, the diversity and configuration of features may record responses to the last major flood event. Vegetation may have a significant role in controlling the rates and types of deposition and erosion on different surfaces. Prior to documenting the range of instream geomorphic units, smaller-scale sand and gravel bedforms are briefly described.

4.2.1 Sand and gravel bedforms

Natural streams are seldom characterized by flat beds. Such a form is unstable, and tends to become deformed to produce a suite of bedforms that adjusts over differing time periods. When shear stress exceeds a critical threshold, cohesionless beds are

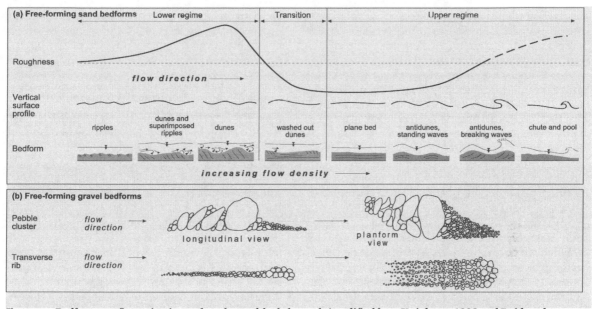

Figure 4.1 Bedform configuration in sand- and gravel-bed channels (modified from Knighton, 1998 and Reid et al., 1992)

The sequence of bedforms in sand-bed streams (a) is dictated by surface water profile and flow intensity (see text). Particle organization in gravel-bed streams (b), such as pebble clusters and transverse ribs, reflects differing flow–sediment interactions.

molded into differing geometric forms dependent upon flow characteristics. In turn, bedform geometry influences flow resistance and the nature/distribution of flow energy in complex feedback relationships (Knighton, 1998). Bedforms reflect local variations in the sediment transport rate, generating orderly patterns of erosional and depositional forms. Sediment transport rates vary across individual bedforms as a result of form-induced accelerations and decelerations in flow, promoting scour in the troughs and deposition towards the crests.

Bed morphology of sand-bed streams adjusts readily to changes in flow and/or sediment supply conditions. Given the small size and low inertia of individual grains, bed material is mobile over a wide range of flows, creating instabilities in the form of ripples, dunes, and antidunes. These lower and upper flow regime forms are classified according to their shape, resistance to flow, and mode of sediment transport (Figure 4.1a, Simons and Richardson, 1966). Lower flow regime conditions comprise plane bed with no motion, ripples, or

dunes. At these stages, form roughness is dominant. Upper flow regime conditions comprise plane bed with motion and antidunes. At these stages, grain roughness is dominant. Bed configuration in the transition zone between these two regimes is characterized by the washing out of dunes as the bed approaches plane bed with motion.

Starting with a flat sandy bed (lower-stage plane bed), some sediment transport can take place over the surface at shear stresses just above the entrainment threshold. However, the bed is deformed at relatively low competent stresses into small wavelets instigated by the random accumulation of sediment and then into *ripples* which are roughly triangular in profile, with gentle upstream and steep downstream slopes, separated by a sharp crest. Rarely occurring in sediments coarser than 0.6 mm, ripples are usually less than 0.04 m in height and 0.6 m in wavelength. These dimensions are seemingly independent of flow depth. With coarser grain sizes, wavelengths tend to be longer, while ripple height is marginally greater. Initiated by the turbulent bursting process, these small bed-forms translate downstream at speeds inversely proportional to their height, reflecting discontinuous movement of bed material load.

As shear stresses increase, ripples are overtaken and eventually replaced by *dunes*, the most common type of bedform. Although superficially similar, they can be distinguished from ripples by their larger height and wavelength, attaining values in excess of 10^1 m and 10^3 m respectively in large rivers. Unlike ripples, dune height and wavelength are directly related to water depth, approximately in the form whereby height is up to one-third of flow depth and wavelength is 4–8 times flow depth (Knighton, 1998). The upstream slope of dunes may be rippled (Figure 4.1). Dunes are eventually washed out to leave an upper-stage *plane bed* characterized by intense bedload transport, which prevents the patterns of erosion and deposition required for the formation of three-dimensional bedforms. As flow intensity again increases, standing waves develop at the water surface and the bed is remolded into a train of sediment waves which mirror the surface forms. These *antidunes* are more transitory and much less common than dunes. They form in broad, shallow channels of relatively steep slope when the sediment transport rate and flow velocity are particularly high. Antidunes can migrate upstream through scour on the downstream face and deposition on the upstream face, move downstream, or remain stationary. They develop under conditions of such rapid flow that the probability of structures being constructed and preserved is very limited.

Bedform features tend to scale to the size of the largest clast. Hence, in gravel-bed and coarser textured streams, a differing array of depositional forms lines the channel bed (Figure 4.1b). Features such as pebble clusters and transverse ribs that form along steep channels with shallow, rapid flow reflect the ability of streams to sort and transport material over a wide range of flow and bed material conditions (Richards and Clifford, 1991). Pebble clusters generally consist of a single obstacle protruding above the neighboring grains together with upstream and downstream accumulations of particles (i.e., their long axis is parallel to flow; Brayshaw, 1984). In contrast, transverse ribs form as sheet-like deposits under highly sediment-charged conditions, and their long axes are trans-

verse to flow (Koster, 1978). Repeated ridges of coarse clasts may be evident, the spacing of which is roughly proportional to the size of the largest particle in the ridge crest.

4.2.2 Sculpted (erosional) geomorphic units

Bedrock and boulder geomorphic units reflect largely nondeformable channel features around which flow and sediment accumulations locally adjust. These features are shaped by antecedent controls such as structural and/or lithological considerations and the impacts of major flood events. Forced morphologies tend to form in reaches with steeper gradients (high transport capacity) and/or lower sediment supply relative to their free-forming counterparts (Montgomery and Buffington, 1997). In most cases, sculpted or erosional forms reflect processes that occur during high-energy conditions. Erosion of bedrock occurs via the chemical action of water (corrosion), the mechanical (hydraulic and abrasive) action of water armed with particles (corrasion), and the effects of shock waves generated through the collapse of vapor pockets in a flow with marked pressure changes (cavitation; Knighton, 1998). The largest clasts are customarily exposed above the water surface and typically have a diameter similar to the depth of the channel (Church, 2002). These features contribute to considerable energy loss during flood events.

A gradient of channel slope, bed material size, and stream power conditions induces a continuum of variants of instream geomorphic units, including *waterfalls (steps)*, *rapids*, *cascades*, *runs*, *forced riffles*, and *pools* (see Table 4.1). Specific conditions under which these differing forms of instream geomorphic units are formed may vary in different environmental settings, reflecting local combinations of factors such as slope, flow, discharge characteristics (or history), range of sediment availability, and bed material caliber, or various forcing conditions such as imposed bedrock steps or constrictions, changes in valley alignment, or loading of woody debris. For these reasons, significant variability has been reported in the range of conditions under which individual instream geomorphic units are formed (cf., Grant et al., 1990; Abrahams et al., 1995; Montgomery and Buffington, 1997, Wohl, 2000; Church 2002;

Table 4.1 Sculpted, erosional geomorphic units.

Unit	Form	Process interpretation
Bedrock step (waterfall) Waterfall / bedrock Step with plunge pool	Locally resistant bedrock that forms channel-wide drops. Transverse waterfalls > 1 m high separate a backwater pool from a plunge pool downstream.	Erosional features formed and maintained as turbulent flow falls near-vertically over the lip of the step. Steps are major elements of energy dissipation. These locally resistant areas may represent headward-migrating knickpoints. Equivalent features may be forced by woody debris.
Rapid Rapid	Very stable, steep, stair-like sequences formed by arrangements of boulders in irregular transverse ribs that partially or fully span the channel in bedrock-confined settings. Rapids in bedrock channels may be analogous with riffles in alluvial systems. Individual particles break the water surface at low flow stage.	Boulders are structurally realigned during high energy events to form stable transverse ribs that are associated with neither divergent nor convergent flow. Typically, 15–50% of the stream demonstrates supercritical flow.
Cascade Cascade	Very stable, coarse-grained or bedrock features observed in steep, bedrock-confined settings. Comprise longitudinally and laterally disorganized bed material, typically cobbles and boulders. Flow cascades over large boulders in a series of short steps about one clast diameter high, separated by areas of more tranquil flow of less than one channel width in extent.	More than 50% of the stream area is characterized by supercritical flow. Typically associated with some downstream convergence of flow. Near-continuous tumbling/turbulent and jet-and-wake flow over and around large clasts contributes to energy dissipation. Finer gravels can be stored behind larger materials or woody debris. During moderate flow events, finer bedload materials are transported over the more stable clasts that remain immobile. Local reworking may occur in high magnitude, low frequency events.
Run (glide, plane-bed) Run	Stretches of uniform and relatively featureless bed, comprising bedrock or coarse clasts (cobble or gravel). These smooth flow zones are either free-flowing or imposed shallow channel-like features that connect pools. They may occur in either alluvial or bedrock-imposed situations. Individual boulders may protrude through otherwise uniform flow.	Plane-bed conditions promote relatively smooth conveyance of water and sediment in these linking features. Slopes are intermediate between pools and riffles.

Table 4.1 *Continued*

Unit	Form	Process interpretation
Forced riffle 	Longitudinally undulating gravel or boulder accumulations that act as local steps. The irregular spacing of these features is dictated by the distribution of bedrock outcrops, woody debris, or hillslope sediment inputs along the river. They tend to occur at wider sections of valley in bedrock-confined systems (e.g., at tributary confluences).	Flow is characterized by high energy turbulence over lobate accumulations of coarse bedload materials, woody debris, and bedrock outcrops. At the lower end of the energy spectrum, riffle–pool spacing in bedrock-confined settings may reflect purely rhythmic hydraulic processes of sediment transport.
Forced pool 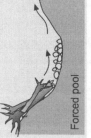	These deeper areas along longitudinal profiles are scour features associated with irregularly spaced bedrock outcrops, woody debris, and forced riffles. A *backwater pool* may form immediately upstream of a bedrock step.	These areas of tranquil flow within high energy settings may accumulate finer-grained materials at low–moderate flow stage, but they are flushed and possibly scoured during extreme events. At the lower end of the energy spectrum, riffle–pool spacing in bedrock-confined settings may reflect purely rhythmic hydraulic processes of sediment transport.
Plunge pool 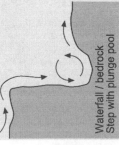 Waterfall / bedrock Step with plunge pool	Deep, circular, scour feature formed at the base of a bedrock step.	As flow plunges over a step, its energy is concentrated and scour occurs by corrosion, cavitation, and corrasion processes. Erosion may be aided by preweakening by weathering.
Pothole 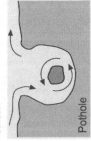 Pothole	These deep, circular scour features occur in areas where flow energy is concentrated. They are commonly associated with weaknesses or structural changes in bedrock.	Potholes are sculpted from bedrock by corrasion (i.e., hydraulic and abrasive action of water). The effectiveness of this process is determined by the volume and hardness of particles that are trapped in the pothole. Abrasion is induced by these particles, which deepen and widen the pothole.

Halwas and Church, 2002). There is considerable overlap in the range of conditions/settings in which individual features form. Hence, interpretations of controls on form–process associations must relate general (theoretical) principles to site specific considerations.

Waterfalls or *bedrock steps* are characterized by falling flow over bedrock or boulder steps that have a near-vertical drop greater than 1 m. *Plunge pools* are circular scour features that form when flow becomes concentrated at the base of waterfalls, steps, or obstacles (Wohl, 2000). The force of the flow induces corrasion and cavitation. *Potholes* are deep, spherical features sculpted into bedrock. They commonly form in areas of bedrock weakness or structural changes. Once initiated, bedload particles trapped within the pothole induce scour by corrasive erosion during turbulent flow, widening and deepening the feature.

Rapids are stair-like arrangements of boulders on steep slopes. Individual particles are numerous enough or large enough to break the water surface at mean annual discharge (Graf, 1979). Rapids form by transverse movement of boulders at high flow stage (recurring perhaps once every few years). A series of ridges of coarse clasts spaced proportionally to the size of the largest clast is produced. Grant et al. (1990) distinguish rapids from riffles by their increased steepness, their greater areal proportion of supercritical flow, and the arrangement of boulders into transverse ribs that span the channel. Rapids in bedrock channels may be analogous with riffles in alluvial systems.

Cascades occur on steep slopes (<0.1 m m^{-1}), and are characterized by longitudinally and laterally disorganized bed material that typically comprises cobbles and boulders (Montgomery and Buffington, 1997). Near-continuous tumbling/turbulent and jet-and-wake flow occurs over and around individual large clasts in a series of short steps about one clast diameter high. These clasts induce significant energy dissipation. Finer gravels can be stored behind larger materials or woody debris. During moderate flow events, finer bed-load materials are transported over the more stable clasts that remain immobile during these flows. Localized reworking may occur in high magnitude, low frequency events.

Step–pool sequences occur on gradients between 0.03–0.10 m m^{-1} (Montgomery and Buffington,

1997). These channel-spanning stair-like features comprise boulder or cobble clasts or woody debris separated by areas of quieter flow in a *backwater pool* upstream from a *plunge pool* downstream. The risers of individual steps are generally made up of several large boulders, or keystones (Zimmermann and Church, 2001). When $D/d \sim 1.0$ and the width of the channel is less than an order of magnitude greater than the diameter of the largest stones within it, keystones form stone lines that define steps (see Chin, 1989, 1999). These stone-lines act as a framework against which smaller boulders and cobbles are imbricated. The tightly interlocking structure of these features results in considerable stability, such that steps are only likely to be disturbed during extreme floods. Mobilization of the keystones typically requires a flood event with a recurrence interval in excess of 50 years (Grant et al., 1990). Given the need for one or more keystones, step development is strongly influenced by local sediment supply and transport conditions. In most cases, steps are randomly placed, reflecting random delivery of keystones to the channel (Zimmermann and Church, 2001; Church, 2002). The small pools between steps provide storage sites for finer grained bedload material, creating a contrast in sediment size which is much sharper than that between riffles and pools. The spacing of consecutive step–pool elements is related to channel size, with average values of about three channel widths (Whittaker, 1987; Chin, 1989). A pseudocyclic pattern of acceleration and deceleration characterizes the flow regime as water flows over or through the boulders forming each step before plunging into the pool below. Such tumbling flow is supercritical over the step and subcritical in the pool. Turbulent mixing results in considerable energy dissipation (Whittaker and Jaeggi, 1982). Further energy is expended by form drag exerted by the large particles that make up the steps. Thus, step–pool sequences have an important resistance role.

Runs are generally uniform and relatively featureless forms with trapezoidal cross-sections. They comprise long stretches of bedrock and coarse clasts, although individual boulders may protrude through otherwise uniform flow. They are typically generated under plane-bed conditions on moderate slopes of 0.01–0.03 m m^{-1} (Montgomery and Buffington, 1997). In general, runs (or glides) have low velocities and low water-surface gradients (McKenney, 2001). However, under plane-bed conditions the volume of coarse sediment inputs exceeds the transport capacity of the channel, such that aggradation induces a relatively homogenous bed profile. These features can form in either bedrock-dominated or fully alluvial settings.

The transition from runs to *riffle–pool* morphology tends to be accompanied by increased sediment supply and/or decreased transport capacity (Montgomery and Buffington, 1997). *Forced pools* and *riffles* are longitudinally undulating features that typically form along confined valleys with slopes >0.01 m m^{-1}. Unlike their free-forming counterparts, these features are generally irregularly spaced. Quiet flow through deeper areas (pools) is often separated by turbulence over lobate accumulations of coarse bedload materials in intervening shallow riffles. The formation of forced pools and riffles may be induced by woody debris accumulations or downstream changes in bedrock resistance, which controls variations in bed topography, valley width, or alignment. Alternatively, sediment input from tributaries or mass-movement inputs from hillslopes may fashion the pattern of riffles and pools. At the lower end of the energy spectrum, riffle–pool spacing in bedrock-confined settings may reflect purely rhythmic hydraulic processes of sediment transport (see below). In these cases, the primary riffles may remain anchored in place or may migrate slowly along the system dependent upon the relative mobility of the material forming the channel bed and the valley configuration. Abrupt changes in valley alignment or confinement may anchor otherwise migratory sediment accumulations (Church, 2002).

The shape of pools may vary markedly along river courses. This is particularly evident in bedrock-controlled reaches, or any local area where forcing elements, such as woody debris or a cluster of large boulders, promote scour. For example, McKenney (2001) differentiates between *bluff pools* and *lateral pools*, both of which have low velocities and low water-surface gradients. Bluff pools are characterized by poorly sorted sand- to boulder-sized bed material, v-shaped cross-sections, and bedrock or coarse talus banks. Lateral pools have gravel- to cobble-sized bed material, asymmetrical cross-sections, and banks

that comprise alluvial materials. In bedrock-controlled reaches, pool morphology is largely imposed by lithologic variability (i.e., measures of hardness) and changes in valley alignment. Any factor that accentuates scour, promotes pool development. Pronounced variability may be evident in pool depth. These features often provide the last remaining waterholes along ephemeral systems. In many settings, shallow elongate pools at low flow stage act as runs (or glides) at moderate flow stage.

4.2.3 Midchannel geomorphic units

Midchannel geomorphic units tend to scale to the dimensions of the channel in which they form. These features have strong relationships with other morphological attributes of rivers, notably channel shape and channel planform. Given the tendency for bed material caliber and slope to decrease and discharge to increase downstream, systematic changes in bed configuration may be expected in that direction. A range of midchannel depositional forms is presented in Table 4.2.

The most common midchannel geomorphic units are accumulations of deposits referred to as bars. These free-forming depositional features are areas of net sedimentation of comparable size to the channels in which they occur (Smith, 1978). Bar form and configuration provide key indicators into formative processes, reflecting the ability of a channel to transport sediment of different caliber. In turn, bars interact with, and influence, the patterns of flow through a reach. Flow divergence produces a zone of low tractive force and high bed resistance, which accentuates sediment deposition. Coarse materials often make up the basal platform of bars (Bluck, 1971, 1976, 1979). Bedload materials stored in bars are frequently reworked as channels shift position. Midchannel forms are more likely to be reworked than bank-attached features as they are often aligned adjacent to, or within, the thalweg zone. Long-term preservation of bars is conditioned by the aggradational regime and the manner of channel movement. These bar forms are more a reflection of sediment supply conditions and channel-scale processes than local fluid hydraulics (Knighton, 1998).

Bars are generally classified by their shape and position, ranging from simple unit bars composed

Table 2 Midchannel geomorphic units.

Unit	Form	Process interpretation
Riffle and pool		

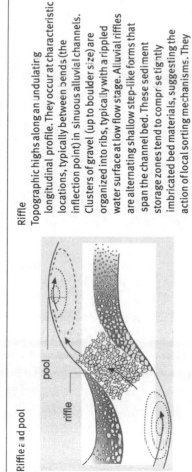

pool

riffle

Riffle

Topographic highs along an undulating longitudinal profile. They occur at characteristic locations, typically between bends (the inflection point) in sinuous alluvial channels. Clusters of gravel (up to boulder size) are organized into ribs, typically with a rippled water surface at low flow stage. Alluvial riffles are alternating shallow step-like forms that span the channel bed. These sediment storage zones tend to comprise tightly imbricated bed materials, suggesting the action of local sorting mechanisms. They induce local steepening of the bed.

Pool

Pools may span the channel, hosting tranquil or standing flow at low flow stage. Alluvial pools are alternating deep areas of channel along an undulating longitudinal bed profile. Pools tend to be narrower than riffles and act as sediment storage zones. These forms tend to occur at characteristic locations, typically along the concave bank of bends in sinuous alluvial channels.

Riffle

Riffles are zones of temporary sediment accumulation that increase roughness during high flow stage, inducing deposition. Concentration of coarser fractions at high discharges (bankfull and above) produces incipient riffles, while lower flows (up to bankfull) may be sufficiently competent to amplify and maintain the initial undulations once they have reached a critical height. In subsequent high discharges, deposition occurs as the resistance of these features induces a reduction in velocity over the riffle surface. At high flow stage the water surface is smooth, as bed irregularities are smoothed out. Riffles are commonly dissected during the falling stage of floods, when the water surface is shallow and steep, and the stepped long profile is maintained. Although very stable, with 5–10% of the stream area in supercritical flow and some small hydraulic jumps over obstructions, riffles may be mobile at and above bankfull stage. Indeed, they may be removed and replaced during extreme floods, as they reform at lower flow stages (velocity reversal hypothesis).

Pool

At high flow stage, when flow converges through pools, decreased roughness and greater bed shear stresses induce scour and flushing of sediment stored on the bed. Subcritical flow occurs at low flow stage, when divergent flow occurs. Pool-infilling subsequently occurs, as pools act as areas of deep, low flow velocity and near-standing water conditions. Pools and riffles are genetically-linked in alluvial rivers. Velocity reversal at high flow stage maintains these features.

Table 4.2 *Continued*

Unit	Form	Process interpretation
Longitudinal bar (medial bar)	Midchannel, elongate, teardrop-shaped unit bar, in gravel- and mixed-bed channels. Bar deposits typically decrease in size downstream, away from a coarser bar head. May contain distinct imbrication.	As flow diverges around the coarse bedload fraction it is no longer competent to transport sediment and materials are deposited in midchannel. Finer materials are trapped in the wake. Alternatively, there is too much sediment for the channel to transport (i.e., exceedence of a capacity limit under highly sediment charged conditions) and material is deposited.
Transverse bar (linguoid bar)	Midchannel unit bar, oriented perpendicular to flow, generally found at points of abrupt channel and flow expansion points in sand-bed channels. They have a lobate or sinuous front with an avalanche face. The upstream section of the bar is characterized by a ramp which may be concave in the center with an arcuate shape.	Formed via flow divergence in highly sediment-charged sandy conditions. Flow moves over the center of the bar, diverges and is pushed up the ramp face. Sediment is pushed over the avalanche face and deposited on the lee side. As a result, the bar builds and moves downstream as a rib.
Diagonal bar (diamond bar)	Midchannel unit bar, oriented diagonally to banks in gravel- and mixed-bed channels. These bars commonly have an elongate, oval, or rhomboid planform. Particle size typically fines down-bar. Commonly associated with a dissected riffle.	Formed where flow is oriented obliquely to the longitudinal axis of the bar. May indicate highly sediment charged conditions or reworking of riffles.
Expansion bar	Coarse-grained (up to boulder size) midchannel bar with a fan-shaped planform. Streamlined ridge forms trail behind obstructions in the channel. Foreset beds commonly dip downstream with a very rapid proximal–distal grain size gradation. Often occur downstream of a bedrock constriction that hosts a forced pool. May be colonized and stabilized by vegetation.	As flow expands abruptly at high flood-stage in high-energy depositional environments, it loses competence and induces deposition. Dissection is common at falling stage. These bars remain fairly inactive between large floods, constraining processes at lower flow stages.

Generally form around a bar core that has been stabilized by vegetation. This induces further sedimentation on the island. Islands are differentiated from bar forms by their greater size and persistence, reflecting their relative stability and capacity to store instream sediments. The pattern of smaller-scale geomorphic units that comprise an island reflects the history of flood events and processes which form and rework the island.

Deposited under high velocity conditions. When the competence limit of the flow drops, the coarsest boulders are deposited, forming obstructions to flow. Secondary lee circulation occurs in the wake of the coarse clasts. Finer boulders and pebbles are subsequently deposited downstream of the core clasts, resulting in distinct downstream fining.

During the waning stages of large flood events, sediments are deposited on top of an instream bedrock ridge. When colonized by vegetation, additional sediment is trapped and accumulates on top of the bedrock core. Over time the bar builds vertically and longitudinally as sediments are trapped in the wake of vegetation.

Formed when transport capacity is exceeded or competence is decreased and bedload deposition occurs across the bed. Generally reflect transport capacity-limited conditions due to an oversupply of sediment. Bedforms are subject to frequent removal and replacement by floods as the sand sheet moves downstream as a pulse.

Vegetated midchannel bar. Can be emergent at bankfull stage. Generally compound forms, comprising an array of smaller-scale geomorphic units. They are commonly elongate in form, aligned with flow direction. They scale to one or more channel widths in length.

Linguoid shaped boulder feature with a convex surface cross-section. Comprise a cluster of boulders without matrix, fining in a downstream direction.

Elongate bedrock ridge over which sediments have been draped and colonized by vegetation. Sediments become finer downstream, and the age structure of the vegetation gets younger.

Relatively homogeneous, uniform, tabular sand deposits which cover the entire bed. May consist of an array of bedforms, reflecting riffle, dune, or plane-bed sedimentation.

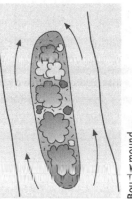

Island

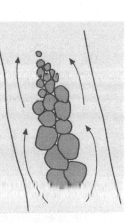

Boulder mound

Bedrock core bar

Sand sheet

Table 4.2 *Continued*

Unit	Form	Process interpretation
Gravel sheet (basal or channel lag) 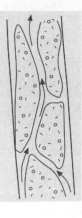	Relatively homogeneous, thin/tabular bedload sheets that are deposited across the bed. Often coarse-grained and poorly sorted. May consist of an array of gravel bedforms such as pebble clusters and ribs.	Deposited under uniform energy conditions in highly sediment charged rivers. Generally indicates transport capacity-limited or competence-limited conditions due to oversupply of sediment. Surficial gravel bedforms are subject to frequent removal and replacement by floods as the sheet moves downstream as a pulse. May represent residual deposits that form a basal lag or a diffuse gravel sheet, reflecting rapid deposition and/or prolonged winnowing. May be armored.
Forced midchannel bar (pendant bar, wake bar, lee bar) 	A midchannel bar form that is induced by a flow obstruction (e.g., bedrock outcrop, boulders, large woody debris, vegetation). The resultant bar form often has a downstream dipping slip face as the bar extends downstream.	Perturbations in flow and subsequent deposition are induced by obstructions. The resultant bar morphology is shaped by the flow obstruction, which forces flow around the obstruction and deposition in its wake in secondary flow structures. Depending on flow stage, these secondary flow structures may locally scour the bed. These bars build in a downstream direction and may become vegetated.
Compound midchannel bar 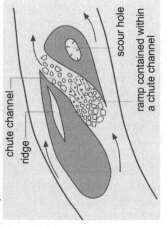	A midchannel bar that comprises an array of smaller-scale geomorphic units. Their variable morphology depends on material texture, flow energy, and the history of flood events that induce formation and subsequent reworking, producing chute channels, ramps, or dissection features. Further deposition may form ridges and lobes. If vegetation colonizes parts of the bar, additional depositional features result, producing an island.	The assemblage of geomorphic units is dependent largely on channel alignment (and associated distribution of flow energy over the bar surface at different flow stages) and patterns of reworking by flood events. Formed initially from the lag deposition of coarser sediments (a unit bar). At high flow stage the bar may be reworked or material deposited around obstructions. At low flow stage, the bar may have finer depositional features deposited on top of the bar platform. The range of bedforms reflects sediment transport across the surface.

of one depositional feature (Smith, 1974) to compound forms that reflect multiple phases of deposition and reworking under a range of flow conditions (Brierley, 1996). As different types of bar tend to develop under particular sets of flow energy and bed material texture conditions, a typical down-valley transition in forms can be discerned, ranging from midchannel to increasingly bank-attached forms (Church and Jones, 1982).

The formation of midchannel bars reflects circumstances in which the coarse bedload fraction can no longer be transported by the flow (exceedence of a competence limit) or there is too much material for the flow to transport, and instream deposition occurs (exceedence of a capacity limit). These conditions tend to be associated with gravel (or coarser) and sand-bed channels respectively. Most midchannel bars are characterized by downstream fining sequences, where the coarsest fraction is deposited at the head of the bar, and finer materials are deposited in the lee by secondary flow currents. Midchannel bars tend to accrete in a downstream direction.

Bed material character and the competence of flow to transport it determine the formation of *longitudinal* bars (Leopold and Wolman, 1957). These features form as flow divides around a tear-drop shaped structure, depositing materials in the lee of the coarser bar head. Subsequent deposition leads to downstream extension and vertical accretion, resulting in an elongate, oval, or rhomboid planform. When flow is oriented obliquely to the long axis of the bar, a *diagonal* feature is produced (Church, 1972). The upstream limb of these bars may be anchored to the concave bank, reflecting a dissected riffle.

In highly sediment-charged sand-bed conditions, flow divergence results in the formation of *transverse* or *linguoid* bars, which extend across rather than down the channel (Smith, 1974; Cant and Walker, 1978). These features have a broad, lobate, or sinuous front with an avalanche face (Church and Jones, 1982). A concavity in the central part of the upstream ramp forms when flow moves over the center of the bar, diverges and is pushed up the ramp face. Sediment falls over the avalanche face, depositing material on the lee side. As a result of this mechanism, these bars extend and move downstream. *Expansion bars* are coarse grained, fan-shaped bars that are deposited in areas

of abrupt flow expansion downstream of a forced pool. Alternatively, the entire channel bed may comprise a homogenous *sand* or *gravel sheet*, where a continuous veneer of sediment moves along the channel.

Areas of channel widening or local slope decrease along confined valleys may induce the development of low-relief, elongate or linguoid-shaped *boulder mounds*. These features form under high velocity conditions by the same mechanism as longitudinal bars, but with much coarser sediments (Zielinski, 2003). Over time, a preferred single channel tends to become established. Following abandonment of a side channel, boulder mounds may evolve into flat, gravel-boulder sheets that are attached to the bank.

Most bars are not simple unit features, but are complex, *compound* features made up of a mosaic of erosional and depositional forms. The specific character of these features reflects river history, differing flow stage interactions, and thalweg shift. Reworking occurs during the rising flood stage and deposition at waning flow stage. On midchannel compound bars, *chute channels* may dissect the bar producing a chaotic pattern of remnant units. *Ridge* features may develop around vegetation. *Islands* typically comprise an array of small-scale units that are scoured or deposited around vegetation. The array of smaller-scale geomorphic units that make up compound features provides key insights into their formation and reworking (see Chapter 5).

Alluvial *riffles* and *pools* are oscillatory bed features, in which patterns of scour and deposition produce a more or less regular spacing between consecutive elements. Riffles and races have high velocities, sorted gravel- or cobble-sized bed material, and high water-surface gradients. Riffles characteristically have low depths and trapezoidal cross-sections. Races (termed coarse runs by Rabeni and Jacobson, 1993) characteristically are deeper than riffles and have u-shaped cross-sections (McKenney, 2001). Concentration and deposition of coarser fractions at high discharges (bankfull and above) produces incipient riffles, while scour occurs in adjacent pools. Flows up to bankfull may be sufficiently competent to amplify and maintain the initial pool–riffle undulations once they have reached a critical size. Riffles tend to have coarser, more tightly imbricated bed materials than adjacent pools, suggesting the action of local sorting mechanisms (Keller, 1971; Lisle, 1979; O'Connor et al., 1986). In general, riffles tend to be wider and shallower than pools at all stages of flow. At low flow, velocity and slope are greater and depth is less over a riffle than in a pool. However, differences in flow geometry and competence are more evenly distributed along a reach at high flows. Indeed, competence may even be reversed so that, contrary to the low-flow condition, it is higher in the pools at those discharges that transport most material in gravel-bed streams (Keller, 1971; Lisle, 1979). In combining high-flow transport through pools and low-flow storage on riffles, such reversal promotes the concentration of coarser material in riffles and the maintenance of the riffle–pool sequence. This pseudocyclic character has a more or less regular spacing of successive pools or riffles that ranges from 1.5 to 23.3 times channel width, with an average of 5–7 times (Keller and Melhorn, 1978).

In low slope, low energy settings with relatively shallow alluvial fills, accretionary forms may develop atop bedrock (e.g., van Niekerk et al., 1999; Wende, 1999). These *bedrock core bars* are characterized by bedrock ridges atop which alluvial materials are deposited during the waning stages of floods. Vegetation cover enhances rates of deposition and vertical accretion of these features, which are common along bedrock-anastomosing rivers.

4.2.4 Bank-attached geomorphic units

The geometry of channel margins reflects a combination of bank erosion processes, as channels rework floodplain deposits or inset features, and depositional processes that generate a range of bank-attached geomorphic units (see Table 4.3). Bank-attached depositional features tend to occur in reaches that are characterized by less sediment-charged, lower-energy conditions relative to reaches in which midchannel features are observed.

Lateral bars are elongate features attached to banks along relatively straight channels. They commonly alternate from bank to bank along a reach. Several platform levels may be evident, separated by steep slipfaces, reflecting lateral accretion, and/or downstream migration. In some

instances, lower platforms become progressively finer-grained, as they form during intermittent stages of flood recession.

Point bars typically have an arcuate shape that reflects the radius of curvature of the bend within which they form. An array of forms may be determined, reflecting bend curvature and bed/bank material texture (Jackson, 1976). Point bars are attached to the inner (convex) bank and are inclined towards the center of the channel, reflecting the asymmetrical channel geometry at the bend apex. They form when helical flow is generated over the bar surface as the thalweg shifts to the outside of the bend at high flow stage. This flow moves sand or gravel bedload by traction processes towards the convex slopes of bends, building the bar laterally. An around-the-bend set of sedimentary structures and grain size trends is commonly observed. The coarsest materials are deposited at the bar head where the thalweg is aligned closer to the convex bank. Further around the bend, the thalweg moves towards the concave bank and finer-grained sediments (a bedload and suspended load mix) are deposited. The most recently accumulated deposits are laid down as bar platform deposits at the bend apex.

Unit point bars comprise one platform, whereas compound point bars have multiple platforms and/or an array of erosional and depositional forms with differing bed material textures that reflect activity at differing flow stages. Compound point bars commonly record multiple phases of bar reworking, bar expansion, lateral migration, and downstream translation. The resulting array of erosional and accretionary patterns reflects the direction and rate of bend adjustment (Hickin, 1974). In some instances, *scroll* bars are deposited in the shear zone between the helical flow cell in the thalweg zone and flow in a separation zone adjacent to the convex bank of a bend (Nanson, 1980). As these features build vertically and the channel shifts laterally, scroll bars become incorporated into the floodplain as lateral accretion deposits. Accretionary *ridges* and intervening *swales* are formed. Swales record the position of the former separation zone. Series of ridges and swales record former positions of the channel and the pathway of migration.

Chute channels may short-circuit the bend, cutting a relatively straight channel from the head of the bar (Brierley, 1991). Enlargement of a chute channel and plugging of the old channel may generate a *chute cutoff* which remains abandoned on the floodplain. High energy flows down a chute channel may form a *ramp*, typically comprising coarse gravels with a steep upstream facing surface, within the chute channel (Blum and Salvatore, 1989, 1994). *Ridges* form where deposition occurs around vegetation. Alternatively, scour may accentuate erosion within a chute channel, leaving a perched ridge shaped feature on the bar surface. These features tend to record the alignment of high flow stage over the bar.

Along tight bends in laterally constrained situations, *concave bank-benches* may form in the upstream limb of obstructed or tight bends. Suspended-load slackwater sediments are deposited in a separation zone at high flow stage. Just like point bars, these features may become incorporated into the floodplain as the bend assemblage translates downstream (Hickin, 1986). Alternatively, *point dunes* may form at high energy flow stages in some laterally constrained bends or on the top of point bars, presenting an alternative to around-the-bend depositional patterns (Hickin, 1969).

In low to moderate sinuosity sand bed channels, oblique-accretion *benches* may form as sand or mud deposits lapped onto relatively steep convex banks. During the rising stage of flood events, bedload materials are deposited atop these step-like features. Suspended load materials cap these deposits at waning stage, forming flood couplets. Similar low energy, falling stage mud drapes are observed along the convex banks of channels characterized by low migration rates and high suspended load concentrations (Nanson and Croke, 1992). When observed on the convex banks of bends, these features are referred to as *point benches*. Bench and point bench features are depositional forms that generally reflect channel narrowing. In contrast, *ledges* are erosional forms that produce a distinct step along channel margins during phases of channel incision and expansion.

Additional forms of bank-attached depositional features may be observed in differing types of setting. For example, open-framework *boulder berms* form a step like feature with a concave cross-section attached to the bank in high energy, boulder bed systems. They may form in the zone of large velocity gradient at bank crests at peak flood stage and are often deposited in one event (Stewart and LaMarche, 1967; Zielinski, 2003). *Channel junction* bars commonly develop as delta-like features, as backwater effects induce slackwater deposition downstream of tributary confluences.

These various bank-attached depositional forms must be complemented by assessment of bank erosional processes to provide a coherent basis with which to appraise variability in channel geometry.

4.3 Bank morphology

Bank morphology records the balance of erosional and depositional processes induced by the alignment and energy of flow at differing flow stages. Bank morphology is also a function of bank composition/texture. Banks comprising sands and gravels are more susceptible to erosion than those with a high silt–clay content. However, the former are also more likely to have bank-attached depositional forms along their margins. Bank erosion processes are outlined in Section 4.3.1. In Section 4.3.2 these insights are combined with analyses of depositional processes to assess the range of bank morphologies.

4.3.1 Bank erosion processes

In contrast to most channel beds, banks tend to have some degree of cohesion because they contain fine-grained material. In vertically stratified (composite) banks, typically characterized by a coarser-grained basal layer overlain by fine-grained alluvium, the strength of less cohesive basal materials controls bank stability and therefore channel width (Klimek, 1974; Andrews, 1982; Pizzuto, 1984). Differential physical properties of cohesive and noncohesive materials result in marked differences in erosion rates, erosion processes, and failure modes. Although fine-grained materials are resistant to fluid shear, they tend to have low shear strength and are susceptible to mass failure. Unlike cohesionless sediment, the erodibility of cohesive fine-grained bank material may vary because of its susceptibility to weakening.

Bank erosion entails two phases, namely detachment of grains from the bank and subsequent

Table 4.3 Bank-attached geomorphic units.

Unit	Form	Process interpretation
Lateral bar (alternate or side bar) 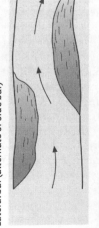	Bank-attached unit bar developed along low-sinuosity reaches of gravel- and mixed-bed channels. Bar surface is generally inclined towards the channel. These bars occur on alternating sides of the channel. They are generally longitudinally asymmetrical, and may not have an avalanche face on the downstream side.	Flow along a straight reach of river adopts a sinuous path. Bar length and width are proportional to these flows. The height of the bar is dictated by flow depth. Bars form by lateral or oblique accretion processes, with some suspended load materials atop (i.e., typically upward fining depositional sequence). They generally migrate in a downstream direction.
Scroll bar 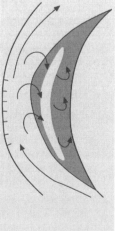	Elongate ridge form developed along the convex bank of a bend. Typically has an arcuate morphology. Commonly develop on point bars.	Formed by a two-dimensional set of flow paths on the inside of a bend. Adjacent to the thalweg, sand or gravel bedload material is moved by traction towards the inner sides of channel bends via helical flow. This is accompanied by a separation zone adjacent to the bank formed at near-bankfull stage and flow alignment shifts adjacent to the bank. These two flow paths converge, leading to the deposition of a ridge-like feature on the point bar surface. Associated with laterally migrating channels, scroll bars reflect the former position of the convex bank. With progressive channel shift and stabilization by vegetation, scrolls develop into ridge and swale topography.
Point bar 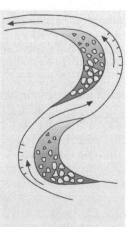	Bank-attached arcuate-shaped bar developed along the convex banks of meander bends. Bar forms follow the alignment of the bend, with differing radii of curvature. The bar surface is typically inclined towards the channel as are the sedimentary structures. Grain size typically fines down-bar (around-the-bend) and laterally (away from the channel). Typically, these unit bar forms are unvegetated.	Result from lateral shift in channel position associated with deposition on the convex bank and erosion on the concave bank. Sand or gravel bedload material is moved by traction towards the inner sides of bends via helical flow. Differing patterns of sedimentation are imposed by the radius of curvature (bend tightness) as well as the flow regime and sediment-load. The coarsest material is deposited from bedload at the bar head, where the thalweg is aligned adjacent to the convex bank (at the entrance to the bend). As the thalweg moves away from the convex bank down-bar, lower energy suspended load materials are deposited in secondary flow circulation cells, as the propensity for deposition is increased. Secondary flow also forces material up onto the face of the bar, building it laterally.

Typically form at high flood stage in reaches where a comparatively minor tributary enters the trunk channel. Flow separation and generation of secondary currents in the backwater zones promote sedimentation in sheltered areas under low flow velocity conditions.

Ridge
Ridge morphology and alignment atop bar surfaces reflect the character of channel adjustment over the bar at high flow stages. Vegetation promotes ridge development with sediment being deposited in the wake.

Chute channel
These features dissect a formerly emergent bar surface. During the rising stage of over-bar flows, scour occurs. If the bar is short circuited, flow energy is concentrated, inducing scour that reworks the bars and forms a chute channel.

Ramp
Under high flow conditions flow alignment over the bar short-circuits the main channel. A relatively straight channel is scoured. Sediment is subsequently ramped up this feature, partially infilling the chute channel with high-energy deposits, such as gravel sheets or migrating dunefields.

Point dune
Produced when high magnitude flow is aligned down-valley rather than around-the-bend, typically in sand-bed streams. Formed at high flood stage, when the thalweg shifts to the inside of the bend (over the point bar). Preserved in the falling-stages when the thalweg switches back along the concave bank of the bend.

Formed at, and immediately downstream of, the mouth of tributaries. These delta-like features have an avalanche face. They generally comprise poorly sorted gravel, sand, and mud with complex and variable internal sedimentary structures. These slackwater deposits are very prone to reworking.

Ridge
Linear, elongate deposit formed atop a bar platform on a midchannel or bank-attached bar. May be curved or relatively straight. Tend to fine downstream. May be formed downstream of vegetation or other obstructions on the bar surface.

Chute channel
Elongate, relatively straight channel that dissects a bar surface. Usually initiated at the head of the bar. A common feature on a range of bank-attached, midchannel bars and islands, leading to the formation of compound features.

Ramp
Coarse-grained, ramp-like form created by partially infilled chute channels. Formed at the upstream ends of bends and rise up from the channel to the bar surface.

Point dune
Dune bedforms that accrete along convex banks, generally atop compound point bars. These have a down-valley alignment, rather than reflecting around-the-bend trends.

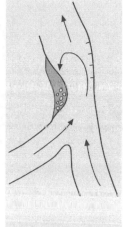

Tributary confluence bar (channel junction bar, eddy bar)

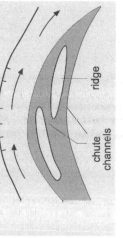

Ridge and chute channels (cross-bar channels)

ridge

chute channels

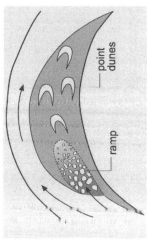

Ramp (chute channel fill) and point dune

point dunes

ramp

Table 4.3 *Continued*

Unit	Form	Process interpretation
Bench and point bench (oblique-accretion bench) 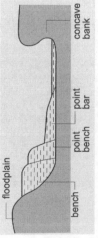	**Bench** A distinctly stepped, elongate, straight to gently curved feature that is inset along one or both banks. Major in-channel sediment storage unit, often situated atop bar deposits. May comprise obliquely attached mud-rich drapes with a convex geometry in suspended load systems, or obliquely and vertically accreted sand deposits in bedload systems. Their sedimentary structure tends to be quite distinct from the floodplain. **Point bench** Distinctly-stepped, bank-attached unit developed along the convex bank of a channel. Has an arcuate planform with a planar surface elevated above the point bar.	**Bench** Formed by oblique- and vertical-accretion of bedload and suspended load materials during small to moderate floods within widened channels. During the rising stage of flood events, bedload materials are deposited atop step-like features. During the waning stages, suspended load materials are deposited as flood couplets atop bedload materials. Oblique accretion benches represent low energy falling-stage suspended load deposition in sand-bed and mud-rich streams. Sediment deposition is often promoted by riparian vegetation. Benches are a major mechanism of channel contraction in over-widened channels. **Point bench** Sediment deposition along the convex bank via vertical and/or oblique accretion of interbedded sands and mud indicates slow lateral migration or lateral accretion within an overwidened bend.
Ledge 	Distinctly stepped, elongate, bank-attached unit. Has a straight to gently curved planform, flanking one or both banks and is flat-topped. Composed of the same materials as the basal floodplain (i.e., sedimentology is laterally continuous from the ledge to the floodplain). These erosional units reflect incision and/or channel expansion.	Formed by channel expansion processes where flows have selectively eroded the upper units of the floodplain as the channel incises and expands. Unpaired ledges reflect lateral shift during incision, whereas paired ledges indicate incision only.
Boulder berm (boulder bench) 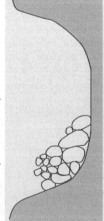	Elongate, bank-attached stepped feature inset against the bank. Can have a convex cross-section. Comprise coarse, boulder bedload materials with limited finer grained matrix.	Formed from bedload deposition in a single event under high velocity conditions. Materials are accreted (or dumped) along the bank where flow velocity decreases substantially. Reworking is restricted to subsequent high velocity events that have the competence to mobilize the boulders.

Associated with flow separation and generation of secondary currents at high flood-stage. Sedimentation occurs in sheltered backwater zones of relatively low flow velocity. Form from flow separation when the primary flow filament continues around a bend. At floodstage, flows separate from the primary flow filament, circulating back around the bend. This is often channeled by a ridge. During the rising stages of flood events, this process may accentuate scour on the surface of the bench. Deposition of suspended load materials subsequently occurs during waning stages.

Development of lateral or compound point bar forms is dependent on channel alignment (and associated implications for the distribution of flow energy over the bar surface at different flow stages) and associated patterns of reworking by flood events. Formed initially from the lag deposition of coarser sediments (a unit bar). At high flow stage, the bar may be reworked or material deposited around obstructions. At low flow stage, the bar may have finer depositional features deposited on top of the bar platform, or a range of bedforms preserved, reflecting sediment transport across the bar surface.

Perturbations in flow and subsequent deposition are induced by obstructions. The resultant bar morphology is shaped by the flow obstruction, which promotes turbulence and deposition in the wake of the obstruction in secondary flow structures. Depending on flow stage, these secondary flow structures may induce local scour around the obstruction.

Bank-attached unit, often with a low ridge across the central portion parallel to the primary channel. Located along the upstream limb (i.e., along the concave bank) of relatively tight bends that abut bedrock valley margins or a flow obstruction. Often inset against floodplain. Comprise slackwater sediments (interbedded sands and mud) and organic materials.

Bank-attached bar that comprises an array of smaller-scale geomorphic units. Generally composed of laterally accreted sand or gravel, but may include silt or boulders. Variable morphology depends on material texture, flow energy, and the history of flood events that form and rework the bar. If a bar is reworked by chute channels, ramps or dissection features may result. Deposition may form ridges and lobes. If vegetation colonizes parts of the bar, additional depositional features result. Forms of bank-attached compound bar include compound point bars and compound lateral bars.

Any bank-attached bar form that is induced by a flow obstruction (e.g., bedrock outcrop, boulders, large woody debris, vegetation). The resultant bar form often has a downstream fining sedimentary sequence.

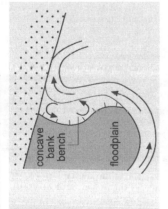

Concave bank bench (convex bar)

concave bank bench

floodplain

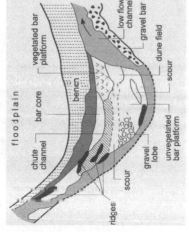

Compound bank-attached bar

floodplain

vegetated bar platform

bar core

bench

chute channel

ridges

scour

scour

gravel lobe

unvegetated bar platform

low flow channel

gravel bar

dune field

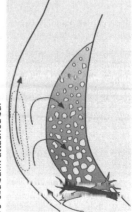

Forced bank-attached bar

entrainment. Flow forces of lift and drag may be entirely responsible for detachment and entrainment, but in most instances, especially in cohesive materials, aggregates and micropeds are loosened and partially or completely detached by weakening or preconditioning processes prior to entrainment (Lawler, 1993). The three most important weakening mechanisms are prewetting, desiccation, and freeze–thaw activity. *Prewetting* influences rates of bank erosion as cohesive materials become more erodible when wet (Wolman, 1959; Hooke, 1979; Thorne, 1982). When very wet, seepage can cause sapping of localized areas of the bank face (Twidale, 1964). Piping may also be evident (e.g., Hagerty, 1991). The role of *desiccation* reflects the nature of clay fabrics that make up cohesive banks (Lawler et al., 1997). Desiccation can encourage higher bank retreat rates. Materials derived from direct spalling of drier upper bank surfaces collect at the foot of the bank and become available for entrainment at higher flow stage. Cracking up and incipient exfoliation of bank surfaces allow flood water around and behind unstable crumbs and ped structures (Lawler, 1992). Slaking refers to "bursting" of bank crumbs and peds during saturation because of a build-up of intracrumb air pressures created by the influx of water into the soil pores during rapid immersion. Finally, in cool and temperate climate settings, *freeze–thaw* processes may be an important agent preconditioning cohesive bank materials for later fluid entrainment (e.g., Lawler, 1988). In some instances, ice lenses can reduce cohesion by wedging peds apart. In larger scale rivers that freeze over during winter, cantilevers of ice attached to the bank and ice rafts may cause serious damage during spring thaw (Church and Miles, 1982).

There are two main types of bank erosion process, namely hydraulic action (also referred to as fluvial entrainment or corrasion) and mass failure (Figure 4.2). *Hydraulic action* refers to grain by grain detachment and entrainment. It is typically associated with banks comprising noncohesive material. Removal of bank material by hydraulic action is closely related to near-bank flow energy conditions, especially the velocity gradient close to the bank and local turbulence conditions, as these determine the magnitude of hydraulic shear. *Fluvial entrainment* occurs when individual grains are dislodged or shallow slips occur

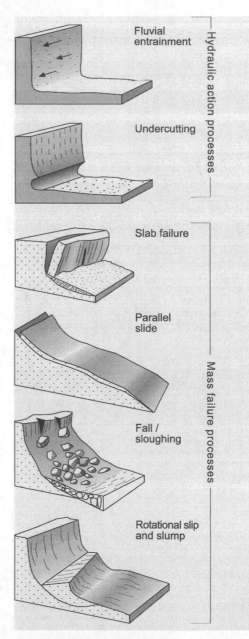

Figure 4.2 Bank erosion processes
Bank erosion processes can be differentiated into variants of hydraulic action and mass failure processes. The dominance of these different processes induces a direct control on bank morphology (see text). Modified from Thorne (1999), © John Wiley and Sons Limited, 2003. Reproduced with permission.

along almost planar surfaces. In some instances, a distinct notch may be left in the bank following a flood event, indicating the peak stage achieved. *Undercutting* occurs when velocity and boundary shear stress maxima occur in the lower bank region. High rates of bank retreat at bend apices are explained by steep velocity gradients and high shear stresses generated within large-scale eddies against outer banks. During the process of undercutting, flow not only entrains material directly from the bank face, but also scours the base of the bank. This leads to oversteepening, and eventually gravitational failure.

The effectiveness of hydraulic action is dictated by the balance between motivating forces that include the downslope component of submerged weight and fluid forces of lift and drag, and resisting forces that include the interparticle forces of friction and interlocking. These effects are especially pronounced along composite banks, where differential erosion may generate overhangs that promote collapse of overlying blocks of cohesive material (e.g., Thorne and Lewin, 1979; Thorne and Tovey, 1981). Fluid entrainment of basal material following collapse is vital to the effectiveness of this mechanism (Pizzuto, 1984). Hence, the stability of the lower bank is crucial to the stability of composite banks in the medium to long term.

Cohesive, fine-grained bank material is usually eroded by entrainment of aggregates or crumbs of soil which are bound tightly together by electrochemical forces, rather than as individual particles. These behavioral traits are heavily dependent upon physical properties of the materials, such as their mineralogy, dispersivity, moisture content, and particle size distribution, and on properties of the pore and eroding fluid, such as temperature, pH, and electrical conductivity (e.g., Grissinger, 1982; Osman and Thorne, 1988; Thorne and Osman, 1988). Entrainment occurs when the motivating forces overcome the resisting forces of friction and cohesion. Hard, dry banks are very resistant (Thorne, 1982). However, wet banks are relatively easy to erode, especially if loosened by repeated wetting and drying or frost action.

The susceptibility of banks to *mass failure* depends on their geometry, structure, and material properties. Deep-seated failures are rare in noncohesive banks, where basal scour, oversteepening, and collapse mechanisms are favored. Along more cohesive banks, weakening and weathering processes reduce the strength of bank material, thereby decreasing bank stability. The effectiveness of these processes is related to soil moisture conditions (Thorne, 1982). Cycles of wetting and drying cause swelling and shrinkage of the soil, leading to the development of interpedal fissures and tension cracks which encourage failure. Seepage forces can reduce the cohesivity of bank material by removing clay particles and may promote the development of soil pipes in the lower bank. The tangential force of the weight of a potential failure block is the primary motivating force. An increase in this force occurs when fluvial erosion leads to an increase in the bank height or bank angle. Catastrophic failure occurs when the critical value of height or angle is reached (Millar and Quick, 1993). Block mass is greatly influenced by moisture content. The switch from submerged to saturated conditions following flood events can cause the bulk unit weight of the soil to double, and can trigger drawdown failures even without the generation of excess porewater pressures. If rapid drawdown does generate positive porewater pressures, friction and effective cohesion are reduced. In extreme circumstances, this can lead to liquefaction (a complete loss of strength and flow-type failures).

Shallow *slips* occur in cohesionless banks, while deep-seated *rotational slip* and *slab failures* are the dominant mechanisms in banks of high and low cohesivity respectively (Thorne, 1982). Rotational slips occur where a curved failure plane and rotational movement leads to slipping of material down the bank face. Retreat of near-vertical banks via slab failure occurs when blocks of sediment topple from the face of the bank into the channel. Other mass failure processes include *fall/sloughing*, where small quantities of material dislodged from the top of the bank accumulate at the base, and *parallel slide*, where slices parallel to the bank slip down the bank face.

The effectiveness of these various bank erosion processes is greatly enhanced by basal scour, which effectively increases bank angle. In composite banks where cohesive materials overlie noncohesive sands or gravels, undercutting of the lower bank by hydraulic action generates an overhang or cantilever in the upper layer, which fails when a critical threshold is reached. During bank

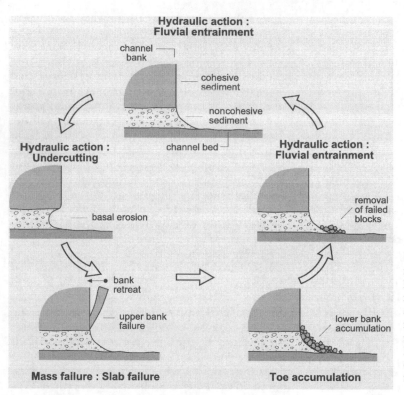

Figure 4.3 A cyclical process of bank retreat along composite banks Hydraulic action and mass failure result in accumulation of materials at the toe of the bank. Until these materials are subsequently reworked, bank retreat is impeded. Each phase of adjustment is characterized by different bank morphologies.

failure and collapse, blocks of bank material slide or fall towards the toe of the bank, where they remain until they are broken down or entrained by the flow. Failed blocks, in turn, may temporarily protect the toe of the bank from erosion. This pseudocyclic process has an important role in controlling bank form, stability, and rate of retreat (see Figure 4.3).

The amount, periodicity, and distribution of bank erosion are highly variable as they are influenced by a multitude of factors. In general terms, bank erosion is accentuated under higher discharge conditions (bankfull stage), but the effectiveness of these flows is determined by bank condition at the time of the event. For example, Wolman (1959) determined that a large summer flood induced little bank erosion on dry banks while lesser winter flows caused considerable bank retreat when acting on thoroughly wetted banks. The effectiveness of weakening, fluvial erosion, and mass failure processes induce considerable variability in rates of bank erosion, instability, and/or retreat. In some instances, rates of lateral

channel change may exceed $1000\,\mathrm{m\,yr^{-1}}$ (see reviews in Hooke, 1980; Lawler, 1993). In a survey of channel migration rates, admittedly completed primarily for disturbed rivers with highly modified vegetation cover, Walker and Rutherfurd (1999) derived a global mean channel migration rate of 3% (range 0.07–25%) of channel width per year, and a median of 1.6% of channel width per year.

Change in one part of a system commonly instigates or accelerates change to the pattern and/or rate of bank erosion in adjacent reaches (Hooke, 1980; Lewin, 1987). Maximum erosion rates tend to be experienced in middle (piedmont) reaches, relating to the peak in stream power and bank textural attributes (Graf, 1982; Lewin, 1987; Lawler, 1992). In general terms, bank materials become finer and more uniform downstream, especially where well-developed floodplains are found. Stream power also declines in a downstream direction, such that the dominant bank erosion process changes downstream from hydraulic action to mass failure. Nanson and Hickin (1986) demonstrated that stream power accounted for 48% of the variability in channel migration rate along various river systems in Western Canada.

Vegetation cover, especially the role of root networks, may reinforce bank materials, thereby increasing resistance to erosion (e.g., Smith, 1976; Abernethy and Rutherfurd, 1998; Brooks and Brierley, 2002). If discharge, slope, bend curvature, bank texture, and bank heights are constant, a river migrating through cleared or cultivated floodplain may erode at almost twice the rate of rivers reworking forested floodplain (Hickin, 1984). Beeson and Doyle (1995) supported these findings, observing that nonvegetated banks were nearly five times more likely to undergo notable erosion compared with vegetated banks.

Vegetated banks tend to have a more open soil fabric and are better drained. Soil is strong in compression but weak in tension. Roots are the reverse, so they reinforce the tensile strength of soils by up to an order of magnitude relative to root-free samples (Lawler et al., 1997). These impacts may be offset, in part, by the additional loading applied to banks by vegetation cover. In large-scale cohesive banks, critical failure surfaces may be well below the root zone. Vegetation structure also influences patterns and rates of flow dynamics adjacent to the banks. Stems and trunks of bank vegetation alter the distribution of near-bank velocity and boundary shear stress (Kouwen and Li, 1979). The spacing (density) and pattern of trees may exert a significant influence on the distribution of form drag, influencing the capacity for detachment and entrainment (Pizzuto and Meckelnburg, 1989). The distribution of woody debris also influences the effectiveness of bank erosion processes (Davis and Gregory, 1994).

4.3.2 Bank morphology: The balance of erosion and deposition

Along any given reach, bank morphology reflects a combination of erosional and depositional processes. Bank-attached geomorphic units typically occur when the thalweg is positioned along the opposite side of the channel, allowing for deposition of materials at the base of the bank. Depositional features along convex banks tend to be flat-topped or gently graded forms that effectively reduce bank angle. These features protect the banks from subsequent erosional activity.

Depending on channel alignment, these units may be longitudinally extensive features, such as benches along low sinuosity reaches, or localized forms such as point bars on meander bends.

Any particular type of bank morphology may reflect a range of differing conditions or circumstances (the principle of equifinality). Hence, significant caution may be required in making interpretations of why banks have adopted a particular morphology. Underlying causes must be appraised with due consideration to a range of possible scenarios. Among the issues that must be appraised are:

• bank position within the reach, and its relation to flow alignment at differing flow stages;
• the balance of erosional and depositional processes operating on or adjacent to banks;
• the sediment mix of materials that make up the bank, and the associated bank sedimentology (including the presence of bedrock);
• the stage of evolution of the bank, as determined by the rate of delivery of materials to the toe of the bank and subsequent rates of removal (see Figure 4.3);
• the primary origin of depositional features that may line the bank (i.e., whether derived from further up the bank, or from upstream sources);
• the combination of fluvial erosion and mass failure mechanisms that erode the bank;
• the aggradational/degradational balance of the reach.

Ongoing interactions among these factors modify bank morphology and the resulting sediment mix, thereby setting the conditions for bank adjustments in response to subsequent formative events. A continuum of variants of bank morphology is evident. Various end member scenarios are portrayed in Figure 4.4. The discussion below provides a cursory overview of some of the factors that may produce these differing forms.

Banks with a homogenous sediment mix commonly have a uniform morphology. This may take a near-vertical form, where fluvial entrainment processes are effective (Figure 4.4a), or various forms of inclined bank (Figures 4.4b–d). In general terms, bank angle is dictated by the sediment mix. Cohesive sediment forms steeper banks, while sandy banks have a gentler angle of repose. Uniform bank morphologies may reflect hydraulic action processes such as fluvial entrainment or

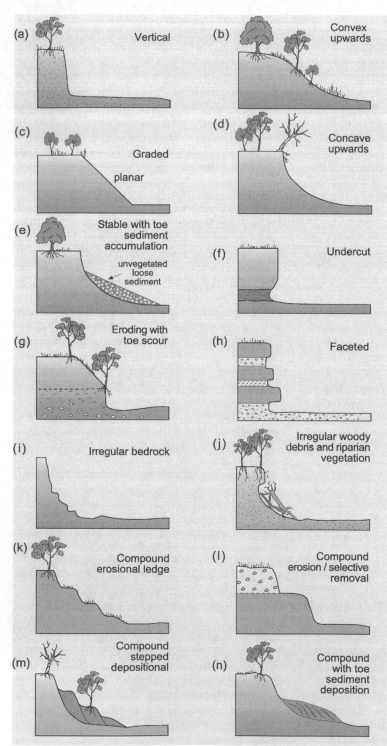

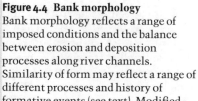

Figure 4.4 Bank morphology
Bank morphology reflects a range of imposed conditions and the balance between erosion and deposition processes along river channels. Similarity of form may reflect a range of different processes and history of formative events (see text). Modified from Thorne (1999), © John Wiley and Sons Limited, 2003. Reproduced with permission.

mass failure processes such as parallel sliding or rotational slips. Along convex-upwards banks (Figure 4.4b), gradual mass movement mechanisms may be inferred. Alternatively, overflow mechanisms from the floodplain to the channel during the waning stages of floods may modify patterns of sedimentation on the bank, especially around vegetation. Graded banks represent a planar condition that may reflect grain-by-grain movement downslope at the angle of repose or parallel sliding down a slip face (Figure 4.4c). A concave-upwards bank profile may arise following removal of rotational slip and slump materials from the base of the bank (Figure 4.4d).

Banks with coarsening upwards profiles, such as sands overlying fine-grained, cohesive sediments, tend to have zones of sediment accumulation along the bank toe (Figure 4.4e). Mass failure processes such as falling and slipping/slumping deliver sediment to the toe of the bank, where it accumulates until it is removed by fluvial entrainment (Figure 4.3). In contrast to banks with bank-attached geomorphic units, their toes comprise material derived directly from the adjacent bank. As such, the sediment composition of the accumulation is similar to that in the bank, whereas materials derived from upstream may have a quite different sedimentary structure, texture, and morphology (see Figures 4.4m and n). In the case of slips and slumps, where entire sections of the bank are removed, the bank structure may be maintained during displacement.

Undercut banks are commonly associated with upward-fining sequences. Commonly, basal gravel lags are overlain by sand or mud deposits. Less cohesive units at the base of the bank are often undercut or eroded by hydraulic action (see Figure 4.4f). Toe scour leaves an overhanging bank of cohesive, finer-grained materials. Mass failure by slab failure or fall results in sediment accumulation at the base of the bank (Figure 4.3). Differing stages of adjustment in this process have quite different bank morphologies. The resultant bank morphology following undercutting reflects the initial condition of the bank prior to commencement of toe scour. For example, Figure 4.4g represents a planar bank that is subsequently subjected to toe scour.

Banks that comprise multiple layers of varying texture tend to have a complex morphology, as coarse materials in the banks are selectively re-worked. This produces a faceted bank with multiple overhangs (Figure 4.4h). Hydraulic action via fluvial entrainment and undercutting commonly acts as a precursor to mass failure by slab failure or fall. In contrast, channels with bedrock margins tend to have an imposed, irregular morphology (Figure 4.4i). Irregular bank morphologies also occur along channels where resisting/forcing elements such as woody debris and riparian vegetation induce local variability in patterns of scour and deposition (Figure 4.4j).

Finally, compound bank morphologies reflect stepped situations that may result from differing sets of erosional and depositional processes. These bank morphologies are commonly found in reaches subjected to bed instability (whether degradation or aggradation). Along incised channels, for example, a ledge feature may be observed (Figure 4.4k). This erosional form may reflect a slot channel inset within the broader trough and is typically associated with phases of channel expansion. Alternatively, this type of bank morphology may reflect selective removal of less cohesive materials from the upper part of a bank profile. For example, reaches that are subjected to valley floor aggradation in response to upstream disturbance events commonly have coarser materials atop the banks. Subsequent channel expansion in the affected reach may selectively erode these less cohesive materials, producing a stepped bank morphology (Figure 4.4l). A compound bank morphology may also be derived from depositionally-induced mechanisms. For example, a stepped morphology may reflect multiple phases of deposition adjacent to the bank, in the form of inset features or benches (Figure 4.4m). Such features are typically associated with phases of channel contraction.

These various scenarios that consider a range of bank morphologies and the array of circumstances that may be responsible for their formation tend to overemphasize the complexity of bank forms. A significant proportion of banks along most reaches reflect relatively simple sets of depositional processes that generate bank-attached bars. These accretionary forms range from lateral and point bars to compound bar features. Resulting bank morphologies have a compound shape (Figure 4.4n). This commonly results in a low-slope, stepped morphology, often with a convex-upwards

Table 4.4 Classification of channel boundary conditions (modified from Knighton, 1998, p. 152).

Primary type	Secondary type	Characteristics
A. Cohesive	A1. Bedrock channels	Generally short reaches with no coherent cover of consolidated material. Common in steep headwater reaches. Bed and bank morphology is largely imposed. Typically have an irregular cross-section.
	A2. Silt-clay sediments	Typically, suspended load systems that have a limited capacity to rework their boundaries and adjust their form. Channels have a low width : depth ratio, and commonly have a symmetrical form. Once entrained, materials are maintained in transport even if flow energy decreases significantly.
B. Noncohesive	B1. Sand-bed channels	"Live-bed" channels that are active over a wide range of discharges, resulting in differing bedform attributes dependent on flow stage and preservation potential. Channels maintain a high width : depth ratio.
	B2. Gravel-bed channels	Mixed-load systems characterized by intermediate width : depth ratios. Bed materials are transported only at high discharge stages. A surface armor protects underlying materials. Banks commonly comprise much finer-grained materials. Channel shape is largely determined by channel alignment and the recent history of formative events. In meandering reaches, for example, the channel is asymmetrical at bends, and symmetrical at points of inflection.
	B3. Boulder-bed channels	Boundary materials are only mobilized during major floods. A surface armor is common. Channel shape is typically irregular.

profile. Vegetation colonization of these various surfaces can induce considerable complexity to the resulting form. This final scenario exemplifies the inherent linkage that must be considered in any examination of bank morphology and associated bank-forming processes, namely how interactions between the bed and banks influence the nature of the channel itself.

4.4 Channel morphology: Putting the bed and banks together

Channels comprise a combination of bed and bank features, reflecting a wide range of formative processes, both contemporary and inherited. Bed and bank processes are not necessarily in phase with each other. For example, as incised channels cut through alluvium they expose older bed and bank deposits that may have been inaccessible to the river for extended periods of time. In general terms, however, more transient bed materials are younger than materials exposed along the banks. It is not unusual for the bed and banks to comprise different material mixes. Indeed, many channels have cohesive banks and a noncohesive bed. Unlike the bed that is typically reworked on a

semiregular basis, bank materials may be an artifact of history. This hiatus between bed and bank materials varies markedly from reach to reach and from system to system, reflecting the aggradational/degradational balance of the channel and the history of formative events.

While bed and bank material texture exert a primary control on channel shape (see Table 4.4), the texture of boundary materials is only one factor that influences channel morphology. Fluctuations in discharge and sediment load fashion the balance between bed aggradation and degradation, and bank erosion and deposition, in any given cross-section or along any reach. The distribution of flow energy at differing flow stages, and flow alignment, can result in a wide range of bed and bank morphologies. Ultimately, however, bed morphology is determined by patterns of sculpted/erosional geomorphic units and mid-channel depositional geomorphic units, while bank morphology reflects the balance of bank erosion processes and the assemblage of bank-attached depositional geomorphic units. A combination of these factors at any given site results in a range of channel shapes (see Table 4.5 and Figure 4.5). However, the same channel shape can result from a range of processes.

Table 4.5 Putting the beds and banks together to assess channel shape.

Channel shape	Bank process	Bed process	Energy distribution
Symmetrical	Erosional (e.g., fluvial entrainment)	Erosional (e.g., headcut formation) or depositional (e.g., sand sheets)	Evenly distributed across channel
Asymmetrical	One erosional (e.g., undercutting), one depositional (e.g., point bar formation)	Depositional (point bar formation along convex bank), and erosional (pool scour along concave bank)	Thalweg along concave bank
Irregular	Imposed condition, or a reflection of erosional (e.g., slumping), or depositional (e.g., bank-attached bar formation) influences	Erosional (e.g., sculpted pools or cascades) or depositional (e.g., midchannel bars and island formation)	Unevenly distributed around alluvial materials or bedrock outcrops
Compound	Depositional (e.g., bench formation) or erosional (e.g., ledge formation)	Depositional (e.g., sand sheet formation) or erosional (e.g., headcut formation)	Dissipated in a nonlinear manner at differing flow stages

A channel with a given width and average depth can be characterized by a wide range of possible shapes depending on the array of midchannel and bank-attached geomorphic units. This reflects the distribution of energy within the channel (which is a function of slope, channel size, and flow alignment), the sediment flux of the reach (i.e., the caliber and volume of available materials, and whether the reach is transport-limited or supply-limited), and process interactions with instream vegetation and forcing elements. In many settings, predictable patterns of geomorphic units are observed. These relationships often reflect the geomorphic effectiveness of the most recent formative flow event. Any adjustment to the sediment flux or energy within a channel that alters bed material caliber and organization or flow characteristics may modify the geomorphic structure of the reach, and hence channel shape. If a channel is *transport-limited*, such that the volume or caliber of bedload material is greater than the capacity or competence of the stream to move it, channels tend to be wide, shallow, and characterized by bed sedimentation and midchannel bars. Conversely, *supply-limited* rivers are able to move all materials made available to them, and channels tend to be narrow and deep with bank-attached depositional forms.

Symmetrical channels tend to be characterized by banks with a uniform or upward-fining cohesive sediments and a near-homogenous bed (Figure 4.5a). Bedload systems tend to have a high width:depth ratio, while suspended load systems tend to have a low width:depth ratio. Channels tend to be relatively free of depositional features other than uniform sheet-like deposits, as flow energy is spread evenly across the bed (Table 4.5). Symmetrical channels commonly occur at the inflection points of bends, along low sinuosity channels, along fine-grained suspended load rivers with cohesive banks, or in incised channel situations.

In *asymmetrical* channels, flow energy is concentrated along the concave bank in bends (Table 4.5). As a result, erosion occurs along one side of the bed, while deposition occurs on the other (Figure 4.5b). Bank erosion via fluvial entrainment or undercutting is common along the concave bank, while bank-attached geomorphic units develop along the convex bank (commonly point bars). These processes promote lateral migration. In partly-confined valleys, discontinuous floodplain pockets and point bars on the convex bank of bends, combined with abutment against bedrock along the concave bank, also induce the formation of an asymmetrical channel shape (Figure 4.5c).

Irregular channels may form under differing sets of conditions. In confined valleys, imposed controls on bed/bank morphology induce an irregular channel shape (Figure 4.5d). Flow energy is distributed unevenly around bedrock or coarse substrate, generating sculpted or erosional geomorphic units (Table 4.5). In more alluvial rivers, midchannel geomorphic units and either erosional or

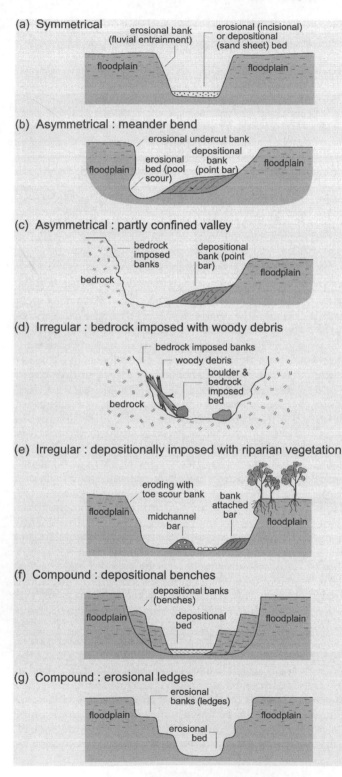

Figure 4.5 Channel shape
Combinations of bed and bank components define channel shape. The hiatus between these components varies markedly from system-to-system, reach-to-reach, and site-to-site. In each instance a combination of erosional and depositional processes may be evident (see text). Elsewhere, channel shape is imposed by bedrock and/or ancient boundary materials and/or other forcing elements (such as riparian vegetation and woody debris).

depositional banks can induce an irregular channel shape (Figure 4.5e). In some instances, an irregular channel shape is inherited from the past and is out-of-phase with contemporary processes. Alternatively, significant heterogeneity is often evident along forested streams, where woody debris and riparian vegetation induce significant local irregularities in channel shape (Figure 4.5e).

Compound channels are commonly associated with macrochannels. Their stepped cross-sectional morphology has the appearance of a smaller channel inset within a larger channel form. They are commonly associated with cut-and-fill activity or rivers that are responding to significant variability in discharge (Table 4.5). Formation of one or more inset levels (i.e., benches) at channel margins may reflect depositional phases associated with channel contraction (Figure 4.5f). Alternatively, channel expansion may be recorded by the formation and/or reworking of ledges (Figure 4.5g).

4.5 Channel size

Alluvial channels adjust their form to convey the water and sediment supplied to them. Various approaches have been developed to characterize "equilibrium" channel dimensions, as determined by mean conditions (Knighton, 1998). For example, regime theory and principles of hydraulic geometry have been used to derive empirical equations that describe relationships between channel width, depth, slope, particle size distribution and flow velocity, and external controls such as catchment area and flow. These principles have been based primarily on analyses of single channel systems in unconsolidated sediments. Local-scale variability in bed and bank materials, the distribution of forcing elements, the role of riparian vegetation and woody debris, and preponderance of other forms of channel configurations introduce a level of diversity that is not captured by these empirical relationships. Hence, application of these principles to describe notionally "characteristic" channel dimensions must be undertaken with caution, especially in environments that differ to those in which the primary data were derived. Consideration must also be given to the geomorphic condition of sites at which data were

collected to derive these empirical relationships. Data gathered from disturbed river systems are unlikely to provide appropriate guidance for management procedures that strive to improve river health.

In general terms, rivers on steeper slopes, or systems that transport large volumes of coarse bedload with divided or braided channels, tend to develop wide, shallow channels with higher width : depth ratios than comparable reaches with meandering or straight planforms (Parker, 1979). Similarly, rivers with a flashier discharge regime and relatively high peak flows tend to develop wider channels. Sand channels with insufficient fine sediment to form resistant banks are particularly sensitive to discharge variability compared to fine-grained systems (Osterkamp and Hedman, 1982).

The role of vegetation also has a significant effect on channel size. Other factors being equal, channels with dense vegetation tend to be narrower and deeper than their sparsely vegetated counterparts (e.g., Charlton et al., 1978; Hey and Thorne, 1986; Millar and Quick, 1993; Montgomery, 2001). Also, as a general rule, the proportion of vegetation occupying a channel cross-section decreases downstream as the channel becomes larger. Zimmerman et al. (1967) suggested that in very small catchments (up to about 2 km²) grass and sedge dominated channels are smaller than channels having similar catchment area (or discharge) that are dominated by trees. However, moving downstream, channels dominated by trees are comparatively smaller than channels with equivalent catchment area but only grass and sedge on the banks and floodplain. An equivalent set of relationships has been described for the geomorphic role of woody debris. The stability of woody debris and its influence on channel forms and processes reflect the relative size of key wood elements compared to channel size (Montgomery and Piegay, 2003). In low order channels, woody debris may induce channel blockage ratios as high as 80%. Moving downstream, woody debris tends to be rotated subparallel to the flow, minimizing the blockage ratio, but maximizing its role in bar accretion and bank toe protection. In wider channels, woody debris may be transported beyond the fall point, and become incorporated into log jams potentially causing local bank erosion, triggering channel avulsion or cutoff development, or promoting island development. Any changes to riparian vegetation and woody debris loading that alter instream and floodplain roughness may modify patterns and rates of depositional and erosional processes within a channel, affecting its morphology and size.

Despite the limitations of regime theory and principles of hydraulic geometry, regionally based applications derived for particular landscape settings that operate under similar hydrologic and sediment (lithologic) conditions, with equivalent riparian vegetation associations, may have considerable application for planning purposes. Empirically derived relationships have been extensively used in the design of river rehabilitation treatments for meandering rivers (e.g., Hey, 1997). Ideally, an equivalent body of work would be developed across the range of natural river diversity, such that design criteria fit the local setting, rather than imposed notions of channel geometry framed in terms of relatively uniform pool–riffle sequences. Each river must be viewed in its landscape context, considering notions of downstream connectivity in flow and sediment regimes, antecedent controls, and local factors that may shape channel morphology and size. These relationships exert a fundamental control on floodplain forms and processes.

4.6 Floodplain forms and processes

Over decades or centuries, rivers transport only a small fraction of the total alluvium stored along their valleys (Knighton, 1998). The bulk of materials stored in *floodplain* or *terrace* (abandoned floodplain) forms between the channel and valley margins is inaccessible to contemporary channel processes. In narrow valley settings, common in headwater situations, floodplains are generally restricted to riparian corridors. These buffer strips act as filters for flow, sediment, and nutrients from adjacent slopes. The functional role of floodplains changes as valleys widen downstream. Interactions with slope processes decrease, and a different assemblage of floodplain forms is observed. In general, floodplains can be separated into proximal (channel marginal) and distal (valley marginal) zones. Within these zones, distinct

packages of landforms may form (Allen, 1965; Lewin, 1978).

Genetic approaches to the classification of floodplains relate river processes to the floodplains they construct (Nanson and Croke, 1992). Various geomorphic parameters can be used to differentiate among floodplain types. Each floodplain type reflects a combination of energy conditions (largely determined by slope and valley width relative to upstream catchment area), the availability of sediment (its caliber and volume relative to the accommodation space along the valley), and the range/history of floodplain forming and reworking processes. A change in one or more of these conditions may alter the dominant mode of floodplain construction.

Floodplains form by a combination of *lateral* (within-channel) and *vertical accretion* (overbank) processes (Table 4.6), and are prone to reworking by various mechanisms (Table 4.7). The type and mix of these processes influence the range and pattern of floodplain geomorphic units found along any given reach. Floodplain geomorphic units are differentiated primarily on the basis of their shape, position, and formative processes. Pronounced differences are evident between floodplains comprised largely of noncohesive alluvium (gravel and fine sand) and those comprised of cohesive alluvium (silt and clay). Significant pocket-to-pocket variability in floodplain forms and processes may be evident along a river (e.g., Ferguson and Brierley, 1999a, b). A summary of form–process associations for various floodplain geomorphic units is presented in Table 4.8.

Lateral accretion occurs when bedload deposits on the convex slope of bends are incorporated into the floodplain as the channel migrates across the valley floor or translates downstream (Figure 4.6a). Patterns of *ridge and swale topography*, which record accretionary pathways of the channel, relate to the radius of curvature of the bend and associated channel sinuosity (see Table 4.8). *Oblique accretion* occurs as sediments are draped along the bank of nonmigrating rivers (Figure 4.6g). As these surfaces build inset floodplains or *benches*, channel contraction occurs.

In general, horizontally-bedded, fine-grained, suspended load materials dominate floodplain sequences beyond the active channel zone. Floodplains that are dominated by *vertically*

Table 4.6 Floodplain forming processes.

Floodplain forming process	Definition
Lateral accretion	Within-channel, bedload materials are deposited as point bars on the convex banks of bends. These materials become incorporated into the floodplain as the channel migrates. Resulting sedimentary structures often dip towards the channel.
Vertical accretion	Accumulation of sediment derived from suspension in overbank flows (typically, fine sand and mud). Overbank deposits commonly comprise vertically stacked beds with flood couplets reflecting the rising and waning stage of flood events. Bioturbation tends to homogenize these materials over time, such that they appear to be massive rather than retaining their primary laminated form. Patterns of sedimentation may be influenced by vegetation cover. In distal areas, silt and clay may remain in suspension in ponds or wetlands for considerable periods. Proximal–distal gradation in material size is commonly observed.
Braid channel accretion	Deposition atop midchannel bars during large flood events promotes the development of stable islands that are beyond the reach of small–moderate flood events. Shifting of the primary channels leads to abandonment of the bars/islands, and their incorporation into the floodplain via infilling of old braid channels with overbank sediments. This process is common along multichanneled systems (e.g., braided rivers).
Oblique accretion	Muddy drapes and sand deposits onlap the channel margin, building vertically over time. Eventually these deposits are incorporated into the floodplain or form an inset floodplain surface. These features comprise oblique accretion (dipping) structures.
Counterpoint accretion	Deposits are laid down as slackwater deposits in a separation zone that forms against the upstream limb of the convex bank of tightly curved bends. These suspended load deposits build vertically, becoming incorporated into the floodplain as the channel translates downstream.
Abandoned channel accretion	Paleochannels formed by meander cut offs or avulsion are infilled by overbank deposits. These features generally comprise fine grained deposits atop the old channel fill. In some instances they act as plugs that influence subsequent patterns of channel adjustment.

Table 4.7 Floodplain reworking processes.

Floodplain reworking process	Definition
Lateral migration	Progressive movement of meander bends across the valley floor. Includes bend extension, translation, and rotation.
Cutoffs	Short-circuiting of a meander bend leaving a billabong or oxbow lake on the floodplain. Can be in the form of meander or chute cutoffs.
Avulsion	Wholesale shift in channel position to a lower part of the floodplain, leaving an abandoned channel.
Stripping	Removal of surface floodplain layers by high energy flows in partly-confined valleys.
Floodchannels	Channels that short-circuit a floodplain pocket at overbank stage, resulting in scour and reworking that forms an elongate, channel-like depression on the floodplain.
Channel expansion	Enlargement of a channel by bank erosion, removing proximal floodplain materials.

accreted fine-grained overbank deposits tend to be relatively flat and featureless. As a river overtops its banks, it loses power due to the greatly reduced depth and energy of the unconfined sheet-like overbank flow. Cyclical flood couplet deposits reflect the rising and falling stages of floods. In many instances, vertical accretion deposits overlie lateral accretion deposits.

Some vertically accreted floodplains have significant topography, typically reflecting patterns

Table 4.8 Floodplain geomorphic units.

Unit	Form	Process interpretation
Floodplain (alluvial flat) and alluvial terrace (fill terrace)	**Floodplain** Lies adjacent to or between active or abandoned channels, confined by valley margin and alluvial ridges. Typically tabular and elongated parallel to active channels, but can be highly variable, ranging from featureless, flat-topped forms to inclined forms (typically tilted away from the channel) to irregularly reworked (scoured) forms. Volumetrically, floodplains are the principal sediment storage unit along most rivers. May be coarse-grained, fine-grained, or intercalated. Floodplains can be separated into *proximal* (channel-marginal) and *distal* (against the valley margin) zones. **Alluvial terrace** Typically a relatively flat (planar), valley marginal feature that is perched above the contemporary channel and/or floodplain. These abandoned floodplains are no longer active. Generally separated from the contemporary floodplain by a steep slope called a terrace riser. Can be paired or unpaired. Often found as a flight of terraces. Terraces may be of great age (e.g., Tertiary terraces are not uncommon). Terraces often confine the contemporary channel, in a manner that is analogous to bedrock valley margins.	**Floodplain** Floodplains are the principal alluvial surface aggrading under the contemporary sediment-load and discharge regime. Floodplain form reflects the contemporary arrangement of out-of-channel sediment build-up and reworking at flood stage. Formed from vertically and/or laterally accreted deposits. Proximal–distal gradation in grain size is common, dependent on the nature of the channel-marginal units and whether they allow deposition of coarse sediments beyond the channel zone. **Alluvial terrace** Initially formed by vertical and lateral accretion under prior flow conditions to form a floodplain. With tectonic uplift, a change to base level, or shifts in sediment-load and discharge regime (linked to climate), downcutting into valley floor deposits results in abandonment of the former floodplain. In many cases, a contemporary floodplain subsequently develops and becomes inset within these terraces. Unpaired terraces reflect lateral shift during incision, whereas paired terraces indicate rapid downcutting only.
Strath terrace	Typically a relatively flat, valley marginal feature that is perched above the contemporary channel or floodplain. These erosional surfaces have a bedrock core, often with a thin alluvial overburden. Strath terraces often confine the channel, analogous to valley margins.	Reflect incision and valley expansion associated with downcutting into bedrock, abandoning terrace surfaces. In many cases, a contemporary floodplain subsequently develops and becomes inset within these terraces. In other cases, where incision occurs with little lateral expansion, a confined valley is formed.

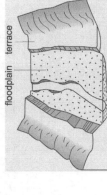

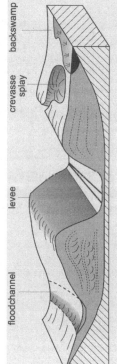

floodchannel   levee   crevasse splay   backswamp

Levee

Raised elongate asymmetrical ridge that borders the channel (i.e., along the proximal floodplain), with a steeper proximal margin. Levees scale in proportion to the adjacent channel. Levee crests may stand several meters above the floodplain surface or be relatively shallow, laterally extensive features. Composed almost entirely of suspended load sediments, i.e., dominantly silt, often sandy.

Levee form is influenced by, and in turn influences, the channel–floodplain linkage of biophysical processes, influencing the lateral transfer of water, sediment, organic matter, etc. Levees are produced primarily from suspended-load deposition at high flood stage. During overbank events, flow energy dissipates when flows spread out over the floodplain. Under these conditions, the flow has insufficient energy to carry its load. The marked reduction in velocity results in coarse sediment deposited on proximal floodplains as levees. Interbedded flood-cycle deposits, termed flood couplets, reflect rising- and falling-stage sedimentation. Finer materials are carried into the distal parts of the floodplain. Highly developed levees along extensive fine-grained floodplains infer a laterally fixed channel zone and well-defined segregation of water and sediment transfer between the channel and floodbasin. As the levee grows, the deposition rate of coarser sediment near the crest is reduced, leading to a generally fining upward sequence of deposits within the levee profile.

Crevasse splay (crevasse channel-fill)

A sediment tongue fed by a crevasse channel breaching the levee crest. Crevasse splays have a lobate or fan-shaped planform, thinning distally away from the levee. The surface may have multiple distributary channels, producing hummocky topography. Composed of bedload material, predominantly sand, sometimes gravel. The crevasse channel fill has a symmetrical, lenticular geometry and low width:depth ratio. Upward-coarsening gradation of grain sizes is common, as is proximal–distal gradation away from the channel.

Crevasse channels breach and erode the levee taking bedload materials from the primary channel and conveying them onto the floodplain at high flood stage. Deposition reflects the rapid loss of competence beyond the channel zone. Flow velocity is sufficient to carry relatively coarse material, which is spread outward onto a fan-shaped area of floodplain that fines away from the levee. The angle of trajectory increases with high levee backslopes and/or decreases with higher flow velocity. Crevasse channel fills represent bedload plugging of old crevasse channels, indicating an aggradational environment. Their formation may be linked to the formation of an alluvial ridge.

Floodchannel (back channel)

Gently curved, subsidiary channel. Entrance height approximates bankfull stage. Commonly observed at valley margins. The depth of the floodchannel tends to increase down-pocket with the basal section of the floodchannel elevated above the low flowchannel (i.e., it lies perched within the floodplain).

Flow alignment along the valley floor short-circuits the channel during high discharge events, steepening the down-valley flow trajectory and inducing scour that forms a floodchannel. At lower flood magnitudes, when the entrance to the floodchannel is not breached, suspended load deposition may occur via backfilling. Channel/valley alignment controls their distribution. Floodchannels do not necessarily lead to meander cutoffs, but may situate future (or past) avulsion channels. Floodchannels may scour and shape distal levee morphology in confined valley-settings.

Table 4.8 *Continued*

Unit	Form	Process interpretation
Flood runner	Relatively straight depression on the floodplain that occasionally conveys floodwaters. Tends to have a relatively uniform morphology.	Acts like a chute during high discharge events, short-circuiting the channel course (i.e., aligned down-valley).
Backswamp (distal floodplain, floodplain wetland, floodpond, floodplain lake)	The distal floodplain, at valley margins, is typically the lowest area of the valley floor. They are major storage units of fine-grained, vertically accreted, suspended load sediments. Morphology is typically fairly flat (or has low relief), with depressions. Ponds, wetlands, and swamps commonly form where lower order tributaries drain directly onto the floodplain.	Forms when the reduction in energy gradient from the proximal to distal floodplain only allows suspended load materials to be transferred to the backswamp. This results in slow rates of fine-grained vertical accretion. A distinct gradation in energy with distance from the channel may result in pronounced textural segregation across the floodplain. Backswamps, wetlands, lakes, and pond features are common in these poorly drained (unchanneled), low-energy, vertically-accreting environments. Naturally colonized by dense aquatic/swamp vegetation that traps fine grained suspended- load sediments promoting cohesive, mud- and organic-rich accumulation. Tend to be highly bioturbated.
Sand wedge	Sandy deposits with wedge-shaped cross-section at channel margins in nonlevee settings. They typically have a scoured basal contact. Basal cross-beds grade to finer-grained flood cycle interbeds.	Sand wedges reflect bedload deposition, thereby differentiating them from levees. They form atop the proximal floodplain in moderate–high energy environments. As flows go overbank, velocity is sufficient to carry relatively coarse material. Energy is spread outward onto a wedge-shaped area of the floodplain, depositing sand.
Floodplain sand sheet	Flat, tabular, laterally extensive sheets in nonlevee settings with massive, often poorly sorted facies. Show little lateral variation in thickness, mean grain size, or internal structure. Surface expression generally conforms to the underlying floodplain. Differentiated from splays by their shape, extensive area, and lack of distal thinning.	Associated with rapid sediment charged bedload deposition on the floodplain during extreme flood events. Competent overbank flows are required to transfer bedload materials onto the floodplain, where they are deposited in sheet-like forms. These planar, homogeneous sequences are common in sandy ephemeral streams. Often formed downstream of transitions from confined to unconfined flows and associated with a break in slope (as on alluvial fans). Sand sheets build the floodplain vertically.

sand wedge

sand sheet

Feature	Description	Process
Paleochannel (prior channel, abandoned, ancestral channel)	An old, inactive channel found on the floodplain. May be partially or entirely filled. Includes more than one meander wavelength (thereby differentiating it from a meander cutoff). Can have a wide range of planforms, from elongate and relatively straight to irregular or sinuous, reflecting the morphology of a former primary channel. Low-sinuosity paleochannels may be overprinted with floodchannels. May have an upward-fining fill comprising a channel lag of coarser material with finer, suspended-load materials atop.	Caused by a sudden shift in main channel position (avulsion), generally to a zone of lower elevation, abandoning a channel on the floodplain. This paleochannel may subsequently fill with suspended-load sediments derived from overbank flooding. They record paleoplanform and geometry of the avulsed channel. If this is markedly different from the contemporary channel, it may indicate a shift in sediment-load, discharge, or distribution of flood power within the system.
Ridge and swale topography	Ridge features represent paleo scroll bars that have been incorporated into the floodplain. Swales are the intervening low flow channels. These arcuate forms have differing radii of curvature, reflecting the pathway of lateral accretion across a floodplain. Ridge and swale topography may indicate phases of paleomigration paths, paleocurvature, and paleowidths of channel bends.	During bankfull conditions the high velocity filament is located along the concave bank of a bend. This thalweg zone contains helical flow that erodes the concave bank of the bend and transfers sediments to the point bar. Eddy flow cells occur in a separation zone along the convex bank. Between these secondary flow circulation patterns there is a shear zone where sediments are pushed up the point bar face to form a ridge (or scroll bar). At bankfull stage this scroll bar accretes vertically. As the channel shifts laterally, the scroll bar becomes incorporated into the floodplain forming ridge and swale topography. Subsequent overbank deposits smooth out the floodplain surface and the former channel position is retained on the inside of the bend.
Valley fill (swamp, swampy meadow)	Relatively flat unincised surface. May have ponds and discontinuous channels or drainage lines. Composed of vertically accreted mud, with possible sand sheets downstream of discontinuous gullies. May comprise organic-rich deposits formed around swampy vegetation.	These sediment storage features are typically formed by flows which lose their velocity and competence as they spread over an intact valley floor, and deposit their sediment load. Vertically accreted swamp deposits are derived by trapping of fine-grained suspended load sediments around vegetation. Mud beds may alternate with laterally shifting floodout and sand sheet deposits.
Floodout	Lobate/fan-shaped sand body that radiates downstream from an intersection point of a discontinuous channel (i.e., where the channel bed rises to the level of the floodplain). Tend to have a convex cross-profile, and fine in a downstream direction. Comprise sand materials immediately downstream of the intersection point, but may terminate in swamps or marshes as fine-grained sediment accumulates downstream.	Formed when a discontinuous channel supplies sediment to an unincised valley fill surface. Sands are deposited and stored as bedload lobes that radiate from the intersection point of the discontinuous channel. At this point there is a significant loss of flow velocity. Beyond the floodout margin, fine-grained materials are deposited in seepage zones. Deposition associated with breakdown of channelized flow may reflect transmission loss and low channel gradient. Floodout lobes shift over the floodplain surface, preferentially infilling lower areas with each sediment pulse.

Table 4.8 *Continued*

Unit	Form	Process interpretation
Meander cutoff (neck cutoff, ox bow, billabong)	A meander bend that has been cut through the neck, leaving an abandoned meander loop on the floodplain. The bends have an arcuate or sinuous planform (generally one meander loop). Generally horseshoe or semicircular in planview, reflecting the morphology of the former channel bend. May host standing water (i.e., oxbow lake or billabong) or be infilled with fine grained materials.	Formed by the channel breaching a meander bend (possibly linked to flow obstruction upstream) or through the development of a neck cutoff during high flow conditions. Represent shortening of stream lengths or decreases in sinuosity of the channel, steepening the water-slope at flood stage. The paleomeander loop subsequently becomes plugged with instream materials. The abandoned meander gradually becomes isolated from the main channel. The loop may infill with fine grained, suspended load materials and develop into a billabong. These features record the paleoplanform and geometry of the channel.
Chute cutoff	Straight/gently curved channel that dissects the convex bend of the primary channel, short-circuiting the bend. This chute then becomes the primary channel. Chute cutoffs have a straighter planform than meander cutoffs. They generally fill with bedload materials.	Represent shortening of stream lengths or decreases in sinuosity of the channel, steepening the water-slope at flood stage. Concentrated flow with high stream powers are able to cut across the bend. With chute cutoff enlargement, the bend may be abandoned, at which point the chute becomes the primary channel. The old channel bend is filled mostly with bedload deposits. Chute cutoffs generally occur in higher energy settings than meander cutoffs.
Anabranch (secondary or flood channel)	Pattern of coexistent multiple-anastomosing channels (repeated bifurcating and rejoining) with low width : depth ratio. These open channels remain connected to the trunk stream(s).	Formed in high flow conditions where the channel avulses to, or reoccupies, another position on the valley floor, but maintains the old channel within a multichanneled network. These channels are dominated by low-energy, suspended-load deposits.

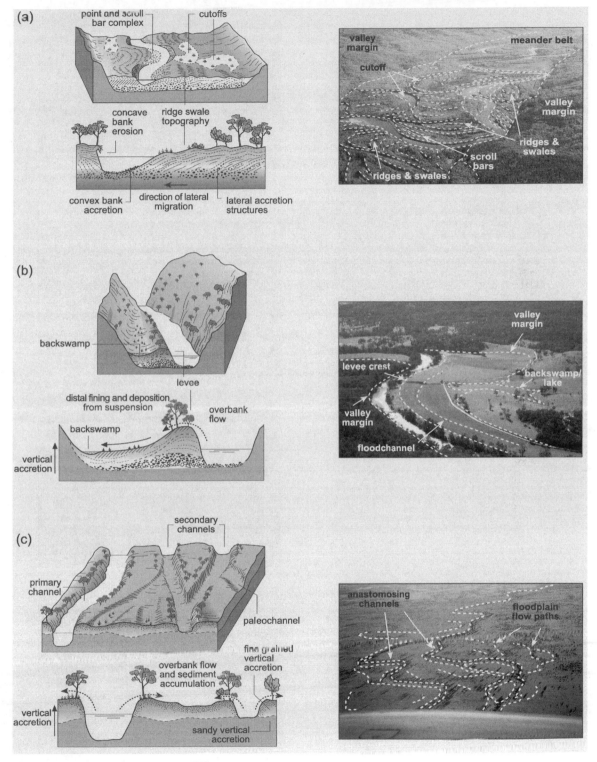

Figure 4.6 Floodplain forming processes

Following principles adopted by Nanson and Croke (1992), seven primary classes of floodplain forming processes may be differentiated, namely: (a) lateral accretion, (b) vertical accretion in a partly confined valley, (c) vertical accretion across a wide plain, (d) abandoned channel accretion, (e) braid channel accretion, (f) counterpoint accretion, and (g) oblique accretion (see text). Cross-sections and block diagrams in a–d reprinted from Nanson and Croke (1992) with permission from Elsevier, 2003.

(d)

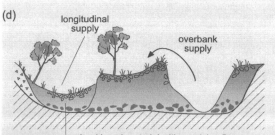

longitudinal supply

overbank supply

coarse grained basal material with overlying fines

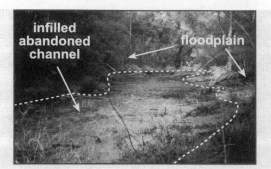

infilled abandoned channel

floodplain

(e)

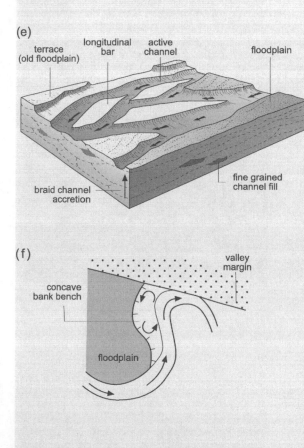

terrace (old floodplain)

longitudinal bar

active channel

floodplain

braid channel accretion

fine grained channel fill

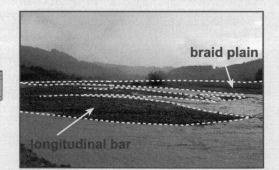

braid plain

longitudinal bar

(f)

valley margin

concave bank bench

floodplain

valley margin

concave bank bench

floodplain

(g)

floodplain   bench

overtop flow and vertical accretion

helical flow and oblique accretion

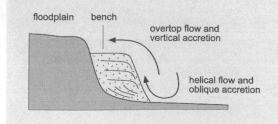

floodplain

lateral bar

bench

Figure 4.6 *Continued*

of reworking. Figure 4.6b displays a sandy flood-plain characterized by vertical accretion atop a *levee*. Levees are raised, elongate, prismatic land-forms with an asymmetrical cross-section. These channel marginal features exert a critical control on the type of floodplain formed. They influence the nature and rate of overbank flows, thereby shaping patterns and rates of sedimentation and re-working of deposits in proximal–distal zones. In general terms, levee dimensions are scaled relative to the size of the adjacent channel. They are best developed at the concave bank of bends, where they commonly form steep high banks (Brierley et al., 1997). A stacked sequence of upward-fining vertical accretion structures can often be identi-fied within the levee. Overbank flows result in distal fining and textural segregation such that coarser materials are deposited on the levee crest and fine-grained suspended load materials are de-posited in *backswamps*. Backswamps occur where overbank (vertical accretion) deposits accumulate slowly in depressions at valley margins, as flow energy becomes increasingly dissipated with dis-tance from the channel. Backswamps generally have a distinctive wetland vegetation association; in some instances peat may accumulate.

The presence of well-developed levees along lat-erally extensive fine-grained floodplains infers that the channel is relatively stable and that segre-gation between the channel zone and the back-swamp is well defined. The greater the height of the levees above the floodplain, the more likely it is that *crevasse splays* and *floodchannels* will form. *Crevasse splays* are narrow to broad, local-ized tongues of sediment that are sinuous to lobate in plan. A crevasse channel that cuts the channel–marginal levee feeds the crevasse splay. Once a crevasse channel is initiated, flood waters may deepen the new course and develop a system of distributive channels on the upper slope of the levee. Crevasse splays generally extend well beyond levee toes onto floodbasin deposits (Farrell, 1987, 2001). Levee construction, and re-striction of a stream to a meander belt, may lead to substantial local elevation of the floodplain sur-face between the levees in the form of an *alluvial ridge*. Perching of the channel above its floodplain enhances the prospect that *avulsion* may occur, bringing about a wholesale shift in channel posi-tion from the present meander belt into the flank-

ing basin. In river systems without levees, espe-cially those with shallow channels, there is con-siderable potential for bedload-caliber deposits to be launched onto the adjacent proximal floodplain in the form of *sheets* or *wedges* (Brierley, 1991).

Most anastomosing rivers are characterized by vertically accreted floodplains produced from draped deposits laid down as flow overtops the banks of multiple channels (Figure 4.6c). Given the mud-dominated nature of materials transported in these rivers, thick, uniform sequences of vertically accreted deposits form flat-topped floodplains. In some cases, these can be organic rich (Smith, 1983).

Vertical accretion deposits also accumulate in *paleochannels* and *cutoffs*. Typically, upward-fining sequences of gravel and fine sands grade into mud and/or swamp deposits. This form of vertical accretion is termed *abandoned channel accretion* (Figure 4.6d). Paleochannels extend over more than one meander wavelength and can have a wide range of planforms, from elongate and relatively straight to irregular or sinuous. Cutoffs are aban-doned meander loops. They are generally horse-shoe or semicircular in planview. In general, the rate of infilling of paleochannels and cutoffs re-flects their antiquity (i.e., the longer the period since their abandonment, the greater the degree of infilling; e.g., Erskine et al., 1992; Hooke, 1995). In some instances, however, the rate of infilling of pa-leochannels and cutoffs reflects their alignment relative to the contemporary channel, and their geometry and position on the floodplain relative to the frequency of overbank flood events (Shields and Abt, 1989; Piégay et al., 2000). Straighter chan-nels are subjected to higher flood velocities and tend to be actively reworked, while more sinuous cutoffs are flooded by backwaters, and tend to be subjected to higher rates of deposition. Areas of consolidated sediment such as clay plugs may exert a marked influence on meander morphology and migratory pathways (e.g., Schumm and Thorne, 1989). If the resistant material is exten-sive, deep scour may develop against the bank, locking the channel into that area for a consider-able period (typically until the bed is overtaken by a more mobile bend from upstream).

A distinct set of geomorphic units occurs along unincised river courses where the *valley fill* accretes vertically over time. *Ponds* tend to be relatively elongate, scour features formed along preferential drainage lines. *Floodouts* are lobate/fan-shaped depositional features composed largely of bedload-caliber materials that radiate downstream from an intersection point of a discontinuous channel (i.e., where the channel bed rises to the level of the floodplain). These deposits are associated with the breakdown of channel flow, reflecting transmission loss and low gradient conditions. They tend to have a convex cross-profile, and fine in a downstream direction. Beyond the floodout margin, selectively sorted fine-grained materials are deposited in seepage zones. Over time, floodout lobes shift over the floodplain surface, preferentially infilling lower areas with each sediment pulse.

Braid channel accretion is observed along some multichanneled rivers (Figure 4.6e). Preferential flow orientation down one of the channels may lead to abandonment of others. Abandoned channels may subsequently be infilled by overbank sediments, and the island may become incorporated into the floodplain (Nordseth, 1973). Another form of vertical accretion is *counterpoint accretion* (Figure 4.6f), where *concave bank benches* formed in secondary circulation cells are incorporated into the floodplain as the channel translates down-valley (Hickin, 1986).

While some floodplains have the appearance of being purely aggradational forms, especially those dominated by vertical accretion processes, others may show considerable evidence of reworking (Table 4.7). Reworking can range from progressive *lateral migration* or downstream translation of trains of meanders, to cutoffs of varying form (*meander* and *chute cutoffs*), to wholesale shifts in channel position (i.e., *avulsion*) that result in abandoned or paleochannels of differing scale and stage of infill across the floodplain, to *floodchannel* or *stripping* mechanisms. *Floodchannels* short-circuit a floodplain pocket by scouring a channel into the floodplain surface. They tend to be relatively straight, depressed tracts of the floodplain that occasionally convey floodwaters. The entrance height of a floodchannel tends to approximate bankfull flood stage, while depth tends to increase down-pocket. The basal section of floodchannels lies above the low flow channel. Following flood events, these features are commonly associated with ponding and deposition of sus-

pended load materials. In some instances they are prone to backfilling. Other forms of floodplain reworking include *stripping* (Nanson, 1986), which involves the removal of entire sections of surficial floodplain material, and *channel expansion* where proximal floodplain deposits are eroded.

Interpretation of various types of depositional and reworking processes provides insight into the form–process associations of floodplain geomorphic units, and the range and history of formative events that produced the floodplain (see Chapter 5). Packages of these differing floodplain features commonly form distinct assemblages that characterize different types of river. For example, they are commonly associated with differing variants of channel planform, as discussed in the following section.

4.7 Channel planform

Channel planform, defined as the configuration of a river in plan view, provides a reach-scale summary of the channel and floodplain characteristics of an alluvial river. Flow patterns and the nature/distribution of physical processes for different planform types are major determinants of channel shape and floodplain character. As such, channel planform provides an excellent initial guide to the capacity and manner by which alluvial rivers adjust their morphology and configuration. A wide range of channel planforms exist, conditioned primarily by available flow energy (and its spatial/temporal variability), sediment caliber and availability, and whether the reach operates as a bedload, mixed load, or suspended load system.

Channel planform is differentiated on the basis of three inter-related criteria, namely the number of channels, their sinuosity, and their lateral stability (Figure 4.7). The *number of channels* is typically differentiated into reaches in which channels are absent or discontinuous, single channeled or multichanneled (Figure 4.7a). Rivers that are capacity- or competence-limited tend to have multiple channels with braided or wandering tendencies. On low slopes, where drainage breakdown occurs, distributary networks develop in the form of anastomosing or anabranching networks. In low energy systems that are unable to incise their

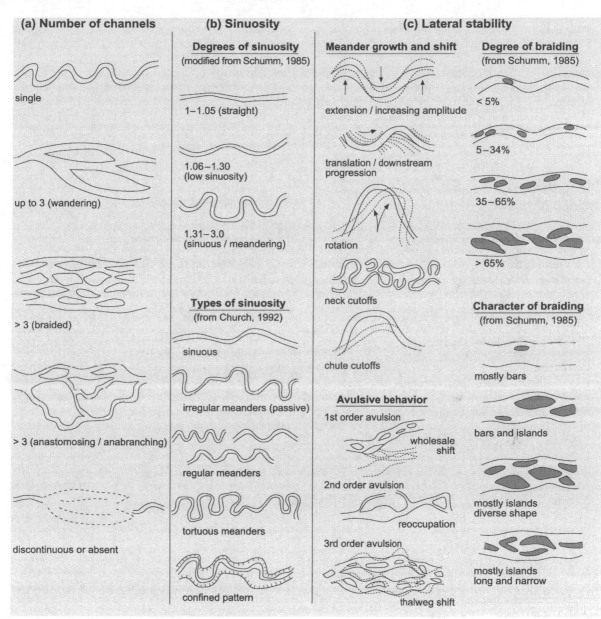

(a) Number of channels

single

up to 3 (wandering)

> 3 (braided)

> 3 (anastomosing / anabranching)

discontinuous or absent

(b) Sinuosity

Degrees of sinuosity
(modified from Schumm, 1985)

1–1.05 (straight)

1.06–1.30
(low sinuosity)

1.31–3.0
(sinuous / meandering)

Types of sinuosity
(from Church, 1992)

sinuous

irregular meanders (passive)

regular meanders

tortuous meanders

confined pattern

(c) Lateral stability

Meander growth and shift

extension / increasing amplitude

translation / downstream
progression

rotation

neck cutoffs

chute cutoffs

Avulsive behavior

1st order avulsion

wholesale
shift

2nd order avulsion

reoccupation

3rd order avulsion

thalweg shift

Degree of braiding
(from Schumm, 1985)

< 5%

5–34%

35–65%

> 65%

Character of braiding
(from Schumm, 1985)

mostly bars

bars and islands

mostly islands
diverse shape

mostly islands
long and narrow

Figure 4.7 Measures of channel planform
Measures of channel planform are based on (a) number of channels, (b) sinuosity, and (c) lateral stability. An alluvial reach comprises a combination of these attributes. Parts modified from Schumm (1985). Reproduced with permission from the *Annual Review of Earth and Planetary Sciences*, Vol 13 © 1985 by Annual Reviews.

valley floor, channels may be discontinuous or absent.

As noted on Figure 4.7b, *sinuosity*, defined as the ratio of channel length to valley length, typically ranges from 1.0 (straight) to 3.0 (tortuous). Meandering is a natural tendency of rivers. Practical experience has indicated that meandering channels have a sinuosity greater than 1.3 (Brice, 1983). Variants of meander types can be differentiated on the basis of their sinuosity. The *degree and type of sinuosity* are dictated by the slope of the river (meandering rivers on lower slopes tend to be more sinuous), the texture of the river (rivers in cohesive material tend to be more sinuous that rivers with sand or gravel substrates), and the type or combination of meander growth and shift forms (i.e., extension, rotation, translation, and cutoffs) (see Chapter 5). In some instances, a meandering configuration may be imposed by antecedent conditions. This referred to as passive meandering, and tends to produce irregular meanders. Rivers that continue to adjust their meandering alignment are referred to as active meandering systems (Richards, 1982).

As alluvial rivers flow through their own deposits, their bed and banks are deformable and prone to lateral adjustment. The *lateral stability of the channel(s)* is defined as the capacity of a river to adjust its position in the valley floor trough. Components of lateral stability include meander growth and shift, the degree of braiding and thalweg shift, and tendency towards avulsive behavior (Figure 4.7c). In many instances, the pattern and rate of lateral stability are fashioned by the vertical stability of the channel, especially the potential for bed incision (degradation).

In most instances, lateral stability is directly related to channel sinuosity. The degree to which *bends grow and shift* provides a key criterion for differentiating among meandering river types (Figure 4.7c). Progressive bend extension, translation, or rotation are considered to be more stable forms of adjustment compared to neck or chute cutoffs. The latter variants induce secondary forms of instability by changing river gradient and length.

A wide range of measures has been used to measure braided indices (see Thorne, 1997). In simple terms, the *degree of braiding* can be expressed as

the percentage of channel length that is divided by islands or bars (Schumm, 1985). This records the extent of thalweg shift along a reach (Figure 4.7c). Bedload multichanneled systems are prone to develop a large number of channels that divide around bars, such that they have a high degree of braiding. Channel stability is enhanced when vegetation colonizes the bars and islands develop. For example, many anastomosing rivers have exceedingly stable channels. Brice (1983) describes reaches as "locally braided" if 5–35% of reach length is divided and as "generally braided" if more than 35% is divided. The *character of braiding* is defined by the assemblage and shape of bars and islands within the braid belt.

Some channels have a tendency towards *avulsive behavior*, making them laterally unstable. Avulsion is defined as a wholesale shift in channel position on the valley floor such that a new or secondary channel results. Leedy et al. (1993) and Nanson and Knighton (1996) describe three orders of channel avulsion (Figure 4.7c). First order avulsion is defined as a relatively sudden and major shift in the position of the channel to a new part of the floodplain. Rivers with coarse bedload (sand and gravel), with noncohesive banks and high stream power are susceptible to this form of avulsive behavior. Second order avulsion is defined as a sudden reoccupation of an old channel on the floodplain. Some fine-grained multichanneled networks, including anastomosing variants with flashy flow regimes, display this form of avulsive behavior. Finally, third order avulsion is defined as relatively minor switching of channels, within a braid train, referred to as thalweg shift (Schumm, 1985).

Combining these various measures of channel planform provides a basis to characterize differing forms of alluvial river morphology, as outlined in the next section.

4.7.1 The continuum of alluvial river morphology

Self-adjusting alluvial rivers flow over, deposit, and rework river-borne deposits, and are able to deform their boundaries both laterally and vertically. Continuous genetic floodplains line both channel margins. A continuum of variants of alluvial channel planform can be differentiated, based on measures of sinuosity, number of channels, and lateral stability (see Figure 4.8).

Discriminating functions used to differentiate among channel planform types have been framed in terms of channel slope, discharge, and particle size (see Lane, 1957; Leopold and Wolman, 1957; Henderson, 1961; Osterkamp, 1978; Bledsoe and Watson, 2001). These empirical relationships outline the bounds within which different planform types operate. They provide a useful basis to predict likely adjustments to changes in controlling variables. For example, channels that sit close to a planform threshold may be particularly sensitive to change. As slope increases for a given discharge, channel sinuosity decreases and the number of channels increases, with associated increases in width : depth ratio, sediment load, and sediment caliber. Hydraulic adjustments are marked by increases in flow velocity, tractive force, and stream power. As a result, bedload transport increases and the lateral stability of the channel decreases. In general terms, this gradation of channel planform types reflects a declining energy gradient from braided through meandering to straight and anastomosing variants (Figure 4.8).

Ultimately, however, the failure of discriminant analysis may be attributed to problems of definition and interdependence of the terms used (neither discharge nor slope are independent), and the fact that functions derived in one region, with one particle size range, cannot necessarily be applied elsewhere (Carson, 1984). The continuum concept presented in Figure 4.8 oversimplifies the boundary conditions within which differing channel planform variants are found in nature. In reality, distinctions between morphologic types are fuzzy and complex (Ferguson, 1987; Knighton and Nanson, 1993; Bledsoe and Watson, 2001). A range

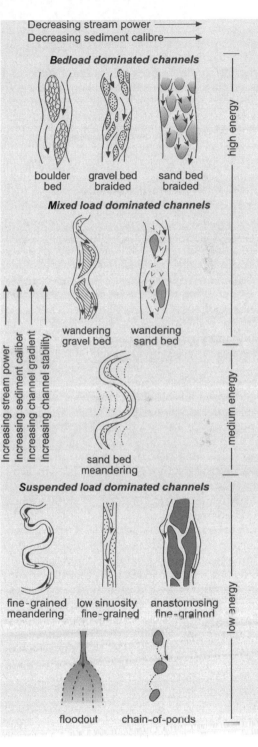

Figure 4.8 The continuum of variants of channel planform (modified from Church, 1992 and Schumm, 1977)
Variants of alluvial river morphology are found along an energy gradient extending from braided rivers through to discontinuous water courses and from bedload through mixed load and suspended load systems. A myriad of intermediary variants can be added to this schematic diagram.

of additional controlling factors affects channel planform, such as riparian vegetation cover, loading of woody debris, antecedent controls, valley alignment/confinement, human disturbance, and off-site impacts. Such extensions are required to explain variants that fall outside conventionally accepted planform types (e.g., discontinuous watercourses), or field examples that do not fit within these general relationships.

Process and historical studies of individual streams suggest that switches between channel planform types may be historically contingent on how intrinsic variables have "primed" a reach for instability and on the state of the channel at the time of impact (Bledsoe and Watson, 2001). For example, Millar (2000) suggested that the meandering-braided threshold for gravel-bed rivers with noncohesive gravel banks is altered by the influence of vegetation on bank stability (i.e., the erosional resistance of the banks). This factor affects the ability of a channel to form a wide-shallow morphology and hence develop braiding tendencies. For given values of discharge and slope, if bank stability is decreased via the removal of bank and riparian vegetation, the meandering-braided threshold is effectively lowered, and braiding is induced along rivers that were once meandering. This relationship also works in reverse with increased forest cover or planting of riparian strips. However, this relationship seems to only hold true for those braided and meandering rivers that are sensitive to change, i.e., they sit close to the meandering-braided threshold. Hence, rivers that sit away from the threshold tend to be relatively resilient to changes in bank and riparian vegetation density. No single threshold function can differentiate among planform variants; rather, a family of threshold curves reflect the sets of conditions within which different rivers operate (Carson, 1984).

In practice, intermediate and transitional forms in the continuum of planform variants are probably more frequently observed than end-member situations (Ferguson, 1987; Thorne, 1997). Dividing lines or threshold conditions that differentiate among planform types are blurred rather than clear. Susceptibility to transformation of channel planform is usually indicated by the range of channel and floodplain forms observed along a reach. If the channel displays only the archetypal

features of a meandering river, it is generally safe to say that it operates some way from notional braided threshold conditions (Thorne, 1997). Because channel planform is frequently transitional, and is shaped by complex sequences of disturbance events, stochastic alternatives to deterministic thresholds may be more useful and realistic for predictive and postdictive work on the effects of planform adjustments (Bledsoe and Watson, 2001).

In addition to these concerns, several philosophical and methodological issues confound reliable appraisals of river morphology framed solely in terms of channel planform. For example, variants of channel planform are commonly not differentiated using discrete criteria. While meandering rivers are defined primarily on the basis of their sinuosity, braided channels are multichanneled (but unstable), and anastomosing river systems are differentiated by their laterally stable multichanneled configurations. Many reaches demonstrate different planform styles at different flow stages; braids can meander, meanders locally braid, and the anabranches of an anastomosing river can be sinuous or straight. Reaches viewed to be intermediate between two or more planform variants cannot *necessarily* be inferred to be an intermediary type of river. Individual channel planform types do not reflect specific geomorphic processes that occur under unique sets of circumstances. Rather, they reflect fluvial adjustment to combinations of inter-related variables, in which limiting factors may impose a particular morphologic response.

Fully adjusting alluvial rivers represent only a part of the spectrum of morphological diversity demonstrated by rivers. Measures used to analyze channel planform are irrelevant to many variants of river morphology, such as bedrock-controlled rivers. Ultimately, the importance of channel planform as a determinant of river type depends on the valley-setting of any given reach. In the sections that follow, form–process associations of differing laterally-unconfined planform types are presented in terms of an energy gradient. These are termed laterally-unconfined rivers rather than fully alluvial, as the role of antecedent sediment stores and bedrock basements may dictate the form–process associations of these rivers to significant degrees. Five key categories of river character are summarized: laterally-unconfined high energy, laterally-unconfined medium energy, laterally-unconfined low energy with continuous channels, laterally-unconfined low energy with bedrock-based channels, and laterally-unconfined low energy with discontinuous channels. These categories, together with their confined and partly-confined counterparts (Section 4.8), are subsequently used as the basis for description of river behavior in Chapter 5.

4.7.1.1 Laterally-unconfined, high energy rivers

Figures 4.9–4.12 provide examples of various laterally-unconfined, high energy rivers. Each type is briefly discussed below.

Boulder-bed rivers Boulder-bed rivers typically occur in areas of local valley widening immediately downstream of confined reaches (Figures 4.9a and 4.10). These high energy settings are character-ized by steep slopes. Floodplains tend to have a convex cross-profile with a fan-like morphology. They thin downstream, as the transport capacity of flows decreases. Single or multichannel systems may develop. These channels tend to have a low sinuosity. They have the propensity to avulse, leaving abandoned channels on the floodplain. Instream geomorphic units include boulder mounds, boulder berms, cascades, rapids, and islands.

Braided rivers Braided rivers are bedload dominated systems in which bars are formed and thalweg shift occurs within a multichanneled configuration (Figures 4.9b and 4.11). When observed in wide alluvial valleys, active channels tend to favor only part of the valley floor at any one time. Avulsion is common. Conditions that promote braiding include moderately steep slopes and flashy discharge regimes that promote high stream

Figure 4.9 **Photographs of laterally-unconfined high and medium energy rivers**
(a) Boulder bed stream (Bemboka River, Bega catchment, New South Wales, Australia), (b) braided river (New Zealand), (c) wandering gravel bed river (Squamish River, British Columbia), (d) meandering river (British Columbia).

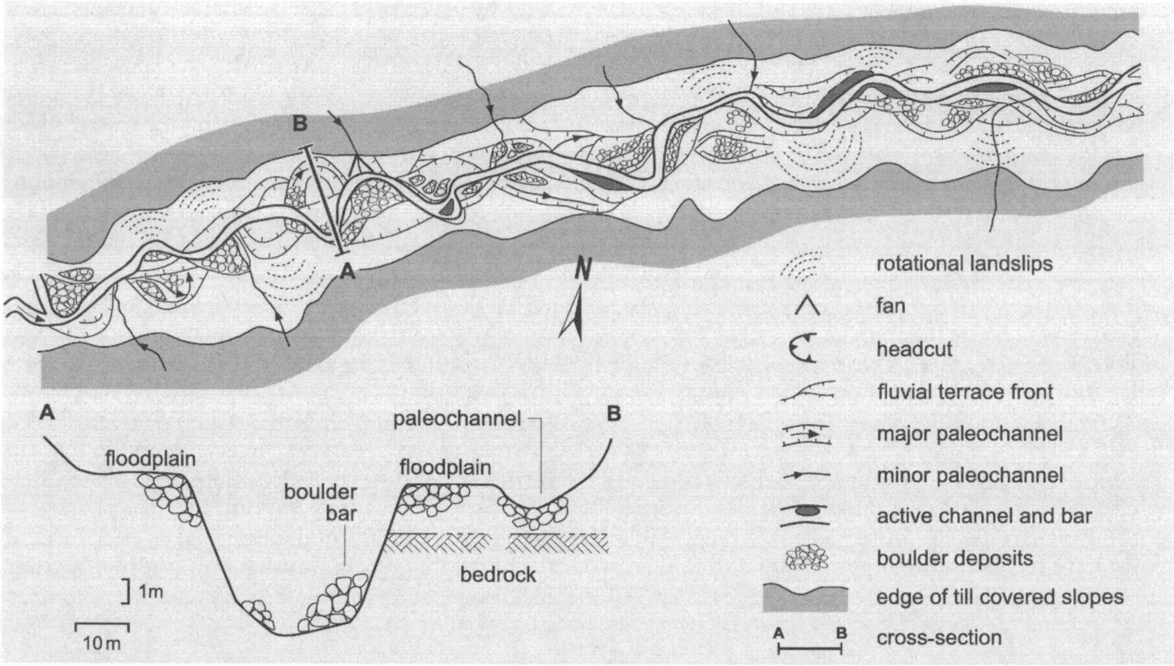
B
A
N
rotational landslips
fan
headcut
fluvial terrace front
major paleochannel
minor paleochannel
active channel and bar
boulder deposits
edge of till covered slopes
A B
cross-section
A
B
paleochannel
floodplain
floodplain
boulder
bar
bedrock
1m
10m

Figure 4.10 Boulder-bed river
Planform and cross-sectional views of a boulder-bed river (from Macklin et al., 1992).

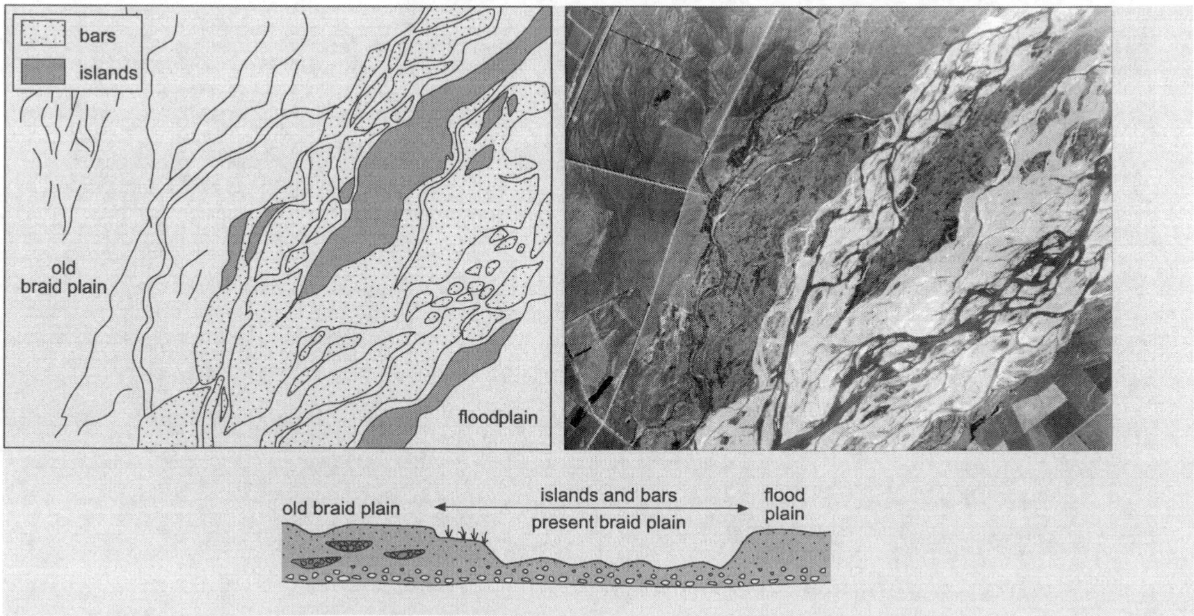
bars
islands
old
braid plain
floodplain
old braid plain
islands and bars
present braid plain
flood
plain

Figure 4.11 Braided river
Planform and cross-sectional views of a braided river (Waimakariri River, New Zealand).

powers, an abundant sediment load that induces high bed mobility, and erodible banks
that enhance the formation of wide, shallow
channels. These rivers tend to be capacity limited
(i.e., the channels cannot transport all sediment
made available to them) or competence limited
(i.e., the caliber of material is too large to be
transported).

The characteristic feature of braided rivers is the
repeated division and joining of channels and the
associated divergence and convergence of flow.
These factors contribute to high channel instability relative to other river types. Bars and islands
separate multiple channels. Bars are transient,
unvegetated, and submerged at bankfull stage,
whereas islands are often stable, vegetated, and
emergent at bankfull stage. Variants of braided
rivers reflect the variability in bed material texture, the level of channel dominance, the degree
of braiding, and the dominance of bars versus
islands.

The floodplains of gravel-bed braided rivers are
formed by migration and avulsion of the primary
braid belt to another section of the valley floor.
This leaves behind areas of accumulated alluvium
comprising multiple stacked bar sequences.
Elevated bars build up in large flood events, and
may become more stable over time. Local channel
incision may result in abandonment and stabilization of adjacent bar surfaces (Reinfelds and
Nanson, 1993). Elevated surfaces may accumulate
overbank fines, typically around vegetation.
Floodplain areas are prone to reworking by lateral
channel shift, avulsion, and reoccupation of old
braid channels.

Sand-bed braided rivers are characterized by
wide, broad floodplains that are reworked by extensive channel widening (e.g., Schumm and
Lichty, 1963; Burkham, 1972). Flow divides
around multiple bars and sand sheets in a wide,
shallow macrochannel. During periods of low precipitation, when the stabilizing effect of vegetation is limited, extreme floods bring about erosion
and channel widening. Vertical accretion, associated with the coalescence of instream islands and
the infilling of secondary channels that separate
the islands from the floodplain, are the primary
mechanisms of floodplain formation. Floodplain
vegetation enhances the accumulation of interbedded sandy units with mud veneers.

Wandering gravel-bed rivers Wandering gravel-
bed rivers (Figures 4.9c and 4.12) are an intermediary form between meandering and braided river
types, and they typically have characteristics of
both. Indeed, they may be observed between these
planform types in a characteristic downstream sequence that reflects a transition in slope and bed
material texture (Brierley and Hickin, 1991). Other
factors that lead to the development of wandering
gravel-bed rivers include coarse sediment inputs
that promote the development of bars and islands
which modify flow alignment, changes in valley
gradient, and periodic formation of log jams
(Church, 1983; Desloges and Church, 1987).
Wandering gravel-bed rivers tend to have fewer
channels and active bars than braided rivers. In
general, a dominant channel can be identified.
Channels are laterally active, with moderately
high width:depth ratios.

In some instances, laterally unstable braided
and anabranching channels are separated by well-
vegetated islands that are sometimes leveed.
Floodplain formation is dominated by lateral
migration of point bars, overbank accretion, and
abandoned channel accretion. Avulsion and
chute channels commonly incise floodplains or
reoccupy paleochannels. In some instances,
nonmigrating gravel-bed channels flow around
well-vegetated islands composed of gravels and
boulders in small, steep basins. Log jams and sediment pulses can induce anabranch formation in
these relatively stable settings (Miller, 1991).

4.7.1.2 *Laterally-unconfined, medium energy rivers*

Variants of laterally-unconfined, medium energy
river types are demonstrated in Figures 4.9, 4.13,
and 4.14. Each type is briefly discussed below.

Meandering rivers Meandering rivers are single
channeled systems with a sinuosity <1.3 and low
width:depth ratios (Figure 4.9d). They tend to
form on low–moderate slopes. Channels have a
relatively low bedload transport capacity (i.e., they
are generally mixed or suspended load rivers).
Channels tend to be wider at the apex of bends,
with an asymmetrical form. Channels migrate
laterally or translate downstream. The rate of
meander shift varies dependent upon the type of

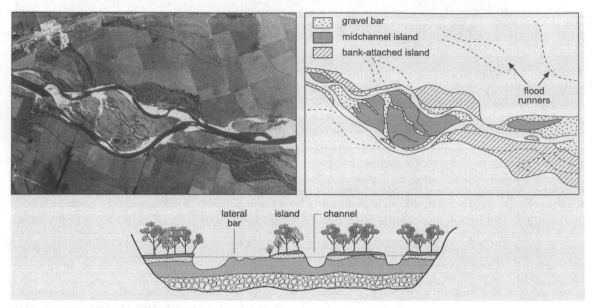

Figure 4.12 Wandering gravel bed river
Planform and cross-sectional views of a wandering gravel bed river. The air photograph is of the Waiau River, New Zealand. Cross-section from Nanson and Knighton (1996), © John Wiley and Sons Limited, 2004. Reproduced with permission.

meandering river and bed material caliber. Bends may develop cutoffs at their necks or grow laterally and shift/translate downstream.

A useful distinction can be made between active and passive meandering (Richards, 1982). Active meandering refers to ongoing bed and bank deformation in self-forming channels (Figure 4.13). Pool–riffle topography is inherently linked to the pattern of bends and crossings in the meandering channel alignment, with pools at bends and riffles at crossings. Riffle spacing (typically five to seven times channel width) is very close to half the meander wavelength (ten to fourteen times channel width), so that there is typically only one pool in each bend, and only one distinct riffle at each crossing (Thorne, 1997). In some instances, however, the meandering process is not fully self-adjusting, such that irregularities are induced to the planform pattern. These instances can be referred to as passive meandering rivers, in which the planform is imposed (Figure 4.14). Fine-grained alluvial rivers commonly exhibit a passive meandering channel alignment. In these cases, there is little evidence for active erosion and the lack of

bedload materials limits the development of point bars. Maintenance of a low channel capacity with cohesive banks minimizes the effectiveness of erosive events, and energy is effectively dissipated at overbank stage.

Meandering reaches of differing sinuosity and textural characteristics may have widely ranging lateral migration rates and associated assemblages of channel and floodplain forms (see Brice, 1964; Church, 1992). The degree to which bends grow and shift provides a means to differentiate among types of meandering river (see Figure 4.8). Differing patterns of geomorphic units provide insight into forms of lateral adjustment. For example, the character of channel movement is recorded by different forms such as paleochannels, cutoff channels, ridge and swale topography, or flat floodplains. Although lateral stability is directly related to channel sinuosity, sinuosity in itself does not provide a measure of lateral stability. Rather, channel sinuosity and the shape of the meander bend influence the form of meander growth. In general terms, meandering reaches on steeper slopes tend to be associated with higher sinuosity and greater lateral

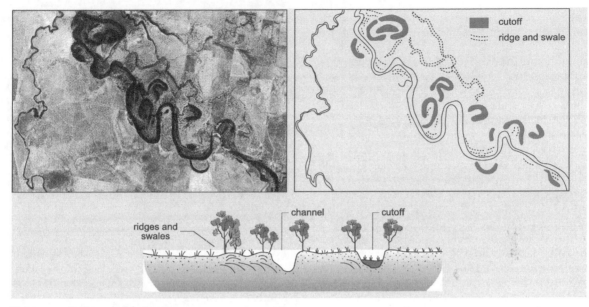

Figure 4.13 Active meandering river
Planform and cross-sectional views of an active meandering river (Murray River, Australia).

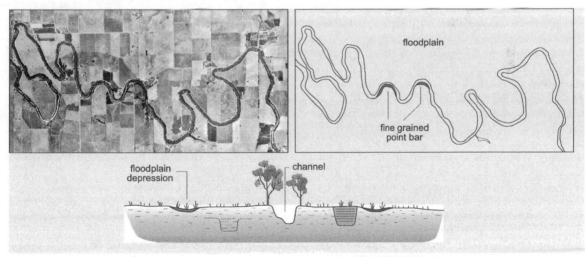

Figure 4.14 Passive meandering river
Planform and cross-sectional views of a passive meandering river (Goulburn River, Victoria, Australia).

instability. Progressive tendencies for bend extension, translation, or rotation are considered to be more stable than neck or chute cutoffs that induce secondary forms of instability, such as changes to floor gradient and length. Rates of bend migration attain a maximum value where the ratio of radius of curvature to channel width ($r_c : w$) approximates 3.0, and decline rapidly for bends with values above and below 3.0 (Hickin and Nanson, 1975, 1984). Increasing rates of bend migration lead to the formation of widely spaced ridges and swales on the floodplain. Discontinuity in rates of bend migration promotes scroll bar development as incipient forms of ridge and swale topography.

Nanson and Croke (1992) differentiated between three main variants of laterally migrating meandering rivers, namely systems with scrolled floodplains, backswamps, and nonscrolled floodplains. Meandering rivers with scrolled floodplains are sand or gravel dominated systems with a characteristic ridge and swale topography. Floodplain scrolls can result from three processes (Nanson and Croke, 1992). Transverse bars may migrate over a point bar that is topped with overbank sediment (Sundborg, 1956; Jackson, 1976). Alternatively, sediments are deposited from suspension within a flow separation cell over a point bar (Nanson, 1980, 1981). These types of scroll bar grow both upstream and downstream, developing a wider spacing as the rate of channel migration increases (Hickin and Nanson, 1975). Finally, undulating ridges and swales are produced when former chute channels that developed between the convex bank and the point bar become incorporated into the floodplain (McGowen and Garner, 1970; Nanson and Croke, 1992; Brierley and Hickin, 1992).

Meandering rivers with backswamp floodplains are characterized by extensive deposits of fine-grained overbank sediment along valley margins (Fisk, 1944, 1947; Farrell, 1987; Woodroffe et al., 1989). Pronounced levees may induce an energy gradient across the floodplain, resulting in reduced rates of vertical accretion with distance from the channel. When flows go overbank, coarser sediments are deposited on the levee crest (Pizzuto, 1987), while finer-grained suspended load materials accrete as flow energy is dissipated over the floodplain. Backswamp, wetland, or floodplain ponds may develop in distal areas. Lateral accretion deposits are confined to the proximal floodplain where there is sufficient energy to rework part of the wide, alluvial plain. This results in a composite floodplain formed by two different sets of processes (Nanson and Croke, 1992).

Meandering rivers with nonscrolled floodplains tend to migrate rapidly when flowing through sandy deposits. Levees are absent. Convex bank deposition builds a point bar that grades into thin, horizontally bedded overbank deposits (Nanson and Croke, 1992). Suspended load system variants are found on low slopes in wide valleys. The lack of bedload material limits the range of instream geomorphic units. Given the cohesive nature of the floodplain sediments, these rivers tend to have low migration rates and floodplains are dominantly built via vertical accretion. Over time, oblique accretion features may slowly build along the convex banks of meander bends. Eventually these features become incorporated into a broad, flat, featureless floodplain. These rivers are prone to avulsion. Meander cutoffs and paleochannels are common.

4.7.1.3 Laterally-unconfined, low energy rivers with continuous channels

Variants of laterally-unconfined, low energy rivers are demonstrated in Figures 4.15–4.17. Each type is briefly discussed below.

Low sinuosity rivers   Low sinuosity or straight channels are generally low energy, suspended load systems with sinuosities <1.3. Given the limited availability of bedload materials, instream geomorphic units are restricted to occasional alternate and transverse bars and various fine-grained depositional forms such as drapes associated with oblique accretion. The high silt–clay content of bank materials promotes low capacity channels with a relatively uniform, deep, and narrow geometry. In these low slope settings, channels are unable to generate sufficient energy to promote bank erosion, and cross-sectional currents are relatively weak. This induces high lateral stability. Floodplains are dominated by vertically accreted silt and clay (e.g., Nanson and Young, 1981; Brakenridge, 1985). Low levees and backswamps may be evident, but in general floodplains are relatively flat.

Sand- and gravel-bed variants of low sinuosity rivers are characterized by alternating lateral and point bars, with occasional midchannel bars and islands and chute cutoffs (Bridge et al., 1986; Bridge and Gabel, 1992) (Figures 4.15a and 4.16). Floodplains comprise abandoned bars and associated features such as scrolls, infilled chute channels, abandoned swampy channels, and narrow levees.

Figure 4.15 Photographs of laterally-unconfined low energy rivers
(a) Low sinuosity river (Bega River, New South Wales, Australia), (b) anastomosing river (Channel Country, southwestern Queensland, Australia), (c) laterally-unconfined bedrock-based river (Sabie River, South Africa), (d) intact valley fill (Barbers Creek, Shoalhaven catchment, New South Wales, Australia), and (e) floodout (western New South Wales, Australia).

Anabranching rivers Particular sets of sedimentary, energy-gradient, and hydraulic conditions promote the operation of river systems as multiple channels separated by areas of floodplain, or as stable vegetated islands or alluvial ridges around which flow divides at discharges up to nearly bankfull. Multiple channels concentrate stream flow and maximize bed-sediment transport (work per unit area of the bed) under conditions where there is little opportunity to increase gradient (Nanson and Knighton, 1996; Jansen and Nanson, 2004). High-energy variants of anabranching river such as wandering gravel-bed rivers have been described earlier. Lower energy variants are stable, multi-channeled, suspended load systems that occur on low slopes. Intervening areas of floodplain or vege-

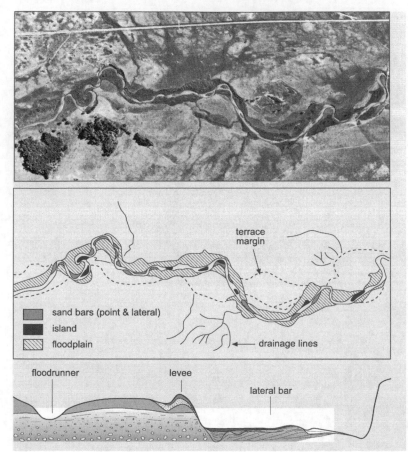

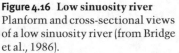

Figure 4.16 Low sinuosity river
Planform and cross-sectional views of a low sinuosity river (from Bridge et al., 1986).

tated islands/ridges are large relative to the size of the channels. As a result, flow patterns in adjacent channel segments are effectively separated. This characteristic sets these rivers apart from braided rivers. Channel capacity is highly variable, and each anabranch may have its own planform. Some channels may meander, braid, or remain relatively straight and eventually rejoin the primary channel. However, given the fine-grained nature of these rivers, the lateral stability of each channel is high, but avulsion is common (Smith et al., 1989; Jain and Sinha, 2004). Mechanisms that promote the development of anabranching channels include erosional excavation of channels within the floodplain via avulsion, and accretional formation of islands and ridges which are subsequently stabilized by vegetation (Nanson and Knighton, 1996).

Multichanneled anabranching rivers with cohesive floodplain are commonly termed anastomosing rivers (Figures 4.15b and 4.17). In these settings, low width : depth ratio channels exhibit little or no lateral migration, but are subjected to changes in flow preference, typically associated with differential rates of channel infilling (a form of avulsion). Floodplains comprise vertically accreted mud and/or organic-rich (peat-like) deposits (Smith and Smith, 1980; Smith, 1983, 1986; Knighton and Nanson, 1993; Nanson and Knighton, 1996). Shallow levees may develop adjacent to the channels. Given the cohesive character of the floodplains, narrow, deep channels are often sinuous but are laterally very stable. Islands that are excised from the floodplain separate channels. Very little bedload is transported in these

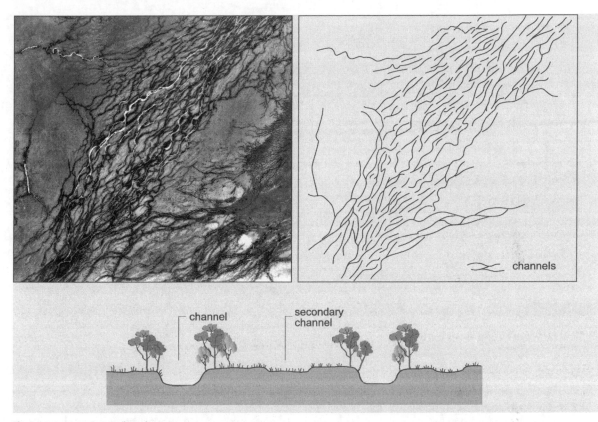

Figure 4.17 Anastomosing rivers
Planform and cross-sectional views of an anastomosing river. Air photograph is of the channel country near Innaminka, southwestern Queensland, Australia. Cross-section from Nanson and Knighton (1996), © John Wiley and Sons Limited, 2004. Reproduced with permission.

suspended load systems. Given their low gradients, anastamosing rivers are found in settings in which unit stream power is very low (<8 Wm^{-2}).

Sand-dominated, island-forming anabranching rivers are characterized by low sinuosity channels that tend to be relatively wide and sometimes contain braided reaches. The channels are dominated by sandy bedload and there is very little fine sediment on the floodplains (Nanson and Knighton, 1996). Vegetation acts as a primary stabilizing force in these noncohesive systems.

Finally, mixed-load laterally active anabranching rivers have meandering, multichanneled systems that migrate laterally across a portion of their floodplain (Nanson and Knighton, 1996). These systems carry a mixed load of sand and mud, and occasionally fine gravel (Nanson and Knighton, 1996; Brizga and Finlayson, 1990). Channel avulsion during major floods forms a laterally active short-lived anabranch. This channel remains relatively inefficient until a higher energy channel is generated.

4.7.1.4 Laterally-unconfined, low energy rivers with continuous bedrock-based channels

In some laterally-unconfined settings, channels may flow within relatively thin veneers of alluvial material atop a bedrock base (Figures 4.15c and 4.18). This situation is especially common in transfer zones of the upper to middle catchment. These low sinuosity bedrock-based channels have

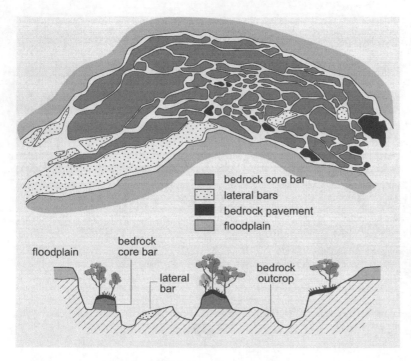

bedrock core bar
lateral bars
bedrock pavement
floodplain

Figure 4.18 Laterally-unconfined bedrock-based river
Planform and cross-sectional views of a laterally-unconfined bedrock-based river (modified from van Niekerk et al., 1999). Planform and cross-sectional views of a laterally-unconfined bedrock-based river. Modified from van Niekerk et al. (1999), © John Wiley and Sons Limited, 2003. Reproduced with permission.

a slot-like (symmetrical) morphology. Depending upon instream roughness, these settings have the capacity to flush bedload caliber materials, while leaving occasional pockets of reworked bedload materials that form shallow bars (typically bank-attached). They tend to be characterized by vertically-accreted fine-grained floodplains. The limited capacity for channel adjustment often promotes the development of a low sinuosity planform. Bedrock steps exert a primary influence on river adjustment. If excess energy is available, channel expansion promotes ledge formation.

In a differing form of transfer reach, bedrock anastomosing rivers may be observed (Figures 4.15c and 4.18). These vertically constrained multichanneled systems are contained within a low sinuosity macrochannel (Heritage et al., 1999; van Niekerk et al., 1999). As the macrochannel is incised into bedrock, it is laterally very stable. Bedrock-based, vegetated ridges (termed bedrock core bars) separate channels. Each anabranch may have a variable morphology, as planform flow paths tend to be related to joint and fracture patterns in the bedrock. As each bedrock anabranch may sit at a different elevation, low flows may be contained within just one or two primary channels. The prominence of bedrock generates considerable local geomorphic and hydraulic diversity. Most bedrock geomorphic units protrude above the water surface at low flow, leaving topographic lows such as backwaters and pools inundated. At high flow stages, the entire macrochannel acts as a flow path. Alluvium is locally deposited atop ridges and in riffles, backwaters, and lee bars. Bedrock distributary channels (anabranches) tend to remain largely sediment-free, as their higher slope and transport capacity promote flushing. The floodplain is rarely inundated given the incised nature of the macrochannel. However, when overbank events do occur, suspended load materials are deposited over the bedrock platform, along the banks, and as a thin veneer beyond the macrochannel. In some settings, long, parallel, sand-dominated ridges separate low width : depth ratio channels (Wende and Nanson, 1998). Vegetation cover has a formative role in the generation and maintenance of these steep-sided ridges. The macrochannel tends to be confined by bedrock, indurated materials, or cohesive mud.

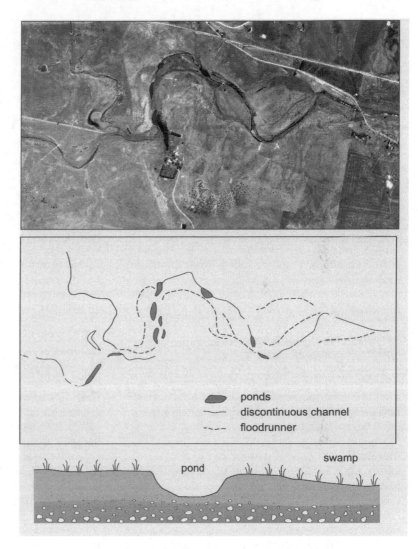

Figure 4.19 Cut-and-fill river
Planform and cross-sectional views of a chain-of-ponds river (Mulwaree River, New South Wales, Australia).

4.7.1.5 Laterally-unconfined, low energy rivers with discontinuous channels

Figures 4.15, 4.19, and 4.20 show examples of laterally-unconfined, low energy rivers with discontinuous channels. Each type is briefly discussed below.

Cut-and-fill rivers with discontinuous channels
Cut-and-fill rivers record spatial and temporal discontinuity of landscape forms and processes. During the fill phase, channels are discontinuous

or absent and unit stream power tends to be very low (< 10 Wm⁻²). Flow energy is dissipated over an intact valley fill surface, resulting in slow rates of vertical accretion (commonly in swamps). When incision occurs (the cut phase), flow is concentrated within the channel and unit stream power increases dramatically. Former phases of channel abandonment may be evident in the bank stratigraphy.

The range of nonchannelized cut-and-fill variants reflects textural characteristics and the character and/or extent of vegetation cover.

Cut-and-fill rivers have been recognized in a wide range of environments. For example, in the humid uplands of southern Australia variants include chains-of-ponds (Figure 4.19), swamplands, intact valley fills (Figure 4.15d), dells, and swampy meadows (Eyles, 1977; Bird, 1982, 1985; Young, 1986; Prosser et al., 1994). Along Mediterranean valleys, equivalent features are called wadis (Vita-Finzi, 1969). In plateau regions of southern Africa, dambos are a form of discontinuous watercourse (Mackel, 1974), while in the semiarid American southwest cut-and-fill systems are termed arroyos (Schumm and Hadley, 1957; Cooke and Reeves, 1976; Graf, 1983a). In general, variants of cut-and-fill river behavior are associated with upper to middle parts of catchments where channels are either absent or discontinuous for a significant proportion of time.

Floodout rivers with discontinuous channels Additional variants of discontinuous watercourse have been described for low-slope settings in semi-arid or arid terrains, where channels may terminate in a floodout or terminal fan (Figures 4.15e and 4.20). In these wide alluvial valleys, the river is unable to maintain a continuous channel or transport its sediment load, and drainage breakdown occurs. Beyond the discontinuous channel, materials "floodout" over the alluvial plain in a fan-like formation. Lobes of sediment accumulate over the valley floor as the location of the floodout shifts over time. Variants of floodout include terminal floodouts where the entire drainage network breaks down (Gore et al., 2000), or intermediate floodouts where anabranching drainage networks terminate prior to reforming downstream (Tooth, 1999).

Floodouts have also been described in temperate environments where bedload materials are deposited atop intact swampy valley fill downstream of discontinuous gullies (Erskine and Melville, 1983; Melville and Erskine, 1986; Brierley and Fryirs, 1998; Fryirs and Brierley, 1998). Bedload caliber materials "floodout" from a gully forming a shallow, fan-shaped feature atop suspended load deposits of the buried swamp.

The continuum of alluvial rivers described in this section outlines end-member situations and is far from complete. Analysis of river character and behavior in any given reach requires appraisal of the form–process associations that characterize geomorphic units along a river, and their interactions (or packages), rather than merely determining which of the variants outlined above is closest to the field situation under investigation (see Brierley, 1996). This approach to analysis of river character and behavior can be extended beyond alluvial variants, as outlined in the following section.

4.8 Valley confinement as a determinant of river morphology

Other than fully self-adjusting alluvial rivers, most rivers flow in valleys in which bedrock exerts some degree of lateral and/or vertical control on their character and behavior. If alluvial deposits on the channel bed are not significantly deeper than the deepest scour holes, such that valley floor slope and structure exert a significant control on channel slope and vertical movement of the channel bed, bedrock confinement represents a major control on river morphology (Kellerhals and Church, 1989). Bedrock rivers lack a continuous alluvial bed and are confined to some degree by valley walls. Hence, bedrock exerts vertical and/or lateral constraints on river forms and processes. Unlike fully alluvial rivers, channel morphology of bedrock rivers reflects interactions between erosive processes and the resistance of the channel substrate (Wohl, 1998).

Valley confinement acts as a primary control on the differentiation of geomorphic process zones along rivers (see Chapter 3). Sediment source, transfer and accumulation zones are influenced by landscape setting and within-catchment position. The capacity for sediment storage or reworking along a reach is influenced by topographic controls such as slope and valley confinement. Bedrock rivers tend to occur in the incisional, degrading parts of landscapes, typically characterized by long-term sediment source or transfer zones. Structural and lithological controls are ubiquitous (e.g., Krzemień, 1999). These reach-scale controls combine with local distributions of forcing elements to impose major constraints on the operation of river forms and processes. In other instances, a particular type of lithology may impose an array of variants of channel morphology, as demonstrated in karstic terrains.

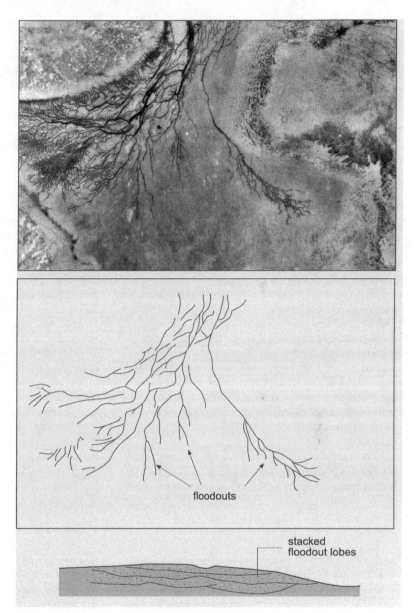

Figure 4.20 Floodout river
Planform and cross-sectional views
of a floodout river (Channel country
near Innaminka, southwestern
Queensland, Australia).

Differences in valley width may control spatial patterns of erosion during geomorphically effective floods. Wide valleys can decrease the peak discharge for a given event, as decreased velocity of overbank flow results in temporary storage of some of the runoff (Woltemade and Potter, 1994). Even for a given discharge, narrow valleys have higher stage, stream power, and shear stresses than

wide valleys (Miller, 1995). These conditions may be reflected in floodplain forming processes and resultant structure (Nanson and Croke, 1992). For example, Ferguson and Brierley (1999a) noted significant variability in floodplain forming and reworking processes in a partly-confined valley, associated with the degree of valley confinement and valley morphology. Significant transitions in instream river character and behavior may accompany changes to valley width (e.g., McDowell, 2001).

A range of schemes has been derived to characterize the role of valley confinement as a control on river morphology. In general, these schemes are based on the distribution of genetic floodplain along river courses (e.g., Lewin and Brindle, 1977; Schumm, 1985; Rosgen, 1994, 1996; see Figure 4.21). Confined or entrenched rivers do not have genetic floodplains, as their channel margins comprise bedrock or ancient alluvium (i.e., terraces) (Figure 4.21). In effect, the entire valley floor acts as a channel. In these settings, the entrenchment ratio, defined as the ratio of flood prone width (i.e., width of the valley over the genetic floodplain) to bankfull channel width, lies between 1.0–1.4 (Rosgen, 1994, 1996).

Given their steep slopes, confined rivers in mountain settings tend to have high flood power and high sediment transport capacities. Channels are strongly coupled to adjacent hillslopes, which act as major sources of sediment. The exposure of bedrock on the channel bed reflects high transport capacity relative to sediment supply. The mobile bedload or suspended load fraction is prone to flushing. Large particles that line the bed exert a significant influence on river character and behavior.

Confined rivers often have deep, narrow cross-sections which encourage the high depths and velocities associated with macroturbulent flow. Given their inherent resistance, such channels change little other than during catastrophic events when peak velocities and depths exceed threshold values required for macroturbulence to develop, such that cavitation occurs (e.g., Baker and Costa, 1987; Kochel, 1988). Under these conditions, bed configurations may bear a striking resemblance to the riffle–pool morphology more commonly associated with alluvial rivers (Baker and Pickup, 1987; Wohl, 1992).

Rivers with discontinuous pockets of genetic floodplain are vertically confined such that bedrock is a key determinant on bed morphology. The extent of lateral confinement influences the potential for floodplain pockets to form (Figure 4.21). Rosgen (1996) defines moderately entrenched rivers as reaches with an entrenchment ratio between 1.41 and 2.2. In the River Styles framework, these are termed partly-confined valleys (Brierley et al., 2002). Differentiation of river types in these settings reflects the position of the channel relative to the valley margin, indicating how often and over what length of river course the channel impinges on the valley margin. In some settings, for example, rivers are confined by terraces and/or bedrock, limiting available space in which discontinuous pockets of genetic floodplain may form. If the width of the genetic floodplain is less than the amplitude of the meandering channel, boxed, or sinusoidal patterns result, in which a meandering channel crosses from one confining margin to another, resulting in discrete pockets of floodplain (Lewin and Brindle, 1977). These valleys tend to be relatively sinuous. The dominant channel adjustment processes include lateral growth, loop extension, downstream translation, and stabilization against confining media (Lewin and Brindle, 1977). In valleys with a more regular alignment, the channel may hug a confining bank for some distance and then suddenly shift to the opposite valley margin, creating discontinuous pockets of floodplain. In these cases, the planform of the river is more readily able to adjust but is pinned by fine-grained floodplains and bedrock valley margins (Brierley et al. 2002).

As valley morphology is a primary influence on the distribution of unit stream power along river courses (e.g., Graf, 1983b; Leece, 1997; Knighton, 1999), it exerts a key control on the transport and deposition capacities within partly-confined valleys. In addition, valley alignment influences the distribution of energy at differing flow stages,

Figure 4.21 Valley confinement as a control on river morphology
Valley confinement records the extent to which the channel abuts the valley margin. Three degrees of confinement are indicated. In each instance, the extent to which the channel abuts the valley margin reflects the alignment of the channel on the valley floor (in part this is determined by the alignment of the valley itself) and dictates whether floodplains are absent (a), discontinuous (b), or continuous (c) (see text).

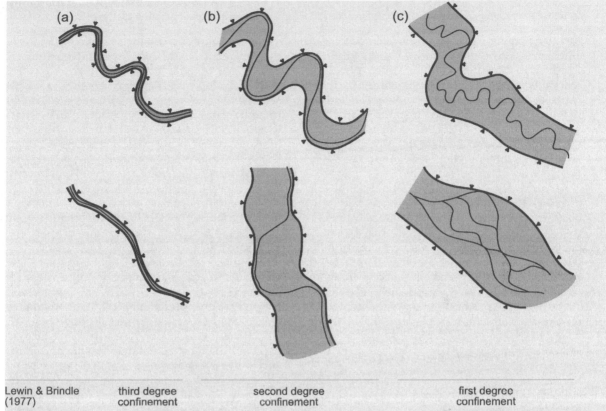

Lewin & Brindle (1977)	third degree confinement	second degree confinement	first degree confinement
Kellerhals & Church (1989)	entrenched	–	alluvial
Schumm (1985)	confined	semi- controlled	alluvial
Rosgen (1994, 1996)	entrenched	moderately entrenched	alluvial
Brierley et al. (2002)	confined	partly - confined	laterally - unconfined

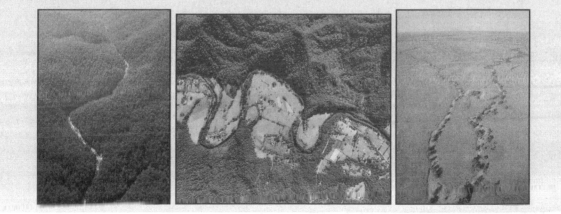

thereby shaping patterns of sediment deposition and reworking along valley floors and on floodplains (Magilligan, 1992; Miller, 1995). Remarkable pocket-to-pocket variability in floodplain sedimentology and structure may be evident, in which individual pockets comprise distinct assemblages of geomorphic units dictated largely by downstream changes in valley configuration and flow alignment over floodplain surfaces (e.g., Ferguson and Brierley, 1999a, b).

An array of river types has been described for rivers with occasional or discontinuous floodplain pockets in confined or partly-confined valley settings. As noted for laterally-unconfined channel planform types, these systems form a continuum, with considerable overlap among variants. In general terms, the energy gradient reflects a combination of slope and valley confinement factors that shape patterns of deposition and the potential for reworking of alluvial stores along river courses.

4.8.1 Confined valley-setting rivers

Variants of confined rivers are shown in Figures 4.22 and 4.23. These rivers types are briefly described below.

4.8.1.1 *Steep headwater rivers or mountain streams*

Mountain streams on steep slopes flush all finer-grained materials, such that large boulders line the bed. However, the stream is only able to mobilize these boulders during extreme flows. Erosion along bedrock-based channels occurs by flaking of rock fragments, block quarrying, and longitudinal groove and pothole development (Wohl, 1998). Although slopes are strongly coupled, rates of sediment supply are highly variable, resulting in irregular patterns of bed adjustments. The short length of slopes in upland watersheds ensures that hillslope materials are often delivered directly to channels, resulting in high sediment delivery ratios. Mass wasting from valley walls and debris flows from tributaries have a stronger influence on channel form in narrow-valley segments than in wider valleys, often adding boulders and woody debris that constrict the channel (Grant and Swanson, 1995).

In general, steep headwater channels have a low width : depth ratio and tend to develop a stepped-bed morphology characterized by fall, cascade, and step–pool sequences (Grant et al., 1990; Abrahams et al., 1995; Montgomery and Buffington, 1997, 1998). The assemblage of these features is dictated by local variability in channel slope and the distribution of forcing elements. Pool size and spacing are determined primarily by the pattern of falls and steps. Cascades, riffles, and runs perform a similar function in situations where coarse basal materials line the channel bed.

Given their high transport capacity relative to sediment supply, most steep headwater rivers function as sediment transport zones that rapidly convey sediment downstream (Montgomery and Buffington, 1997). Sediment supply, lithology, woody debris, and local slope have significant effects on the ability to discriminate bedrock

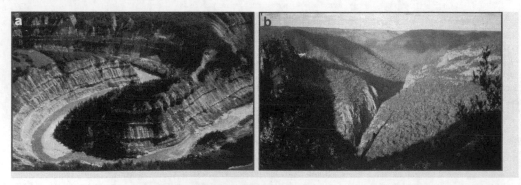

Figure 4.22 Photographs of rivers in confined valley settings
Variants of rivers found in confined valley settings (a) gorge (Colorado, United States), (b) gorge (Shoalhaven River, New South Wales, Australia).

(a) Gorge

(b) Confined valley with occasional floodplain pockets

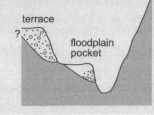

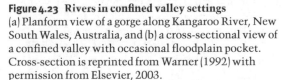

Figure 4.23 Rivers in confined valley settings
(a) Planform view of a gorge along Kangaroo River, New South Wales, Australia, and (b) a cross-sectional view of a confined valley with occasional floodplain pocket. Cross-section is reprinted from Warner (1992) with permission from Elsevier, 2003.

from alluvial reaches within headwater settings (Montgomery et al., 1996; Massong and Montgomery, 2000). Some of the factors affecting sediment supply are highly variable in time, controlled features such as sediment-trapping woody debris and recent landslides (Hogan et al., 1998). Given the episodic delivery of sediment from debris flows and generation of sediment waves, headwater basins may be sediment starved most of the time (Benda and Dunne, 1997a, b). However, pulsed sediment input does not necessarily cause pulsed output (Lancaster et al., 2001). Rather, small basins may absorb the sporadic and abrupt inputs of sediment from debris flows and disperse the material in place (Lisle et al., 1997, 2000). As a result, output may produce a relatively smooth fluvial sediment transport signal, as indicated by Massong and Montgomery (2000).

Along more open channels in which bedrock has a local-forcing rather than a dominant role on channel morphology, a range of instream geomorphic units may be evident, including an array of bar types. In some cases, plane-bed channels may develop, comprising well-sorted cobbles or a homogenous bedrock surface. These rivers are commonly straight with a rectangular to trapezoidal cross-sectional form and low width : depth ratios. Channels with plane bed features may function as either sediment source or sediment storage zones depending on the flow : sediment ratio, the degree of bed armoring, and frequency of bed mobility (Montgomery and Buffington, 1997). This reflects sediment supply- and transport-limited conditions respectively.

4.8.1.2 Gorges or canyons

Gorges and canyons typically form following headward retreat of large knickpoints through a landmass, creating a deep, narrow valley with walls that are only slightly modified by slope processes (Nott et al., 1996; Seidl et al., 1996) (Figure 4.22 and 4.23a). Many gorges preferentially erode weakened rock along joint planes. While slot canyons tend to have straight channels, some gorges have a superimposed meandering outline. The valley floor slope typically comprises alternating pools, rapids, and steps. Plunge pools form as scour features downstream of steps, waterfalls, or localized inputs of coarse sediment from hillslopes or small tributaries. Rapids comprising coarse boulders form from local influx of materials by tributaries or mass movement. These features exert a clear impact on the longitudinal profile. Bedrock steps may represent secondary knickpoints that act as local base level controls. If subjected to significant sediment supply, beds can be temporarily characterized by bars or stores trapped behind forcing elements or flow obstructions such as woody debris.

4.8.1.3 Confined rivers with occasional coarse textured floodplains

Flash flooding in steep, bedrock-confined valleys may mobilize very coarse bedload and induce catastrophic erosion (e.g. Stewart and LaMarche, 1967; Baker, 1977). These valleys tend to comprise bedrock steps, scour pools, and runs, with occasional bars or thin, shallow floodplain pockets. The latter features are typically composed of poorly sorted gravels and boulders. These are preserved behind bedrock spurs or accumulations of woody debris (Figure 4.23b). Floodplain pockets may have coarse grained levees, sand and gravel splays, chute channels, scour holes, and abandoned channels covered by thin overbank deposits of fine alluvium (Nanson and Croke, 1992). Floodplains formed by a combination of lateral, vertical, and abandoned channel accretion are prone to stripping or reworking by chute cutting or channel avulsion. Abandoned channels can be infilled by coarse sediments (Baker, 1977).

4.8.2 Partly-confined valley-setting rivers

Various end-member situations of river types in partly-confined valleys are shown in Figures 4.24 and 4.25, and are discussed briefly below.

4.8.2.1 Partly-confined rivers with vertically accreted sand and silt floodplains

Floodplain pockets preserved behind bedrock spurs in meandering or irregularly-shaped partly-confined valleys comprise vertically accreted sand and silt materials (Nanson, 1986; Ferguson and Brierley, 1999a). These may be gravel-based rivers where lag materials line the valley floor and flood-plains are largely silty (e.g., Nanson, 1986), or sand-bed variants with sand and silt floodplains (e.g., Ferguson and Brierley, 1999a) (Figures 4.24a, and 4.25a and c). As the channel impinges against bedrock along much of its course, it is laterally stable. Large compound point bars and point benches characterize the inside of bends, with bedrock induced pool–riffle sequences along the bed. Floodplains are dominantly flat, but accentuated floodchannels, levees, and crevasse splays may be observed (Ferguson and Brierley, 1999a). During high magnitude, low frequency events, catastrophic erosion may strip floodplain pockets down to the basal gravel lag (Nanson, 1986). This periodic destruction may reflect progressive build up of a levee that increasingly concentrates stream power within the channel zone during high energy flood events, prior to extreme events that strip the floodplain.

In valleys with a more regular alignment, low sinuosity channels may be pinned along one valley margin before suddenly shifting to the opposite valley margin. These planform-controlled variants of partly-confined rivers also have vertically accreted silty floodplains (Brierley et al., 2002). The cohesive nature of floodplain sediments ensures that the channel remains pinned in place. Floodplains tend to be relatively flat, but shallow levees may line the channel margin. Occasional floodchannels may short-circuit floodplain

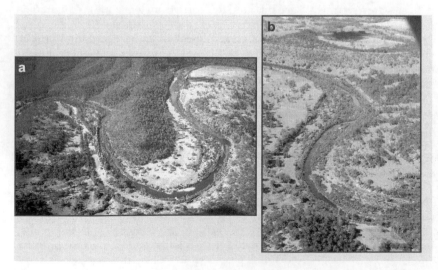

Figure 4.24 Photographs of rivers in partly-confined valley settings
(a) A vertically accreted sand and silt floodplain variant, and (b) a laterally accreted sandy with ridge and swale topography floodplain variant. Both examples are from the Clarence Catchment, NSW, Australia.

pockets, but stripping does not tend to occur. Bedrock-induced pools and midchannel bars may be evident.

4.8.2.2 Partly-confined rivers with laterally accreted sandy floodplains

In sinuous valleys flowing through easily erodible sedimentary rocks, discontinuous floodplain pockets form between bedrock spurs along the convex margin of the valley (Figures 4.24b and 4.25b). These have been referred to as ingrown meandering rivers (Brakenridge, 1985). Asymmetrical valley cross-sections have steep bedrock cliffs on the outsides of bends and gently sloping bedrock slip faces with point bar and point bench deposits on the insides of bends (Brakenridge, 1985; Ferguson and Brierley, 1999a). Bedrock-lined pools and riffles characterize the channel zone. As these rivers migrate laterally, they leave bedrock ledges that represent the former position of the channel along the convex banks of bends. Alluvium deposited over these ledges forms a series of overlapping lateral accretion deposits, point benches, and levees. Occasional floodchannels may short-circuit the floodplain pocket along swales. In some instances, where bedrock is less of an influence on bed morphology, ridge and swale topography may record the meander pathway down and across the valley floor, developing from an initial low sinuosity course to a pattern of arcuate forms (Ferguson and Brierley, 1999a). As the channel incises and translates or moves laterally, progressively higher surfaces are sometimes abandoned as terraces. When the bend eventually impinges on the concave or downstream valley margin, migration ceases and floodplains can be transformed into vertically accreting forms.

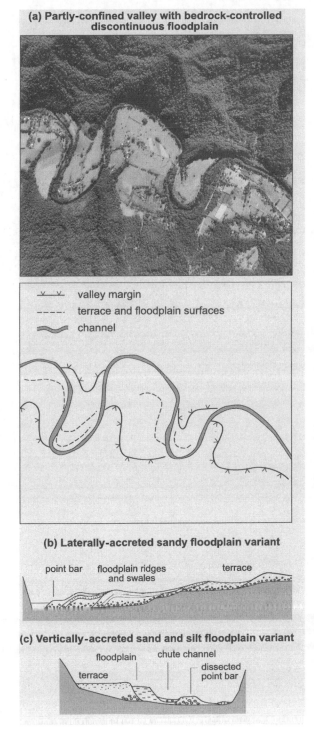

(a) Partly-confined valley with bedrock-controlled discontinuous floodplain

ᵛ ᵛ ᵛ   valley margin

- - - - -   terrace and floodplain surfaces

~~~~   channel

(b) Laterally-accreted sandy floodplain variant

point bar   floodplain ridges   terrace
            and swales

(c) Vertically-accreted sand and silt floodplain variant

floodplain   chute channel

terrace   dissected
              point bar

Figure 4.25 Variants of rivers found in partly-confined valley settings
Planform and cross-sectional views of various partly-confined river types. (a) Planform map of a vertically accreted discontinuous floodplain river (Bellinger valley, NSW). (b) Cross-section of a vertically accreted sandy floodplain (modified from Brakenridge (1984). (c) Cross-section of a vertically-accreted sand and silt floodplain. Reprinted from Warner (1992) with permission from Elsevier, 2003.

4.9 Synthesis

In this chapter, a hierarchy of inter-related attributes has been shown to produce a remarkable diversity of river character. Seven primary categories of river morphology have been described, reflecting the continuum from confined through partly-confined to laterally-unconfined rivers. This continuum reflects the relative balance of erosional and depositional processes in any given setting. However, this balance is not time invariant, resulting in pronounced variability in the extent and frequency with which different types of river adjust their morphology. Indeed, how a river behaves is more important for management applications than descriptions of how a river looks. Insights into river system dynamics and associated behavioral regimes are appraised in Chapter 5.

CHAPTER 5

River behavior

The Northwest's great river, Celilo Falls revealed, is a convection, not a collection; purest verb, not noun . . .

David James Duncan, 2001, p. 6

5.1 Introduction: An approach to interpreting river behavior

Rivers are never static. Disturbance is the norm, as reaches constantly adjust one way or another. Indeed, regeneration and replenishment are critical components in the maintenance of vibrant aquatic ecosystems. Natural disturbance events, such as thunderstorms, cyclones, fires, tectonic uplift, volcanic activity, etc., and associated landscape responses, vary markedly from place to place. River adjustments reflect cumulative responses to recent events, lagged responses to previous events, and off-site responses to events that occurred elsewhere. Responses to human disturbance, whether advertant or otherwise, lie atop this natural diversity. The site-specific nature and timing of these factors, and patterns of connectivity within any catchment, result in pronounced diversity of river responses to past and ongoing disturbance events. River management strategies that strive to "work with" the inherent natural variability of river systems, respecting the inherent diversity of river forms and processes and their capacity to change, must recognize that nonlinear nature adjustments are a functional and desirable part of longer-term evolutionary trends.

In this chapter, a conceptual approach to appraisal of system dynamics is developed. The approach considers what attributes of a reach are able to adjust, the timeframe over which adjustments take place, and the geomorphic consequences of those adjustments. An assessment is made to determine whether ongoing adjustments are part of the "expected" behavioral regime for that type of river, such that the reach sustains a characteristic morphology and associated set of process attributes, or

whether a fundamental shift in form–process associations is underway and the reach is evolving to a different type of river. Marked differences in types, patterns, and rates of geomorphic adjustment are expected for differing reaches in differing settings. River behavior is considered in this chapter, while river change is analyzed in Chapter 6.

River behavior is defined as *adjustments to river morphology induced by a range of erosional and depositional mechanisms by which water moulds, reworks, and reshapes fluvial landforms, producing characteristic assemblages of landforms at the reach scale.* Reach behavior is considered over timeframes in which boundary conditions have remained relatively uniform, such that flow and sediment load inputs and outputs are near consistent, and a characteristic set of attributes is maintained. In making these assessments, system responses to differing forms of disturbance are appraised, outlining the capacity for adjustment of the reach and associated recovery times.

Any sense of system dynamic requires an appreciation of system evolution. Evolution takes a multitude of forms, operating at rates that range from the catastrophic to the inexorably slow. Physical consequences are manifest in different ways in different systems. In differentiating behavior from change, timeframes of adjustment must be appraised in a landscape- or catchment-specific manner. Critical to this assessment is a sense of the natural or inherent stability of the system. A stream is considered to be unstable if it exhibits abrupt, episodic, or progressive changes in position, geometry, gradient, or pattern that are anomalous or accelerated (Shields et al., 2003). Highly dynamic streams are considered to be geomorphologically stable if consistent patterns and rates of behavior are demonstrated over the medium–long term (i.e., hundreds or thousands of years). For example, a stream may experience rapid rates of lateral migration and bank retreat, but these attributes may be "natural" or "expected" for the setting.

A practical approach to differentiation of behavior from change must be generic, such that it can be applied to any given type of river in any given setting. It must also be flexible, enabling relevant geomorphic attributes to be appraised for the system under investigation. For example, river adjustments in a gorge are quite different to those experienced along an anastomosed alluvial river. From this premise, analyses of river behavior and change are framed in terms of the boundary conditions within which any reach operates. Imposed boundary conditions are determined by the landscape setting, reflecting the regional geology and topography (valley width, slope, and relief). These considerations, in turn, fashion the volume and caliber of material that is made available to the river and the way in which energy is utilized along river courses. These interactions control the distribution of processes that erode, transport, and deposit materials.

The landscape setting determines the *potential range of variability* of any reach, which summarizes the range of process activity that is possible for that setting. Variability in the operation of biophysical fluxes enables the river to adopt a range of morphologic variants. Hence, river form adjusts within a particular range, influenced by contemporary *flux boundary conditions* and historical considerations. Collectively, these factors determine the *natural capacity for adjustment* of the river, the bounds of which are set by the potential range of variability for that setting (Section 5.2). At any given time, the range of behavior may reflect a small or large proportion of the potential range of variability. Hence, any reach could demonstrate a range of river types within the imposed boundary conditions. These notions are conceptualized in a "river evolution diagram" (Section 5.3).

The natural capacity for adjustment is defined as *morphological adjustments of a river in response to the changing nature of biophysical fluxes that do not bring about a wholesale change in river type, such that the system maintains a character-* *istic state (i.e., morphology remains relatively uniform in a reach-averaged sense).* In other words, river adjustments are inevitable as biophysical fluxes are altered in response to modifications in impelling and/or resisting forces, but the reach-scale configuration of geomorphic attributes is maintained. The prevailing set of biophysical fluxes determines the likelihood that a characteristic river morphology will be maintained over any given interval of time. If a reach is subjected to a significant change in biophysical fluxes or other boundary conditions, such that a wholesale shift in the capacity for adjustment of a river brings about a different set of form–process relationships, *river change* is said to have occurred. Appraisal of what drives the type, pattern, and rate of river change is considered in terms of environmental change and natural disturbance events in Chapter 6, and in response to human impacts in Chapter 7.

Just as there are multiple attributes of river morphology and incredible diversity in the range of river character and behavior, as outlined in Chapter 4, differing morphological attributes are able to adjust in differing settings (Table 5.1). The likelihood that adjustments will take place varies for different types of river. This reflects the *degrees of freedom* of the river. Each degree of freedom (bed character, geomorphic units, channel morphology, and channel planform) records the ability of a certain component of the river system to adjust (Hey, 1982; Kondolf and Downs, 1996; Montgomery, 1999). In many instances, forms of adjustment are mutually inter-related. For example, changes in channel morphology have a direct impact on forms of floodplain adjustment and resultant planform. Deriving a coherent framework with which to interpret and explain the range of river forms and adjustments in any given setting, referred to as the behavioral regime of a reach, is the primary aim of this chapter.

The key tool that is used to interpret river behavior in this book is analysis of reach-scale assemblages of channel and floodplain geomorphic units. Distinct assemblages of channel and floodplain geomorphic units provide key insights into river character *and* behavior at the reach scale. They reflect both contemporary form–process associations, as flow stage relations produce and rework differing features, and reach history, as interpreted

Table 5.1 Scales of river adjustment.

Feature	Spatial scale (m)	Nature of adjustment	Timeframe of adjustment (years)
Bed material organization and sedimentary bedforms	$10^{-1}-10^1$	Adjustment in grain size and/or distribution, and the associated nature and pattern of hydraulic features such as ripples, dunes, particle clusters, etc. This may reflect dissection and reworking of sand/gravel forms, infilling of pools, patterns of scour, development of an armor layer, and local headcuts.	$10^{-1}-10^1$
Geomorphic units	10^0-10^3	Adjustments to the presence/absence, abundance, and distribution of channel and floodplain forms such as bar types, pools, and riffles, levees, backswamps, etc.	10^0-10^2
Channel geometry	10^0-10^3	Adjustments to the nature, pattern, and/or rate of erosion and deposition on the channel bed or banks are marked by modifications to the pattern of instream geomorphic units, thereby bringing about alterations to channel capacity, shape, and width : depth ratio. Adjustments may include: • Bank erosion that promotes channel migration or expansion. • Bench or ledge formation (reflecting channel contraction and expansion respectively). • Channels may degrade, aggrade, widen, shift at both banks, or shift laterally. • Altered channel–floodplain relationships, related to adjustments to channel geometry.	10^0-10^2
Channel planform	10^2-10^6	Adjustments in the ability of the channel(s) to shift position on the valley floor. This may be exemplified by alterations to channel multiplicity, channel alignment (i.e., sinuosity, meander pattern or wavelength, bend radius of curvature), lateral stability of channel(s), or floodplain character (as measured by the assemblage of floodplain geomorphic units). These modifications are marked by the presence of active cutoffs, floodchannels, crevasse splays, sand sheets, avulsion channels, floodplain stripping, etc.	10^1-10^3

from patterns of reworking and the nature/distribution of remnant features, such as ridge and swale topography, abandoned channels, or terraces. While the former issue is a critical concern in assessment of river behavior (this chapter), the latter issue guides interpretation of river change (Chapter 6).

The nature and rate of river adjustments vary at different spatial and temporal scales (Table 5.1). Observed patterns of small-scale bedforms are determined primarily by conditions experienced during the most recent flow event, almost regardless of its magnitude, as these transient features are readily reworked. The resulting surface expression reflects the waning stage of the last bed deforming flow. At a coarser scale, geomorphic units are reworked and evolve over longer timeframes. In some instances these features may be destroyed, but the reach-averaged assemblage of forms remains roughly consistent over time. Viewed in this way, river *behavior* equates to adjustments around a characteristic assemblage of geomorphic units over timeframes of tens to hundreds of years. Local redistribution of erosional and depositional processes modifies channel geometry, but a reach-averaged morphology is retained. Similarly, local adjustments to channel planform may ensue, noted by alterations to channel multiplicity, bend migration, occasional cutoff development, modifications to patterns of floodplain deposition and reworking, etc. However, in terms of river behavior, the suite of morphological attributes along a reach is roughly equivalent over timeframes of tens to thousands of years. River adjustments at the planform scale alter the distribution and extent of geomorphic units, but the range of units observed along the reach remains near-consistent. As such, the characteristic river structure and function is retained. For example, localized planform adjustments, such as the formation of a few cutoffs along a meandering river, do not change river structure and function at the reach scale, and are considered to be part of the natural capacity for adjustment. Channel geometry may be locally modified, but systematic reach-scale adjustments are unlikely. Indeed, reach-scale changes to channel geometry may reflect the condition of the reach, rather than a fundamental change to the type of river. Similarly, adjustments to bed material organization and bedform-scale features reflect recent flow events, rather than being indicative of changes to the behavioral regime of a reach.

By definition, adjustments to broader scale attributes highlighted in Table 5.1 bring about modifications at smaller scales, whether in terms of their nature/extent, or their pattern/distribution. Indeed, these adjustments are an integral part of the behavioral regime of the river, and associated notions of naturalness. However, if fundamental shifts in river structure and function at the planform scale mark a discernible alteration to the assemblage of geomorphic units, such that a *change* occurs to the type/pattern of geomorphic units and channel geometry, a new river type results. For example, if a braided river is transformed into a meandering river, or the sinuosity of a meandering river is reduced from 2.2 to 1.3 leaving a series of cutoff channels, adjustments to river structure and function are accompanied by changes to the types of geomorphic units found along the reach and the resulting channel geometry. These changes are often accompanied by alteration to channel bed slope. The resulting arrangement of the river has a modified balance of erosional and depositional forms. The altered river structure modifies the distribution and extent of flow energy at differing flow stages, resulting in differing proportions of bedforms and patterns of bed material organization. For example, a smoother channel with a straighter alignment and less roughness may promote the development of higher energy bedforms relative to the previous channel geometry.

River behavior is controlled by the balance of sediment supply and the relative energy that is available to transport or deposit that material. This balance is influenced by the tectonic setting within which a river operates, and the climatic regime. Tectonic setting is a key control on the landscape setting (i.e., the relief and slope) and sediment supply. The climatic regime controls discharge variability and the nature of vegetation cover. In many instances, river character and behavior are shaped by antecedent controls or landscape history, such as inherited (geological) controls on slope and valley width (topography/relief), or patterns/volumes of sediment stores deposited in the past (e.g., reworked glacially derived materials). Elsewhere, reaches are adjusting to off-site impacts or lagged responses to disturbance, or virtually instantaneous responses to a major flood event. Of key concern in management terms is determination of situations in which the manner and rate of behavioral attributes that shape river morphology are "expected" given the particular setting.

River behavior adjusts to any factor that changes the boundary conditions under which rivers operate. Landscape forming events may be recurrent and sustained (e.g., in monsoonal climates) or irregular, chaotic, and unpredictable (e.g., in arid settings). If a system is close to a threshold condition, seemingly small perturbations may provoke profound responses. Elsewhere, negligible responses to disturbance events may reflect inbuilt resilience of the system. Patterns and rates of channel adjustment vary in different environmental settings. For example, channels in sparsely vegetated semiarid catchments are relatively unstable, not prone to display characteristic forms, are likely to be subjected to significant adjustment during extreme floods. As such, there is considerable variability in the certainty with which patterns and rates of morphological adjustments can be predicted in differing settings.

The approach to river analysis adopted in this book merges top-down thinking, framed in terms of within-catchment position and landscape setting, with bottom-up thinking, framed primarily as a constructivist approach to analysis of rivers that operates at the geomorphic unit scale (Brierley, 1996). Three scales of reference are considered in this chapter, building on equivalent discussion in Chapter 4. Analysis of river dynamics commences with appraisal of bedform-scale adjustments in response to textural and flow energy relationships over timescales of 10^{-1}–10^1 years (Section 5.4). This is followed by analysis of channel shape, as interpreted by lateral and vertical adjustments along the banks and bed respectively (Section 5.5). Packages of instream geomorphic units provide critical insights into channel behavior over timescales of 10^0–10^2 years (Section 5.6). In general terms, these forms vary along an energy gradient reflecting reach slope (i.e., available energy) and valley confinement (which determines the capacity of the channel to adjust its position and geometry). Broader reach-scale attributes of river behavior are outlined in Section 5.7, where adjustments at the planform scale are presented. This highlights the myriad of ways in which some channels are able to adjust across (or along) the valley floor over timeframes of 10^1–10^3 years. Analysis of floodplain geomorphic units highlights how valley confinement induces differing capacity for adjustment and associated diversity of floodplain types in confined, partly-confined, and laterally-unconfined settings. Scales of river adjustment, and the unifying theme offered by analysis of geomorphic units across the spectrum of river types, are discussed in Section 5.8. Findings from this chapter are summarized in Section 5.9.

5.2 Ways in which rivers can adjust: The natural capacity for adjustment

The diversity of boundary conditions under which rivers operate, along with the continuum of flow, sediment caliber, slope, and vegetation associations, ensure that there is considerable variability in what attributes of river morphology are able to adjust and how readily adjustments can occur for different types of river (Table 5.1). In this book, this is referred to as the *natural capacity for adjust*

ment. Reaches in different valley settings are able to adjust their morphology in quite different ways (Table 5.2). Rivers that have a significant natural capacity for adjustment can readily modify their bed character, channel morphology, geomorphic unit assemblage, and channel planform. These systems are able to respond quickly to relatively small triggering events, and are considered to be sensitive to adjustment. For example, laterally-unconfined rivers have significant capacity to rework and mold sediments stored on the valley floor (e.g., sand-bed alluvial rivers). Rivers with limited natural capacity for adjustment may not elicit a morphologic response to a perturbation. Reaches that are able to absorb the impacts of disturbance events are considered to be resilient to adjustment. For example, rivers in confined valley-settings have limited capacity to adjust their bed character, channel morphology, and planform given their imposed bedrock character. Thus, a resilient river would adjust only slightly in response to a disturbance event that would cause significant displacement in a sensitive system (Kelly and Harwell, 1990). Assessment of the natural capacity for adjustment, framed in terms of the inherent character and behavior of a given type of river, provides a basis to predict the likelihood that differing forms of adjustment will occur. Identification of landforms or landscapes that are sensitive to change, and insights into the proximity to threshold conditions at which change is likely to occur, are important considerations in the design and implementation of preventative conservation programs and appropriate treatment strategies (Schumm, 1991).

Stark differences in the nature and extent of possible adjustments may be demonstrated by different types of river in differing valley settings, as shown schematically in Figure 5.1. In this diagram, the various arrows portray the degree to which lateral, vertical, and wholesale adjustments to bed character, channel morphology, and channel planform are likely to occur. Vertical adjustment records the likelihood that the channel bed will incise or aggrade, while lateral adjustment reflects the ability of the channel banks to adjust (i.e., via lateral migration, channel expansion, or contraction). Combinations of these adjustments are marked by modification to the assemblage of instream geomorphic units (i.e., midchannel and

Table 5.2 The natural capacity for adjustment of rivers in different valley settings.

Valley setting	Bed character	Channel morphology	Channel planform	Natural capacity for adjustment (band width) and river sensitivity
Confined	Grain size, sorting, and hydraulic diversity are constrained by bedrock, restricting adjustments to local reworking of transient bedload fluxes.	Channel size, shape, and bank morphology are imposed by bedrock or ancient materials. Bank erosion is negligible. Local slope and forcing elements such as woody debris induce the pattern of geomorphic units, such as the spacing of step–pool sequences.	No potential to adjust the number of channels, sinuosity, or lateral stability. Geomorphic units are largely imposed forms. Riparian vegetation is not a significant control on geomorphic structure.	Limited (narrow band) Resilient
Partly-confined	Bed often constrained by bedrock. Gravel-bed rivers have well-segregated point bars, riffles, etc. that induce significant hydraulic diversity. Surface–subsurface textural variability may be significant. Bed adjustments are dependent on material availability and the history of bedload transporting events.	Channel width and shape are adjustable where floodplain pockets occur; otherwise they are constrained by bedrock or ancient materials along the valley margins. Bank erosion is restricted to areas where floodplain pockets occur. Instream geomorphic units adjust locally where space permits.	Local potential for lateral or downstream translation of bends, but largely constrained by bedrock. Floodplain pockets may be prone to scour, stripping, and reformation. Adjustments are restricted to areas where floodplain pockets occur.	Localized (relatively narrow band) Moderately resilient
Laterally-unconfined, high-energy with continuous channel(s)	Grain size, sorting, and hydraulic diversity may be constrained by coarse sediments that armor the bed. Transient bedload fluxes induce significant local adjustments. When adjustment occurs, it tends to be dramatic, as it is driven by infrequent, high magnitude events.	Channel size and shape can adjust laterally and vertically over the valley floor. Moderate potential for bank erosion. Largely bedload dominated geomorphic units.	Significant potential for adjustment to the number, sinuosity, and lateral stability of channels. May be considerable variability in floodplain geomorphic units, with significant potential for floodplain reworking.	Moderately significant (moderately wide band) Moderately sensitive

Laterally-unconfined, medium-energy with continuous channel(s)	Mobile bed is subject to recurrent shifts in character, composition, and hydraulic diversity as channel geometry and planform adjust. Surface-subsurface variability may be significant. Bed adjustments are dependent on material availability and the history of bedload transporting events.	Channel size and shape can adjust laterally and vertically over the valley floor in these mixed load systems. Significant potential for bank erosion. Riparian vegetation and woody debris may be significant controls on channel shape and geomorphic units. High potential for reworking of erosional and depositional geomorphic units.	Significant potential for adjustment to the number, sinuosity, and lateral stability of channels. Floodplains are formed by vertical or lateral accretion, and reworked by various processes, resulting in a wide range of floodplain geomorphic units.	Significant (widest band) Very sensitive
Laterally-unconfined, low-energy with continuous channel(s)	Limited hydraulic diversity with little potential to adjust given the cohesive sediments.	The capacity for channel size and shape to adjust laterally and vertically is constrained by cohesive banks along these suspended load systems. Little variability in geomorphic unit assemblage given the lack of bedload material.	Moderate potential for adjustment to the number, sinuosity, and lateral stability of channels. Floodplains are dominated by fine-grained vertical accretion deposits. Localized reworking occurs, largely by avulsion. Little variability in floodplain geomorphic units.	Localized (relatively narrow band) Moderately resilient
Laterally-unconfined, low-energy with bedrock-based continuous channel(s)	Limited variability as a thin veneer of bedload materials adjusts over the bedrock channel bed.	Imposed bed condition. Potential for bank erosion and adjustments to channel geometry are dependent upon floodplain composition and channel alignment. Suspended load systems have limited capacity to adjust their form.	Highly variable, dependent upon planform type. Suspended load systems are prone to avulsion, but have limited capacity to modify the array of geomorphic units given their limited bedload.	Localized (relatively narrow band) Moderately resilient.
Laterally-unconfined, low-energy with discontinuous channels	Valley floor texture dependent on sediment supply. Hydraulic diversity is low. Potential for sediment lobe deposition in swamps and floodouts.	Channels absent or discontinuous. Vegetation can induce significant resistance.	Relatively simple geomorphic structure with little potential for adjustment. However, headcuts may impose dramatic adjustments to river morphology.	Limited (relatively narrow band) Moderately resilient (in this state), but very sensitive if subjected to incision.

Limited adjustment potential (resilient)

Localized adjustment potential (moderately resilient)

Significant adjustment potential (sensitive)

bank-attached features). In some instances, wholesale shift in channel position on the valley floor (e.g., via avulsion, thalweg shift, or cutoff formation) may alter channel planform and the assemblage of floodplain geomorphic units.

Gorges are found in confined valley settings (Table 5.2). Channel configuration in a gorge is ostensibly stable, with no potential for lateral adjustment. Vertical adjustment is restricted to local redistribution of materials around coarse substrate (Figure 5.1a). The geomorphic unit structure of the bed reflects the geomorphic effectiveness of infrequent high magnitude flood events. These may be the only events that are able to mobilize coarse bed materials. Extreme floods may sculpt erosional geomorphic units. The natural capacity for adjustment is limited.

Along a partly-confined valley with bedrock-controlled discontinuous floodplains, vertical adjustment is limited as bed level stability is imposed by the bedrock valley floor (Figure 5.1b; Table 5.2). The channel bed comprises a mix of erosional geomorphic units (e.g., bedrock pools) and depositional forms (e.g., gravel riffles and point bars). Lateral adjustment via channel expansion is restricted to areas adjacent to floodplain pockets. Local channel expansion and contraction may result in a range of bank-attached geomorphic units, such as gravel point bars, benches, or ledges. There is limited capacity for wholesale adjustment to channel planform, but channels may rework floodplain pockets as flow short-circuits bends within these partly-confined valley settings. This river is considered to have localized capacity for adjustment. Adjustments may be restricted to a single pocket for any given event or series of events.

Transfer reaches that are characterized by laterally-unconfined, bedrock-based channels are able to adjust laterally, but have limited capacity for vertical adjustment (see Table 5.2). In the example shown in Figure 5.1c, a low sinuosity bedrock-based river is unable to incise its bed. Bedload materials are limited in this suspended load system, so the potential for bed aggradation is small. Cohesive banks resist bank erosion, inhibiting the capacity for adjustments to channel geometry and channel alignment.

Braided rivers are laterally-unconfined, high-energy systems (Table 5.2). These rivers have significant natural capacity for adjustment in both vertical and lateral dimensions (Figure 5.1d). They are also prone to wholesale adjustment via thalweg shift, as channels switch position over the valley floor, leaving behind abandoned braid plains, paleochannels, and islands. Given the highly sediment-charged nature of these aggradational environments, significant variability may be evident in the assemblage of instream geomorphic units, including a wide range of midchannel bars and islands. Each channel has significant potential to independently adjust via expansion and contraction processes.

Sand-bed meandering rivers in laterally-unconfined, medium-energy settings can adjust in both vertical and lateral dimensions and may be prone to wholesale adjustment (see Table 5.2; Figure 5.1e). Stacked point bar sequences reflect lateral migration processes. A point bar–pool–riffle morphology tends to be maintained along the channel. Lateral migration may result in ridge and swale development on the floodplain. Cutoff formation, abandonment of meander bends, or channel avulsion may result in wholesale adjustment in channel position on the valley floor. Hence, this type of river has significant capacity to adjust.

Anastomosing rivers are found in laterally-unconfined, low-energy settings (Table 5.2). Although these rivers are able to adjust in both vertical and lateral dimensions, and may be subjected to wholesale shifts in channel position on the valley floor, rates of adjustment are slow because of their suspended-load nature (Figure 5.1f). Vertical adjustment occurs as channel belts build within wide plains. Instream geomorphic units tend to be limited to pools and runs because of the limited availability of bedload caliber materials. Lateral expansion, contraction, or migration of channels is limited by the cohesive nature of the banks and the low-energy conditions under which this type of river operates. Hence, these rivers are considered to be moderately resilient to adjustment. On occasions, channel avulsion may bring about wholesale adjustment in channel position on the valley floor. Paleochannels are abandoned and subsequently infill.

Cut-and-fill rivers are found in laterally-unconfined, low-energy settings, typically in uplands. Channels may be continuous or discontinuous, dependent on the stage of adjustment (Table 5.2). The fill stage represents the phase of

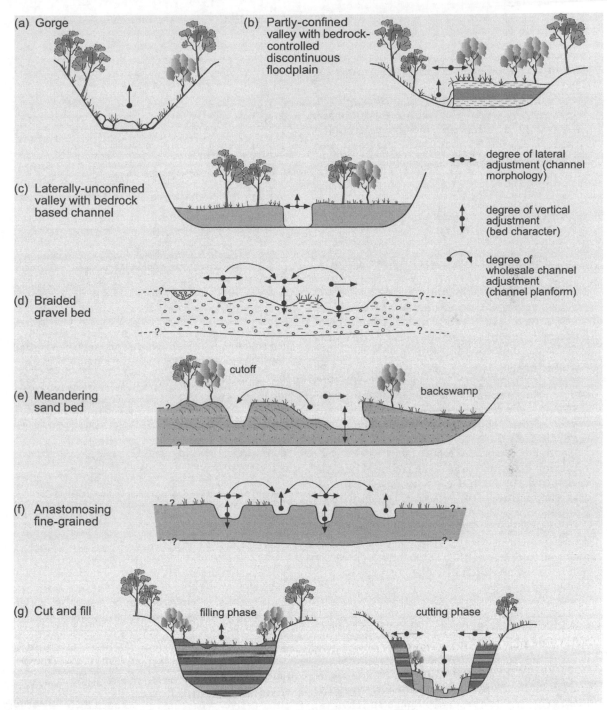

Figure 5.1 The natural capacity for adjustment of rivers in different valley settings
Different types of river have different capacities to adjust, whether vertically, laterally, or through wholesale shifts in channel position. The primary form of adjustment, and the timeframe over which adjustments take place, vary for different types of river. In general terms, as valley confinement is reduced, the ease of adjustment increases. Bedrock confinement constrains the capacity for adjustment in many settings (examples a–c). In laterally-unconfined settings, the ease of adjustment decreases from bedload (example d), through mixed load (example e), to suspended load (example f) situations. In example g, the capacity for adjustment varies markedly at the fill and cut phases of discontinuous watercourses.

adjustment when the channel is either absent or discontinuous on the valley floor (Figure 5.1g). Over timeframes of hundreds or thousands of years these valley floors are subjected to slow, pulsed, yet progressive aggradation. During this fill phase, there is limited capacity for adjustment. Should an erosional threshold condition be exceeded, this river has significant capacity to adjust both vertically and laterally (initially channel expansion, but subsequently contraction). Eventually, the system reverts to infilling the incised channel via aggradation.

5.3 Construction of the river evolution diagram

A conceptual tool called the *river evolution diagram* is presented here as a basis to interpret the range of river character and behavior in different landscape settings. Application of this tool provides an understanding of the type and extent of adjustments that are expected, or should be considered to be appropriate, for the given type of river. These insights into system dynamics enable management activities to be framed in terms of the "natural" behavioral regime of a given river type.

There are three core components to the river evolution diagram, namely the potential range of variability, the natural capacity for adjustment, and the pathway of adjustment (Figure 5.2). Components of the diagram are defined in Table 5.3. A five step procedure is applied to construct a river evolution diagram (Figure 5.3). Various examples that summarize the range of river character and behavior in different valley settings are presented in Figure 5.4.

5.3.1 Step One: Imposed boundary conditions and the potential range of variability

In Step One (Figure 5.3), *imposed boundary conditions* are appraised in terms of valley setting, slope, and lithology at a particular position in the catchment (Figure 5.2). Over geomorphic timeframes, these conditions are effectively set. These considerations determine the energy conditions under which rivers operate, as determined by upstream catchment area, slope, valley confinement, and sediment caliber. Geological setting influences

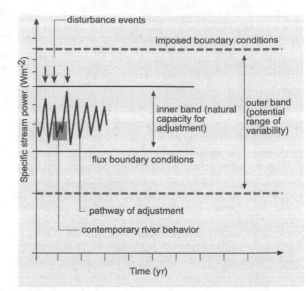

Figure 5.2 Components of the river evolution diagram In this conceptual framework that examines how rivers adjust over time, energy settings are determined by imposed boundary conditions (outer band), and prevailing flux boundary conditions (i.e., flow and sediment regimes; inner band). When subjected to differing forms of disturbance events, the river adopts a pathway of adjustment (the jagged line within the inner band). This records the pattern and rate of morphological variability that is characteristic for that type of river (see text). If changes to flux boundary conditions are experienced, a change in river type may occur. This is marked by a shift in the position of the inner band (upwards to a higher energy state and vice versa).

landscape relief and the range of material textures that are available to the river (i.e., whether it is bedrock, boulder, gravel, sand, or mud-dominated). Areas of mixed lithology typically make a range of particle sizes available (hence a wide outer band), while areas of more uniform lithology (e.g., sandstone) have a more restrictive range (i.e., a relatively narrow outer band).

The *potential range of variability* defines the range of river types that can potentially form within the imposed boundary conditions. The range of formative stream powers and resulting range of river morphologies determine the width of the outer band of the river evolution diagram. This reflects the maximum range of formative energy

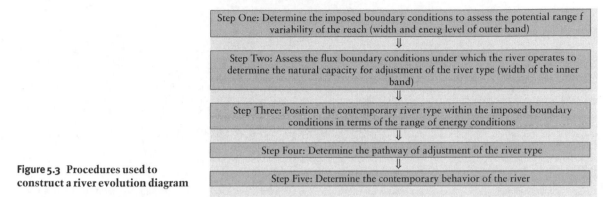

Figure 5.3 Procedures used to construct a river evolution diagram

Step One: Determine the imposed boundary conditions to assess the potential range f variability of the reach (width and energ level of outer band)

⇓

Step Two: Assess the flux boundary conditions under which the river operates to determine the natural capacity for adjustment of the river type (width of the inner band)

⇓

Step Three: Position the contemporary river type within the imposed boundary conditions in terms of the range of energy conditions

⇓

Step Four: Determine the pathway of adjustment of the river type

⇓

Step Five: Determine the contemporary behavior of the river

Table 5.3 Definition of components of the river evolution diagram.

Component	Definition
Specific stream power	Stream power provides a summary of the capability of energy to perform geomorphic work along a river. Total stream power is calculated as the product of discharge acting in any given cross-section multiplied by channel slope. When calculated as the energy acting on a given area, it is referred to as unit stream power. The latter term is used in the river evolution diagram, as it conveys the mutual interactions between available energy and the manner of river adjustment at any given site. It is represented on the y-axis using a logarithmic scale. Geomorphic work reflects the ability of a flow to induce adjustment in bed character, channel morphology, the assemblage of geomorphic units, and channel planform, without inducing change for a particular river type.
Time	Represented on the x-axis of the river evolution diagram using a linear scale. Defines the timeframe over which the full suite of behavior occurs for a particular river type.
Outer band	Reflects the *potential range of variability* in the types of rivers that can form under a certain set of *imposed boundary conditions* (i.e., valley-setting, slope, and lithology).
Inner band	Reflects the *natural capacity for adjustment* for a particular river type which represents the degree to which vertical, lateral, and wholesale change can occur for a river type. The width of the inner band is defined by the *flux boundary conditions*, i.e., the range of flow and sediment fluxes and vegetation dynamics that dictate the potential extent of adjustment in the assemblage of geomorphic units, channel planform, channel morphology, and bed character of the river type.
Pathway of adjustment	Defined by the frequency and amplitude of system responses to disturbance events. The shape of the pathway reflects the variability in the trajectory and timeframe of recovery in response to disturbance events. This records the behavioral regime of a river. In some instances, rivers may adjust among multiple states.
Disturbance event	Formative events that induce geomorphic adjustments to a river type. The size of the arrows represents the relative magnitude of the event that induced adjustment.
Contemporary river behavior	Adjustments that take place under contemporary flux boundary conditions while maintaining the river type.

conditions under which a range of river types operate for that specific landscape setting at that position in the catchment.

Stream power is considered to provide the most appropriate measure with which to differentiate among variants of river settings (i.e., the y axis of the river evolution diagram) as it reflects both the amount of energy that is available to be utilized in any given setting (total stream power) and it refers to the manner with which energy is used, as

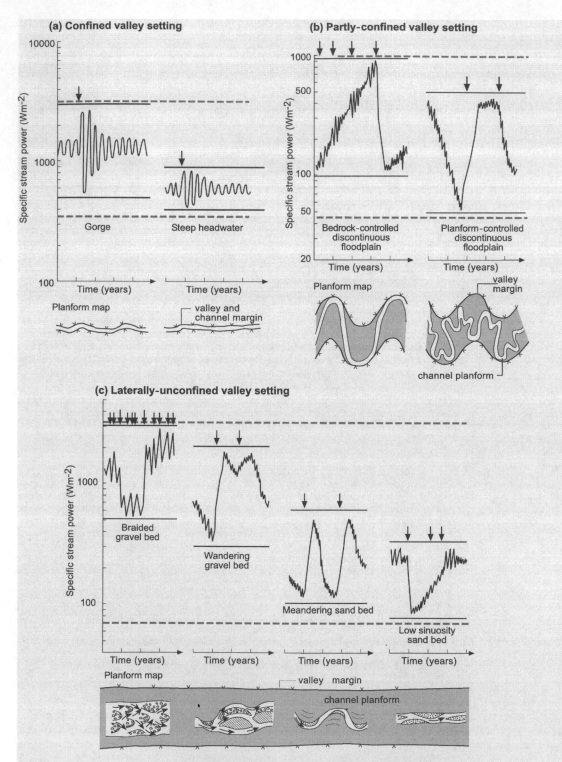

Figure 5.4 Schematic examples of the river evolution diagram in differing valley settings
Stream power estimates are derived from available literature: (a) confined valley setting (based on Costa and O'Connor, 1995), (b) partly-confined valley setting (based on Nanson and Croke, 1992; Miller, 1995; Ferguson and Brierley, 1999a, b), (c) laterally-unconfined valley setting (based on Nanson and Croke, 1992). See text for details.

determined by channel capacity (and active channel width; specific or unit stream power). Adjustments to channel geometry modify the use of energy, thereby altering the position of differing river settings (and associated channel configurations) on the river evolution diagram. It is recognized explicitly that adjustments in other external variables may alter the width of the inner band, or its position within the potential range of variability. For example, an influx of sediment may alter various attributes of river morphology, including channel capacity, thereby modifying formative unit stream power conditions. These mutual adjustments accentuate the underlying role of stream power as the most appropriate single determinant of river character and behavior.

Examples of river evolution diagrams for rivers in differing valley settings are portrayed in Figure 5.4. A broad valley, with a relatively steep slope in a granitic catchment has a wide band, as the valley setting is laterally-unconfined, there is considerable range in the energy conditions under which the river operates, and materials of differing caliber are available to be moved (Figure 5.4c). As such, a wide range of river morphologies and associated process domains may be adopted in this setting. A partly-confined valley with a lower slope within a metasedimentary catchment will have a narrower band, as moderate energy conditions, valley confinement (i.e., less space to adjust), and the mixed texture of the sediment load produce a restricted range of river morphologies (Figure 5.4b). These situations contrast significantly with, say, a narrow, steep valley in a volcanic terrain, which is represented by a narrow band, as the confined valley setting and the uniform sediment load impose particular river morphologies under a narrow range of high-energy conditions (Figure 5.4a). The position of different rivers within the imposed boundary conditions on Figure 5.4 reflects an energy gradient from high-energy variants (on the left) to low-energy variants (on the right).

5.3.2 Step Two: Flux boundary conditions and the natural capacity for adjustment

The width of the inner band represents the contemporary range of flux boundary conditions within which the reach operates (Figure 5.2).

Combinations of these factors, operating within the imposed boundary conditions, determine the range of river types and behavioral states that could be observed in that setting. The prevailing flux boundary conditions may be quite different to those experienced in the past. Hence, different types of river with differing character and behavioral regime may be observed within the same set of imposed boundary conditions.

The characteristic form for a given river type is not a static configuration or structure; rather, it reflects an array of potential adjustments among the assemblage of geomorphic units, channel geometry, channel planform, and bed material organization as determined by the contemporary range of flow, sediment, and vegetation conditions. These considerations determine the *natural capacity for adjustment*, as shown by the width of the inner band on the river evolution diagram (Figure 5.2). The potential extent of adjustments is measured in terms of the range of formative unit stream powers that induce adjustments to various attributes of river morphology, without resulting in river change. Rivers with significant natural capacity to adjust have wide inner bands. Those with limited natural capacity to adjust have narrow inner bands.

In appraisals of river behavior outlined in this chapter, the river evolution diagram is framed in terms of contemporary flux boundary conditions viewed over timeframes in which a characteristic set of form–process associations has become established along the reach, such that a particular type of river is evident. This timeframe varies markedly from setting to setting and for different types of river. For some river types, the "natural" behavioral regime may comprise differing states. In these instances, transitions between states in response to breaching of internal (intrinsic) threshold conditions are considered to be part of the natural capacity for adjustment for that type of river. Examples include cut-and-fill rivers, partly-confined valleys prone to floodplain stripping, meandering rivers that adjust their slope following generation of cutoffs, or various types of river subjected to avulsion or changes in channel multiplicity. In general terms, the width of the inner band that conveys possible states varies with the ease of adjustment of the river. Sensitive rivers have wider bands than resilient rivers, reflecting the inherent range in the degrees of freedom within which rivers operate.

As each reach adjusts to disturbance events, the nature and extent of response may vary markedly. In terms of the behavioral regime of a river, the type and extent of adjustment do NOT result in the adoption of a different river character and behavior. This latter circumstance describes river change, as discussed in Chapter 6. In a sense, the prevailing flux boundary conditions determine the type of river that is observed today, and its natural capacity for adjustment. The *natural capacity for adjustment* determines the range of *behavior* that any particular type of river may experience, while the *potential range of variability* determines the range of types of river that may be found in any given landscape setting (i.e., within its imposed boundary conditions), thereby providing a measure of the possible states that the river could adopt if *change* occurred.

In this book, a natural river is defined as one that dynamically adjusts so that its geomorphic structure and function operate within a range of variability that is appropriate for the type of river, and the range of flux boundary conditions under which that river type operates. Natural or expected river character and behavior is viewed in terms of the range of processes and associated forms that occur within the bounds determined by the inner band on the river evolution diagram (Figure 5.2). This natural state is considered in the absence of human disturbance.

The natural capacity for adjustment varies markedly for differing types of river, over differing timeframes, reflecting a combination of factors, such as:

1 *The variability of sediment mix at any given point along a river.* This may reflect local considerations that determine the relative balance of, say, gravel, sand and finer particles, or the influx of materials from upstream.
2 *The flow regime.* Some rivers are adjusted to relatively uniform flow conditions in which mean annual floods are the primary determinant of river form. In these situations, the inner band is relatively narrow. However, if the system is adjusted to significant flow variability, the inner band is likely to be wider.
3 *Riparian vegetation and woody debris.* These components of flow resistance vary markedly

from setting to setting, potentially exerting a significant influence on the natural capacity for adjustment of certain types of rivers.
4 *System history.* In some instances, longer-term climate-induced changes to the nature and pattern of sedimentation on the valley floor may impose constraints on contemporary system behavior (e.g., gravel terraces or fine grained cohesive banks that line river courses), thereby imposing a narrow band to the natural capacity for adjustment.

5.3.3 Step Three: Placing rivers within the potential range of variability

Step Three in construction of the river evolution diagram entails positioning of the river within the potential range of variability, based on its prevailing energy conditions (Figure 5.2). If the contemporary river operates under relatively high-energy conditions, the inner band is situated high in the potential range of variability (Figure 5.4). Alternatively, if contemporary energy levels are low (relative to the range of conditions that can be experienced under the imposed boundary conditions), the inner band is placed towards the bottom of its potential range of variability. The width of the inner band reflects the range of energy conditions experienced under prevailing flux boundary conditions. Its placement within the outer band reflects the relative extent of those energy conditions (i.e., whether the inner band is positioned high or low within the outer band).

5.3.4 Stage Four: The pathway of adjustment

In assessing the types and extent of adjustment that define the range of expected character and behavior of a given river type, responses to differing forms of disturbance must be appraised. Collectively, these adjustments define the *pathway of adjustment* on the river evolution diagram (Figure 5.2). The behavioral regime of any given type of river, as defined by the natural capacity for adjustment, encompasses ongoing adjustments to alterations in flux boundary conditions. Reaches may operate at different positions within their natural capacity for adjustment as pulse disturbance events of differing magnitude and frequency alter water and sediment regimes and vegetation associations (Chapter 3). If a press disturbance breaches threshold conditions, positive feedback mechanisms may drive the system to a different state, possibly inducing a change in river type (Chapter 6). These considerations determine the pathway of adjustment of a reach, as marked by modifications to the arrangement and abundance of geomorphic units, adjustments to the organization of material on the channel bed, and local alterations to channel planform. Within the inner band of the river evolution diagram, system responses to disturbance events may be indicated by oscillation around a characteristic form, or adjustments among various characteristic forms.

The form of the pathway of adjustment summarizes system response to disturbance events, indicating how any given river type is able to accommodate adjustments to flow and sediment transfer conditions. In essence, the pathway of adjustment integrates all components of adjustment, describing the morphologic and behavioral adjustments to ongoing variability in the nature, extent, and sequence of disturbance events on the one hand (i.e., impelling forces), and the capacity of the system to absorb change on the other (i.e., the effectiveness of response mechanisms as conditioned by resisting forces along the reach).

As noted in Table 5.2, river responses to disturbance events reflect reach sensitivity, measured here as the ease with which the river is able to adjust its form. This provides a measure of the capacity of the system to accommodate the impacts of disturbance events via mutual adjustments, such that the river is able to sustain a characteristic form. The behavioral regime of certain river types may entail fluctuation among various states, reflecting breaching of intrinsic thresholds. Disturbance events are indicated schematically on the river evolution diagram by arrows on the edge of the inner band (Figures 5.2 and 5.4). The frequency and sequence of disturbance events are conveyed by the spacing of arrows, while the size of the arrow indicates the relative magnitude of the event.

The form of the pathway of adjustment is defined by its amplitude, frequency, and shape (Figure 5.5a). *Amplitude* reflects the extent of adjustment in response to a disturbance event. *Frequency* reflects the recurrence with which disturbance events drive geomorphic adjustments.

The *shape* of the pathway of adjustment reflects the trajectory of response to disturbance events. Variants include progressive adjustments in a particular direction, oscillations around a mean condition, or jumps between characteristic states. The spacing of disturbance events that drive adjustment varies in differing settings, influencing the river type and its sensitivity to adjustment. In behavioral terms, however, the collective response to disturbance events does not drive the system outside its natural capacity for adjustment.

The pathway of adjustment summarizes system responses to sequences of disturbance events of varying magnitude and frequency. Examples of differing forms and timeframes of system recovery that determine the shape of the pathway of adjustment are shown in Figure 5.5. The type and timeframe of response depend partly on whether the disturbance induces adjustments that reinforce or counteract existing tendencies. Recovery time may be highly variable, reflecting the condition of the system at the time of the impact, as influenced by the recent history of events, among many considerations. Disturbance responses may be instantaneous or delayed (i.e., lagged responses). Their consequences may be short-lived or long-lasting. Combinations of disturbance responses, and the resulting shape of the pathway of adjustment, can be simple (temporally uniform) or complex (temporally variable).

If the geomorphic response is damped out, and the previous state is restored after a short recovery time, the pathway of adjustment has a jagged shape reflecting minor adjustments away from a characteristic form. This form of adjustment is exemplified by cutoff formation along a meandering river (Figure 5.5bA). Elsewhere, progressive adjustments may promote shifts to an alternative characteristic form, with an altered nature and/or level of activity, but adjustments remain within the natural capacity for adjustment for that river type. In this case, steps along the pathway of adjustment record shifts among multiple characteristic states. Intervening flatter areas record minor modifications around one of these states. These types of rivers are prone to cyclical patterns of threshold-induced adjustments, such as avulsion (Figure 5.5bB), incision, and aggradation (Figure 5.5bC), and floodplain stripping (Figure 5.5bD). Reaches

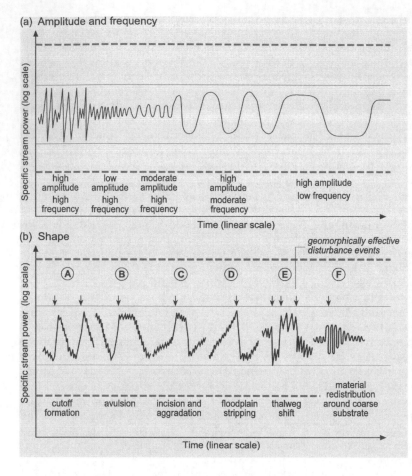

Figure 5.5 Components of the pathway of adjustment as used in the river evolution diagram
Because different rivers adjust in different ways, significant variability is evident in the form and rate of adjustment that may be experienced. These notions are summarized as the pathway of adjustment on the river evolution diagram. Three components are considered in appraisal of these pathways of adjustment, namely amplitude, frequency, and shape. System response to disturbance events ranges in amplitude and frequency as shown in figure (a). Examples of differing shapes of adjustment that reflect different types of geomorphic activity are indicated in figure (b). These issues are discussed more fully in the text.

that are prone to abrupt adjustments also have a cyclic pattern of adjustment with short recovery times. However, this pathway reflects recurrent (tight) oscillations around a characteristic form, as exemplified by thalweg shift in a braided river (Figure 5.5bE) or redistribution of bedload material around coarse substrate in a gorge (Figure 5.5bF).

Building on the examples used to demonstrate the potential range of variability in Section 5.2, various schematic applications of the river evolution diagram are presented in Figure 5.6. The natural capacity for adjustment for a gorge is relatively narrow, as adjustments maintain a uniform state over timeframes up to 10^3 years (Figure 5.6a). These deeply etched bedrock rivers are resistant to change, and demonstrate very short periods of disturbance response, such that adjustments are bare-

ly discernible over the short to medium term (< 10^2 years). As the river has limited capacity to adjust, it is characterized by a low amplitude, high frequency pathway of adjustment within a narrow inner band.

Rivers in partly-confined valley settings may be prone to floodplain stripping (Figure 5.6b). Although this type of river has relatively limited capacity for adjustment, and is considered to be resilient to change, it demonstrates stepped adjustments over timeframes of 10^3–10^4 years. Such adjustments include channel expansion, floodplain building, and floodplain reworking via floodplain stripping (see Nanson, 1986). This is induced by the breaching of an energy threshold within the partly-confined valley. The pathway of adjustment reflects different phases of response to disturbance events, as the river adjusts between

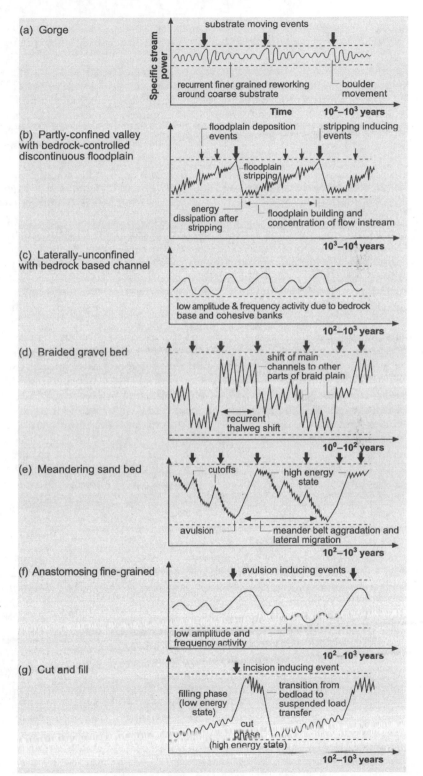

Figure 5.6 Evolution diagrams for different types of river
Forms and timeframes of geomorphic adjustment vary markedly for different types of river. This diagram, which only conveys the inner bank of the river evolution diagram, indicates different pathways of adjustment in response to disturbance events for different types of system. Profound differences in the inferred ranges of variability are indicated for the different systems. There is also marked variability in the timeframes over which adjustments take place.

phases of progressive floodplain aggradation and short periods of catastrophic erosion. During the aggradation phase, disturbance events tend to have a lower amplitude and lower frequency as periods of floodplain inundation decrease. Eventually, catastrophic events bring about floodplain stripping during a short phase of adjustment that is characterized by high amplitude, moderate frequency responses to disturbance.

Bedrock-based laterally-unconfined rivers tend to act as transfer reaches, sustaining an approximate balance between sediment input and output with a relatively thin veneer of deposits over the valley floor (e.g., Heritage et al., 2001). On the river evolution diagram, a low sinuosity variant of this river with cohesive banks is characterized by low amplitude, low frequency adjustments over timeframes of 10^2–10^3 years (Figure 5.6c). The river oscillates around a relatively stable form and configuration.

Braided rivers have a significant capacity to adjust, with a wide inner band (Figure 5.6d). Frequent disturbance events induce recurrent reworking of bedload material via thalweg shift, flow stage adjustment, and local adjustments to bed level over timeframes of 10^0–10^1 years (e.g., Williams and Rust, 1969). The pathway of adjustment is characterized by low amplitude, high frequency responses to disturbance with short recovery times.

A meandering sand bed river in a laterally-unconfined valley setting has a wide natural capacity for adjustment (Figure 5.6e). These sensitive reaches have significant capacity to adjust both vertically and laterally. Progressive channel migration builds the meander belt over time (e.g., Brooks and Brierley, 2002). As sinuosity increases, the energy of the system decreases. Cutoff channels may induce phases of disturbance response as the channel readjusts its slope to the reduced sinuosity, typically over timeframes of 10^1–10^2 years. The river is then subjected to progressive adjustments as the characteristic meandering form is maintained. Over longer timeframes, meandering sand-bed rivers may be prone to avulsion, as they adjust their course beyond the meander belt and sediments accumulate elsewhere on the valley floor. Following avulsion, the river reestablishes its meander belt via lateral migration and vertical accretion. Hence, this type of river is characterized by a stepped pathway of adjustment, with a wide range of disturbance responses of varying amplitude and frequency.

Low energy alluvial rivers tend to be moderately resilient to adjustment. Although these rivers have a wide natural capacity for adjustment that includes modifications to channel morphology and shifts in channel position on the valley floor, cohesive channel boundaries induce progressive rather than dramatic adjustments. In the anastomosing example presented here, the pathway of adjustment is characterized by disturbance responses with high amplitude but low frequency, as occasional avulsion events alter channel multiplicity (e.g., Nanson et al., 1988; Figure 5.6f).

Cut-and-fill rivers show significant natural capacity for adjustment, as they oscillate between two characteristic states (e.g., Cooke and Reeves, 1976; Figure 5.6g). During the aggradation phase, discontinuous channels are quite resilient to adjustment. Eventually, however, exceedance of a threshold condition may promote dramatic incision and formation of a continuous channel. During the incision phase, the system responds more dramatically to disturbance events, as the energy and sensitivity of the system are enhanced. Over time, the channel infills, producing an intact valley floor once more. Typically, cut-and-fill cycles occur over timeframes of 10^2–10^3 years. Responses to disturbance events vary during these different phases, with low amplitude and low frequency responses during the fill stage, but high amplitude and high frequency responses during the cut stage.

5.3.5 Step Five: Contemporary river behavior

The final component in construction of the river evolution diagram, Step Five, entails determination of the *contemporary behavior of the river* (see Figure 5.2). The contemporary river can sit anywhere on the pathway of adjustment for the river type. Appraisal of river behavior is based on how the river adjusts its form to contemporary flux boundary conditions. In some instances, former flow and sediment conditions may impose constraints on the contemporary range of river character and behavior. For example, it may take the system a considerable period of time to adjust to a major flood event that mobilized the coarsest bedload fraction if more frequent, lower magnitude events are unable to do so.

At any point in time, a river can operate anywhere within its natural capacity for adjustment. As each river type has a distinct set of form–process associations, its character and behavior adjust to a given set of disturbance events within a certain range of responses. Ongoing interactions form and rework geomorphic units. The operation of flow and sediment fluxes and prevailing vegetation conditions shape the present character and behavior of the river. Assessment of river behavior is framed in terms of the period of time over which flux boundary conditions have remained relatively uniform such that a characteristic river form results, with a particular assemblage of geomorphic units, bed material organization, and channel planform.

In the sections that follow, adjustments to disturbance events are appraised using the scalar approach outlined in Chapter 4.

5.4 Bed mobility and bedform development

In bedload or mixed load systems, changes to flow depth and associated energy conditions induce bedform adjustments, redistributing materials over the channel bed. Capacity for adjustment is high, as bedforms are modified on an event by event basis. Oscillations in bed configuration record variations in sediment flux resulting from adjustments to flow geometry, the distribution of flow energy across the channel bed, and the availability and caliber of sediment (see Chapter 4). High flow regime forms generated during the rising stage of flow events are reworked and replaced by low flow regime forms during the waning stage. Alternatively, dependent upon sediment availability, the bed may be scoured at high discharges on the rising stage and filled to approximately the preflood level on the falling stage. Considerable spatial variation may be noted in this process. Through their role as a determinant of instream resistance, bedform adjustments modify hydraulic variables such as velocity and depth, thereby influencing the local sediment transport rate. Channel geometry and flow alignment exert a significant influence on these relationships. Hence, if channel geometry and the character/pattern of geomorphic units are altered, changes to flow depth and the distribution of flow energy modify the nature and pattern of bedforms and associated bed material organization.

Sand-bed systems are especially prone to adjustment, given the ease of mobility of bed material, while gravel-bed and coarser fractions require higher energy (less-frequent) flows to initiate motion, especially if the bed is armored. The threshold for bed adjustment is much greater in boulder bed and bedrock streams, where extreme flows are required to mobilize the larger clasts or pluck materials from the channel bed. In suspended load systems, resistance to erosion depends more on the strength of electrochemical bonds between cohesive, silt-clay materials than on the physical properties of the particles themselves. The capacity for bedform adjustment is limited as high velocity flows are required to entrain these materials. As a result, suspended load rivers tend to have relatively planar beds.

Although bedform features are transient forms, exerting only a minor influence on longer-term patterns of geomorphic adjustment, they may be critical considerations in appraisal of low flow river behavior and habitat diversity along a reach. For example, replacement of a heterogeneous bed by a homogeneous sand slug may represent significant loss of habitat. More subtle adjustments may be equally devastating in ecological terms. For example, influxes of fine-grained sediments may choke the interstices between gravels, requiring flushing flows to reaerate the bed (e.g., Pitlick and Wilcock, 2001).

5.5 Adjustments to channel shape

Variability in bed and bank material texture along a river influences the capacity for channel adjustment and resulting channel shapes. The inherent strength and stability of the bed and banks determine the sensitivity of the channel to adjust, whether vertically (the depth dimension) or laterally (i.e., channel width). The composition of materials that make up the bank influences the effectiveness of bank erosion mechanisms (i.e., lateral adjustments). Floodplain character, and the ease with which materials can be reworked at flood stage, exert additional controls on river morphology, influencing the capacity for lateral expansion or channel migration.

As bedform adjustments largely reflect transient sets of contemporary processes, they have little effect on the gross geomorphic structure or behavioral regime of the river. In contrast, bank sediments record former depositional conditions and/or events that were responsible for floodplain formation. In some instances, materials that were deposited under a former depositional regime may constrain contemporary channel size and shape. For example, the prevailing flow regime may no longer be able to mobilize coarse boulders activated under extreme events in the past. Alternatively, ancient fine-grained, cemented materials that record suspended load deposition on floodplains and/or terraces under a former climatic regime may limit the capacity for adjustment of the contemporary channel.

Channel geometry records the balance of erosional and depositional processes that shape the bed and bank. These considerations vary at different positions along a river, reflecting flow energy, sediment availability, and landscape history. Adjustments to channel shape may be locally variable along a reach, reflecting adjustments to flow alignment, channel position on the valley floor, accentuated scour/erosion, influx of depositional units, or forcing elements such as bedrock outcrops, instream/bank vegetation, and woody debris. Channel geometry adjusts to accommodate flow and sediment fluxes. Key differences in river character and behavior are observed in aggradational and degradational environments.

Channel shape is influenced by the energy available to erode or deposit materials of different caliber along the bed and/or banks at different flow stages. In general terms, channel-forming events are most effective at or near bankfull stage when the capacity to perform geomorphic work is maximized (e.g., Leopold et al., 1964). Identification of formative (bankfull) stage is a reach-specific exercise. Depending on patterns of instream sedimentation, or the stage of evolution, bankfull stage may vary over time for any given reach. However, adjustments to bed morphology may occur at flow stages less than bankfull, as instream geomorphic units are formed, reworked, and reorganized.

Variability in bank morphology reflects a balance between bank erosion processes and the formation/deposition of bank-attached geomorphic units. Flow alignment, as determined by thalweg position at different flow stages, dictates the distribution of flow energy adjacent to banks, thereby influencing this balance and the manner/rate of adjustment to bank morphology. Undermining of noncohesive deposits at the toe of composite banks may occur at low–moderate flow stages. Deposits may be draped against the banks, or bank-attached features may form, at these flow stages. Bank-attached geomorphic units typically form in low-energy areas, away from the thalweg, and include features such as point bars, lateral bars, and benches. These features protect the bank from erosion and produce a compound bank morphology.

Midchannel geomorphic units that determine the shape of the channel bed range from sculpted (imposed) to free-forming variants. The balance of erosional and depositional processes is determined primarily by available flow energy and the sediment transport regime of the river (i.e., whether it is a suspended load, mixed load, bedload, or bedrock-dominated system). The nature and extent of deposition and reworking are determined by the frequency of inundation of differing surfaces. As noted in Chapter 4, midchannel forms reflect an energy gradient from step–pool sequences to cascades to runs and a range of depositional features including riffles, islands, and midchannel bars.

Five key types of channel adjustment affect channel shape (Figure 5.7). The specific combination of lateral and vertical components varies in differing settings and over time. Lateral adjustment processes include lateral migration, channel expansion, and channel contraction. Lateral migration describes progressive channel movement across the valley floor. Although multiple forms of planform adjustment may occur, including bend rotation, extension, and translation, an asymmetrical channel geometry is maintained as the convex bank accretes and the concave bank erodes. The type of adjustment reflects the radius of curvature of the bend, floodplain sedimentology, and the presence of obstructions (Hickin, 1983). Channel expansion refers to enlargement of the channel in response to erosion of one or both banks, without the compensatory effect of equivalent deposition. Channel expansion via bank erosion processes tends to be episodic or catastrophic, rather than progressive. An array of bank geometries and bank-attached erosional geomorphic units, such as ledges, may form. Finally, channel contraction

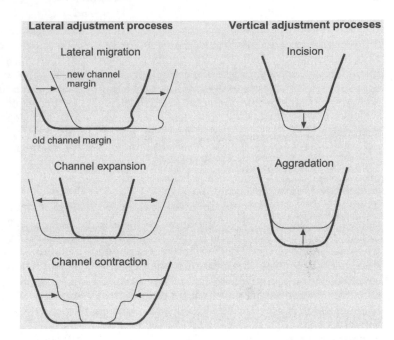

Lateral adjustment proceses

Lateral migration

new channel margin

old channel margin

Channel expansion

Channel contraction

Vertical adjustment proceses

Incision

Aggradation

Figure 5.7 Forms of adjustment to channel shape
Channels can modify their shape through combinations of lateral and vertical adjustment, reflecting the balance of erosional and depositional processes acting on the bed and banks.

can occur along one or both banks depending on flow alignment. Deposition occurs when the thalweg is aligned away from the bank, allowing bank-attached geomorphic units to form in the lower energy segment of the flow, contracting the channel. Alternatively, channel contraction is often associated with incision.

Vertical adjustments reflect bed degradation (i.e., channel incision) or channel aggradation (typically marked by the accumulation of bedload materials, but suspended materials may accumulate in swampy channel environments along discontinuous watercourses). Channel incision occurs where the bed is destabilized, or scoured, resulting in bed lowering and channel deepening. Elsewhere, degradation entails accentuated erosion of sculpted geomorphic units. Incision is commonly initiated by a reduction in the availability of bed material such that erosive energy increases. Galay (1983) distinguished between downstream-progressing and upstream-progressing degradation, the first being associated with a decrease in bed material load or increase in water discharge, and the second with a fall in base level. Upstream-progressing degradation generally proceeds at a much faster pace than its downstream counter-

part, because one increases and the other decreases slope as the degradation head migrates. In contrast, channel aggradation results in shallower channels, generating an array of depositional geomorphic units that are typically characterized by midchannel forms.

Although channel geometry is one of the most readily adjusted attributes of river morphology, especially in the width dimension, patterns, and rates of adjustment vary markedly from reach to reach and from type to type. The ultimate determinant of adjustment tendencies is whether the balance and distribution of erosional and depositional processes are maintained along the reach. For example, the extent to which lateral adjustments brought about by channel migration or expansion modify channel geometry depends on the accompanying pattern and rates of deposition of geomorphic units. Erosion at one site may induce deposition downstream, potentially adjusting channel geometry in both locations.

Channel width and depth do not adjust independently. In many instances, adjustment to channel depth via incision or aggradation triggers secondary adjustments in channel width and geometry. These patterns of adjustment may be accentuated by positive feedback mechanisms following exceedence of critical threshold conditions. For example, degradation may promote bank instability via undercutting or exceedence of critical bank height. As incision continues, banks become higher and are increasingly oversteepened. Indeed, in some instances, channels become so deep that hydraulic action only impacts on lower parts of the bank, and mass failure mechanisms become the dominant form of bank erosion (e.g., Brooks et al., 2003). Increase in channel depth is commonly followed by dramatic increases in width, resulting in enormous increases in channel capacity. As a consequence, flow energy is increasingly concentrated within the enlarged channel, whereas previously it was dissipated over the floodplain. In contrast, aggradation is typically associated with channel expansion and an accompanying increase in the width : depth ratio of the channel. As flows are increasingly dispersed over the wider channel bed, the bed level continues to aggrade and channel depth is further reduced. This induces more frequent overbank flow, further dissipating energy and inducing deposition of coarser materials atop the floodplain.

Mutual adjustment processes among bed and bank forms result in genetically-related assemblages of instream geomorphic units along different river types (see Section 5.6). These packages are determined primarily by the sediment transport regime of the river and the distribution of flow energy within the channel. For example, suspended load systems in low energy, low gradient environments are unable to maintain the transport of coarser materials, which tend to accumulate as bank-attached geomorphic units. Vertically accreted, fine-grained floodplains are prominent features of these aggradational environments. Resulting banks are cohesive, inhibiting bank erosion and lateral adjustment. In contrast, bedload systems tend to be high energy, sediment-charged systems that entrain, transport, and deposit coarser materials. The mobile bed is subjected to phases of degradation and aggradation, dependent upon the prevailing flow–sediment balance, but the long-term tendency is generally aggradational. Depositional, midchannel geomorphic units dominate. Floodplains comprise a range of geomorphic units, including some coarse-grained, noncohesive features that are readily reworked. This pres-

ents significant potential for lateral and vertical adjustment. Finally, mixed load systems tend to be characterized by a wide range of channel and floodplain geomorphic units. The bed and banks comprise contrasting sediment mixes. Within-channel sorting of materials leads to the formation of bank-attached, coarse-grained geomorphic units, while the finer fraction is deposited on the floodplain. These rivers often have a composite bank sedimentology, in which a coarse basal fraction is overlain by interbedded coarse- and fine-grained fractions. Composite or faceted bank morphologies are common. Selective reworking of coarser lenses presents significant potential for lateral adjustment.

Differing combinations of lateral and vertical adjustment processes, and associated assemblages of instream geomorphic units, are key determinants of channel shape (Table 5.4). In general, *symmetrical channels* have a relatively uniform distribution of flow energy across the channel bed. When observed along low sinuosity rivers or at the point of inflection between meander bends, this channel form is indicative of negligible lateral and vertical channel adjustment. In contrast, many incised rivers have symmetrical channels that are prone to lateral expansion.

In general terms, *asymmetrical channels* are associated with laterally adjusting (i.e., migrating) channels in which flow energy is concentrated along the concave bank of a bend. Secondary flow circulations promote deposition of bank-attached (point) bars on the convex bank. Asymmetrical channels are also commonly observed in partly-confined valley settings, where discontinuous pockets of floodplain line one bank (with point bars or point benches common), while the other bank hugs the valley margin.

In *compound channel* situations, a smaller channel is typically inset within a broader macrochannel. This is commonly observed in landscapes that are subjected to significant flow variation. The stepped cross-sectional morphology of compound channels comprises a suite of geomorphic units that reflect either phases of channel expansion and/or contraction, or record river activity at different flow stages. For example, while benches record channel contraction via depositional processes, ledges are erosional forms that are indicative of channel expansion. Elsewhere,

Table 5.4 Channel adjustment processes and channel shapes for differing river types.

River type (sediment transport regime)	Vertical adjustment	Lateral adjustment	Dominant instream geomorphic unit assemblage	Resultant channel shape
Bedrock-confined rivers				
Steep headwater with bedrock/boulder bed (bedload)	Negligible	Negligible	Erosional, sculpted dominated (step–pool–cascade–rapid assemblage)	Irregular
Gorge with bedrock/boulder bed (bedload)	Negligible	Negligible	Erosional, sculpted dominated (forced pool–run assemblage)	Irregular
Partly-confined rivers				
Partly-confined valley with bedrock-controlled discontinuous floodplain (mixed load)	Negligible	Expansion and contraction	Bank-attached depositional dominated (bench–point bar–pool–riffle assemblage)	Compound or asymmetrical
Laterally-unconfined, high-energy rivers				
Braided gravel bed (bedload)	Aggradation	Expansion and contraction	Midchannel depositional dominated (longitudinal bar–run–pool assemblage)	Macrochannel is irregular, individual channels are symmetrical
Wandering gravel bed (mixed load)	Aggradation	Expansion, contraction, lateral migration	Dominated by midchannel and bank-attached depositional forms (longitudinal bar–run–pool–island–lateral bar–point bar–riffle assemblage)	Largely irregular, but asymmetrical in bends
Braided sand bed (bedload)	Aggradation	Expansion and contraction	Midchannel depositional dominated (transverse bar–island assemblage)	Macrochannel is irregular, individual channels are symmetrical
Low sinuosity sand bed (bedload)	Negligible	Negligible	Midchannel depositional dominated (run–sand sheet–lateral bar assemblage)	Symmetrical
Laterally-unconfined, medium-energy rivers				
Meandering gravel bed (mixed load)	Negligible	Lateral migration	Bank-attached depositional dominated (compound point bar–riffle–pool–lateral bar assemblage)	Asymmetrical in bends, symmetrical at inflection points

Table 5.4 *Continued*

River type (sediment transport regime)	Vertical adjustment	Lateral adjustment	Dominant instream geomorphic unit assemblage	Resultant channel shape
Meandering sand bed (bedload)	Aggradation	Lateral migration	Bank-attached depositional dominated (point bar–lateral bar–run assemblage)	Asymmetrical in bends, symmetrical at inflection points
Laterally-unconfined, low-energy rivers				
Low sinuosity fine grained (suspended load)	Negligible	Negligible	Bank-attached erosional dominated (lateral bar–pool–run–ledge assemblage)	Symmetrical
Meandering fine grained (suspended load)	Negligible	Negligible	Bank-attached erosional dominated (point bar–run–pool–ledge assemblage)	Asymmetrical in bends, symmetrical at inflection points
Anastomosing fine grained (suspended load)	Negligible	Negligible	Bank-attached erosional dominated (point bar–lateral bar–run–pool–ledge assemblage)	Symmetrical in low sinuosity sections, asymmetrical in bends
Laterally-unconfined, low-energy rivers with bedrock-based channel				
Anastomosing bedrock-based (suspended load)	Negligible	Negligible	Erosional, sculpted dominated (bedrock core bars–forced pool assemblage)	Irregular
Laterally-unconfined, low-energy rivers with discontinuous channels				
Intact valley fill (suspended load)	Aggradation	N/a	Depositional dominated (swamp-floodout assemblage)	N/a
Channelized fill sand bed (bedload)	Incision and aggradation	Expansion and contraction	Bank-attached depositional dominated (bench–lateral bar–run assemblage)	Symmetrical to compound
Channelized fill fine grained (suspended load)	Incision	Expansion	Bank-attached erosional dominated (ledge–run assemblage)	Symmetrical to compound

compound channels may reflect long-term river evolution recorded by terraces that may have formed under different environmental conditions to those experienced today. Terraces perched above the contemporary river system may reflect changes to the flow–sediment regime, tectonic uplift, or isostatic rebound, among many considerations.

Irregular channels do not have a clearly defined shape that has been molded by a particular set of flow–sediment interactions. Rather, channel shape is locally variable, reflecting site-specific characteristics. In some instances, local lithological variability may impose differing patterns of sculpted forms (e.g., in a gorge). In steep headwater rivers, flow may adopt irregular patterns around boulders. In both these instances, there is negligible vertical and lateral adjustment around imposed geomorphic units. Elsewhere, an irregular channel shape may indicate that the river has yet to become fully adjusted to the prevailing flow and sediment conditions, such that a chaotic pattern of depositional forms is found. For example, the irregularly shaped macrochannel of braided rivers reflects formation and dissection of midchannel bars.

These observations indicate that channel geometry, in itself, is not directly indicative of formative processes and the likely pattern (or rate) of geomorphic adjustments. To gain insights into these components of river behavior, assemblages of instream geomorphic units must be analyzed.

5.6 Interpreting channel behavior through analysis of instream geomorphic units

Assemblages of instream geomorphic units typically reflect the operation of genetically linked sets of processes under certain types of conditions at characteristic locations. These processes are determined by the transport regime of the system (i.e., whether it is bedload, mixed load, or suspended load) and the distribution of energy as determined by flow alignment at differing flow stages. Resulting patterns and rates of erosion, sediment transport, and deposition fashion the arrangement of geomorphic units along a reach.

Interpretation of assemblages of geomorphic units provides fundamental insights into the bal-

ance of erosional and depositional processes along a reach, the structural diversity of the river, and channel–floodplain relationships. Sediment availability and its relation to the discharge regime influence processes and patterns of material reworking along channels, and the resulting mix of sculpted versus midchannel versus bank-attached forms along a reach. Particular assemblages of these features are commonly observed for different types of river. Adjustments among these components are key determinants of channel morphology, in turn shaping (and reflecting) the manner of channel adjustment on the valley floor. If the energy balance or sediment flux is altered, such that the assemblage of geomorphic units is modified, channel change has occurred (Chapter 6). However, ongoing adjustments to the assemblage of geomorphic units along a reach, such that the reach retains a characteristic form with an equivalent mix of erosional and depositional processes (operating at differing places within the reach), are core attributes of reach behavior.

As noted in Table 5.4, differing assemblages of instream geomorphic units tend to result in characteristic channel shapes along bedrock, bedload, mixed load, and suspended load rivers. In this section, assemblages of instream geomorphic units are interpreted to assess channel behavior for the seven classes of river defined at the landscape setting scale of the nested hierarchical framework presented in Chapter 4.

5.6.1 Channel behavior along bedrock-confined rivers

Bedrock confined rivers have limited ability to adjust in either vertical or lateral dimensions. A continuum of bedrock/boulder sculpted geomorphic units is found along a slope-induced energy gradient (e.g., Montgomery and Buffington, 1997; see Figure 5.8). A characteristic pattern observed along high slope mountain streams is marked by downstream transition from step–pool sequences to cascade to rapid assemblages, while runs and forced pool–riffle assemblages are common along lower sloped gorges. Given the bedrock-dominated character of these rivers, channel shape is irregular. Patterns of geomorphic units are influenced by the distribution of forcing elements such as bedrock outcrops and woody debris.

Figure 5.8 Channel behavior in bedrock-confined rivers
The behavior of bedrock-confined rivers is dictated by sculpting of bedrock/boulder geomorphic units. A continuum of these features is found along a slope-induced energy gradient. (a) shows a sequence of step–pool sequences along a steep section of the Sangainotaki River (Three Steps of Waterfall), Japan, while (b) shows a run–lateral bar complex along a lower slope section of the same river. (c) depicts the assemblage of geomorphic units formed on a bank-attached compound bar along the upper Kangaroo River, New South Wales, Australia, and (d) shows a cascade–pool–run complex in a gorge near Launceston, Tasmania, Australia.

In these settings, geomorphic units are generated, locally redistributed, and reshaped during high magnitude, low frequency flow events. Bedrock may be sculpted through corrasive action, resulting in the formation of potholes and plunge pools. All but the coarsest fractions are flushed downstream. In a sense, this forced morphologic condition constrains river behavior at anything other than extreme flow stages. Low flow stage events exert negligible impact on the geomorphic structure of these rivers. Bedrock and boulder geomorphic units dissipate flow energy. Fine-grained materials that locally accumulate in pools and behind obstructions are flushed by subsequent events.

5.6.2 Channel behavior along rivers in partly-confined valley settings

Bedrock exerts a dominant influence on channel behavior along partly-confined rivers, limiting the potential for vertical adjustment. Lateral adjustment varies markedly, with potential for channel contraction or expansion adjacent to discontinuous pockets of floodplain, whereas in other sections the channel is set in place. Flow alignment relative to valley configuration, the distribution of floodplain pockets, and the location of bedrock outcrops are key determinants of energy dissipation within the channel. Given the confined nature of these channels, high-energy instream geomorphic units dominate, the distribution of which is influenced by local channel bed slope. These bedload-dominated channels tend to be relatively narrow and deep. Composite banks may facilitate channel expansion through undercutting and slumping at higher flow stages. Dissipation of energy at channel margins promotes the formation and reworking of bank-attached geomorphic units, especially adjacent to floodplain pockets (Figure 5.9). These bank-attached features can be erosional or depositional forms. Hence, the assemblage of instream geomorphic units typically comprises a mix of depositional forms such as compound bank-attached bars (point and lateral) and benches and/or erosional forms such as ledges, forced pool–riffle sequences, and bedrock steps.

At low flow stage, flow is confined to forced pool–riffle sequences. Fine-grained materials accumulate in pools. As flow stage increases and flow remains confined to the channel, energy increases and a range of instream geomorphic units are formed and reworked. As flows overtop bar surfaces, flow alignment within the channel shifts, reworking bar surfaces. Chutes and ramps may form as flow short-circuits the bend (e.g., McGowen and Garner, 1970; Gustavson, 1978). During waning stages, fine-grained sediment may accumulate around vegetation as ridges on compound bars. At bankfull stage and beyond, both instream and floodplain features are formed and reworked. Riffles are reworked and pools are scoured. In some instances, high-energy flows may sculpt bedrock, typically by corrasion. Lateral bars and inset features may form at channel margins, comprising bedload and suspended load materials. These geomorphic units produce a compound channel shape, with a series of steps in the form of flat topped, elongate benches or inset features that reduce channel dimensions.

5.6.3 Channel behavior along laterally-unconfined high-energy rivers

Laterally-unconfined, high-energy rivers are formed on high slopes and tend to be bedload dominated. Noncohesive banks are easily reworked, re-sulting in wide, shallow, single or multichanneled networks with braiding and wandering tendencies (Table 5.4). These mobile channels are also prone to vertical adjustment, exemplified by net aggradation of braid and alluvial belts (Figure 5.10). The landscape settings in which these rivers are found typically induce significant sediment delivery to channels, prompting the development of mid-channel bars. While the macrochannel has an irregular shape, individual channels tend to be symmetrical (or asymmetrical at bends). Thalweg shifting induces recurrent reworking of bedload in these dynamic systems. The presence of midchannel bars indicates that bed material is too coarse to be carried (i.e., the system is competence limited), or the volume of material is too great to be transported (i.e., the system is capacity limited). The latter scenario generally prevails along sand-bed rivers, where highly sediment charged channels tend to be characterized by transverse bars (or ribs) or have a planar bed dominated by sand sheets (e.g., Smith, 1970). These conditions are common along braided and low sinuosity sand-bed rivers.

The downstream gradation of bar types along braided or wandering gravel-bed rivers reflects energy conditions and sediment availability (Church and Jones, 1982). Riffles and runs are commonly observed between bars, while pools form at points of flow convergence. Bars build downstream and vertically as materials are deposited around a coarse bar head. With continued deposition in the lee of the bar head, downstream fining results. Individual bars show a range of accretionary patterns, reflecting long-term aggradation and downstream and lateral migratory tendencies. The placement of recently deposited platforms that comprise less coarse material guides insight into patterns of accretion (e.g., Williams and Rust, 1969). Once established, bar position and shape have a significant effect on flow alignment and the formation of adjacent geomorphic units. At moderate–high flow stages, bars are submerged and flow alignment (i.e., thalweg position) shifts to a down-valley orientation. Bars are subjected to dissection and modification, producing compound features with a range of platforms, chute channels, and ridges. The nature and pattern of these features provide insight into geomorphic adjustments at differing flow stages and the history of flow events. If vegetation becomes establishes on these surfaces, islands are formed.

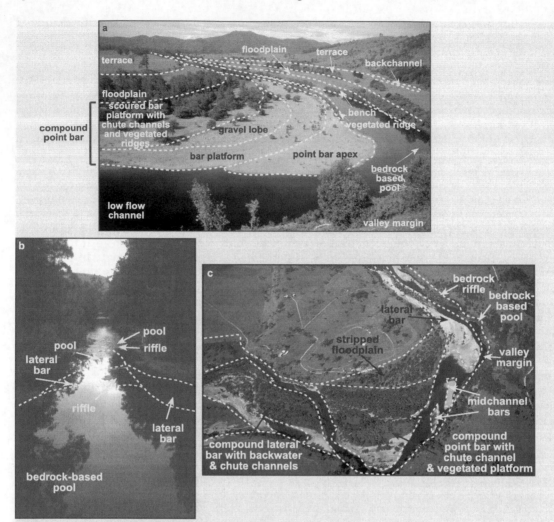

Figure 5.9 Channel behavior in a partly-confined valley setting
Significant variability in the assemblages of geomorphic units found along partly-confined valley setting reflects the degree to which bedrock confines the channel and aligns flow at different flow stages. In general, high-energy instream and floodplain geomorphic units are formed. (a) shows a range of geomorphic units on a compound point bar on a bedrock-confined bend of the Macleay River, New South Wales, Australia. (b) shows bedrock-induced pool–lateral bar–pool complexes along the Kangaroo River, New South Wales, Australia. (c) shows a series of stepped floodplain-terraces, providing evidence of stripping, along the Clarence River, New South Wales, Australia.

5.6.4 Channel behavior along laterally-unconfined, medium-energy rivers

Laterally-unconfined, medium-energy rivers are formed on moderate slopes and tend to be mixed load or bedload dominated systems. Bank-attached forms are observed more frequently than in higher-energy settings, where midchannel geomorphic units are dominant. Noncohesive or composite banks are readily eroded and reworked, giving the channel significant capacity to adjust its form both laterally and vertically. Lateral migra-

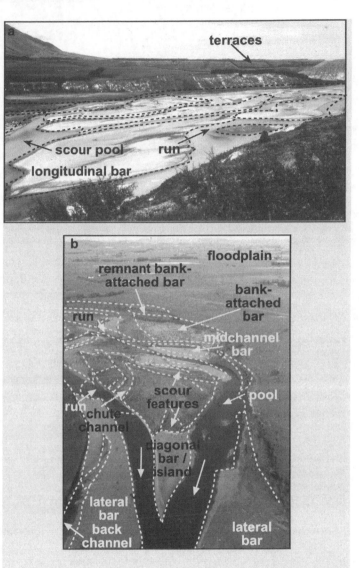

Figure 5.10 Channel behavior in laterally-unconfined, high-energy rivers
Laterally-unconfined, high-energy rivers on steep slopes tend to be characterized by wide, shallow, bedload-dominated, multi-channeled systems with braiding and wandering tendencies. (a) depicts the braided Rakaia River, New Zealand with an assemblage of longitudinal bars, runs, and shallow scour pools. Terrace–floodplain surfaces are evident in the background. (b) A wandering gravel bed reach of the Waiau River, New Zealand, characterized by compound midchannel and lateral bars.

tion results in a meandering channel planform (Figure 5.11). Along each bend, flow is deflected from the convex to the concave bank, resulting in deposition of point bars along convex banks and erosion of the concave bank. Bank-attached forms are reworked less frequently than midchannel bars in high-energy situations. A genetically linked assemblage of point bars, pools, and riffles is generat-ed. Alternating deeps (pools) and shallows (riffles) are characteristic of both straight and meandering channels with bedload or mixed load transport regimes. A regular pattern of scour and deposition reflects alternation of convergent and divergent flow along the channel, combined with secondary circulation currents (Keller and Melhorn, 1973). Surface flow convergence at the pool induces a de-

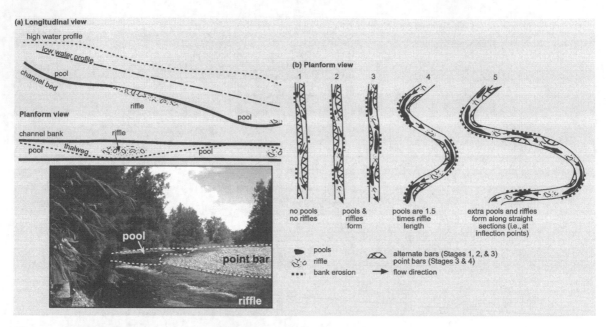

Figure 5.11 Channel behavior in laterally-unconfined, medium-energy rivers
Laterally-unconfined, medium-energy rivers are found on moderate slopes and tend to be mixed load or bedload dominated systems. They are characterized by bank-attached bars, such as point bars, with pool–riffle sequences. A self-sustaining process maintains these features (Keller, 1971). The photograph is from the Gloucester River, New South Wales, Australia.

scending secondary current which increases the bed shear stress and encourages scour, while surface flow divergence at the riffle produces convergence at the bed and thereby favors deposition. Once initiated, bed perturbations interact with flow to generate conditions necessary for the maintenance of riffle–pool sequences. The hypothesis of velocity (Keller, 1971) or shear stress (Lisle, 1979) reversal highlights that there is greater flow competence in pools than over riffles for a certain range of flows. This provides a self-sustaining mechanism by which patterns of erosion and deposition selectively transport and deposit material of differing caliber across the bed, maintaining the pool–riffle–point bar complex.

At low flow stage, broken water is evident over riffles, while pools trap finer-grained sediments in relatively still water. As flow stage increases, point bars on the insides of bends push the thalweg, and concentrate flow energy, along the outside of a bend. This initiates scour along the concave bank where a pool develops and an asymmetrical chan-

nel is formed. At the inflection point between bends, flow is distributed evenly over the channel bed as the thalweg is positioned along the central axis of the channel. Gravel accumulations in riffles are generally lobate in shape and frequently slope alternately first towards one bank and then towards the other, so that the flow tends to follow a sinuous path even in a straight channel. The presence of lateral bars in straight reaches enhances the meandering tendencies of the thalweg and reinforces the point bar–pool–riffle assemblage. Flow deflection and patterns of scour/sedimentation dictate channel geometry. This, in turn, influences the distribution of flow energy within a reach, fashioning the pattern of sedimentation in areas of lower flow energy. In a meandering situation, point bars form on the insides of bends when secondary flow circulations push sediments towards the convex bank, where they are deposited. As flow stage increases, this process is reinforced by deposition of sediments around-the-bend and atop the bar surface. Depending on flow alignment and erosive potential, a range of erosional and depositional geomorphic units such as chute channels, ramps, and ridges can form on these bar surfaces, producing a compound point bar (e.g., Brierley, 1991).

5.6.5 Channel behavior along laterally-unconfined, low-energy rivers

Laterally-unconfined, low-energy rivers that form on very low slopes can be characterized by low sinuosity single-channeled rivers, but multichanneled networks are more commonly observed. In these suspended load situations, banks comprise vertically accreted silt and clay. These materials

limit the ability of channels to adjust laterally, such that a trench-like configuration, with a low width : depth ratio, is adopted. Symmetrical channels form along low sinuosity variants of these rivers, while asymmetrical channels occur along meandering variants. The range of instream geomorphic units is limited because of the lack of bedload caliber materials. Drapes are common, especially along bank-attached features. In some settings, clay-pellets may act as sand-sized bedload materials (Maroulis and Nanson, 1996). A typical assemblage of geomorphic units comprises occasional lateral or point bars, ledges, suspended load benches, scour pools, and runs (Figure 5.12). Stepped banks and compound channel geometries,

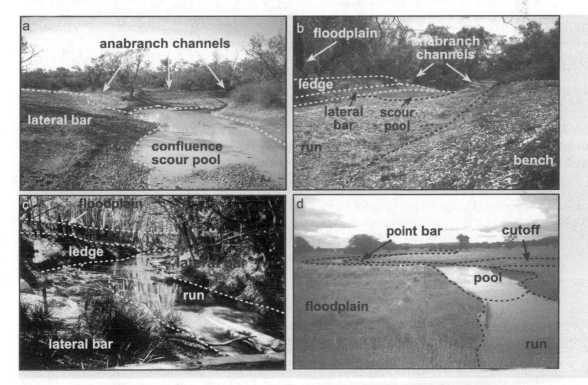

Figure 5.12 Channel behavior in laterally-unconfined, low-energy rivers with continuous channels
Laterally-unconfined, low-energy rivers are suspended load systems on very low slopes. They can be single or multichanneled networks. Cohesive, fine-grained bank materials limit the capacity of these rivers to adjust. The range of instream geomorphic units tends to be relatively simple. (a) and (b) show the assemblage of lateral bar–pool, and ledge-bench features found at the confluence of anabranches along Cooper Creek, Queensland, Australia. (c) and (d) show the assemblages of instream geomorphic units in single channeled fine-grained meandering rivers. (c) shows a relatively simple lateral bar–run complex along Bucca Bucca Creek, northern New South Wales, Australia, while (d) shows a pool–run sequence along Wingecaribee River, Southern Highlands, New South Wales, Australia. In the latter instances, floodplains are relatively flat, but cutoffs and paleochannels may form when the channel avulses.

as such, may reflect erosional and/or depositional scenarios that occur at differing flow stages (or the history of flow events). Shallow runs commonly characterize sections of planar bed between pools. Run–pool assemblages are common along meandering fine-grained rivers, low sinuosity fine-grained rivers, and anastomosing rivers.

5.6.6   Channel behavior along laterally-unconfined, bedrock-based rivers

Laterally-unconfined, bedrock-based rivers are found on low slopes, under low-energy conditions and are suspended load dominated (e.g., Heritage et al., 2001). The key difference between these rivers and their fully alluvial counterparts is that their beds are confined by bedrock. Sculpted bedrock forms are prominent (Figure 5.13). Channel shape tends to be irregular in these single or multichanneled networks. Lateral adjustment is limited because of the cohesive, fine-grained nature of the floodplain.

At low flow stage, flow paths reflect the distribution of scour features in bedrock. Suspended load materials line the bedrock pools and adjacent surfaces. At high flow stage, bedrock sculpting may occur. Unlike other bedrock-based rivers (e.g., confined and partly-confined rivers), the position, type, and distribution of bedrock geomorphic units is influenced less by slope (given the low slope conditions) and more by the resistance to erosion of the underlying bedrock. Weak strata are eroded to produce pools, while more resistant elements tend to be dominated by runs and bedrock platforms.

In anastomosing variants, flow divides around bedrock core bars and ridges, accentuating scour in adjacent secondary channels. During waning stages, suspended load materials drape the surfaces of bedrock ridges. Once colonized by vegetation, sediment trapping is enhanced, and the bedrock core bar builds vertically (Wende and Nanson, 1998).

5.6.7   Channel behavior along discontinuous watercourses

In general, discontinuous watercourses are found in wide, unchannelized valleys where low-energy conditions promote the dissipation of floodwaters and the deposition of suspended load materials.

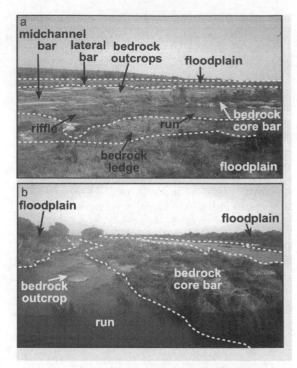

Figure 5.13   Channel behavior in laterally-unconfined, low-energy rivers with continuous, bedrock-based channels
Laterally-unconfined, bedrock-based rivers on low slopes tend to be suspended load systems. The assemblage of instream geomorphic units along these vertically confined rivers is dominated by sculpted forms. Bedrock core bars form as sediments are deposited along instream bedrock ridges during the waning stages of flood events. (a) and (b) show these features along the Sabie River, South Africa.

Instream geomorphic units are seldom evident other than in discontinuous gullies, though swamps, floodouts, pools, and ponds may be observed (see Figure 5.14). At low flow stage, flow is confined to depressions or preferential drainage lines on the surface of the valley fill. Suspended load deposits accrete in these depressions. At higher flow stages, the entire valley floor may be covered with a sheet of water. Any bedload materials are rapidly deposited as flow energy is dissipated across the valley floor, resulting in the formation of a floodout (e.g., Brierley and Fryirs, 1998). If

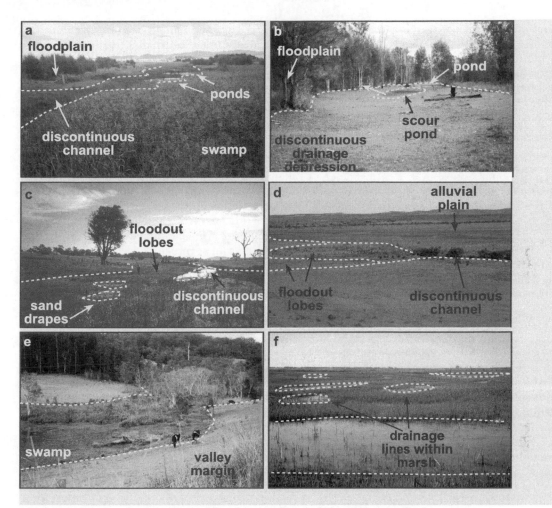

Figure 5.14 Channel behavior in laterally-unconfined, low-energy rivers with discontinuous channels
Discontinuous watercourses are generally found in wide, unchannelized valleys. Low-energy conditions dissipate floodwaters over the valley floor. Geomorphic units include ponds, drainage depressions, floodouts, and swamps, as shown for various systems in New South Wales, Australia: (a) Mulwaree Ponds, Wollondilly Catchment, (b) Six Mile Swamp Creek, Richmond Catchment, (c) Frogs Hollow Creek, Bega catchment, (d) near White Cliffs, (e) Bellinger catchment, and (f) Macquarie Marshes.

enough energy is created, ponds may scour along preferential drainage lines.

If these rivers incise, an entrenched channel with a symmetrical or compound form is produced. Concentration of energy within the channel produces an array of geomorphic units, including bank-attached and midchannel features. If bank materials are relatively cohesive and resist-

ant to change, the stepped channel cross-sectional morphology may reflect erosion of flat topped, elongate forms at channel margins (i.e., ledges). These are common along channelized fill rivers with a suspended load transport regime. Where a sand substrate dominates, deposition adjacent to the bank produces benches. Both of these features are formed under high flow stage conditions, when erosion and deposition occur along the channel margin.

The combination of form–process associations in channel and floodplain compartments, and their interactions, defines the summary behavioral regime of any given reach. Inevitably, this interaction has limited meaning for some systems (i.e., those without floodplains). This very distinction emphasizes the importance of valley confinement, and the associated distribution and character of floodplain pockets, as a critical determinant of river types. Adjustments to channel position on the valley floor, and associated insights into longer-term river behavior, are considered in the following section.

5.7 Adjustments to channel position on the valley floor

Implicit to the scalar theme applied to characterize river morphology and appraise behavioral regimes in this book is the notion that adjustments to channel position on the valley floor (i.e., planform-scale river behavior) influence morphological adjustments at the scale of channel geometry, the associated assemblage of instream geomorphic units, and the range of bedform-scale features that are likely to be observed on the channel bed. Mutual adjustments among these morphological attributes describe the behavioral regime of any given reach or river type. Adjustments to channel position on the valley floor are influenced by the energy available to form and rework floodplains and alignment of flow beyond bankfull stage. Planform adjustments are recorded by the assemblage of floodplain geomorphic units. Interpreting the form–process associations of these units, and their history of formation and reworking, provides insight into long-term river behavior. Assessment of the capacity of the channel to adjust its position across the valley floor entails analysis of adjustments to channel multiplicity, sinuosity, and lateral stability. In this section, adjustments to various combinations of these attributes are shown using examples of different types of river.

Adjustment to channel planform is valley setting dependent. Obviously, this characteristic is irrelevant for confined, bedrock-controlled reaches, as the channel is stable and floodplains are absent.

In partly-confined valley settings, accommodation space allows the channel to locally adjust adjacent to discontinuous floodplain pockets. However, bedrock features and terraces limit the potential for planform adjustment, but existing sediment stores are prone to reworking because of the high-energy setting. Valley configuration dictates flow alignment at overbank stage, often resulting in short-circuiting of the floodplain pocket (Miller, 1995). A limited set of planform adjustments and floodplain responses result. Capacity for adjustment to channel planform is maximized in laterally-unconfined valley settings. A wide range of floodplain forms characterizes these areas of long-term aggradation. Floodplain formation may entail a mix of progressive lateral and vertical accretionary processes, or braid accretion and abandoned channel infilling, along with various forms of reworking. In some settings, catastrophic adjustments such as channel avulsion form part of the natural behavioral regime. Bedload or mixed load systems tend to be characterized by lateral accretion or channel abandonment processes, whereas vertical accretion is dominant in suspended load systems.

The assemblage of floodplain geomorphic units represents a mix of formation and reworking processes (Chapter 4). Analysis of channel-marginal geomorphic units guides interpretation of channel–floodplain linkages, as they influence flow and sediment fluxes from the channel to the floodplain. In some cases, floodplain pockets record phases of formation and reworking at differing flood stages, reflecting patterns of flow alignment and energy. This may result in significant pocket-to-pocket variability in the geomorphic unit assemblage for the same type of river (e.g., Ferguson and Brierley, 1999a). Assessment of the summary mix of these attributes can be used to characterize the behavioral regime of these types of river. In the discussion that follows, packages of floodplain geomorphic units are interpreted to assess forms of planform adjustment for different types of floodplain.

5.7.1 Flat-topped, vertically accreted floodplains

Flat-topped floodplains are typically characterized by simple assemblages of geomorphic units. Two

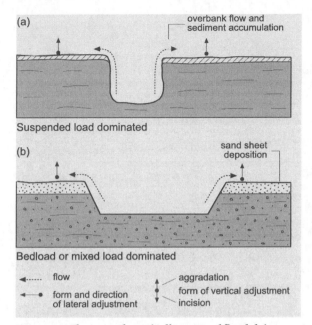

(a)

overbank flow and
sediment accumulation

Suspended load dominated

(b)

sand sheet
deposition

Bedload or mixed load dominated

◄······ flow

●──● form and direction
of lateral adjustment

↕ aggradation

● form of vertical adjustment

↕ incision

Figure 5.15 Flat-topped, vertically accreted floodplains
Flat-topped vertically accreted floodplains may be
subjected to incremental rates of accumulation in
suspended load settings (a). In contrast, bedload or
mixed load systems may experience intermittent sand
sheet deposition (b).

variants can be differentiated. In the first instance,
fine-grained materials vertically accrete in suspended load rivers with stable channels (Figure
5.15a). Equivalent looking forms may also be produced in more active systems in which bedload
deposits are deposited atop the floodplain as relatively continuous sand sheets (see Figure 5.15b).
The latter variant is prone to extensive reworking
via channel expansion or floodplain scour.

5.7.2 Progressive, lateral channel migration with ridge and swale formation

In laterally-adjusting rivers, patterns of migration
can be unidirectional or multidirectional. In the
latter instance, combinations of bend extension,
translation, and rotation produce bends of varying configuration (Hickin, 1974). Series of
accretionary ridges and intervening swales record
the migration pathway, marking former positions
of the channel (Sundborg, 1956). Ridge and swale

topography is genetically related to scroll bars that
form on the insides of bends (Figure 5.16a). During
bankfull conditions, the high velocity filament is
located along the concave bank of a bend. Patterns
of helical flow erode the concave bank of the bend
and transfer sediments to the point bar. At bankfull stage this scroll bar accretes vertically. As the
channel shifts laterally, the scroll bar becomes
incorporated into the floodplain. Over time, a
number of ridges with intervening swales can be
formed. Until overbank deposits smooth out the
floodplain surface, the hummocky appearance of
former channels is retained on the inside of the
bend.

As bend curvature increases, the potential for
thalweg scour increases, leading to greater bank
erosion, more rapid rates of bend migration, and
wider spacing of ridges and swales (Hickin and
Nanson, 1975, 1984). Bend migration reaches a
maximum when the ratio of radius of curvature (r_c)
to channel width (w) (i.e., r_c/w) approximates 3.0.
Beyond this stage, the bend migrates along a different erosional axis or is cutoff.

The focal point of bank erosion in any given
bend reflects channel alignment (i.e., sinuosity)
and flow stage. Commonly, maximum flow velocity impinges on the concave bank progressively
further downstream, increasing the sinuosity and
generating tortuous meanders (Bridge, 2003). At
low flow stage, high flow velocity occurs towards
the upstream end of the bend apex, inducing unidirectional bend extension (Figure 5.16b). At high
flow stage, concentration of erosion downstream
of the bend apex promotes bend translation (Figure
5.16c). Varying phases and patterns of erosion at
different flow stages promote multidirectional
bend migration, with various combinations of extension and translation leading to bend rotation
and the development of meander lobes (Figure
5.16d). Accentuation of bends increases channel
sinuosity.

The downstream translation or rotation of bends
may be hindered if the migration path comes into
contact with obstructions such as bedrock valley
margins or cohesive sediments at the margin of a
meander belt (Figure 5.16e). This produces an irregular meandering pattern. If obstructions on the
outsides of bends promote the development of
secondary flow circulation at flood stage, erosion
along the outside of the upstream limb of the bend

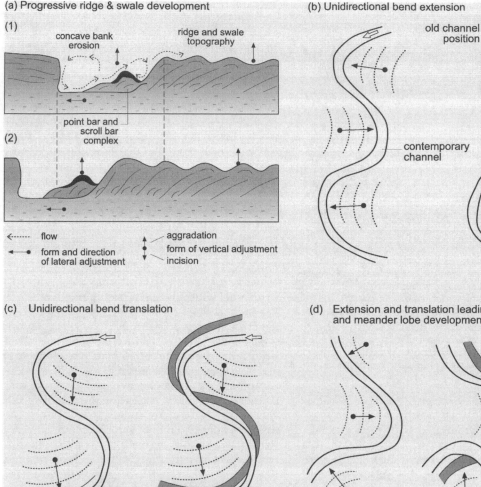

(a) Progressive ridge & swale development

(b) Unidirectional bend extension

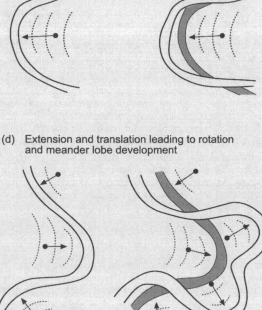

(c) Unidirectional bend translation

(d) Extension and translation leading to rotation and meander lobe development

(e) Translation and rotation leading to counter point accretion in a partly confined valley

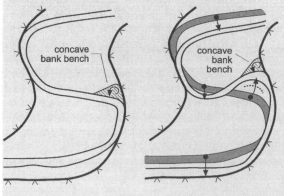

Figure 5.16 Ridge and swale floodplains and lateral channel migration patterns
A range of accretionary patterns on the inside of bends results in differing configurations of ridge and swale topography. (a) depicts the flow mechanisms involved in ridge and swale development while (b), (c), (d), and (e) depict (in plan view) various forms of bend development including bend extension, translation, rotation, and counterpoint accretion.

and counterpoint accretion create concave bank benches that progressively translate downstream (Hickin, 1986).

5.7.3 Meander cutoffs (billabong or oxbow development) and chute cutoffs

Meander cutoffs form whenever a meandering stream shortens its course by cutting off a bend. Slope is locally increased, enhancing prospects for future instability. Abandoned channels are left on the floodplain. Neck cutoff is the primary mechanism of meander loop abandonment. Such cutoffs occur late in the development of the loops, as a result of tightening of a bend (i.e., accentuated sinuosity) via extension and translation (Figure 5.17a). A new channel erodes the narrow neck of land be-

tween two loops (Figure 5.17b), or the loop is captured by the next bend upstream. Cutoff formation is a form of lateral instability that starts as a progressive adjustment, but can be catastrophic at the time of the cutoff. Bedload sediment rapidly plugs the ends of the abandoned channel to produce an oxbow lake (or billabong) (Figure 5.17c).

Chute cutoffs occur where part of a bend is short-circuited at high flow stage by erosive flows that are aligned down-valley rather than around the bend, generating a relatively straight channel. Differing degrees of infilling of paleochannels and meander cutoffs reflect varying rates of sedimentation from overbank flows (e.g., Erskine et al., 1992; Piégay et al., 2000). Abandoned channels may contain clay plugs that subsequently resist lateral migration of the channel. Differing forms of abandoned channels with differing dimensions, alignment, and degrees of infilling provide insight into river history.

5.7.4 Floodplain stripping

Floodplain scour or stripping is particularly pronounced in partly-confined valley settings. It occurs when the concentration of flow energy at high flow stage is sufficient to remove alluvium from floodplain pockets (Nanson, 1986; Warner, 1997). The resultant morphology varies depending on the position of the floodplain pocket on the valley floor, and hence the alignment of flow over the pocket. Two common morphologies result, namely stepped, flat-topped floodplains which comprise terraces or inset floodplain geomorphic units, and levee–floodchannel floodplains. Vertically accreted mud tends to form a flat-topped morphology, whereas vertically accreted sands are common in the levee–floodchannel variant. In the accretionary phase, the floodplain builds vertically. The resultant morphology may be relatively flat or inclined towards the valley margin (Figure 5.18a). If the rate of floodplain accretion is greater than channel bed accretion, progressively larger floods are required to produce overbank deposits (Figure 5.18b). Hence, in general, the thickness of bedding decreases vertically. Over time, flow energy becomes increasingly concentrated within the channel. In a high magnitude flood event (or sequence of moderate events), flow strips the floodplain to basal materials, whether bedrock or gravel

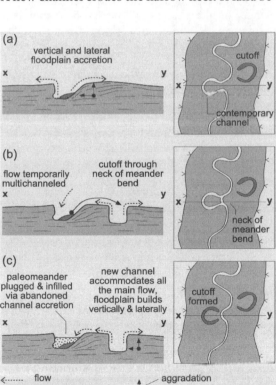

Figure 5.17 Cutoff formation
During high magnitude events, narrow necks on bends may erode, ultimately leading to channel abandonment and the formation of cutoffs. This figure depicts the series of events that lead to cutoff formation.

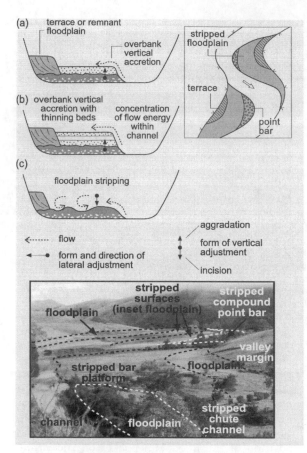

Figure 5.18 Floodplain stripping
Progressive vertical accumulation of floodplain deposits increases flow energy within the channel over time. As noted in the text, this may ultimately lead to stripping of the floodplain adjacent to the channel. This figure depicts the series of events that lead to floodplain stripping in a partly-confined valley setting. The photograph is from the Upper Manning River, New South Wales, Australia.

lag (Figure 5.18c). Planform adjustment along these rivers is dominated by channel expansion and lateral instability, while maintaining a single, low sinuosity channel within the partly-confined valley.

5.7.5 Accentuated levee–floodchannel complexes

In some partly-confined valleys, pronounced levees may be elevated up to 10 m above the low flow channel of bedload-dominated rivers. The levee tends to have an accentuated asymmetrical morphology, with a pronounced slope from the levee crest to the channel and notable inclination towards a floodchannel adjacent to the valley margin. In sandy substrates, fall velocities are reached quite readily (Pizzuto, 1987), and levees build vertically during overbank events as materials are deposited preferentially on the levee crest (Figure 5.19a). Over time, the incline from the levee towards the valley margin becomes more pronounced (Figure 5.19b). With further accentuation of the levee crest, the alignment of floodplain flows during high discharge events is accentuated along the valley margin where it short-circuits the channel. As a result, scour is induced along the valley margin and a floodchannel is formed (Figure 5.19c). Hence, unlike the levee, the floodchannel is a product of floodplain reworking processes. The depth of the floodchannel tends to increase down-pocket (e.g., Ferguson and Brierley, 1999a). In many instances, the basal section of the floodchannel is elevated above the low flow channel (i.e., it lies perched within the floodplain). At lower flood magnitudes, when the entrance to the floodchannel is not breached, suspended load deposition may occur via backfilling as the exit (at the downstream end of the pocket) is the lowest entrance to the floodchannel. Reworking and scour in the floodchannel and accretion atop the levee accentuate the pronounced lateral relief of the floodplain.

5.7.6 Levee–backswamp–crevasse splay formation

Levee–backswamp–crevasse splay complexes are common in laterally-unconfined valleys with wide, continuous floodplains. In contrast to their counterparts in partly-confined valley settings, the levees in this setting tend to be relatively low features that dip gently to a backswamp at the distal valley margin (e.g., Fisk, 1944, 1947; Figure 5.20a). This reflects the smaller grain size and lack of valley confinement. The backswamp sits at a lower elevation than the levee, effectively disconnected from bedload movement in the channel. However, if high magnitude events breach the levee, and sufficient bedload materials are moved from the channel to the floodplain, crevasse splays

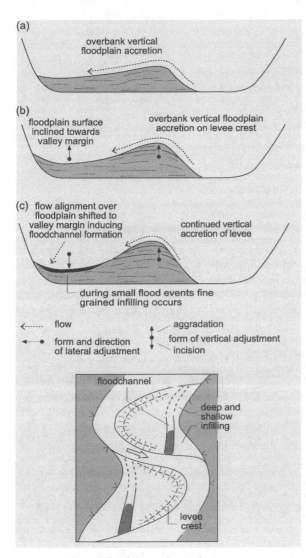

(a)
overbank vertical
floodplain accretion

(b)
floodplain surface
inclined towards
valley margin

overbank vertical floodplain
accretion on levee crest

(c) flow alignment over
floodplain shifted to
valley margin inducing
floodchannel formation

continued vertical
accretion of levee

during small flood events fine
grained infilling occurs

<----- flow
●──● form and direction
of lateral adjustment

aggradation
form of vertical adjustment
incision

floodchannel

deep and
shallow
infilling

levee
crest

Figure 5.19 Levee–floodchannel complexes
Some floodplains in partly-confined valley settings have
distinct proximal–distal topographic variability
accentuated by levee–floodchannel complexes. Over
time, floodchannels may scour or be partially infilled.

may be produced on the proximal floodplain slope
(Figure 5.20d). As energy is dissipated over the
floodplain, it forms a discrete splay-like feature
that thins and become finer grained distally.

Levee formation occurs via vertical accretion
on the levee crest during overbank flood events.

Given the relatively high-energy conditions at
the channel marginal, the coarser fraction is
deposited first (Figure 5.20b). The reduction in
energy from the proximal to the distal floodplain
results in textural segregation, as suspended load
materials slowly accrete in the backswamp. These
floodplains either comprise a mix of lateral and
vertical accretion deposits where a meander belt is
formed on the proximal floodplain, or they are
dominated by vertical accretion deposits (Figure
5.20c).

5.7.7 Avulsion

Wholesale abandonment of channels and adoption
of a new channel course via avulsion processes is
recorded through preservation of paleochannels
that remain perched on the floodplain (Figure
5.21). Unlike localized meander cutoffs, these paleochannels tend to be relatively long sections of
river. Varying degrees of infilling may be observed.
Generally, an area of floodplain is preferentially
reworked at high flow stage. This is typically
associated with less resistant vegetation, lower
elevation, or steeper slopes (Figure 5.21a).
Wholesale shift in channel position leaves a
perched/elevated paleochannel (Figure 5.21b). As
the new channel develops and builds the adjacent
floodplain, the paleochannel is infilled via overbank flows (Figure 5.21c). Individual paleochannels may have different geometries and planform
configurations, ranging from relatively straight
to tortuously meandering, from laterally stable
reaches to reaches with significant capacity to adjust. These features preserve a record of paleoriver
planform and channel shape/dimensions.

Avulsion is a form of lateral instability that induces catastrophic adjustments to river planform
(Schumm, 1985). The position of the channel on
the valley floor adjusts in response to aggradation
of a meander belt, fine-grained alluvial plain, or
braid belt (see Figure 5.21). Coarse bedload rivers
with noncohesive banks and high stream power
are susceptible to relatively sudden and major
shifts in channel position (first order avulsion;
Figure 5.21a, b, c). Accentuated rates of floodplain
accretion in a meander belt result in the channel
becoming perched, prompting avulsion to a lower
position on the valley floor. Secondary floodplain
channels in fine-grained vertically-accreted multi-

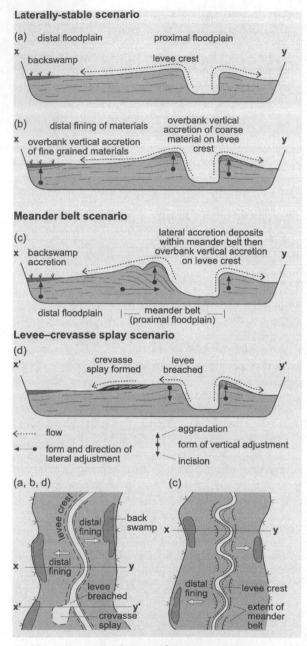

Laterally-stable scenario

(a) distal floodplain   proximal floodplain

x backswamp   levee crest   y

(b) distal fining of materials   overbank vertical
                                                   accretion of coarse
x overbank vertical accretion   material on levee   y
   of fine grained materials   crest

Meander belt scenario

(c)   lateral accretion deposits
                                                  within meander belt then
x backswamp   overbank vertical accretion y
   accretion   on levee crest

distal floodplain |— meander belt —|
                         (proximal floodplain)

Levee–crevasse splay scenario

(d)
x'   crevasse   levee
                 splay formed   breached   y'

←------ flow   — aggradation
                                        form of vertical adjustment
←—•— form and direction of
      lateral adjustment   incision

(a, b, d)   (c)

levee crest
           distal   back
           fining   swamp   x   y
x distal
   fining   y
                                  distal   levee crest
           levee   fining
           breached
x'   y'   extent of
           crevasse   meander
           splay   belt

Figure 5.20 Levee–backswamp formation
In relatively wide, laterally-unconfined settings, backswamps may develop in distal areas of floodplains. These features may be texturally segregated from the channel, comprising an array of fine-grained suspended load deposits or even peat development. Three scenarios of channel–floodplain connectivity are presented: (a) and (b) laterally-stable scenario, (c) meander belt scenario, (d) levee–crevasse splay scenario.

channeled networks with flashy flow regimes are prone to reoccupation at differing flow stages (second order avulsion; Figure 5.21d). Finally, thalweg shift refers to relatively minor switching of channels within a bedload-dominated braid train (third order avulsion) (Figure 5.21e). Multiple channels adjust around midchannel bars and islands. The formation and build up of midchannel bars and islands during large flood events leaves them elevated beyond the range of small–moderate flood events, eventually inducing avulsion and channel abandonment.

5.7.8 Cut-and-fill processes

A distinct set of geomorphic units may be evident along unchannelized river courses, in which the valley fill accretes vertically over time as energy is dissipated over a wide, swampy surface (e.g., Prosser et al., 1994; Fryirs and Brierley, 1998; Figure 5.22a). These valley fill deposits range from organic-rich mud through to sand. The fill may grade in slope to the valley margin, with wetlands evident. The entire valley floor may comprise a swamp. Elsewhere, discrete ponds may be observed. Discontinuous channels provide key insights into prospective changes. If the valley fill becomes dissected, and a discontinuous gully develops, floodout deposits are spread across the valley floor in the form of a shallow fan. Once initiated, incision proceeds rapidly via headcut erosion, producing a continuous channel (Figure 5.22b). This is quickly followed by channel expansion (Schumm et al., 1984). Initially, the incised channel tends to be characterized by a near-homogeneous sand sheet, as large volumes of material are released (Figure 5.22c). Eventually, a diverse array of within-channel geomorphic units may be found, promoting the development of a compound channel. During the filling phase, coarse bedload materials vertically accrete within the incised trench. Ultimately, fine-grained deposition may be reinstigated in swampy environments. If the channel reincises, one set of geomorphic units becomes inset within another. This may include benches, inset features, and terraces. The lateral stability of the (dis)continuous channels varies over time. While the intact fill is relatively stable, incised channels may experience significant adjustment in both vertical and lateral dimensions.

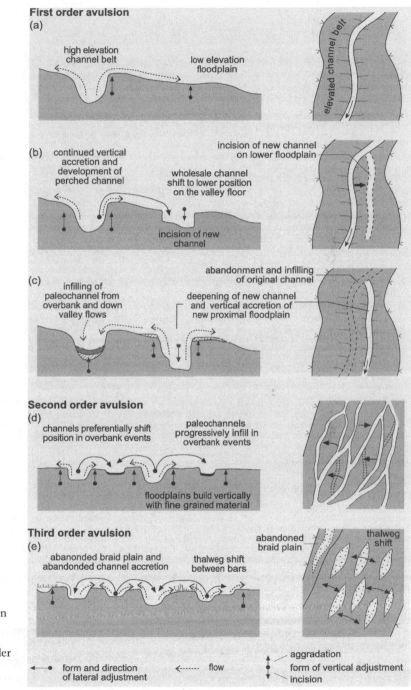

Figure 5.21 Avulsion
Wholesale shifts in channel position can occur in a range of different scenarios. These are referred to as first order (a), (b), and (c), second order (d), and third order (e) avulsion (see text).

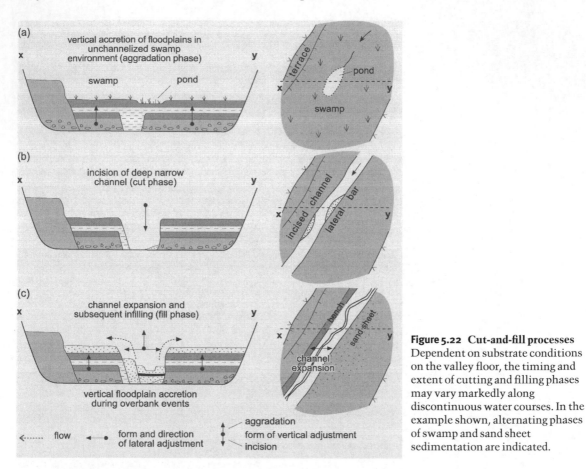

Figure 5.22 Cut-and-fill processes
Dependent on substrate conditions on the valley floor, the timing and extent of cutting and filling phases may vary markedly along discontinuous water courses. In the example shown, alternating phases of swamp and sand sheet sedimentation are indicated.

5.8 Use of geomorphic units as a unifying attribute to assess river behavior

Differing attributes of river morphology provide differing insights into forms of river adjustment. The most dramatic adjustments are often related to planform adjustments at the reach scale. While local adjustments to sinuosity brought about by development of a cutoff, channel shifting of a braided river, or channel abandonment in an anastomosing channel network characterize the behavioral regime of a river, considerable potential exists in these settings for wholesale changes to channel configuration (see Chapter 6). However, planform-scale adjustments are not pertinent for all rivers, particularly those operating in confined and partly-confined valley settings where the capacity for adjustment is limited.

Hence, reach-scale analyses of channel planform adjustments do not provide a unifying theme with which to appraise alterations to river character and behavior across the range of river diversity. Similar limitations are evident in analyses of channel geometry. Channel shape and size may be highly sensitive to local-scale factors, and they are not relevant attributes with which to analyze discontinuous watercourses. Indeed, channel geometry is perhaps a better guide to geomorphic condition of a reach rather than being viewed as an innate attribute of river character and behavior. Finally, bedform-scale adjustments are indicative of flow conditions and sediment availability during the last flow event, but they do not necessarily provide a good indicator of how and why the channel, or the river itself, is adjusting.

Adjustments to the assemblage of geomorphic units along a reach provide the most consistent, insightful, and reliable attribute of river morphology with which to analyze system responses to disturbance events and related notions of river evolution. Geomorphic units are evident along all river types. They provide a record of form–process associations and the distribution of erosion and deposition along a reach. Changes to their assemblage and pattern indicate adjustments to river structure and function. The presence, character, and distribution of channel and floodplain geomorphic units can be interpreted to guide the history of preservation, indicating how a reach has changed over time. Working from the geomorphic unit assemblage as a starting point, direct linkages can be made to associated changes to channel planform, channel geometry, and bedform-scale features, whenever such attributes are pertinent to the type of river under investigation.

Analyses of geomorphic units provide a basis to appraise river adjustments at various scales of interaction. For example:

1 The assemblage of geomorphic units, channel alignment, and associated channel geometry (shape and size) influence the distribution of flow energy at any given flow stage, thereby determining the pattern of bedform-scale features and their likely preservation potential.

2 Adjustments to within-channel geomorphic units record the local balance of erosional and depositional processes that determines channel shape and size at any given cross-section. Alterations to channel geometry are recorded by differing types of midchannel and bank-attached bar forms or various sculpted (erosional) geomorphic units. Adjustments to channel capacity may alter the relationships between channel and floodplain forms and processes, modifying the periodicity and geomorphic effectiveness of phases of floodplain inundation.

3 Adjustments to the morphology and assemblage of floodplain geomorphic units may record a shift in the balance of depositional and reworking (erosional) processes or alteration to the type of floodplain forming processes (e.g., the balance of lateral and vertical accretion processes).

4 When combined, analyses of channel and floodplain geomorphic units enable interpretation of channel planform adjustments.

Just as modifications to these relationships provide critical guidance into the behavioral regime of a river, alterations to the assemblage of geomorphic units along a reach provide pivotal insights into river change, as considered in Chapter 6.

5.9 Synthesis

Assessment of river behavior appraises the range of mechanisms by which water moulds, reworks, and reshapes fluvial landforms, producing characteristic assemblages of landforms at the reach scale. Reach behavior is considered over timeframes in which boundary conditions, and associated flow and sediment fluxes, have remained relatively consistent. Differing types of river, in different settings, are characterized by marked variability in their expected ranges of behavior. Notions of "natural" river behavior encapsulate progressive, dynamic adjustments that may include catastrophic or negligible responses to differing forms of disturbance event. Adjustments around a characteristic state for a given interval of time are part of the natural capacity for adjustment for a given type of river. To effectively characterize the behavioral regime of a particular river type, mutual adjustments among a range of parameters must be analyzed at a range of spatial and temporal scales, focusing on adjustments to bed material organization (sediment transport), channel shape (instream processes), and channel planform (floodplain processes). Analysis of channel and floodplain geomorphic units provides an integrative tool with which to interpret river behavior at differing flow stages across the range of river types.

The river evolution diagram provides a conceptual tool that can be applied to interpret the range of river behavior that is evident across the spectrum of morphological complexity. This approach to analysis of river dynamics differentiates river behavior from river change, in that change results in a shift in river type, resulting in a new character and behavior (Chapter 6). This provides a basis to place human disturbance to river forms and processes in context of natural variability, as discussed in Chapter 7.

CHAPTER 6

River change

. . . . (T)ime is not a process in time.
Richard Chorley and Barbara Kennedy, 1971, p. 251

6.1 Introduction

Analysis of river change at the reach scale, viewed in context of changes to catchment-scale linkages, provides a critical basis to develop proactive river management programs. Appraisal of the pathway and rate of river evolution is required to assess whether ongoing adjustments are indicative of long-term trends or whether they mark a deviation in the evolutionary pathway. Such insights guide interpretation of the likelihood that the direction, magnitude, and rate of change will be sustained into the future. To perform these analyses, it is important to determine what components of a river system are likely to change over any given timeframe, and what the consequences of those changes are likely to be. As noted in Chapter 5, different types of river exhibit considerable variability in their behavioral regime. However, all river systems are subject to change.

Variability in the configuration and history of each system, the multitude of factors that drive change, and the range of system responses to disturbance events, make it difficult to isolate geomorphological principles with which to evaluate river changes in a coherent and consistent manner. Just because a particular type of river in a given system responds to an event of a given magnitude in a certain way, it does not mean that an equivalent type of river in an adjacent catchment will respond to a similar event in a consistent manner. Even if particular cause and effect relationships are well understood, the complex, chaotic, and nondeterministic nature of some systems may result in surprising responses to disturbance events. In ecological terms, unique responses may underpin biotic interactions that are the very components that should be sustained and/or enhanced in management programs that address concerns for healthy aquatic ecosystems. In accordance with these notions, disturbance-based paradigms form primary bases for inquiry in both geomorphology and ecology (e.g., Wu and Loucks, 1995; Phillips, 2003).

Instinctively, human attention is drawn to landscapes that are subject to change. Observations of bank erosion, river responses to flood events, anecdotal records of river adjustments, or analyses of historical maps and air photographs provide compelling evidence of the nature and rate of river changes (Table 6.1). Fluvial geomorphologists apply an array of techniques to record and interpret underlying causes of these changes. This chapter builds on the river evolution diagram presented in Chapter 5 to develop a set of procedures with which to interpret river change. River change is defined as *adjustments to the assemblage of geomorphic units along a reach that record a marked shift in river character and behavior.* In this chapter, analysis of Late Holocene river change frames the assessment of "natural" variability (Section 6.2). In Section 6.3, an appraisal is made of what components of rivers are able to change for different types of rivers. A series of intellectual constructs with which to interpret the complexity of patterns and rates of river changes, building on the river evolution diagram, is outlined in Section 6.4. Section 6.5 provides a synthesis of controls on the nature and rate of river change, differentiating between driving forces that initiate change and system characteristics that resist change. Temporal perspectives of river change are discussed in Section 6.6. These spatial and temporal themes are used to assess sys-

Table 6.1 Sources of evidence of river change (modified from Knighton, 1998, p. 264).

Direct observations	• Instrument records (rarely continuous) – typically applied over intervals from minutes to years • Photographic records • Ground surveys – such as repeated field surveys of cross-sections • Local anecdotal knowledge
Historical records	• Historical records, such as explorers and survey notes, paintings, bridge plans, newspaper articles, etc. • Archival maps, such as portion plans • Vertical aerial photographs • Historical photographs • Remote sensing and satellite imagery
Sedimentary evidence	Long-term records of river changes are largely derived from complex and generally incomplete sedimentary records. This entails analysis of: • Surface forms and paleochannels • Subsurface forms using techniques such as Ground Penetrating Radar • Exploratory sedimentary data (borehole, core-log, and trench data) • Bed material caliber, bedforms, paleocurrent indicators and the architecture of depositional units
Dating techniques	(1) *Relative methods* • Relative height • Organic remains • Artifacts (archaeological remains, especially pottery shards) (2) *Absolute methods* • Radioactive isotopes (^{14}C, ^{210}Pb, ^{137}Cs, etc.) • Dendrochronology • Thermoluminescence dating
Inferential reasoning	Ergodic hypothesis (space for time substitution)

tem vulnerability, susceptibility, and sensitivity in Section 6.7.

6.2 Framing river evolution in context of Late Quaternary climate change

Analyses of long-term river evolution and system responses to environmental changes provide an important context with which to interpret contemporary river character and behavior. Landscape history may exert a profound legacy on contemporary river forms and processes, whether viewed in terms of long-term controls on accommodation space and sediment availability, or responses to recent disturbance events that may shape how the system responds to subsequent events. For example, Pleistocene environmental change and neotectonic activity have left a strong imprint on basin physiology, sediment sources, and supply which, in turn, influence present-day channel and floodplain characteristics and activity rates

(Church and Ryder, 1972; Church and Slaymaker, 1989; Macklin and Lewin, 1997).

The nature, extent, and timing of climate changes vary markedly across the Earth. During the Quaternary Era (i.e., the last 2 million years), climatic changes have induced, and in turn been affected by, the repeated expansion and contraction of continental ice sheets in mid- and high-latitude regions, and related variations in precipitation regime in lower-latitude regions. Annual temperature between full glacial and interglacial conditions varied by more than 15 °C, with marked fluctuations in precipitation (Lamb, 1977).

Throughout the Quaternary Era, shifts in climate and vegetation cover, and associated changes to flow regime and material caliber and availability, have brought about significant changes to river morphology. While recognizing the imperative to consider regional or catchment-specific histories of geomorphic adjustments in the Late Quaternary, it is instructive to consider the nature and extent of disturbance and associated system responses for certain areas. For example, Knox (1995) identified three primary phases of fluvial activity in the northern hemisphere over the past 20,000 years:

• *20–14 ka BP.* During this glacial period, the Earth was colder and drier than at present, with extensive ice sheet and permafrost development in high and middle latitude areas and greater aridity over much of the tropics. Continental ice sheets dominated temperate latitudes, reaching their maximum extent by around 18 ka BP. Reduced vegetation cover generally increased runoff and clastic sediment supply. Erosion rates in river systems that drained former glacial margins were around an order of magnitude greater than contemporary rates, resulting in extensive valley floor aggradation. Large volumes of glacial meltwater and sediment were contributed to proglacial areas, where extensive braided river systems developed.

• *14–9 ka BP.* Major climatic changes saw the transition from glacial to postglacial conditions. The initial phase of deglaciation was characterized by very high rates of sediment supply. However, rates progressively subsided as vegetation cover increased. Meltwater sources and increasing precipitation maintained high river discharges. Over time, the reduced sediment load coupled with relatively high discharges brought about a transition from an aggrading to an erosional regime, and an associated shift from braided to meandering in many temperate valleys. In presently unglaciated river basins of moderate to high relief, beyond the influence of postglacial sea-level rise, rapid valley floor incision occurred during this period.

• *9–0 ka BP.* From 9–4 ka BP temperatures were higher than today across much of the northern temperate zone. Rates of incision slowed, or were reversed. An episode of climatic deterioration became most marked at around 3–2.5 ka BP. Various phases of river instability have been discerned over the past 9 ka, the most recent of which has been related to the Little Ice Age.

Notable contrasts in geomorphic responses to climate change were experienced in areas remote from direct impacts of glacial activity during the Late Quaternary. For example, remarkable changes in flow-regime during the Late Quaternary (past ≈ 100 ka) resulted in dramatic changes to river morphology in systems in southeastern Australia that were not affected by headwater glaciation (Nanson et al., 2003). These adjustments have been manifest primarily through adjustments to the flow regime, and the capacity of rivers to transport materials of differing caliber, rather than accompanying adjustments to sediment availability (as brought about by glacial activity). Various Late Pleistocene phases of alluvial activity were very powerful, and were quite unlike the relative inactivity of contemporary rivers in the region (Page and Nanson, 1996).

In general terms, large shifts in climate bring about river changes. However, the sensitivity of some alluvial channels may be such that relatively modest climatic changes can trigger major episodes of fluvial adjustment (Knox, 1993, 1995). This depends, in large part, on proximity to a threshold condition. Geomorphic responses to disturbance depend on the sensitivity of the system at the time of the event. Interpretation of catchment-specific patterns and rates of geomorphic response to long-term climate changes presents a critical basis with which to analyze contemporary river character and behavior.

6.3 The nature of river change

Landscape adjustments are conditioned by inherited (endogenic) controls such as slope, topography, and sediment type, and exogenous controls, such as climatic change, changes in sediment availability, and changes to vegetation cover. The distribution of forcing factors that drive river changes may be local (e.g., a landslide), regional (e.g., a storm), or continental (e.g., glaciation). Adjustments to the balance of impelling and resisting forces along a reach promote geomorphic change. If threshold conditions are breached, the reach may adopt a fundamentally different configuration. The nature and rate of river change are shaped, in large part, by the capacity for adjustment of the type of river under investigation, especially the ease with which the channel is able to adjust its position on the valley floor. These considerations are determined by the boundary conditions under which the reach operates, and system responses to flux boundary conditions. Hence, the capacity for adjustment of a gorge is much less than for an alluvial sand-bed river. If changes to the operation of fluxes drive the system to adopt a dif-

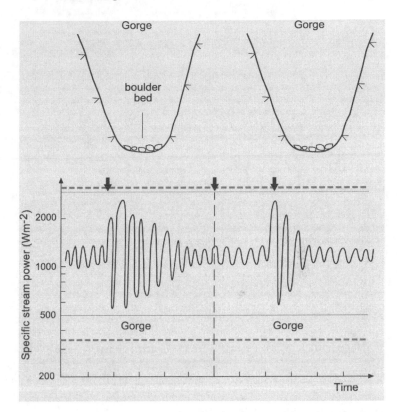

Figure 6.1 Pathways of river evolution within a gorge
The resilience of gorges is such that the system responds quickly to disturbance events, with negligible adjustments to river morphology.

ferent morphology and associated set of processes (i.e., the behavioral regime has changed), the reach has been subjected to metamorphosis (Schumm, 1969). These notions of river change are outlined schematically for a range of scenarios, with accompanying representations of the river evolution diagram, in Figures 6.1–6.6. Each scenario is discussed below.

Figure 6.1 represents an imposed river configuration such as a gorge. Coarse basal materials, typically cobbles or boulders, are the only materials that are retained in these settings for any length of time. Pebbles and finer materials are stored in transient forms or are flushed through these reaches. The distribution of erosional and depositional geomorphic units is fashioned primarily by local-scale variability in flow energy (determined primarily by local variability in slope and valley width), the volume and caliber of bed material, and variability in erosional resistance of the bedrock. The nature and rate of material influxes from upstream, tributaries, and valley marginal

slopes determine the suite of depositional forms, and the ease/frequency with which they are reworked. The range of erosional and depositional forms varies little in a reach-averaged sense over timeframes of tens, hundreds, or even thousands of years.

Schematic cross-sections indicating the type of river changes that can occur in a partly-confined valley are shown in Figure 6.2. In its previous configuration, the river had a ridge and swale floodplain, reflecting progressive lateral accretion as the channel migrated across the valley floor. When the channel became pinned against the valley margin, floodplain forms and processes were transformed to a vertically accreting system, with distal fining marked by the presence of a backswamp.

The capacity for river adjustment, and hence the likelihood that river change will be experienced, is greatest in laterally-unconfined valleys in aggradational environments. Considerable variation in river forms and processes is evident in these settings, fashioned largely by the sediment mix (sand

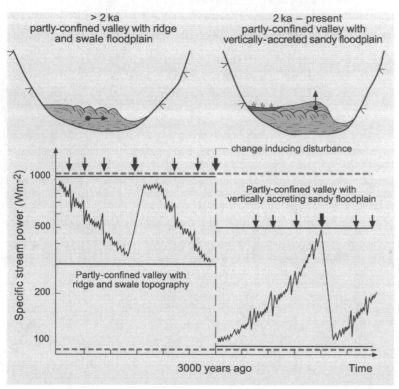

Figure 6.2 Transition from a laterally-migrating river to a vertically-accreting river in a partly-confined valley setting
Once lateral accretion of the Tuross River (South Coast, New South Wales, Australia) had pushed the channel against the valley margin in the period following postglacial adjustments to sea level (i.e., it attained its maximum sinuosity), the river was transformed into a lower energy system characterized by vertical accretion processes (Ferguson and Brierley, 1999a, b).

being the most readily transported and hence altered medium) and the range of formative events that drive system behavior. Channel and flood-plain forms and processes mutually adjust in these alluvial settings. Various scenarios that exemplify the types of river change that may be experienced are presented in Figures 6.3–6.6.

Adjustments to flow and sediment fluxes in the postglacial period commonly brought about changes from a braided to a meandering channel planform (e.g., Kozarski and Rotnicki, 1977; Starkel, 1991b; Figure 6.3). In the early postglacial interval, abundant sediment, highly variable flows, and negligible vegetation cover promoted the development of braided rivers. These bedload-dominated systems were characterized by a wide range of midchannel bar forms, with progressive shifting and abandonment of channels of varying size. Over differing time periods, but typically by the Mid-Holocene, many of these braided rivers had been transformed into mixed-load meandering systems, characterized by laterally-migrating sin-

gle channels with point bars and associated in-stream geomorphic units, and an array of laterally and vertically-accreted floodplain forms. This transition was brought about by progressive reduction in sediment availability in the postglacial era, reduced variability in discharge, and progressive encroachment of vegetation onto the valley floor.

A different set of Late Quaternary river changes was experienced by alluvial systems in Australia that were not subjected to the impacts of glaciation *per se*. However, climatic changes induced significant adjustments to river courses, over a range of timescales. In the example shown in Figure 6.4, a mixed load laterally migrating channel has been transformed into a slowly migrating suspended load river with a much smaller channel capacity. This transition reflects a decline in fluvial activity driven by changes to the discharge regime (Page and Nanson, 1996). Equivalent sets of climatically induced changes to prevailing flow and sediment fluxes brought about a different set of morpho-dynamic changes to other rivers in the region. In

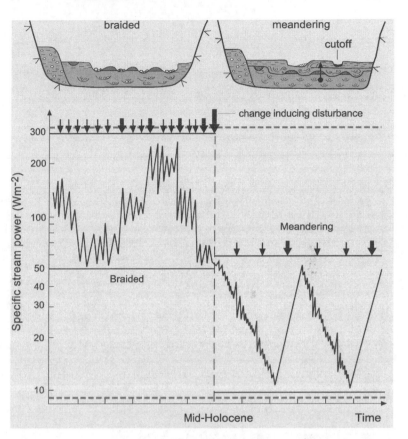

Figure 6.3 Transition from a braided to a meandering river in a laterally-unconfined valley setting
Many rivers subjected to high sediment loads in formerly glaciated landscapes were characterized by braided planforms. As vegetation became established and flow regimes became less variable in the Holocene, the energy of these braided systems diminished and they were transformed into meandering rivers. These meandering systems operated under much lower energy conditions and were characterized by greater degrees of inherent channel resistance. However, they were subjected to occasional phases of neck and/or meander cutoff or shifts in position on the valley floor.

Figure 6.5, an extremely stable low sinuosity sand-bed river with a vertically accreted floodplain has replaced a gravel-bed braided river (Nanson et al., 2003). Equally dramatic changes to river forms and processes in the Late Quaternary have been experienced in many low gradient upland settings of southeastern Australia. In Figure 6.6, a very thin veneer of gravel materials with a braided configuration lines the valley floor at the last glacial maximum (20–15 ka BP). Cool climatic conditions and negligible vegetation cover promoted removal of other valley floor deposits. Amelioration of climate and encroachment of vegetation onto the valley floor promoted the accretion of fine-grained materials and development of discontinuous watercourses. Throughout the Holocene, extended phases of valley floor accretion alternated with short phases of incision, resulting in a cut-and-fill type of river (Prosser et al., 1994).

The dramatic changes to river morphology and associated process domains outlined in Figures 6.2–6.6, admittedly over very long periods of time, occurred within the same valley setting. It was changes to flux boundary conditions, and the imprint of river history, that shaped the transition in river type. In each example, changes to the assemblage of channel and floodplain geomorphic units provide a critical guide with which to frame interpretations of river change. In the section that follows, these notions are added to the river evolution diagram.

6.4 Framing river change on the river evolution diagram

In Figures 6.1–6.6, the imposed boundary conditions, as shown by the outer band of the river evolution diagram, are viewed to be consistent over time. The width of the outer band increases from

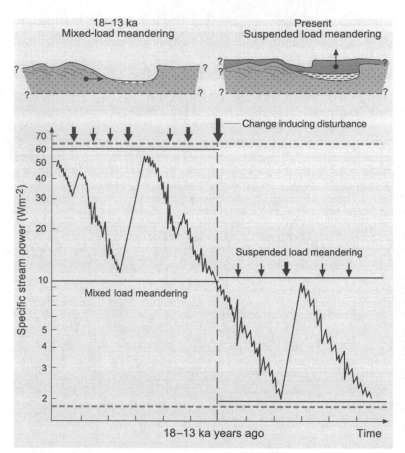

18–13 ka
Mixed-load meandering

Present
Suspended load meandering

Figure 6.4 Transition from a mixed load meandering rivers to a suspended load meandering rivers in a laterally-unconfined valley setting On the wide plains of western New South Wales, Australia, rivers have adjusted to much lower energy regimes in the period since the Last Glacial Maximum. This has been marked by a transition from mixed load meandering rivers that were prone to progressive lateral migration, to suspended load systems that are characterized predominantly by vertical accretion. This diagram is based on phases and timescales of river evolution along the Murrumbidgee River documented in Page and Nanson (1996).

the confined through the partly-confined to the laterally-unconfined setting, as the potential range of variability increases. Rivers can more readily adopt differing morphologies and behavioral attributes if there is space for the channel to adjust on the valley floor, and channel configuration is less imposed. A similar degree of variability is evident in the width of the inner band on these figures. This reflects the natural capacity for adjustment as determined by flux boundary conditions. The width of this inner band represents the range of states that the river can adopt while still being considered to be the same type of river (i.e., a consistent set of core geomorphic attributes that reflect river character and behavior is retained). In a sense, this is a measure of the sensitivity of the river, as it records the ease with which the river is able to adjust. As indicated for the potential range of

variability, the width of the inner band is usually greatest in laterally-unconfined settings.

Each of the evolution diagrams in Figures 6.1–6.6 conveys how rivers change in response to disturbance events, indicated by the arrows at the top of the inner band. The spacing of the arrows indicates their frequency, while the size of the arrow indicates their magnitude. The shape of the pathway for adjustment, shown by the jagged line within the inner band, has a different form for different types of river. In most instances, disturbance events promote river adjustments but the reach remains within the inner band (i.e., perturbations fall within the natural capacity for adjustment). River adjustment within the inner band may breach intrinsic threshold conditions, marking a shift in the way energy is used (either concentrated or dispersed). Typically this reflects an

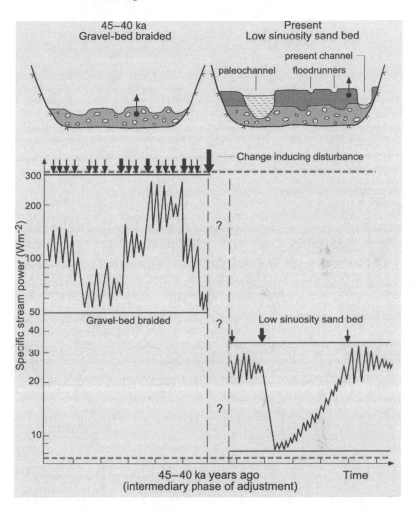

Figure 6.5 Transition from a gravel-bed braided to a low sinuosity sand-bed river in a laterally-unconfined valley setting
Profound changes to river morphology have characterized the Late Quaternary evolution of the Hawkesbury-Nepean River, New South Wales, Australia. This marks a transition from a gravel bed meandering system to a low sinuosity sand-bed river that is effectively inset within a gravel braid plain. Much uncertainty remains regarding the transitional forms of adjustment between these markedly different river configurations (and associated energy conditions). Phases and timescales of river evolution outlined in this diagram were extracted from Nanson et al. (2003).

adjustment in the character or distribution of resisting forces (e.g., bed resistance, form resistance, resistance induced by riparian vegetation or woody debris). These internal adjustments alter the assemblage of erosional and depositional landforms on the valley floor, yet fall within the behavioral regime of the river (see Chapter 5). If the frequency of high magnitude events that are able to cause further disturbance is less than the time required for recovery, the river is able to retain its original character and behavior, with a slightly modified state. This is shown by a shift in vertical position of the inner band for the partly confined and laterally unconfined valley settings in Figures 6.1–6.6.

In some instances, changes to the prevailing flux boundary conditions may result in changes to the formative processes that generate, sustain, and adjust river morphology. Such a scenario is highlighted by the shift in position of the inner band that conveys the capacity for adjustment shown on the right hand side of the river evolution diagrams shown in Figures 6.2–6.6. Reaches now operate within a different inner band, with altered energy conditions. The accompanying transition in the pathway for adjustment marks a change in form–process associations along the valley floor, such that there is a change in river morphology. The new configuration represents a different type of

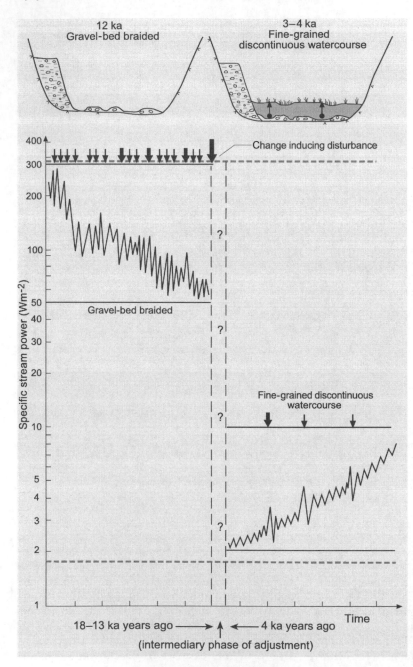

Figure 6.6 Transition from a gravel-bed braided river to a fine-grained discontinuous watercourse in a laterally-unconfined valley setting As noted for the lower Hawkesbury River (Figure 6.5), a marked reduction in energy conditions and river morphology has characterized the Holocene evolution of Mulloon Creek, New South Wales, Australia. In this instance, the river has been transformed from a gravel bed braided river to a discontinuous water course characterized by swamp sedimentation (Johnston and Brierley, subm.).

river, with a different appearance (character) and set of formative processes (behavior). For simplicity, shifts in river type following disturbance events are indicated as near-instantaneous responses in these figures, recognizing explicitly that lagged responses are likely in many instances. Inevitably, there may be some overlap in the position of former and contemporary bands, and various geomorphic units may be evident in both situations. However, the range of geomorphic units in the two bands differs, reflecting a change to the character and behavior of the reach.

The shift in the position of the inner band on Figures 6.2–6.6 can be induced by a press disturbance that exceeds an extrinsic threshold. This usually reflects alteration to flux boundary conditions, whereby adjustments to flow and sediment transfer regimes (i.e., impelling forces) drive river change. In this case, the time that is required for recovery following perturbation is longer than the recurrence interval of disturbance events (Brunsden, 1980, 1996). Effectively, the previous configuration of the river was unable to cope with changes to the magnitude and rate of stress applied to the landscape. Rare floods of extreme magnitude, or sequences of moderate magnitude events that occur over a short interval of time, may breach extrinsic threshold conditions, transforming river character and behavior.

Dependent on the subsequent set of form–process associations adopted by the river, the natural capacity for adjustment may widen or contract as the new type of river adjusts to different flux boundary conditions. The position of the inner band within the potential range of variability (the outer band) indicates whether the change in river type marks a transition to a higher energy state (an upwards adjustment) or a lower energy state (downward adjustment). Changes to the amplitude, frequency, and shape of the pathway of adjustment within the inner band indicate how the river responds to pulse disturbance events of varying magnitude and frequency.

The relationship between the potential range of variability and the natural capacity for adjustment for a gorge is shown in Figure 6.1. Disturbance events that have the capacity to induce changes in other settings are unable to bring about significant geomorphic change along confined rivers, as the inherent resilience of the system is too strong. Perturbations to the flow and sediment regime are accommodated by instream adjustments to hydraulic resistance characteristics, such as the nature and distribution of bedforms, dissipating flow energy. Changes to river character and behavior are negligible and the river type remains the same. The assemblage of erosional and depositional geomorphic units along the reach is likely to remain consistent over timeframes $< 10^2$ years (at least).

A different pattern of responses to changes in external stimuli may be experienced in partly-confined valley settings, where the potential range of variability is somewhat broader than in confined valleys (Figure 6.2). This enables a greater range of possible river morphologies to develop. Antecedent controls and prevailing flux boundary conditions shape the contemporary configuration of the river. In the example shown, ridge and swale features developed along the inside of a bend of a floodplain pocket. However, when the channel had migrated as far as possible, the pattern of sedimentation on the floodplain was transformed to that of a vertically accreting silty floodplain. This transition to the assemblage of channel and floodplain geomorphic units likely occurred over decades or hundreds of years. As a result of these changes, the natural capacity for adjustment shifted and a new, lower energy river type developed. This is indicated on the river evolution diagram by a downward shift in the position of the inner band (the natural capacity for adjustment) within the outer band (the potential range of variability). In addition, the range of river behavior has been reduced (i.e., the width of the inner band has narrowed; note the logarithmic scale).

The potential range of variability and the natural capacity for adjustment are at a maximum in laterally-unconfined valley settings. Changes are shown from a braided configuration to a meandering mixed load system (Figure 6.3), from a mixed load meandering to a suspended load meandering river (Figure 6.4), from a gravel-bed braided to a low sinuosity sand bed river (Figure 6.5) and from a braided to a fine-grained discontinuous watercourse (Figure 6.6). These changes reflect a change in both the impelling forces that promote change (i.e., less variability in flow, less coarse-sized material on the valley floor, etc) and internal system adjustments that modify the pattern and extent of resistance. The adjustments shown indicate an increase in resistance that increases the capacity of the system to trap finer grained materials, thereby aiding the transition to a single-channeled or discontinuous channel configuration. Increased stability enhances prospects for vegetation development on the valley floor. A major shift in the assemblage of geomorphic units ensues, resulting in altered patterns of midchannel and bank-attached geomorphic units, and processes of floodplain formation and reworking. Channel geometry and bedform assemblages are transformed as well.

Rivers are more sensitive to change in laterally-unconfined valley settings relative to partly-confined and confined valleys. While smaller-scale pulse disturbances events result in oscillation within the natural capacity for adjustment, a larger or more intense perturbation (press disturbance) may push these reaches over a critical threshold, bringing about a fundamental shift in river character and behavior. In all cases, the natural capacity for adjustment has shifted down within the potential range of variability. This reflects the adoption of a lower energy river type within the same landscape setting. This transition is especially pronounced in Figure 6.6. In all cases, the natural capacity for adjustment is narrower, reflecting a reduction in the range of behavior. In many cases, the pathway of adjustment has also changed significantly. For example, the transition from braided to meandering configurations shown in Figure 6.3 is marked by a switch from tight chaotic oscillations reflecting recurrent reworking of materials on the channel bed to a jagged shape that reflects the occasional formation of cut-offs and subsequent readjustment.

River change can result from alteration to impelling forces, alteration to resisting forces, or both. Resulting adjustments modify the nature, intensity, and distribution of erosional and depositional processes along a reach. Ultimately, changes to the pattern of erosion and deposition are the process manifestations of adjustments to flow and sediment fluxes. A range of considerations determines the likelihood that differing forms of river change will occur at any given place (Section 6.5) over any given timeframe (Section 6.6).

6.5 The spatial distribution of river change

Landscape setting, as shaped by geological history and lithological considerations, exerts a primary influence on river forms and processes. Of particular concern here is how tectonic and structural controls shape landscape relief and degree of dissection. These considerations fashion sediment storage on slopes and valley floors and the propensity for sediment replenishment. Sediment generation, in turn, is largely determined by the underlying lithology, its response to the prevailing weathering regime, and rate of uplift. Sediment sources vary markedly from river to river. Rivers in mountainous areas are often fed by coarse sediments from discrete sources, such as landslides, whereas rivers in lowland areas typically carry fine sediment supplied from diffuse sources, including the surrounding watershed and upstream reaches.

Lithological controls dictate the nature of weathering break down products. The rate of breakdown, relative to the effectiveness of reworking processes, determines the character and extent of sediment stores. In some situations the sediment mix is largely inherited from former conditions, often producing coarse lag deposits or cemented fine-grained materials. Prevailing climatic conditions determine the nature and rate of weathering processes, the range and effectiveness of denudational processes, and the efficiency of mechanisms that redistribute sediments. Climatic conditions also influence riparian vegetation composition and cover, and hence resistance along a reach.

Geological considerations influence the drainage network pattern, valley configuration, shape of the longitudinal profile, and connectivity of landscape compartments. These factors determine, among other considerations, the caliber and volume of materials supplied to rivers. Ultimately, landscape setting determines the potential energy in any given landscape, how that energy is likely to be used (i.e., concentrated or dispersed), and what materials are available to act as erosive tools, to be transported along channels, and to create depositional forms along river courses. The juxtaposition of landscape compartments fashions the distribution of geomorphic process zones (i.e., source, transfer, and accumulation zones), exerting a primary control on the nature and distribution of river changes in any given catchment.

In terms of assessments of river change, two key issues emerge here. First, relief and valley confinement are primary controls on the impelling forces that drive river change. Second, sediment availability and patterns of sediment storage in landscapes dictate the nature of river change over the short–medium term ($<10^3$ years). In its simplest sense, river change refers to the adjustment in the balance of erosional and depositional processes along a reach (i.e., changes to sediment fluxes).

River change can only occur if sediment stores are available to be reworked and restored to produce differing arrays of landforms along valley floors. Measures of river sensitivity reflect the ease with which morphological attributes are able to adjust in different settings, as determined by the natural capacity for adjustment of a reach. While bedrock reaches are remarkably resilient to change because of their imposed character and behavior, alluvial reaches are often sensitive to change, especially bedload or mixed load variants. In large part, the natural capacity for adjustment reflects the balance between the volume, caliber, and location of sediment stores along a reach, and the recurrence of forcing events that remobilize or rework these deposits. Differing units may act as stores that are reworked over intervals $<10^1$ years, or sinks that may have residence times extending beyond 10^3 years (e.g., Fryirs and Brierley, 2001).

Tectonic setting, and particularly the rate of uplift, exerts a first order control on differentiation of sediment transport- and supply-limited landscapes. Other factors influence this broad-scale balance of erosional and depositional processes, not least of which is the legacy of landscape history as a determinant of patterns of sediment storage. Excess energy in supply-limited landscapes ensures that sediment stores along the valley floor are efficiently flushed through the system. Impelling forces and erosional processes are dominant. The balance is tilted the other way in transport-limited landscapes, where there is insufficient energy to mobilize sediment stores, such that impelling forces and erosional processes are balanced by resisting forces and depositional mechanisms. The nature and consequences of river change may be quite different in supply- and transport-limited landscapes. In transport-limited settings, regular or recurrent events shape river adjustments, with relatively high potential for recovery. In supply-limited landscapes, the opposite effect may be observed. Disturbance responses are infrequent, but the capacity for recovery is inherently limited by the availability of sediments that can replenish losses following erosion (e.g., Fryirs and Brierley, 2001; Brooks and Brierley, 2004).

Valley confinement and slope are key physical controls on river morphology, fashioning the availability and concentration (or dissipation) of flow energy along valley floors. While valley width is effectively set over timescales $<10^3$ years, river morphology is potentially highly sensitive to changes in channel bed slope. Sensitivity is fashioned primarily by the nature and extent of alluvial deposition. Available energy and sediment availability vary markedly in sediment source, transfer, and accumulation zones, impacting profoundly on the balance of erosional and depositional tendencies and resulting river adjustments. Analysis of the distribution of stream power along long profiles provides a useful initial guide in interpretations of patterns of river change (e.g., Fonstad, 2003; Reinfelds et al., 2004). The use of available energy is conditioned, in large part, by valley width, which determines the extent to which available energy is concentrated or dispersed over narrow or wide valley floors respectively. Equally important is the way in which the channel has adjusted on its valley floor, in terms of its size and alignment. These considerations dictate channel–floodplain relationships and associated dispersion of flow energy, thereby shaping patterns and rates of erosional and depositional activity.

In source zones, where there is excess energy and erosion is dominant, an array of bedrock geomorphic units is evident. The specific nature and distribution of these forms reflects local variation in slope, which in turn reflects long-term rates of bedrock erosion as determined by the hardness or resistance to erosion of the local lithology and the available flow energy. Regardless of the specific nature of these interactions, the capacity for adjustment of the river is negligible over timeframes of relevance to river management and limited change results.

At the other end of the spectrum, reaches with limited excess energy on the valley floor (i.e., low stream power conditions) tend to be characterized by sustained aggradation. If the base level to which the river naturally adjusts is effectively set, erosion is largely restricted to lateral adjustments. As noted in Chapter 4, a bewildering array of river morphologies may be adopted in these accumulation zones. Of critical concern in assessment of the type and likelihood of adjustment in these settings is what controls local base level. In coastal settings, base level is set by sea level. Although glacially induced changes to sea level during the Late Quaternary exceeded 100 m, and sea-level adjustments in some regions subjected to continental glaciation continue to experience the consequences of isostatic rebound, most rivers have had sufficient time to adjust to the prevailing sea level. However, rates of infilling of lowland basins are highly variable, and a wide range of river morphologies is evident. Any change to sea level may alter the balance of fluvial and coastal processes, potentially resulting in profound changes to river morphodynamics (see Woodroffe et al., 1993). Hence, incremental rises in sea level associated with global warming may exert a wide range of adjustments to the pattern and rate of depositional and erosional processes along lowland plains.

In other instances, bedrock steps along long profiles reduce upstream slope and promote the development of alluvial reaches atop relatively shallow valley fills (e.g., Fryirs, 2002; Tooth et al., 2002). Adjustments to channel sinuosity modify channel bed slope in these reaches. Given their shallow fill, these laterally-unconfined reaches are commonly incised to bedrock, though floodplains line each channel margin. As such, the propensity for vertical channel adjustments is limited, but lateral adjustments may be readily accommodated.

In addition to slope, a key control on the capacity for river adjustment is the caliber of materials, and whether the channel acts as a bedload, mixed load, or suspended load system. Alluvial rivers with coarse bedload (i.e., boulder) or fine-grained suspended load boundaries tend to be relatively resilient to change because of their imposed and cohesive textures respectively. In contrast, alluvial rivers with sand or gravel bedload or mixed load boundaries are more sensitive to change because of the ease with which these materials can be reworked and transported.

In source or accumulation zones, river adjustments record the dominance of erosional or depositional processes respectively. However, any change to the balance of impelling and resisting forces in transfer zones may promote an adjustment to river morphology. Hence, river adjustments in these settings are brought about by changes to the types of processes that are taking place, as prevailing fluxes can alternate between phases of erosional or depositional dominance. As such, these sections of river may be subjected to pronounced variability in their process domain. The extent of change that may be experienced in

these reaches, and their inherent sensitivity to change, are at a maximum. For example, bed level instability may present a critical trigger for river adjustment, altering the aggradational–degradational balance of the reach. Indeed, any factor that promotes incision and subsequent headcut retreat promotes dramatic and often catastrophic changes to river morphology and the resulting array of geomorphic processes. In a generalized sense, these adjustments record differing phases of valley floor aggradation (i.e., filling phases) and degradation (i.e., cut phases).

An additional layer of sensitivity is often added to transfer reaches, in that they commonly occur in areas of stream power maxima along longitudinal profiles, as noted in Chapter 3, and they have a bedload or mixed load composition. Changes to these reaches commonly entail transformations in channel planform and associated adjustments to the range of geomorphic units (e.g., the dominance of erosional or depositional forms along the reach) and the resulting channel geometry. For example, phases of channel expansion and contraction commonly accompany adjustments at different stages of the incisional process (Schumm et al., 1984, 1987). This exerts a major influence on patterns and rates of floodplain inundation, and resulting geomorphic adjustments to floodplain forming and reworking processes.

Geological controls on patterns and rates of river change, as influenced by landscape configuration, material availability, and the catchment-specific distribution of geomorphic process zones, must be appraised in light of the climatic setting within which the catchment is located. Much of the "natural" unsteadiness in fluvial systems is caused by climatic fluctuations which, through their influence on vegetation patterns, water balances, and process activity affect flow regimes and the production, supply, and transport of sediment. Climatic controls determine the type, distribution, and effectiveness of geomorphic processes that rework and redistribute materials. This influence is manifest in a range of ways. For example, the temperature regime is a major determinant of the distribution and effectiveness of various geomorphic processes (e.g., ice-related mechanisms, aeolian activity, etc). The amount, distribution, and intensity of precipitation are critical determinants of flow regime, differentiating between perennial and ephemeral streams, flow events of varying magnitude and frequency, and associated implications for the effectiveness of geomorphic events. Climate also influences vegetation distribution, thereby exerting an important secondary control on the nature and extent of resisting forces in landscapes. The ways in which these climatic influences impact upon the spatial distribution of river changes are discussed in turn.

The balance between precipitation input and evapotranspiration determines the amount of excess water that generates runoff that drives the geomorphic work performed along rivers. While topographic and geologic controls shape the sediment side of impelling forces that influence river character, behavior, and change, climatic controls determine the flow regime and associated magnitude–frequency relationships that bring about river adjustments and/or change. Climatic factors also influence the volume and recurrence of sediment inputs from slopes. The frequency and magnitude of landslides, debris flows, earth flows, debris torrents, and other mechanisms are major considerations in determining whether the river operates as a supply limited or transport-limited system. Other key factors to be considered here are the volume of sediment input from upstream, and the relative effectiveness with which flow events rework materials on the valley floor. Mechanisms and rates of sediment delivery from slopes vary markedly in arctic, temperate, tropical, Mediterranean, or arid settings. Obviously the terrain in these differing climate settings also shapes the effectiveness of sediment delivery processes.

Flow regime is also a critical determinant of river change. The frequency of formative discharge events varies markedly from reach to reach. Profound variability in discharge extremes is experienced across the planet (McMahon et al., 1991). Extreme floods are much more pronounced in monsoonal and semiarid areas compared to humid-temperate or tropical rivers. They are also more pronounced in headwater settings relative to downstream reaches.

Magnitude–frequency relations vary markedly from system to system, in terms of the nature of formative events (e.g., discharge variability and flood periodicity, especially extreme events) and the geomorphic effectiveness of events (as determined by the inherent sensitivity of the system).

Of particular importance is the duration of events beyond a critical threshold stage (e.g., an erosional threshold), as these are the intervals that drive landscape adjustment (e.g., Costa and O'Connor, 1995). In a very general sense, humid-temperate alluvial rivers have been observed to adjust to bankfull-stage events that occur with a recurrence interval of a few years (Leopold et al., 1964). In essence, these rivers are subjected to less variability in flood discharge, enabling them to adjust more readily to prevailing flow conditions. In other settings, however, this relationship may be wildly different (e.g., Pickup, 1984). For example, tropical rivers are adjusted to recurrent high magnitude events. In monsoonal settings, seasonal variability of wet–dry phases is immense, but the interannual variability is generally relatively small. In contrast, rivers in Mediterranean or arid/semiarid areas have less reliable flows. Long-term variability in flow, and the extraordinary magnitude of extreme flows, may be the primary determinant of river morphology and resulting river changes. The geomorphic effectiveness of high magnitude events is accentuated along ephemeral rivers in Mediterranean/semiarid relative to perennial rivers in humid-temperate settings as the resisting role offered by vegetation cover is reduced.

Although climatic setting exerts a first order influence on the way in which river adjustments are driven by differing magnitudes of flood event that occur with a differing frequency, there is an inordinate degree of system-to-system variability. The geomorphic makeup of each reach reflects its own history and sequencing of flood events. Viewed in this light, derivation of prescriptive management strategies that are framed in terms of formative flows of a given magnitude and frequency is foolhardy, as it negates the inherent variability or river systems. Patterns and rates of river responses to flood events are fashioned not only by the natural capacity for adjustment of differing settings, they are also influenced by the condition of the landscape at the time of the flood, the duration of the event, and the recent history of events, as discussed in Section 6.6.

The mediating role of vegetation cover as a determinant of erosion rate varies in differing climatic settings. In simple terms, a relationship can be drawn between the frequency and intensity of rainfall events that drive erosion (i.e., impelling forces), and the nature and extent of ground cover that resists erosion by protecting materials at the ground surface. Sediment yield is maximized under semiarid conditions (annual rainfall of around 250 mm), as ground cover is sparse yet there are sufficient rainfall events to generate erosion (i.e., each event tends to be relatively efficient; Langbein and Schumm, 1958). Although ground cover is reduced under drier conditions, runoff is also reduced so sediment yields may be exceedingly low. As annual precipitation increases beyond 250 mm, ground cover increases, whether in the form of grassland, scrub, or forest conditions. Although runoff events occur more frequently than in semiarid areas, their erosive potential is reduced. A second peak in the rate of sediment yield may be experienced in tropical environments, where the volume and intensity of runoff may be so profound that they effectively override the resisting force played by vegetation cover. These relationships partially account for the differential sensitivity of landscapes in different climate settings when subjected to changes in ground cover, whether in response to natural bushfires or effects of human-induced land-use change (see Chapter 7). Disruption to ground cover may greatly enhance erosivity, such that rates of sediment yield rise asymptotically with mean annual runoff until they approach a maximum level (Dunne, 1979). Alternatively, an increase in precipitation in a semiarid area brought about by climate change may result in an increase in ground cover or vegetation density, ultimately reducing the amount of work done per unit of energy applied.

Any factor that alters the interplay between these various controls on sediment availability and runoff enhances the prospects for river change. The likelihood that change will occur reflects the capacity for adjustment of any given reach. Geographic location is a critical consideration when comparing geomorphic observations or measurements (Schumm, 1991). It influences the potential for change, as it records upstream or downstream factors that may promote adjustments. Similar looking landscapes may have differing trajectories of adjustment. The key considerations in making such appraisals are evolutionary assessments on the one hand, and interpretation of factors that may promote adjustments on the other. Analysis of position in the landscape

must be framed in relation to what is upstream and downstream and the threats or limiting role that off-site considerations may have in shaping the nature and rate of change in any given reach.

Two simple examples highlight the significance of geographic location. First, any downstream factor that lowers base level promotes bed level instability. Ensuing headcuts promote river adjustments upstream, unless restricted by a bedrock step. Second, changes to upstream rates of sediment delivery influence river character and behavior downstream. Grossly accelerated rates of supply may generate sediment slugs, the impacts and longevity of which depend upon reach-specific considerations such as the character and behavior of the affected reach, the volume and caliber of sediment input, subsequent flow events, the energy conditions of the reach, etc. Such adjustments may promote significant changes to river forms and processes. Alternatively, river metamorphosis may occur if sediment supply rates are reduced. This may arise in response to lower rates of erosion upstream, exhaustion of available sediment stores, fewer transporting events, or imposition of barriers or buffers that impede downstream conveyance of sediment. Equivalent sets of adjustment may arise if the caliber of material transported is changed.

In summary, many factors must be considered in explaining the nature of river changes at differing positions in landscapes in differing environmental settings. Just as there is profound variability in the form and extent of change that may be experienced in different river systems, so may profound differences be evident in the rate at which change occurs. The likelihood that changes will occur, and associated rates of system response, are considered in the following section.

6.6 Temporal perspectives of river change

At the beginning of this chapter an overview of environmental changes in the Late Quaternary provided the long-term context with which to appraise patterns and rates of river change. These are vital considerations in understanding adjustments to flow and sediment fluxes, and vegetation associations, that frame changes to the position of the inner band (i.e., flux boundary conditions) in the river evolution diagram. In general terms, the post-glacial period has been characterized by declining energy conditions along valley floors, prompting the oft-quoted transition from braided to meandering rivers in a particular subset of settings (see Figure 6.3; e.g., Kozarski and Rotnicki, 1977). An equivalent reduction in fluvial activity has been reported over the Late Quaternary in landscapes such as southeastern Australia that were not subjected to the impacts of continental glaciation, reflecting climate-induced reduction in the flow regime (e.g., Figure 6.4–6.6; e.g., Nott et al., 2002; Nanson et al., 2003). In both these instances, profound changes to flux boundary conditions have brought about dramatic river changes. These changes are noted on the river evolution diagram by a repositioning of the inner band to a lower position within the potential range of variability. Effectively, these rivers operate today under lower energy (unit stream power) conditions relative to their state in the Early Holocene.

Timescales of geomorphic change in the examples above reflect adjustments over thousands of years. However, river change may be virtually instantaneous. This typically records the breaching of a threshold condition, whether internal to the system or driven by external events such as catastrophic floods. These events may bring about dramatic erosion in some reaches, or overwhelming deposition in others. Alternatively, contemporary river adjustments may record an extension to longer-term evolutionary tendencies that have operated over timeframes of millions of years, such as responses to knickpoint retreat in escarpment-dominated landscapes. The nature and rate of these adjustments reflect the landscape setting and the inherent sensitivity to change of the reach under investigation.

Magnitude–frequency relationships, and the effectiveness of disturbance events of a given size and recurrence interval, vary markedly for different types of river in differing environmental settings (Wolman and Gerson, 1978). The relationship between magnitude of external forcing process and the resultant morphological response may vary over time. For example, if the recurrence interval between major floods was shortened, such that notionally extreme events occurred more often, the propensity for river changes would be enhanced, and profound changes would likely occur in more sensitive or vulnerable settings. In some instances, landforms have become resistant to the occurrence of events of a certain magnitude and frequency. This is referred to as event resistance (Crozier, 1999). For example, sediment exhaustion may ensure that extreme events may be relatively ineffective in geomorphic terms (e.g., Fryirs and Brierley, 2001). Alternatively, if the landscape is primed to change (i.e., it is close to a threshold condition), relatively small events may bring about profound responses. This is referred to as event sensitivity.

The geomorphic efficiency of large events is inhibited by their duration, especially the period beyond critical erosional threshold conditions (e.g., Huckleberry, 1994; Costa and O'Connor, 1995). Whether or not a disturbance event has a large or lasting effect on a river system depends partly on whether a threshold is exceeded. The breaching of threshold conditions that drives a reach beyond its former range of variability to a new process domain is often facilitated by *positive feedback* mechanisms. In some instances, internal adjustments progressively approach a threshold condition, such that a relatively minor forcing event can bring about dramatic responses. This is typically associated with situations in which resisting forces are effectively diminished over time, such that the system becomes increasingly vulnerable. Hence, the condition of the landscape at the time of a given event determines its efficiency/effectiveness (e.g., Brooks and Brierley, 2000).

The effectiveness of floods of a given magnitude is also dependent on preceding events, or event ordering (Beven, 1981). Event effectiveness is likely to be greater if a large event rather than a small one precedes it, but even a small event will be more efficient if it follows a preceding event quickly (in most instances). Thresholds for sediment movement may shift with a change in climate and/or vegetation (e.g., Wolman and Gerson, 1978). Rivers can only move those materials made available to them, so any preceding event that modifies sediment availability may impact upon the effectiveness of subsequent events. Some landscapes may be subjected to lagged responses, as evidenced when influxes of sediment of a particular caliber or volume exert an influence on river adjustments for some period into the future. In a sense, these are the temporal equivalent to off-site impacts.

The influence of past flood events on river form varies both with their absolute magnitude (more extreme floods having a longer-lasting effect) and with their relative magnitude, as expressed by the ratio of individual flood peaks to the mean annual flood (Stevens et al., 1975). Where sensitivity is high and response times are short, historical influences are less important determinants of contemporary river character and behavior. Profound variability in the manner and rate of river response to disturbance reflects, in part, sediment availability, and associated potential for geomorphic recovery. In settings where extreme events are the key agent of channel form adjustments, the adoption of a differing morphology following catastrophic disturbance is so pronounced that the system is unable to recover to its previous state over an extensive timeframe, say hundreds or thousands of years (e.g., Stevens et al., 1975). These landscapes may be out of phase with contemporary processes, such that notable adjustments to river forms and processes can only be brought about by infrequent, catastrophic floods. In many instances, this type of river character and behavior reflects a circumstance in which boundary materials reflect previous intervals of river activity. Elsewhere, there may be insufficient time for recovery between disturbance events, such that the system never attains a characteristic state.

Antecedent controls may exert a pervasive influence on contemporary river forms and processes. The timeframe of lagged responses is influenced by the residence time of differing sediment stores. This may range from the impacts of recent (extreme) events to previous climate phases that mobilized significant volumes of material. For example, many rivers are influenced by a legacy of sediment supply fashioned by tectonic activity or reworked glacially modified sediment stores that line river courses (Baker, 1978; Church and Slaymaker, 1989). Elsewhere, sediment supply limitations may extend back over millions of years within low relief landscapes (e.g., Australia).

Ultimately, timeframes of river adjustment or change reflect the prevailing balance of fluxes operating along a reach, framed in light of the capacity of the reach to change its form (i.e., its sensitivity). Any factor that impedes or accentuates flow or sediment fluxes may alter the geomorphic effectiveness of floods, thereby promoting

change. Alternatively, changes to resisting factors may inhibit or enhance the likelihood of change. For example, if roughness is reduced following clearance of riparian vegetation or removal of woody debris, there is greater opportunity for accelerated geomorphic responses to the next formative event. In contrast, incursion of exotic vegetation may choke a river system, initially promoting metamorphosis but then inhibiting the prospects for change (other than accelerated rates of aggradation). A similar degree of variability may be evident in the capacity for systems to recover following disturbance events. The time taken by a system to recover varies. This is dependent not only on the initial amount and extent of displacement, but also on climatic conditions, especially as they relate to the subsequent flow regime and vegetation cover. While alluvial rivers in humid regions tend to recover quickly from the effects of major floods, the slow rate of vegetative regeneration in more arid areas inhibits rapid recovery (Wolman and Gerson, 1978; Bull, 1991).

6.7 Appraising system vulnerability to change

Adoption of a future focus for management activities requires appraisal of the likelihood that river change will take place, along with interpretation of the likely direction of change, its extent, and its off-site implications. Understanding why these adjustments take place in the manner that they do provides the basis to develop strategies to manage the underlying causes of change, rather than their symptoms. Among the many concepts that must be appraised in assessment of these considerations are notions of the susceptibility, vulnerability, and sensitivity of a system to change.

In many instances, river responses to disturbance do not operate under simple cause and effect scenarios. Rather, different parts of the landscape adjust to disturbance events in quite different ways, dependent on reach condition at the time of the event. Environmental change and perturbing influences do not always promote geomorphic responses, and, if they do, responses may be quite different in different parts of the landscape. The notion of *complex response* indicates that there may be pronounced variability in the nature and rate of system responses to the same external stimuli, resulting in stark differences in the pattern of spatial and temporal lags. Hence, relatively uniform climate changes across a region may lead to different patterns of erosional response, the effects of which may be propagated through systems in different ways. Responses to impacts in one place may dampen or buffer effects elsewhere. Alternatively, small disturbances in one part of a system may set up chain reactions that breach threshold conditions elsewhere, resulting in chaotic responses. Resulting alluvial chronologies are temporally and spatially out of phase from system to system.

A measure of the inherent diversity of river behavior and the propensity for river change can be gained through consideration of the response of different types of river in differing environmental settings to disturbance events of a given magnitude. Among the continuum of responses that may be observed are the following situations:

• *No response may be detected*, as systems absorb the impacts of disturbance. A landscape that can tolerate considerable variation in controlling factors and forcing processes can be considered to be relatively stable. For example, gorges are resilient to adjustment or change. Alluvial systems with inherent resilience may demonstrate limited adjustment or changes over timeframes of 10^3 years, whether induced by the cohesive nature of valley floor deposits, or the mediating influence of riparian vegetation and woody debris. In some settings, landscapes may be so stable and insensitive, or so slow to react, that they are adjusted to a previous set of controlling conditions and forcing processes.

• *Effects may be short-lived or intransitive* (Chappell, 1983). In general terms, humid alluvial channels often show little change following disturbance events, or they have the capacity to recover rapidly. For example, extreme floods may engender only localized and short-lived responses (e.g., Costa, 1974; Gupta and Fox, 1974; Gomez et al., 1997; Magilligan et al., 1998).

• *Part of progressive change*. Rivers may respond rapidly at first after disruption, but at a steadily declining rate thereafter, such that recovery pathways have a basically exponential form (e.g., Graf, 1977; Simon, 1992, 1995). For example, in high energy steepland settings, where slopes and channels are effectively coupled, channels adjust

among various states conditioned by the balance between sediment input from slopes and tributary systems and phases of sediment flushing on the valley floor (e.g., Madej, 1995; Kasai et al., 2003, 2004). Elsewhere, floodplain pockets in partly-confined settings may be stripped by sequences of flood events over annual or decadal intervals, with subsequent build up of the floodplain over periods of thousands of years (Nanson, 1986).

• *Change may be instantaneous*, as breaching of threshold conditions prompts the adoption of a new state or even a new type of river. These effects tend to be long-lasting or persistent (i.e., transitive; Chappell, 1983). In many instances, threshold-driven changes reflect extrinsic controls induced by impelling forces (e.g., extreme events or sequences of moderate-magnitude events over a short interval). For example, emplacement of large quantities of very coarse material during extreme floods may induce a situation in which channels are unable to recover their original morphology, as subsequent smaller flow events do not have the competence to transport this material (e.g., Stewart and LaMarche, 1967; Wohl, 1992; Zielinski, 2003). These settings tend to retain an extended memory of extreme floods (i.e., a flood-dominated morphology), as the time between major floods is shorter than the recovery time. Extreme disturbance may result from massive inputs of sediment, such that the channel becomes overloaded (e.g., Simon, 1992; Simon and Thorne, 1996). Alternatively, geomorphic recovery may be constrained along sediment starved rivers (e.g., Fryirs and Brierley, 2001; Brooks and Brierley, 2004).

• *Change may be lagged*. Off-site impacts of major disturbances may induce a lagged response in downstream reaches (e.g., conveyance of a sediment slug). In some instances, geomorphic recovery is delayed by subsequent climatic conditions, the history of flood events, and associated vegetation adjustments (e.g., Schumm and Lichty, 1963; Burkham, 1972; Friedman et al., 1996).

• *Equivalent responses may occur via differing pathways of adjustment*. For example, either gradual or catastrophic adjustments can bring about a transition between braided and meandering channel planform types (Graf, 1988). A gradual transformation can be achieved when control variables (stream power and bank resistance) change in such a way that the system follows a smooth path. Alternatively, the system may change abruptly following exceedance of an extrinsic threshold.

River character and behavior in any given reach reflect cumulative responses to a range of disturbance events. Evolutionary investigations must consider long time periods in order to identify the critical processes involved in river change. Spatial and temporal controls on river adjustment and change must be meaningfully integrated to provide insights into system vulnerability.

In this book, *vulnerability* refers to the potential of a reach to experience a shift in state within its natural capacity for adjustment or to be transformed to a different type of river. Vulnerability can result from the breaching of either an intrinsic or an extrinsic threshold. If an extrinsic threshold is breached, a transition to a different type of river may occur. Exceedance of an extrinsic threshold in a vulnerable landscape results in a new assemblage of landforms with a new process regime. Only a minor perturbation in the external environment is required to produce a major response (Schumm, 1991). However, breaching of an intrinsic threshold equates to a behavioral adjustment that forms part of the natural capacity for adjustment for a vulnerable type of river. These rivers operate among two or more states. The degree of vulnerability depends upon the stage of adjustment, and how close to an intrinsic threshold the reach sits. Under circumstances of incipient instability that arise as threshold conditions are approached, minor disturbance events may bring about changes to a new state (e.g., Begin and Schumm, 1984). This is indicated on the river evolution diagram by a stepped pathway of adjustment with flat areas. For example, breaching of intrinsic thresholds is a natural part of the behavioral regime of cut-and-fill landscapes. These settings are characterized by discontinuous watercourses atop significant sediment stores in valley fills. Gradual increases in valley floor slope associated with progressive aggradation may lower threshold conditions for incision over time (Schumm and Hadley, 1957; Patton and Schumm, 1975; Schumm et al., 1984). Eventually, a relatively trivial event on the oversteepened valley floor may incise the valley fill (e.g., Brierley and Fryirs, 1998). The development of meander cutoffs provides a further example of this type of instability, as selective cutoffs

may return a river to a more stable sinuosity in which there is a better balance between transporting ability and gradient (Knighton, 1998). Finally, floodplain stripping in partly-confined valleys represents a shift in state for the same type of river. In these settings, progressive build up of sediments on the valley floor increasingly concentrates flow within the channel until a threshold condition is breached and stripping occurs (Nanson, 1986).

In contrast, *susceptibility* refers to the ease with which a system or reach is able to adjust within its natural capacity for adjustment (cf., Brunsden and Thornes, 1979; Schumm, 1991). This is part of the behavioral regime for these types of river. Susceptible landscapes retain a characteristic identity as they form and reform under a given process regime. Relatively rapid responses follow recurrent disturbance events. Frequent perturbations result in minor adjustments to river character and behavior without inducing a shift in state or change to a different type of river. At first glance, continual adjustment may be perceived as a form of instability, but this is not always the case, as a susceptible river is not necessarily vulnerable to adjustment or change. For example, this type of adjustment is characteristic of some high energy fluvial systems, such as braided or wandering gravel bed rivers, where bars and various channels are recurrently reworked (or destroyed), only to be replaced by similar landforms. In contrast, a nonsusceptible system may be subjected to progressive adjustment. This may be characteristic of a fine-grained meandering river or anastomosing river, for example, where gradual responses follow recurrent perturbations.

Measures of *river sensitivity* to change are considered to reflect the sum of susceptibility and vulnerability for any given river type. This reflects the ease with which adjustment can take place (i.e., the way in which the reach has adjusted its form to resist change) and the proximity to threshold conditions. This provides a measure of the response potential of a river to external influences and smaller-scale disturbance events. Morphological responses to disturbance events are likely to be more pronounced along more sensitive reaches. River sensitivity is dictated to some degree by the within-catchment position of the reach and patterns/rates of geomorphic linkages (i.e., off-site impacts). In this book, *sensitive rivers* are readily able to adjust to perturbations but are prone to dramatic adjustment or change. They are effectively threshold breaching rivers. Conversely, *resilient rivers* have an inbuilt capacity to respond to disturbance via mutual adjustments that operate as negative feedback mechanisms. In this scenario, long-term stability is retained because of the self-regulating nature of the system, which mediates external impacts. These rivers are readily able to adjust to perturbations, without dramatic adjustment or change in form–process associations. The ability of a system to absorb perturbations, such that disturbance events do not elicit a morphological response, is referred to as the buffering capacity. In systems with large buffering capacity and/or with large thresholds to overcome, considerable time lags may be experienced between the timing of the disturbance event and morphological response.

Assessments of river sensitivity (susceptibility and vulnerability) can be grounded through consideration of various geomorphic principles in a stepwise procedure, as summarized in Table 6.2. Dependent on their potential range of variability and associated capacity for adjustment, different types of river respond to disturbance in different ways (Stage 1). These broad-scale considerations determine the nature, extent, and pattern of geomorphic units along the valley floor (Stage 2). The makeup of these features records the distribution of sediment stores, and the ease with which they are likely to be reworked. These features, in turn, provide insights into the prevailing balance of erosion and deposition along the reach, and whether the reach operates as a source, transfer, or accumulation zone. The sensitivity of a river to change is influenced by the nature, extent, and distribution of resisting forces along the valley floor. As noted in Chapter 3, these are determined in large part by valley setting (e.g., forcing elements such as points of confinement, valley alignment, bedrock steps, etc.), river morphology (e.g., channel alignment, channel geometry, bed material organization, etc.), and the presence, abundance, and geomorphic role of riparian vegetation and woody debris. Any factor that reduces the effectiveness of these resisting forces enhances the prospects of adjustment in state or river change. For example, if instream roughness is reduced, positive feedbacks may be set in train, bringing about river metamorphosis.

Analysis of reach evolution is required to interpret changes to the balance of impelling and resisting forces along the reach (Stage 3). At different stages of geomorphic adjustment, a river may have enhanced vulnerability to change. For example, the sequence of flood events may affect the vulnerability of the river. The critical factor that determines the geomorphic effectiveness of any given event, regardless of its magnitude and frequency, is the position of the reach within its evolutionary sequence and the condition of the landscape at the time of the event.

In order to develop an understanding of variability in the pattern, extent, rate, and consequences of geomorphic responses to disturbance, local, or reach-specific factors must be framed within a broader, system-wide appraisal of the boundary conditions that determine how the reach operates (Stage 4). Changes to one part of a river system may result in secondary impacts elsewhere, with profoundly different responses and residence times in other reaches. Indeed, in some instances, propagatory influences from a minor disturbance somewhere in a catchment may result in exceedance of some threshold condition elsewhere, possibly transforming river character and behavior. For example, any factor that promotes bed level instability may trigger incision and channel expansion in upstream reaches. Accelerated rates of sediment transfer may result in river metamorphosis downstream. The nature and consequences of lagged responses to off-site disturbance are system-specific. In general terms, however, disturbance effects are accentuated in lowland (alluvial) reaches at the bottom end of catchments, which accumulate the cumulative effects of upstream changes. If these effects are sufficiently severe, such that a change in boundary conditions occurs, river metamorphosis may eventuate.

Finally, specific indicators of threshold conditions that are likely to result in dramatic changes to river forms and processes must be interpreted for different types of river (Stage 5). This has been referred to as threshold spotting (Newson, 1992b) or identification of thresholds of probable concern (Rogers and Biggs, 1999; Mackenzie et al., 1999). Examples include indicators such as the degree of bank erosion (relative to what is expected for that type of river), the degree of vegetation cover, or the presence of a headcut, sediment slug, or thin neck

Table 6.2 Guiding principles for interpreting system sensitivity and vulnerability.

Stage and principle	Action
(1) Identify the type of river and its valley setting.	Assess the potential range of variability and contemporary capacity for adjustment.
(2) Appraise the balance of erosional and depositional landforms (geomorphic units) along the reach.	Interpret the ease with which various features are likely to be reworked, based on their composition (e.g., texture, cohesivity) and position along the reach (e.g., proximity to the thalweg). Interpret whether the reach presently operates as a sediment source, transfer, or accumulation zone.
(3) Interpret reach evolution as a basis to interpret how the balance of impelling and resisting forces along a reach adjusts over time.	Analyze the formative factors that drive geomorphic change, such as the dominant discharge, the role of extreme events, and the history of events. Interpret the type of events that have shaped the contemporary geomorphic state of the reach. The condition of the landscape at the time of any given event is a critical factor in appraisal of geomorphic effectiveness and how sensitive a river is to change. Hence, due regard must be given to the nature and distribution of resisting forces along the reach, highlighting any factor that may make the system prone to accentuated responses. In some instances, contemporary forms and processes may be imposed by events from the past. Due regard must also be given to the impelling forces that drive change.
(4) Based on analysis of the trajectory of change, interpret what the system is evolving towards. Identify limiting factors and pressures that may modify the prevailing balance.	River change can be driven by an increase in impelling forces, a decrease in resisting forces, or a combination of these factors. Interpretation of system responses to past events provides guidance into these relationships. Location for time substitution can be used to predict the likely future trajectory of change (see Chapter 3). In many settings, past events induce impacts that may have remarkable persistence. Detailed consideration of the spatial context of the reach under consideration is required to identify potential off-site and lagged impacts, noting any upstream or downstream factors that are likely to modify the pattern or rate of geomorphic adjustment.
(5) Identify threshold conditions that guide interpretations of indicators of change.	Thresholds of probable concern are likely to be specific to a particular type of river. In some instances, direct tools can be applied, whether empirically based (e.g., r_c/w link to lateral migration) or theoretically based (e.g., notions of critical bank height) to aid in threshold spotting. However, in many cases system-specific knowledge of the nature of adjustment is required to provide fundamental insight into how prone a river is to change or adjustment.

cutoff. A range of empirical and theoretical tools can be used in this analysis. However, system-specific knowledge on river evolution, the linkage of geomorphic processes in a catchment, and the resultant trajectory of change provides fundamental insight into how close to intrinsic or extrinsic thresholds a reach sits. These insights provide a basis to appraise likely system responses to disturbance events.

River changes must be viewed in context of the boundary conditions within which any reach operates, and associated disturbance events. The nature, rate, extent, and consequences of river changes vary markedly in differing environmental settings. The inherent complexity of river systems ensures that disturbance events of similar charac-

ter, magnitude, and duration do not necessarily bring about uniform response, in terms of either the extent or rate of change. For example, the volume and caliber of sediments stored along a reach exert a considerable influence upon prospective geomorphic adjustments. Propagation of disturbance responses reflects system-specific configuration. While geomorphic responses of a system to headcut incision typically record upstream-progressing degradation and downstream aggradation, the consequences of these adjustments may vary markedly from system to system. If the downstream reach is confined or partly-confined, or the volume of released sediment is relatively small, impacts will likely be restricted to local bed aggradation and transitory infilling of pools. However, impacts are likely to be much greater if a sediment slug reaches an alluvial reach, where responses may include channel contraction, accelerated rates of floodplain sedimentation, and smothering of the channel bed. Endeavors to understand the contemporary character, behavior, and evolutionary tendencies of rivers must add a further complicating factor to this inordinate complexity of river systems, namely responses to human-induced disturbances. These considerations are appraised in the following chapter.

CHAPTER 7

Geomorphic responses of rivers to human disturbance

We have been bequeathed a legacy of ageing river engineering projects whose objectives were simply designed, whose effectiveness are uncertain and which were planned in ignorance of their long-term physical and environmental impacts on the river system.

Phillip Williams, 2001

7.1 Introduction: Direct and indirect forms of human disturbance to rivers

Change is an integral and natural part of all ecosystems. However, human disturbance has introduced a source of change that is foreign to the geomorphic and biotic conditions of river systems. Human disturbance has modified the nature and rate of river adjustments, altering the spatial and temporal distribution of river forms and processes. Diversity of river character, dynamics and behavioral regimes, and propensity for river change were outlined in Chapters 4–6. In this chapter, a further layer of complexity is added through examination of the nature, rate, and impacts of human disturbance to rivers.

Exploitation of natural resources is vital for human survival. As social, economic, and cultural circumstances have evolved, so too have the mechanisms and extent by which these resources have been utilized. Endeavors to exploit these opportunities to meet societal demands have demonstrated remarkable ingenuity and creativity. Over the last 5000 years, human modifications have been the dominant form of disturbance to the fluvial environment, exerting a greater influence than adjustments caused by climate changes, although extreme "natural" events continue to be a significant cause of change (Brookes, 1994; Knighton, 1998). The intensity of development of land and water resources has been particularly pronounced over the past 100–500 years (Petts, 1989).

Many human activities have a landscape context, requiring specific localities for certain forms of practice. For example, towns are typically located adjacent to water sources, preferential trade routes had concerns for terrain, and agricultural developments sought out certain soils. Hydraulic civilizations developed along some of the world's great rivers, marking a turning point in social organization (Wittfogel, 1956). Profound variability in population and resource pressures, tied to the contemporary nature and extent of river regulation and management activities, shapes opportunities for ongoing and future developments. Across much of Western Europe, for example, floodplain areas settled initially for agricultural purposes have now been largely replaced by extensive urban and industrial developments, with associated flood management problems. Alternatively, the race to dam any significant body of moving water in the western United States has resulted in profound scarcity of river systems for which conservation programs truly relate to natural conditions (e.g., Reisner, 1986; McCully, 1996).

In most instances, human endeavors have sought to control and stabilize rivers. While responses to human disturbance may be reversible or at least can be stopped or reduced in some instances, elsewhere impacts are irreversible, marking a change in the direction of long-term evolutionary trends. These considerations have major implications for river structure and functioning, conditioning how landscapes work today, and constraining how rivers are likely to behave in the future. Across much of the world, the nature and rate of human impacts on river forms and processes are in a transition phase. Most reaches that are either readily navigable or present substantive opportunities for water development programs have already been exploited. In many parts of the world, regeneration of land cover or forest regrowth has followed the wholesale clearance of forest cover. Altered water, sediment, and nutrient fluxes have brought about secondary adjustments to river forms and processes (e.g., Newson, 1992a; Liebauldt and Piégay, 2002). Likely future river character and behavior must be predicted in light of these societally induced changes.

Human modifications to biophysical attributes of river systems can be direct or indirect (Park, 1977, 1981; Table 7.1). While most direct modifications are intended, indirect modifications are inadvertent. *Direct* modifications to the channel bed and/or banks have typically taken the form of resource development activities (e.g., water supply, power generation, gravel extraction) or structural engineering works designed to alleviate the effects of flooding. Clearance of riparian vegetation cover and removal of woody debris have generally accompanied these activities. *Indirect* human impacts refer to adjustments brought about as secondary responses to changes outside the channel that modify the discharge and/or sediment load of the river. These impacts relate primarily to changes in ground cover that modify the nature, balance and interaction of water and/or sediment fluxes. In general terms, impacts of indirect catchment-scale changes predate those associated with direct human modifications to river courses.

Although these impacts may appear to be less dramatic than direct disturbance responses, their effects are often more ubiquitous and far-reaching, with considerable lagged and off-site impacts. It is often exceedingly difficult to differentiate river responses to direct human disturbance at the reach scale from indirect human impacts at the catchment scale.

Human impacts on river character can only be reliably interpreted if longer-term controls on river evolution are understood. Hence, appraisals of system response to human disturbance must be made in context of inferred adjustments that would have occurred under natural disturbance regimes. Whatever the form of disturbance, whether a site-specific direct impact such as dam construction or indirect disturbance such as vegetation clearance, effects can be transmitted long distances from their source. Ultimately, however, management must consider cumulative responses to disturbance, interpreting how these adjustments will shape likely future patterns and rates of river changes.

This chapter is structured as follows. Direct and indirect human impacts on river forms and processes are considered in Sections 7.2 and 7.3 respectively. Spatial and temporal variability in river responses to human disturbance, and their cumulative nature, are discussed in Section 7.4. Finally, implications for rehabilitation activities are emphasized, demonstrating how notions of reversible or irreversible changes are integrated into the river evolution diagram (Section 7.5).

Table 7.1 Forms of human disturbance to river courses (modified from Brookes, 1994).

Direct channel changes	Indirect catchment changes
River regulation	*Land-use changes*
• Water storage by reservoirs and water diversion schemes (e.g., for irrigation)	• Changes to ground cover, including forest clearance, afforestation, and changes in agricultural practice (e.g., conversion of grazing land to arable land and emplacement of agricultural drains and irrigation channels)
Channel modifications	• Urbanization and building/infrastructure construction, including stormwater systems
• River engineering. Channelization programs include flood control works, bed/bank stabilization structures, and channel realignment	• Mining activity
• Sand/gravel extraction and dredging programs	
• Clearance of riparian vegetation and removal of woody debris	

7.2 Direct human-induced changes to river forms and processes

Schemes that set out to stabilize and regulate river systems typically endeavor to fix a channel with a given size and configuration in a given position. Such aspirations reflect concerns for potential impacts of river change, and are typically applied with little concern for the inherent diversity of the natural world. Change is natural. Innately, the river will adjust. Once a river is fixed in place, the cost of keeping it there rises inexorably. Protection of societal, economic, and institutional infrastructure has ensured that this course of action must be maintained. However, in many parts of the world, bold initiatives are being taken to let rivers run free once more. Equally important are programs to conserve remaining minimally constrained rivers in their notionally natural state, as far as practicable.

Although various forms of direct human disturbance are typically interlinked in river management programs, individual components are considered separately in this section. Emphasis is placed on dams and interbasin transfers (Section 7.2.1), channelization programs (Section 7.2.2), sand and gravel extraction (Section 7.2.3), and impacts of changes to riparian vegetation cover and woody debris loading (Section 7.2.4).

7.2.1 Dams and reservoirs

The dominant form of direct human-induced disturbance to river courses reflects schemes that have endeavored to control and regulate their flow, and associated concerns for water supply, whether for agricultural (irrigation), commercial/industrial, or residential purposes. Enormous efforts have been undertaken to make dry lands wetter and wet lands drier, ensuring that water is available for human purposes (Cosgrove and Petts, 1990). The clinical efficiency with which engineering programs have achieved this task is, in itself, a testimony to human ingenuity. The extent of these programs is staggering. The global volume of freshwater trapped in reservoirs now exceeds the volume of flow along rivers.

Given the critical importance of security of water supply for both human consumption and agricultural developments, dam construction has played a major part in pivotal societal changes, such as the development of hydraulic civilizations (Wittfogel, 1956). Indeed, dams have been constructed for more than 5,000 years. The pace of construction quickened dramatically after the Second World War, and each year more than 200 large dams are completed (Gregory, 1995). A recent decline in the rate of dam development and associated water transfer projects across much of the western world reflects the lack of remaining reasonable opportunities. This factor, among other considerations, marks a shift towards a mature water economy, in which concerns for efficient water use have replaced prospects for generation of further water supply facilities (Smith, 1998). This situation has yet to be reached across much of the developing world, where pressures for dam megaprojects continue to be seen as part of nation-building exercises.

Individual dams commonly form part of integrative water supply or hydroelectricity programs, such as interbasin transfer schemes, wherein water is transferred across system boundaries, thereby accentuating water storage and flow in some systems while diminishing flow elsewhere. The function of dams may vary markedly, ranging from water supply facilities to flood control impoundments. In some instances, dams may operate as flow-through structures. A common basis for all water supply programs, however, is the disruption they place upon patterns and rates of flow. Flow regulation is inimical to natural variability. Impacts range across various timescales, whether measured in terms of instantaneous flow releases over seconds and minutes, from season to season, or over annual or decadal timeframes. By definition, flow regulation reduces the extremes of flow, substantially lowering flood peaks and modifying base flow conditions. The variability of flow that drives various geoecological processes is anathema to the regularity of supply that constitutes the raison d'être for dam construction. Dams not only disrupt the longitudinal continuity of flow along rivers; they also act as major barriers to sediment transfer.

Collectively, disruption to water and sediment transfer mechanisms impacts directly on river structure and function both upstream and downstream of the control structure (Figure 7.1). The reduction in channel gradient following elevation of base level upstream of dams reduces the transporting capacity of flow as it enters the reservoir. This

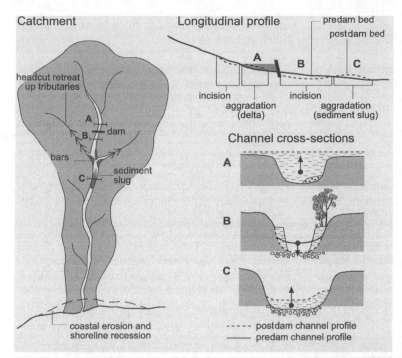

Figure 7.1 Geomorphic impacts of dam construction on river character and behavior
Dam construction traps sediment in a delta, creating an accumulation zone at the entrance to the reservoir (point A).
Suspended load sediments drape the former channel at this point, which now lies beneath the reservoir. At point B,
immediately downstream of the dam, reduced bedload and increased erosive potential of the "hungry river" have
induced bed incision following dam closure. A slot channel has been produced, and the bed has become armored. Inset
floodplains that line the compound channel have been colonized by dense, rapidly growing weeds. The original
floodplain is increasingly decoupled from the channel because of changes to the flow regime and morphological
adjustments. The channel at point B has become an area of net sediment loss following dam closure. Sediments
released from this zone have accumulated downstream at point C, where the channel has contracted through the
formation of lateral bars. Accelerated rates of bedload sediment supply to this reach cannot be sustained from
upstream because of the armoring effect at point B, and reworking of sediments is likely. These effects are
progressively propagated further downstream. Off-site impacts of dam construction may include incision of tributary
streams and altered morphodynamics at the coastline.

promotes delta development at the backwater
limit, reducing the water storage capacity of the
reservoir. Although aggradation takes place rapidly initially, its upstream extent may be limited or
long delayed, dependent upon sediment supply
conditions (Leopold and Bull, 1979). Reservoirs
make excellent sediment traps, commonly retaining more than 90% of the total load and the entire
coarse fraction (i.e., all bedload sediment and all
or part of the suspended load). In large reservoirs,
trap efficiency is commonly greater than 99%
(Williams and Wolman, 1984).

Downstream impacts of dam construction reflect lowered flood peak magnitudes and marked
reductions in sediment load (Williams and
Wolman, 1984). Reduction in sediment concentration, coupled with a reduction in high flows, can
reduce the total sediment load to a fraction of
predam values. The impact of sediment reductions
on downstream channels can vary widely, depending on the amount of reservoir storage, the dam
operations, and the location of the dam relative
to sediment sources (Brandt, 2000; Pitlick and
Wilcock, 2001).

Water releases have the energy to move sediment but little or no sediment load is available to them. This "hungry water" is able to expend its energy on erosion of the channel bed and banks (Kondolf, 1997). Typically, the channel incises, and decreases in bankfull cross-sectional area of over 50% are not uncommon (e.g., Petts, 1979; Andrews, 1986). Where there is no sediment supply immediately downstream of a dam, and bed materials are relatively fine-grained, bed degradation may be experienced (Williams and Wolman, 1984). In extreme circumstances, basal scour may potentially undermine the structure itself (Komura and Simons, 1967). In general terms, downstream channel contraction and degradation continue until development of bed armor or reduction in the energy slope stabilizes the channel. Under conditions of notable degradation, progressive reductions in slope and increases in channel roughness may modify hydraulic conditions such that the rate of degradation diminishes. An initial decrease in channel width following dam closure may be followed by a widening phase as the bed becomes armored and relatively more resistant (Williams and Wolman, 1984; Xu, 1990, 1996). However, if impacts on the flow regime are sufficiently dramatic, such that substantive flows no longer occur, no incision may be observed regardless of bed material texture (Kondolf, 1997). Alternatively, channel capacity may be reduced as fine-grained bars and berms are deposited at channel margins. This may be accompanied by a change in channel planform, such as an increase in channel sinuosity or a decrease in channel multiplicity. If vegetation encroachment occurs, channels may become increasingly stable (e.g., Kondolf, 1997; Erskine et al., 1999; Steiger et al., 2001).

Bed incision and channel narrowing inevitably entail a timelag, as materials are removed and/or redeposited. Changes to sediment transfer regimes following dam closure are manifest over timeframes ranging from 10 to over 500 years (Petts, 1984). Since many dams were constructed in the twentieth century, it may be a century or more before the river adjustment process is fully realized, especially in downstream reaches. Exponential decay functions used to describe incisional responses of channels to dam closure are far from uniform. Around 50% of the total change may be

achieved in the first 5% of the adjustment period (Knighton, 1998).

Large distances may be required before the river regains, by boundary erosion and tributary inputs, the same sediment load that it transported prior to dam construction, and in some instances it may never do so (Pitlick and Wilcock, 2001). Downstream degradational impacts following dam closure may extend over hundreds of kilometers, at rates extending up to tens of kilometers per year (e.g., Galay, 1983; Williams and Wolman, 1984). Variability in response reflects pattern of flow releases and bed material characteristics. Offsite impacts of flow regulation can be especially pronounced at the interface with other landscape compartments. Downstream-progressing degradation along the trunk stream can induce upstream-progressing degradation along tributaries, promoting accelerated deposition at tributary confluences (e.g., Howard and Dolan, 1981; Petts, 1984; Brierley and Fitchett, 2000). The effects of river regulation tend to diminish with distance downstream, as nonregulated tributaries make an increasing contribution to the flow. In some instances, the pattern and rate of morphodynamic interactions may be altered a considerable distance from the control structure, as exemplified by accelerated erosion and shoreline recession at the coastal interface (e.g., Kashef, 1981; Kondolf, 1997; Brierley and Fitchett, 2000).

Dams also induce changes to thermal regimes, water quality, and biogeochemical fluxes, impacting on habitat availability and viability along the trunk stream and secondary channels (e.g., van Streeter and Pitlick, 1998). Altered base flows and associated adjustments to the water table may result in the loss of refugia in isolated pools, increased predation by terrestrial animals, changes to riparian vegetation cover, and increased outbreaks of algal blooms.

Many dams no longer fulfill the purpose for which they were initially constructed. Indeed, in many instances, rates of sedimentation were underestimated, such that infilling of the reservoir compromised or negated some of the core functions of the dam prior to its completion. New, larger structures may make the original structures redundant. This realization, along with increasing awareness of the geoecological consequences of dams, has resulted in many calls for dam removal, and cautious steps are underway to make this happen. An array of geoecological consequences will accompany dam removal, relating to altered flow and sediment fluxes, and accompanying changes to river structure and function. These concerns require careful planning to appraise the long-term viability and sustainability of dams and other forms of control structures along river courses. In rare instances, collapse of dams may lead to devastating consequences, in societal and geoecological terms.

7.2.2 Channelization programs

As many settlements are established along valley floors, concerns for flood control and hazard reduction to support infrastructure development have prompted calls for the training of river courses. These issues, along with widespread efforts at drainage improvement, erosion prevention, and maintenance of navigational arteries are the primary purposes of channelization practices (Brookes, 1988). Most streamlines in urban and peri-urban areas have been channelized via concrete lining and piping of flow. Swampy areas have been extensively drained for agricultural purposes. A range of structural measures can be applied to stabilize channel bed and bank conditions (Table 7.2). In contrast to dam construction, which essentially represents a point disturbance with off-site impacts, channelization activities are applied over varying lengths of river.

Initial endeavors at river clearing and engineering date back to the Roman era (e.g., Herget, 2000). However, systematic channelization programs that set out to address human concerns for navigation and flood control only began in earnest in the seventeenth century (Brown, 1997; Petts, 1989). Since then, channelization programs have extensively modified tens of thousands of river kilometers (Brookes, 1985). These activities typically transform a heterogeneous system into a homogeneous one, with resulting loss in the complexity of instream geomorphic structure and associated changes to flow interactions and habitat availability (e.g., Rhoads and Herricks, 1996; Toth, 1996).

Artificial cutoff and realignment programs may increase the efficiency with which the channel is

able to convey flow and sediment in the short term, initially enhancing prospects for flood control and navigability. However, local adjustments to bed slope trigger secondary adjustments that may have negative and very costly consequences. For example, levee construction deepens flows, potentially increasing rates of bed erosion (James, 1999). Since channelization involves manipulation of one or more of the dependent hydraulic variables of slope, depth, width, and roughness, feedback effects promote adjustments towards a new characteristic state (Brookes, 1988). Geomorphic response times following the emplacement of river engineering works depend on the type of works installed and the extent to which they alter flow and stream power, sediment supply, and vegetation cover. The time taken to attain this new characteristic state may be anything up to 1,000 years (Brookes, 1988).

Channelization may induce instability not only in the "improved" reach but also upstream and/or downstream (Figure 7.2). Impacts are particularly pronounced in response to channel slope modifications or straightening programs that increase local bed steepness and hence erosive potential (e.g., Winkley, 1982). The character of bed level adjustment depends on reach position relative to the area of maximum disturbance (Simon, 1989a, b, 1992, 1994). This zone acts as a fulcrum with net degradation upstream and net aggradation downstream. Degradation tends to be at a maximum immediately upstream of the area of maximum disturbance, as the upstream progression of headcuts accentuates unit stream power. This effect declines in severity upstream. Rapid upstream progression of headcuts may compromise the integrity of infrastructure such as bridges (Brookes, 1989). Incision and basal erosion increase bank height and bank angle until a critical state is reached, promoting mass failure and channel widening. Bed incision and subsequent channel expansion may increase channel capacity by several hundred percent (Brookes, 1989, 1994; Jaeggi, 1989). Over time, armoring may increase resistance, inhibiting further bed erosion. Degradation and widening provide an effective means of energy dissipation as systems adjust to channelization (Simon, 1992). Bed level change (aggradation or degradation) may alter the biological and chemical

Table 7.2 Geomorphic impacts of channelization procedures (modified from Knighton, 1998, p. 312).

Methods	Purpose	Description	Impacts
Straightening (realignment)	Flood protection (flow evacuation); infrastructure development	River is shortened artificial by cutoffs	Gradient is steepened as flow follows a shorter path. Flow velocity and transport capacity are increased. Degradation ensues, progressing upstream as a headcut. Bed and bank erosion increase sediment load to the reach downstream, ultimately flattening its slope and promoting aggradation.
Resectioning (overwidening)	Increase conveyance capacity to reduce overbank flooding	Widening and/or deepening of the channel	Widening reduces velocity and unit stream power, thereby lowering sediment transporting capacity, promoting bench deposition.
Levee and floodwall construction	Flood protection, confine floodwaters, maintain irrigation channels	Raise channel banks, increasing channel capacity	Reduces floodplain inundation and hence sedimentation rates, inducing profound changes to wetland ecosystems. May "trap" floodwaters in extreme events. Alternatively, concentration of flow may promote bed incision.
Channel stabilization and bank protection works	Control bank erosion	Use of structures such as paving, gabions, steel piles, subaqueous mattressing, dikes, and jetties	Alters channel width and roughness components, with secondary implications for bed incision and subsequent sediment release, thereby adjusting channel bed slope. May promote sedimentation adjacent to the bank, potentially increasing flooding if channel capacity is reduced.
Clearing and snagging	Aid flood passage and navigation capacity	Removal of obstructions from the river	Decreases resistance and increases flow velocity, thereby promoting bed degradation, subsequent widening, and marked increase in channel capacity.
Dredging	Maintain navigable channels	Sediment removal from the bed to deepen the channel, especially along the thalweg in lowland reaches	May promote degradation through lowering of base level, enabling knickpoints to migrate upstream, thereby contributing additional sediment to the dredged reach. Deepening may also promote bank collapse and promote upstream-progressing degradation within tributaries.
Weir and lock emplacement	Regulate slope for navigation	Channel spanning structures	Alter bed slope, reducing conveyance of sediment. Modify river structure, promoting elongate pools in place of hydraulic diversity.

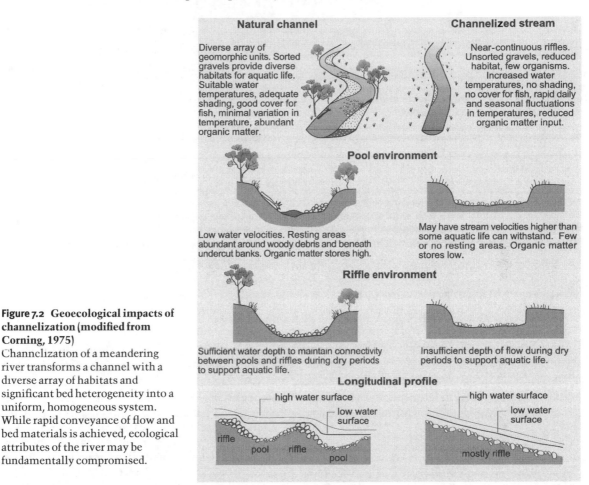

Figure 7.2 Geoecological impacts of channelization (modified from Corning, 1975)

Channelization of a meandering river transforms a channel with a diverse array of habitats and significant bed heterogeneity into a uniform, homogeneous system. While rapid conveyance of flow and bed materials is achieved, ecological attributes of the river may be fundamentally compromised.

functioning of channel and floodplain zones. Geomorphic responses to bed incision exert a significant impact on hydraulic and habitat conditions along river courses, the lateral connectivity of channels and floodplains, and vertical changes to substrate conditions and associated hyporheic zone processes. Drops in water-table levels may bring about detrimental effects to agriculture on adjacent alluvial lands.

Greater concentration of flow within the channel may accelerate the transmission of flood waves and accentuate flood peaks, relative to the period prior to channelization (e.g., Wyzga, 1993, 1996). Not only are effects of flood hazard transferred elsewhere, their extent and consequences may be exaggerated! Effects of flood alleviation and land drainage via combinations of channel widening,

dredging, and straightening are particularly marked on rivers that transport significant amounts of bed material load, as they are able to respond very quickly to imposed changes. On rivers with low bed material loads, responses are less dramatic. Transfer of excess load to reaches downstream of the area of maximum disturbance may result in accelerated channel aggradation and/or bank accretion, reducing channel capacity. This secondary response progressively works its way upstream over time. In areas of reduced velocity that promote the deposition of sediments, response times for stabilization and channel contraction may be enhanced by revegetation of sediment stores. Increased flow resistance promotes accelerated rates of sedimentation. Changes to the nature and extent of riparian vege-

Table 7.3 Geomorphic influence on riparian vegetation recovery patterns following channelization. Modified from Hupp and Simon (1991). Reprinted with permission from Elsevier, 2004.

Stage	Geomorphic attributes	Ecological attributes	Duration
I: Premodified	Aggrading bed and banks, meandering channel, low convex, upward banks, minimal mass wasting	Mature, diverse riparian communities, 100% cover	Stable
II: Construction	High gradient straight channel, linear banks	All woody vegetation removed, 0% cover	Short <1 yr
III: Degradation	Active bed degradation, minimal mass wasting, linear banks	Channel adjustments exert little influence on woody vegetation, which is generally high and dry, 100% cover	1–3 yr
IV: Threshold	Active bed degradation, active mass wasting, concave upward banks, severe instability	Bank failure removes most woody species, herbaceous weeds present, 0–5% cover	5–15 yr
V: Aggradation	Aggrading channel bed, mild mass wasting, significant bank accretion, multiple thalweg, diverse bank forms	Initial active revegetation at same site as initial bank accretion, 10–50% cover	50 yr
VI: Recovery	Meandering channel, low banks and gradient, convex upward banks, general mild aggradation, point bar development	Diverse bank vegetation growing down and into water, 90–100% cover	Stable

tation cover may exert a significant role in channel recovery (Table 7.3; Hupp and Simon, 1991; Hupp, 1992).

Although engineering works generally result in a reduction of the sediment flux, local areas may experience accelerated rates of erosion and sediment transfer associated with river bed scouring, bank erosion, and increased tributary sediment supply (e.g., Bravard et al., 1999). Accentuated sediment loads may result in build-up of deposits, especially in lowland basins. Generation of sediment slugs may diminish habitat availability, presenting barriers for fish passage. If these responses impede human activities, dredging may be undertaken to maintain a navigable channel. Such actions not only act against the natural depositional tendency of the river, they may also trigger bed level instability, promoting the generation of headcuts that may extend some distance upstream.

7.2.3 Gravel/sand extraction

Gravel and sand extraction can take the form of instream (wet mining) where sediment is extracted from instream bar and bed surfaces, or open flood-plain pits. Instream mining may involve extensive clearing, diversion of flow, stockpiling of sediment, and excavation of deep pits (Kondolf, 1994, 1997; James, 1999). In many instances, sediment extraction has been applied without due regard for sustainable rates of bedload transport (i.e., replenishment), such that the bed and floodplain have effectively been mined. By removing sediment from the channel, the preexisting balance between sediment supply and transport capacity is disrupted. Typical responses include lowering of the streambed, local increases in slope and flow velocity upon entering the pit, and adjustments to channel geometry (Figure 7.3). Once bed armor is destroyed, enhanced bed scour may generate headcuts in oversteepened reaches, and hungry water erodes the bed downstream (Peiry, 1987; Kondolf, 1997).

As geomorphically effective sediment transporting events are infrequent in many gravel-bed rivers, instream mining activities may operate for several years without obvious effects upstream or downstream (Kondolf, 1998a, b). However, responses may be manifest during high flows many years later. Headcuts may propagate upstream for

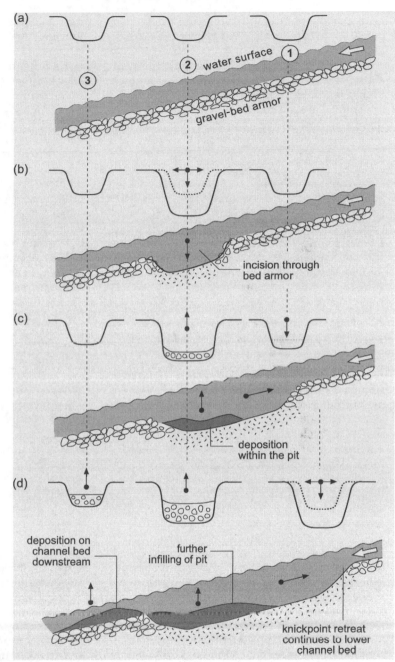

Figure 7.3 Geomorphic impacts of instream gravel mining
(a) In the preextraction condition, the river's sediment load and the force available to transport sediment are continuous through the reach. (b) Excavation of an instream pit breaks the bed armor and instigates a headcut at the upstream end of the pit. Initially, the pit traps sediment, interrupting the transport of sediment through the reach. Downstream, the river retains the capacity to transport sediment but has no sediment load. (c) Headward extension of the headcut acts to maintain bed surface slope. Hungry water erodes the downstream end of the pit, as incision expands both upstream and downstream. (d) Sediments released following the upstream progression of the headcut and associated channel expansion partially infill the incised and expanded trench of downstream zones in the form of bars and benches. This results in a compound channel form. Modified from Kondolf (1994) and reprinted with permission from Elsevier, 2003.

kilometers on the main river and tributaries, potentially undermining bridges and weirs and exposing aqueducts, gas pipelines, and other utilities buried in the bed. Incision may also be accompanied by coarsening of bed material, as smaller, more mobile fractions are transported first. Undercutting of banks promotes channel expansion. Planform changes may ensure, typically entailing the adoption of a low sinuosity, single channeled river. Enhanced rates of downstream sediment delivery may promote channel aggradation and instability. These adjustments alter the availability and viability of aquatic habitat, groundwater levels, and riparian vegetation associations.

Removal of sand and gravel via floodplain mining also represents a nonrenewable exploitation of resources. However, if managed effectively, with clear separation of geomorphic activity from the channel zone, pits may be stabilized and left open once mining activities are completed, creating habitat for local flora and fauna in large open-water ponds. However, if not carefully managed, the pit may be captured by the channel, resulting in upstream and downstream propagation of incision and consequent bed coarsening, channel widening, and destabilization of the banks (e.g., Kondolf, 1997).

7.2.4 Geomorphic responses of rivers to clearance of riparian vegetation and removal of woody debris

As noted in Chapter 3, river change is brought about when the preexisting balance between impelling and resisting forces along a reach is unsettled. Adjustments that modify the water and/or sediment regime reflect changes to either the *driving forces* that promote change, or the *resisting forces* that inhibit change. Riparian vegetation and woody debris have key roles in various feedback linkages that influence channel capacity, hydraulic roughness, channel slope, sinuosity, sediment transport rates, bank strength, and floodplain evolution. Riparian vegetation cover and the loading of woody debris are perhaps the most readily manipulated form of channel resistance. Whether induced by environmental (climatic) changes, bushfire activity, or as a consequence of human activity (mechanical removal or grazing), changes to riparian vegetation cover are likely to bring about significant adjustments to river structure and function. Although human disturbance to riparian vegetation represents a direct human action, its consequences have been largely unintended.

As widespread disturbance to riparian zones occurred before records were kept across much of the world, insights into the nature, pattern, and rate of river adjustments to clearance of riparian vegetation and removal of woody debris remain inferential, rather than directly proven. Regardless of environmental setting, contemporary river morphodynamics across most of the globe have adjusted to conditions in which riparian vegetation and woody debris are either absent or highly altered. Variability in geomorphic response to vegetation removal reflects the type of river and the role played by riparian vegetation and woody debris as determinants of form–process associations, the inherent capacity of a river to adjust, within-catchment position (and related scalar considerations), and the sequence of driving factors (i.e., floods) that promote change (e.g., Hupp and Osterkamp, 1996; Gurnell et al., 2002).

River responses to clearance of riparian vegetation and/or woody debris are likely to be greatest in those settings where vegetation exerts greatest influence on river morphology, namely sand-bed alluvial rivers. For example, the presence of an intact riparian forest maintained a low capacity, slowly meandering sand-bed channel throughout the Holocene, at least, along various river courses in southeastern Australia (Brooks and Brierley, 2002). However, removal of riparian vegetation and woody debris resulted in catastrophic incision and channel expansion within a matter of years (Brooks et al., 2003). In their intact state, there was sufficient inherent roughness in these riparian landscapes that thresholds for geomorphic change were virtually unattainable and geomorphologically effective floods were unable to bring about river metamorphosis. Even following major floods, there was sufficient capacity for geomorphic recovery, such that a characteristic state was maintained. However, once the inherent resilience of these valley floors was altered by wholesale clearance of riparian forests, accompanied by the removal of woody debris, subsequent events initiated channel metamorphosis. In one river system, direct human disturbance to the riparian vegetation cover tied to a desnagging program brought about a 700% increase in channel capacity, a 360% increase in channel depth, a 240% increase in channel slope, and a 150-fold increase in the rate of lateral channel migration within a few decades (Brooks et al., 2003; Figure 7.4). Once triggered, the enlarged channel capacity not only reduces channel resistance; it also increases energy concentra-

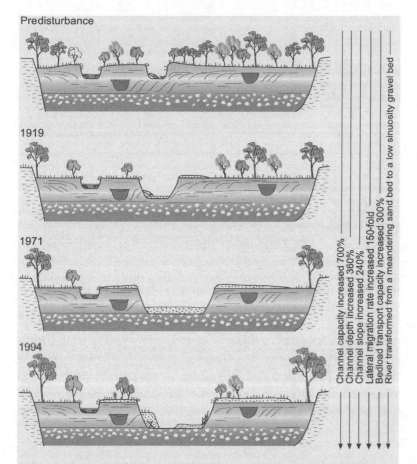

Figure 7.4 **Geomorphic changes following clearance of riparian vegetation and removal of woody debris along the Cann River, Victoria, Australia**

Clearance of riparian vegetation and removal of woody debris induced dramatic changes to the geomorphic structure of the Cann River in East Gippsland, Victoria (see text; Brooks et al., 2003). Partial clearance of riparian vegetation had occurred by 1919, but a desnagging program in the decade prior to 1971 was primarily responsible for channel metamorphosis. The near-instantaneous reduction of vegetative roughness elements lowered threshold conditions that determine bed level stability and critical bank height, such that the channel became highly sensitive to change. Flood events that brought about minor perturbations under intact vegetation conditions were much more geomorphologically effective under altered boundary conditions. Exceedance of threshold conditions brought about fundamental shifts in river character and behavior via incision, straightening, and channel expansion. The progressively enlarging channel increasingly concentrated flow energy at flood stage. The channel became increasingly decoupled from its floodplain.

tion. The capacity of woody debris to increase roughness and stabilize instream sediment may significantly enhance geomorphic river recovery following disturbance (e.g., Cohen and Brierley, 2000).

In general terms, loss of riparian vegetation increases bank erosion and promotes channel widen- ing and shifting, or bed degradation. Removal of floodplain vegetation can cause the water table to drop leading to secondary salinization (Burch et al., 1987). Piping, tunnel, and gully erosion may dam- age infrastructure such as roads, railways, and bridges. Alternatively, influxes of exotic vegeta- tion can smother a channel bed, inducing excess resistance and promoting aggradation. These adjustments alter habitat diversity and the nature and rate of biogeochemical fluxes.

Somewhat ironically, maximizing flow resistance along a reach by increasing instream roughness using wood and riparian vegetation, tied to strategies that increase the structural heterogeneity of a river, now form a major aim of many river rehabilitation programs. In situations where riparian forests can be regenerated, fast-growing trees should be reestablished initially, such that large trees quickly become available to create key pieces for log jams (Collins and Montgomery, 2001). Placement of key pieces or constructed jams can be important in interim decades (e.g., Abbe et al., 1997; Brooks et al., 2001, 2004). Over time, slower-growing, more durable species become dominant, in a form of "restoration succession," working towards self-sustaining riparian forests over timeframes of 50–100 years (Collins and Montgomery, 2001). Geoecological benefits of such initiatives extend well beyond concerns for geomorphic river structure and function, aiding a host of ecosystem attributes such as the temperature regime (shade), hydraulic diversity around instream plants, habitat for invertebrates, input of organic matter, nutrient processing, food web processes, etc.

7.3 Indirect river responses to human disturbance

Indirect impacts on river forms and processes refer to human actions that modify components of the landscape which, in turn, alter flow and sediment transfer regimes, thereby affecting river character and behavior in ways that were unplanned and/or unforeseen. These adjustments entail spatial and temporal lags of varying intensity. Three variants of indirect human disturbance to rivers are considered here, namely: deforestation and subsequent land-use changes (Section 7.3.1), urbanization (Section 7.3.2), and mining activities (Section 7.3.3).

7.3.1 River responses to forest clearance and afforestation programs

It has long been recognized that human-induced changes to ground cover bring about indirect changes to river forms and processes (e.g., Marsh, 1864). Significant human-induced changes to forests have occurred for at least the last 5,000 years (Williams, 2000). In historic times, humans have reduced global forest cover to about half its maximum Holocene extent, and eliminated all but a fraction of the world's aboriginal forests (Montgomery and Piégay, 2003). Today, forests cover around one-third of the Earth's land surface.

Alterations to ground cover, and subsequent land-use changes, impact directly on the balance of resisting and impelling forces that control water and sediment fluxes. Removal of protective forest cover increases the sensitivity of soils to erosion and reduces the rainfall threshold for erosion initiation. Such changes increase the sensitivity of runoff and sediment yield to climatic events. Changes to ground cover induce changes to flood hydrographs by reducing infiltration capacity, thereby shortening lag times and increasing flood peaks. The effects of deforestation upon catchment hydrology are dependent upon the dominant plant species that is removed and the climate, especially the rainfall regime (Knox, 1977, 1987; Sahin and Hall, 1996; Newson, 1997). Yield changes are most pronounced in smaller catchments in areas of high rainfall. Increased rates of runoff, as flow becomes more concentrated and peaked, increase the capacity for sediment transfer, bringing about an exponential increase in the total volume of sediment mobilized on slopes. The extent of slope–channel coupling in differing landscape settings results in significant variability in the efficiency with which these materials are transferred to river systems, and associated geomorphic responses.

In general terms, runoff and sediment yields are high from cultivated and heavily grazed rangeland and relatively low for forests and ungrazed rangeland. Any changes that eliminate or reduce vegetative cover are likely to increase sediment discharge proportionately more than water discharge. For example, sediment loads increase markedly following clearcutting of steep slopes, as much of the prelogging drainage pattern is obliterated by mass movement and debris torrents (Beechie et al., 2001). Timber production increases the frequency of landslides, rates of runoff, and associated erosion rates. These responses increase sediment loads in rivers, thereby enhancing their instability. Surviving channels are choked with huge volumes of sediment and logging trash. Replanting of logged areas stabilizes slopes and reduces runoff, erosion, and sediment yield. Hence, following logging-induced forest clearance, sediment is initially mobilized in great amounts, but yields then decline as secondary vegetation becomes established. Selective logging practices markedly reduce these impacts. Average sediment supply rates from immature forests (i.e., stands < 20 years old) are four times greater than from mature forests (Beechie et al., 2001). Landslide rates from roads are roughly 45 times the rate from mature forests, and landslide rates from clearcut areas are roughly 4 times that of mature forests (Beechie et al., 2001). Sediment yield is strongly dependent on the degree of disturbance to the soil by skidding, hauling, road construction, etc., rather than the removal of trees *per se*. Specific impacts on bedload and suspended load yields are dependent on whether sediment sources on slopes are directly linked to channels. Drainage network extension may be promoted via gully development, and accentuated runoff may accelerate rates of bank erosion (e.g., Madej, 1995; Nolan and Marron, 1995; Madej and Ozaki, 1996).

In response to these collective adjustments, river sediment loads following forest clearance may be increased by an order of magnitude or more (e.g., Milliman et al., 1987). Sediment overloading accentuates the tendency for downstream channels to become wider, shallower, less sinuous, and more braided. Rates of vertical accretion on floodplains may be accelerated by an order of magnitude or greater (e.g., Trimble, 1974, 1977; Costa, 1975; Knox, 1987, 1989; Starkel, 1991a; Brooks and Brierley, 1997). A schematic representation of sediment responses to land-use change is presented in Figure 7.5. Following initial increases in sediment yield associated with forest clearance and cropping, a phase of partial reforestation and application of conservation measures is commonly observed (Wolman, 1967). If sediment supply declines sufficiently, yet flows remain effective, channel entrenchment may ensue, possibly leading to the formation of new floodplains inset within older ones (Costa, 1975; Miller et al., 1993; Trimble, 1974).

In some instances, land management practices may directly influence water and sediment fluxes. For example, emplacement of agricultural drains may modify the flow network (either enhancing or diminishing local patterns of flow). Alternatively, tillage may accelerate the delivery of fine-grained materials to rivers. In many places, improved conservation practices and reforestation initiatives have drastically reduced rates of soil erosion and hence sediment loading of rivers, thereby improving water quality (e.g., Archer and Newson, 2002). Notable reductions in streamflow following afforestation are first detected around 3 years after planting, with notable reductions evident after 5–8 years (Bosch and Smith, 1989). Responses to afforestation depend on vegetation cover and climate, causing different rivers to adapt to afforestation in different ways. Secondary responses of afforestation may include incision into previously eroded sediments such that channels become narrower (e.g., Trimble, 1974, 1983; Davies-Colley, 1997; Bravard et al., 1999; Liebaultd and Piégay, 2002). However, much of the eroded sediment may remain as colluvial sheetwash deposits on hillslopes or as alluvium in floodplains and channels.

7.3.2 Urbanization

Relative to forest clearance, impacts of urbanization tend to be relatively localized. The spread of impervious surfaces and installation of efficient sewerage and stormwater systems increases the area of low or zero infiltration capacity and increases the speed of water transmission in channels or surface water sewers. This increases the volume of runoff for a given rainfall and gives rise to a flashier runoff regime with shorter lag times and higher peak discharges. Smaller, more frequent floods are much more affected than extreme floods. Hollis (1975) showed that 20% urbanization enhances the 1 year event 10-fold, while the 1 in 2 year event is only doubled or trebled. Increases in discharge volume and peakedness increase flow velocity and hence competence, potentially leading to accelerated erosion, an increase in channel slope, and deposition of sediment downstream. Geomorphic consequences vary markedly dependent upon the form, extent, and intensity of modifications to streams and the ground cover, the type of urban development, and the physical and climatic setting of the city.

In terms of sediment yield, urbanization can be considered as a two-phase post-disturbance response (Figure 7.6). When large amounts of soil are exposed in construction phases, runoff has the capacity to induce extensive erosion. Sediment con-

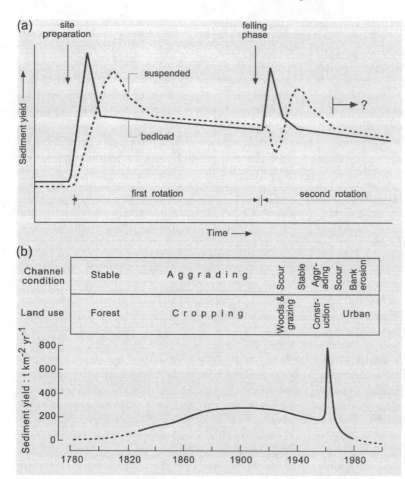

Figure 7.5 Effects of forest rotation and land-use change on river sediment yields
(a) Schematic representation of changes in upland stream sediment yields over a forest rotation. Forest clearance promotes an initial pulse in bedload materials, while the pulse in suspended load is delayed. Re-establishment of ground cover following site preparation reduces sediment yield following the initial pulsed event. Subsequent forest clearance in the second rotation results in a similar, but more subdued pattern of response. From Leeks (1992) in Billi et al. (1992), © John Wiley and Sons Limited, 2003. Reproduced with permission. (b) The effects of land-use change on sediment yield in the Piedmont region, USA (from Wolman, 1967). Increase in sediment yield following forest clearance eventually reaches a plateau. Transformation of agricultural land into urban areas promotes a sharp peak in sediment yield during the construction phase. Ultimately, sediment yields may return to levels that are similar to, or lower than, those associated with a predisturbance condition.

centration and yield increase by one and two orders of magnitude respectively (Wolman and Schick, 1967). In this initial phase, channel capacity is reduced due to local aggradation caused by increased sediment supply to the channel. Sediment yields decline after the construction phase because of the decreased availability of sediment sources in response to newly concreted and asphalted surfaces. Resulting levels may be lower than those associated with forested basins. This factor, along with more peaked floods as overland flow is directed rapidly to the channel via stormwater sewers, accentuates erosion and promotes channel enlargement (Wolman, 1967).

7.3.3 Mining

Mineral extraction for fossil fuels (e.g., coal, lignite, and peat), metals (e.g., gold, silver, lead, zinc, and copper), and aggregates (e.g., alluvial sand and gravel) has exerted a profound impact on river systems in various parts of the world. These activities disrupt the hydrological regime (through vegetation removal and drainage modification), accelerate slope erosion, and increase sediment delivery to rivers. As a consequence, they induce long-term, large-scale instability, with significant off-site impacts (e.g., Lewin et al., 1983; Lewin and Macklin, 1987; James, 1991; Knighton, 1991). In extreme situations, large open-cut mines may remove entire hills or even mountains, infilling intervening valleys.

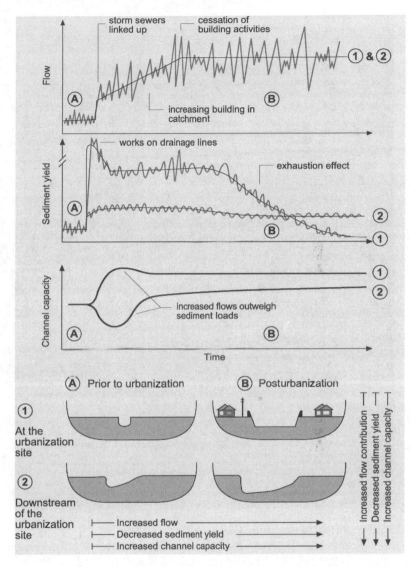

Figure 7.6 Hypothetical trends in channel adjustment following urbanization
Changes to flow and sediment yields following urbanization are accompanied by adjustments in channel capacity, both at the site of disturbance (1) and in reaches downstream (2). Patterns of geomorphic adjustments have slightly different trends, with lagged responses in the latter instance in the transition from a preurbanization phase (A) to a posturbanization phase (B). Modified from Roberts (1989) in Beven and Carling (1989), © John Wiley and Sons Limited, 2003. Reproduced with permission.

Patterns and rates of system response to mining activities depend upon the type of river under consideration, the sensitivity of downstream reaches, and the specific form and extent of mining-induced disturbance. In some instances, alluvial mining activities may simply turn over the channel bed, impacting on the bed material mix and enhancing removal of the finer fraction. Heavy equipment may accentuate the packing of bed material. Elsewhere, extensive disruption may occur through excavation of channel and floodplain ma-

terials and disposal of waste materials (Figure 7.7). Patterns of river response depend on the extent of material input and its texture relative to the natural load. Lewin and Macklin (1987) differentiate between *passive dispersal*, in which mining waste is transported along with the indigenous load without major disruption to river morphology, and *active transformation* in which the increased sediment load induces metamorphosis. This commonly takes the form of an aggradation–degradation cycle. Extensive build up of sediment

Figure 7.7 Mining-induced impacts on rivers
Mining impacts on river character and behavior can take many forms. Photographs (a) and (b) show waste disposal from a copper mine along the King River in Tasmania, Australia. Photograph (c) shows enhanced slope failures in tailings dams at the Porgera mine site in Papua New Guinea. Equivalent waste disposal procedures at the Ok Tedi mine, also in Papua New Guinea, have locally induced more than 100 m of bed aggradation downstream of mine sites (photographs (d) and (e)). Accelerated rates of sedimentation have reduced significant off-site responses through contraction of channel capacity and increased floodplain sedimentation, resulting in loss of wetlands and impacting negatively on cultural/spiritual values (and navigability, ironically impacting on mining operations).

during the *aggradation phase* promotes channel widening and the adoption of a multichannel planform (e.g., James, 1993). As mining-induced supply diminishes, the *degradation phase* prompts the channel to narrow and reinstigate a single channel, producing a series of terraces (e.g., Macklin and Lewin, 1989).

Since both the supply of material from mining and the occurrence of competent discharges are likely to be intermittent, the coarser waste tends to move downstream in the form of slugs rather than continuously (Lewin and Macklin, 1987). Finer fractions move more rapidly through the system, being deposited some distance from source on floodplain surfaces or in the accumulating basin. The nature and extent of off-site and lagged responses reflect site-specific circumstances, as they are influenced by the history of the mine and the sequence of formative flood events. Effects of contamination following deposition of metalliferous fines and other toxic substances can exert an indirect impact on river forms and processes over extensive lag periods. For example, toxic waste products may retard riparian vegetation development, and associated hydraulic roughness, for centuries.

7.3.4 Geomorphic responses to past river engineering endeavors

The mindset with which river management practices were applied in the past emphasized concerns for river stability, focusing attention on measures to "train" or "improve" river courses. Many efforts were overengineered, to ensure security in flood protection or related goals (Jaeggi, 1989). Endeavors to nullify river instability often prompted a greater sense (or faith) that engineering strategies could be applied to rectify or control problems. This mentality promoted greater exploitation of riparian corridors, increasing societal needs for the stability of river systems, thereby increasing the costs for ongoing maintenance. These strategies were typically applied in a piecemeal manner, with little consideration given to basin-wide perspectives or off-site impacts (e.g., Bravard et al., 1997). As a consequence, many activities not only failed to achieve their intended goals; they also had unforeseen, undesirable effects. In some instances, engineering works have promoted or enhanced river instability, inducing local increases in bedload, uncontrolled aggradation, and channel widening (e.g., Leeks et al., 1988; Bravard et al., 1999). This lack of foresight and appropriate planning marked reactive thinking that was concerned with the immediacy of issues, rather than a reflective, proactive approach that sought to address the underlying causes of problems. Protection of infrastructure now demands the continuance of measures that are unsympathetic to the needs of healthy aquatic ecosystems.

In recent years there has been increased awareness of the limited economic feasibility and high maintenance requirements of these programs, with their misplaced faith in techno-fix solutions using hard-core engineering strategies (e.g., Brookes, 1988; Williams, 2001). This has resulted, in some areas, in a shift in ethos from one of controlling the environment to one of working with nature, finding compromises between requirements of human activities and environmental needs. Efforts are being made to check and heal the damage caused by past management practices, ensuring that new developments proceed in an environmentally sensitive way. Alternative methods attempt to minimize the adverse physical and ecological consequences of conventional engineering practices by incorporating "natural" channel tendencies (Keller and Brookes, 1984; Gore, 1985). *Soft engineering* approaches that entail a lesser degree of structural manipulation, such as riparian vegetation management, emplacement of woody debris, and use of flexible materials (geotextiles), have induced positive responses to rivers in both physical and ecological terms.

In stark contrast to floodplain protection measures that sought to increase the efficiency of flow conveyance along river courses, many contemporary river rehabilitation programs seek to redress secondary problems of bed/bank instability by maximizing the resistance to flow along a reach by increasing channel and/or floodplain roughness. The "living river" concept seeks to recreate riparian corridors within which the channel is able to adjust. Putting roughage back into river systems is achieved primarily through wood emplacement and riparian vegetation plantings. Maintaining roughness through stock exclusion and other programs is equally important.

Appropriate measures with which to address concerns for river rehabilitation must build on system-specific appraisals of human impacts on river character and behavior. This requires that the various forms of human disturbance outlined in this section are appraised individually and collectively, in both spatial and temporal terms.

7.4 Spatial and temporal variability of human impacts on rivers

Although impacts of fire have seemingly been an agent of landscape change that has been modified by human activity for hundreds of thousands of

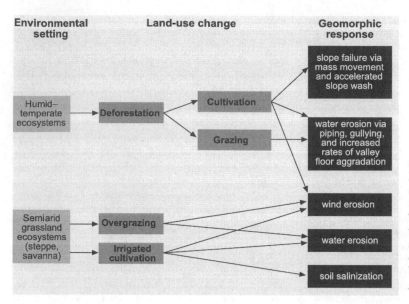

Figure 7.8 Geomorphic responses to land-use changes in humid and semiarid regions (modified from Starkel, 1987)
The nature of agricultural exploits varies in differing environmental settings, prompting differing forms of geomorphic response expressed by adjustments to differing geomorphic processes.

years, demonstrable evidence for human transformation of the environment commenced with the Neolithic agricultural "revolution" in the Near East around 10,000 years ago. Equivalent impacts commenced much later elsewhere across the planet. Indeed, the transition from hunter-gatherer communities in the New World has occurred at various intervals since the sixteenth century. Phases of agricultural intensification resulted in severe environmental degradation. For example, dramatic erosion of the Mediterranean area followed deforestation of much of the Mediterranean region around 5,000 years ago (Marsh, 1864; Thomas, 1956).

In developed areas of the Old and New Worlds, erosion and sedimentation problems associated with deforestation and agricultural expansion preceded the dramatic deterioration of aquatic ecosystems during the Industrial Revolution, when rivers were used as conduits for waste products. Unfortunately, in most developing countries, rapid urbanization, industrialization, and agricultural intensification have been accentuated over the past century or so, inducing simultaneous perturbations in fluxes of water, sediment, and contaminants along river courses (Macklin and Lewin, 1997). Simplified pathways of geomorphic responses to land-use change in

different environmental settings are presented in Figure 7.8.

Geomorphic impacts of changes to ground cover vary for differing landscape compartments, reflecting local relief, length of slope, aspect, degree of landscape dissection, upstream catchment area, etc. Just as important is the configuration of the catchment which shapes the connectivity of slope and channel systems and the capacity of river courses to store and convey materials downstream. Atop this landscape imprint, the magnitude, frequency, and effectiveness of formative geomorphic events induce stark contrasts in landscape responses to disturbance of ground cover, and resulting consequences in both geoecological and societal terms.

Several considerations dictate the nature of landscape responses to changes in ground cover. First, disturbance events are only able to mobilize and convey sediments that are made available to them. Hence, the nature and pattern of sediment stores condition what types and volumes of material are available to be moved by disturbance events. These sediment stores also influence patterns of water retention and associated rates of runoff generation. Second, the nature of ground cover – its density, structure, extent of canopy cover, root networks, and a host of other factors, influence the role that it has as a resisting force that impedes erosion. These interactions determine potential increases in erosion upon the removal of vegetation cover. Third, landscape relief exerts considerable control on disturbance response as it determines the potential energy at any given site. Erosion potential is clearly greater in steep, uplifting terrains relative to low relief settings. The nature and effectiveness of geomorphic processes is fashioned largely by slope angle and length, influencing the preponderance of differing slope mass failure mechanisms that deliver sediments to rivers, such as debris flows, debris torrents, earthflows, landslides, or creep mechanisms. Fourth, once sediments have been mobilized from any given part of the landscape, the capacity for subsequent erosion and sediment transfer is determined by the rate of sediment regeneration on slopes. This is influenced largely by the lithology (its hardness and resistance to erosion) and the weathering regime. Once more, topography greatly influences the capacity for sediment storage, while the frequency of formative events determines the timeframe over which accumulation continues, relative to erosion and down-slope or downstream transfer. Finally, coupling relationships in any given landscape influence the conveyance of water and sediment, thereby affecting the extent and timeframe of system response to disturbance. In strongly coupled landscapes, responses tend to be catastrophic in the short term, then prone to sustained phases of recovery. In largely decoupled landscapes, disturbance responses may not be dramatic initially (or at least they are relatively localized), but responses are sustained over extensive timeframes, and prospects for recovery may be limited. These spatial considerations that fashion indirect human-induced changes to river courses are further complicated by the variable manner, rate, and extent of vegetation clearance, and the recurrence of subsequent land-use changes.

Human-induced changes to river morphology do not directly modify the physical *processes* that occur along rivers. Fundamental hydraulic and geomorphic processes such as the mechanics of sediment transport, bank erosion, and suites of depositional bedforms, continue to be driven by energy relationships, sediment availability, and related

considerations. However, human disturbances modify the spatial *distribution* (pattern, extent, and linkages) and *rate* (accelerated or decelerated) of these processes, often inducing profound changes to river morphology, whether advertently or otherwise. Differing forms of human disturbance vary in terms of their spatial and temporal distribution and extent, their intensity, and their recurrence (i.e., whether they are one-off events or sustained impacts). Spatial and temporal components of river response to human disturbance are considered in Sections 7.4.1 and 7.4.2 respectively. The cumulative nature of human-induced changes to river courses is discussed in Section 7.4.3.

7.4.1 Spatial ramifications of human-induced disturbance to river courses

The impacts of human disturbance on river courses have varied enormously. Some landscapes are primed for change and inherent balances can be unsettled quite readily. Profound responses may be induced at a site or reach, and effects propagated efficiently through the system. Other landscapes may be subjected to intense disturbances with relatively inconsequential geomorphic adjustments. The nature and rate of river response to human disturbance reflects the sensitivity of any given reach to change. The volume and caliber of sediment stored on the valley floor, and the shift in energy regime associated with disturbance, fashion the rate and extent of reworking and hence geomorphic change. For example, if sand materials are protected by resisting elements along the channel, profound responses are likely if human disturbance reduces flow resistance by removing these resisting elements (e.g., Brooks et al., 2003; Brooks and Brierley, 2004). The rate and extent of disturbance response are likely to be less marked along rivers with cohesive banks that are less sensitive to change. However, if dramatic adjustments are recorded in the latter instance, the capacity for recovery may be limited because of the scarcity of bedload-caliber materials that build recovery features and the ease with which finer grained materials are flushed through the system. Profound river adjustments have been evidenced along virtually all alluvial reaches which have been "developed" for human purposes.

The second spatial characteristic of profound significance in interpreting river responses to human disturbance is the juxtaposition of a reach relative to other parts of the catchment that have been subjected to change. Ultimately, changes to flow and sediment fluxes in any given reach affect the balance of these factors elsewhere. Disconnection of sediment stores by upstream barriers, and hence depletion of sediment supply, may produce hungry rivers that promote incision. Any factor that lowers base level promotes bed degradation and associated upstream-progression of headcuts along the trunk stream and affected tributaries. Alternatively, excess sediment loading from upstream may overload reaches downstream. The connectivity of reaches, and associated tributary–trunk stream linkages, are key determinants of the pattern, extent, and consequences of human disturbance to river courses.

The extent and intensity of differing forms of human disturbance have a marked geographic selectivity. Different types of rivers and their adjacent landscapes present different opportunities for human exploitation. For example, zones of bedrock confinement are frequently located in inaccessible areas of catchments, the land is less arable, and hence may remain relatively undisturbed. However, these areas present opportunities for dam construction. Environmental and societal considerations determine the viability of such developments, and their associated functions for hydro-power generation, irrigation schemes, or multipurpose applications. Agricultural opportunities reflect the availability of suitable land, under appropriate climatic conditions. For example, many alluvial reaches have considerable agricultural potential, and these are the very settings where the capacity for geomorphic adjustment is at a maximum. In most cases, the arable alluvial flats adjacent to these rivers were the most sought after land selections and became the first areas in catchments to be developed, initially for agriculture, and subsequently for trade, urban development, etc. These rivers typically occur in lowland basins where river responses to on-site and cumulative catchment disturbance events have been particularly profound.

The spatial extent of differing forms of direct human disturbance varies markedly. Although weirs or dams are point-source forms of disturbance, considerable off-site effects may be experienced because of changes to the longitudinal connectivity of the river system and associated base level adjustments. Most sand and gravel extraction enterprises constitute similar localized impacts, though system responses may be dramatic. Channelized rivers tend to be reach-based, but they may be system-wide in urban settings. In contrast to these site- or reach-specific forms of disturbance, human modifications to riparian vegetation cover have been virtually ubiquitous. Indeed, contemporary river morphodynamics across much of the globe have adjusted to conditions in which riparian vegetation and woody debris are either absent or highly altered. These direct modifications to river courses have been compounded by human-induced changes to ground cover and associated secondary consequences for flow and sediment regimes.

In many places, a reversal in trend in the extent of clearance of ground cover is evident, as forests are allowed to naturally regenerate or afforestation programs are emplaced. This exerts significant changes to water budgets and associated sediment loadings. If ground cover is quickly reestablished and not subsequently disturbed, previous runoff and erosion rates may be reestablished. However, if ground cover is subject to recurrent phases of agricultural land use, series of smaller perturbations will continue, whether induced by natural or human events (or their combination), creating an irregular pattern to resulting runoff and sediment yields. Off-site geomorphic responses are shaped, in part, by the degree to which initial disturbance responses can be sustained. However, depending on the nature and rate of system adjustments, responses to afforestation are not necessarily the converse of those associated with deforestation. The cumulative, interconnected sequence of impacts and system adjustments, and whether they bring about irreversible river change, are critical considerations for management applications (see Section 7.5).

7.4.2 The rate/intensity and extent of human-induced disturbance to river courses

Impacts of inadvertent human disturbance are often delayed until well after the original activity has ceased. Such responses depend on landscape sensitivity to change and the proximity of the system to threshold conditions. The effects of direct changes are usually more rapid and are not so dependent on the crossing of thresholds. The time it takes for a system to recover following disturbance depends upon the extent of displacement, the subsequent flow regime, and the availability of sediment to drive recovery processes. In systems with large buffering capacity and/or with large thresholds to overcome, there may be considerable time lags between perturbation and morphological response. The nature and level of perturbation may be such that some landscapes are capable of withstanding external disturbance, while others fall apart. Profound variability in the manner and rate of river responses to disturbance, and associated potential for geomorphic recovery, reflect, in part, sediment availability. In relative terms, the capacity of river systems to respond and subsequently recover following disturbance may be enhanced in transport-limited landscapes compared to supply-limited landscapes. Transport-limited settings tend to be prone to progressive disturbance because of their readily available sediments, while supply-limited rivers tend to be prone to more dramatic adjustments during infrequent high magnitude events. However, in the former instance, sediments are readily available to be reworked, whereas in the latter instance the timeframe for recovery may be prolonged once sediments are evacuated from a reach (e.g., Fryirs and Brierley, 2001; Brooks and Brierley, 2004).

It is not only the abundance and spatial extent of geomorphic responses along rivers that has been affected by human activity, but also the type or rate of geomorphic process activity. Certain processes and landscape responses now happen more often in more places than they did prior to human disturbance. Alternatively, certain processes may no longer occur in areas where they once were common, or they may occur less frequently within a smaller geographic range. For example, contemporary upland landscapes of the American Southwest and southeastern Australia are characterized by cut phases of cut-and-fill cycles, whereas fill phases were dominant throughout most of the Holocene (Cooke and Reeves, 1976; Prosser and Winchester, 1996). While incision was localized and generally of a limited extent in the past, today it is near-ubiquitous and has brought about an ex-

tent of change that is more profound than has been inferred throughout the Holocene. In addition, it is likely that contemporary channel incision, expansion, and floodplain stripping are far more widely distributed, and more severe, than at any stage in the Holocene.

Since the Neolithic period (around 5,500 years ago), human activity has had a significant impact on floodplain forest systems. Initially, forest clearance likely occurred in a piecemeal manner, in accord with changes in population density, settlement history, and land-use control (Brown, 1997). Relatively low rates of population growth in the mid-Holocene were mirrored by gradual land-use changes, building out from fragmented areas of forest clearance. Major episodes of woodland regeneration likely occurred (Brown, 1997). Intensive human land use was initially localized and transient, and progressive forest clearance occurred over millennia rather than a few centuries. Montgomery and Piégay (2003) refer to manipulation of riparian vegetation cover along European streams during this period as riparian gardening. Although sedimentation rates were undoubtedly increased, especially for fine-grained suspended load materials, profound adjustments to river morphology seem to have occurred much later than this initial phase of vegetation disturbance, associated primarily with direct human modification to river courses (Brierley et al., in press). Seemingly, profound river metamorphosis was induced by direct human intervention via channelization programs, which began in earnest across most large European rivers in the seventeenth century, rather than indirect responses associated with riparian forest clearance over preceding millennia. Interestingly, this phase was coincident with the removal of woody debris from channels to minimize navigation problems and for flood mitigation purposes (Triska, 1984; Gregory et al., 1993; Gippel 1995). These channel disturbances profoundly increased the geomorphic effectiveness of flood events.

In contrast, human-induced disturbances to riparian forest cover and woody debris loadings along river courses in New World settings brought about fundamental changes to river character and behavior within a matter of decades of European settlement. In New World societies, human endeavors sought to exploit the best lands, and use

Figure 7.9 River responses to forest clearance on the North Eastern Cape of North Island, New Zealand Rapid clearance of forest cover and the introduction of sheep brought about dramatic adjustments to river character and behavior, as indicated by bank erosion on this photograph taken immediately following the impact of Europeans in this part of New Zealand. Photograph titled "Sheep crossing Mata River, East Coast, New Zealand, Puketoro Station" by Frederick Hargreaves. Reproduced with permission from the Tairawhiti Museum, Gisborne, New Zealand.

them intensively. Inevitably, subsequent patterns of use have depended on environmental opportunities and societal needs. Eventually, only areas of increasing marginality remained. Initially, riparian zone clearance was a priority of pioneer settlers, as these were the most fertile and well-watered lands (Figure 7.9).

In colonial settings, clearance of riparian vegetation and removal of woody debris were facilitated by far more efficient tools than were available at the time of Old World forest clearance and land-use changes (Crosby, 1986; Lines, 1991; Flannery, 1994; Diamond, 1997). Development of agricultural systems in the New World was more extensive, widespread, and synchronous than equivalent endeavors in the Old World. The intensity of these activities brought about rapid landscape changes, marked by pronounced river metamorphosis within the first generation after colonization (e.g., Knox, 1972, 1977, 1987, 1989; Brierley et al., 1999; Brooks et al., 2003). Geomorphic changes to river courses in the New World were so profound and the lag time between disturbance and metamorphosis so short (typically a few decades), that once critical responses were initiated, it was exceedingly difficult for systems to recover. The systematic nature of riparian vegetation clearance, along with the fact that vegetation regrowth was seldom possible (whether

associated with stocking rates or management practice), almost entirely negated the opportunity for recruitment of woody debris and associated geomorphic recovery mechanisms. In some instances, recovery was further inhibited by sediment exhaustion, such that geomorphic changes experienced over a remarkably short interval induced river responses that are effectively irreversible over timeframes of centuries or even millennia (e.g., Fryirs and Brierley, 2001; Brooks and Brierley, 2004). Once threshold conditions were breached, a completely different set of river forming processes was established under altered channel/catchment boundary conditions. Ramifications of these adjustments were spatially variable because of the differing connectivity of biophysical processes in differing catchments, and differing patterns of sediment storage.

Many rivers have been adjusting to clearance of natural vegetation over hundreds or thousands of years. In some areas of the world a reversal is underway, and river courses have an enhanced vegetation cover relative to conditions in the past (e.g., Liebauldt and Piégay, 2002). Regardless of the nature, extent, and direction of human disturbance, each system has its own cumulative memory. Perturbations build upon each other, making it difficult to isolate specific cause and effect relationships and predict future consequences.

7.4.3 Cumulative responses of rivers to human disturbance

Catchments comprise complex, interactive landscapes. Their unique configuration and history fashion catchment-specific patterns and rates of biophysical fluxes and associated responses to disturbance. Given the complexity of biophysical linkages and the cumulative nature of disturbance impacts, along with profound variability in the inherent sensitivity of individual river systems to adjustment, it is often difficult to isolate cause–effect relationships that directly link changes in river morphology to discrete underlying factors. Within any given catchment, not all landforms may have responded to the last external influence in the same way, resulting in considerable complexity in patterns and/or rates of landscape responses to disturbance events. As impacts following disturbance are conveyed through a system, consequences may be manifest for some time, possibly up to thousands of years. Although patterns, rates, and consequences of geomorphic river responses to disturbance are catchment-specific, disturbance effects are accentuated at the bottom end of catchments, where the cumulative effects of upstream changes are manifest.

Geomorphic responses of rivers to human disturbance are dictated by the timing, sequence, and magnitude of flood events. If no formative events occur, no geomorphic changes are likely to be recorded following human disturbance. When floods do occur, their consequences may be so severe that they are often referred to as natural disasters, conveniently forgetting the multiple forms of disturbance that may have sensitized the system to change. However, it is not the floods that are unusual; indeed, they are an integral part of the process regime that drives the natural variability of river courses. Rather, it is the exaggerated nature, rate, and extent of system responses to disturbance events that are so alarming, as the geomorphic effectiveness of floods is intensified. Changes to the boundary conditions within which floods act, especially marked reductions in resisting elements along the channel bed and/or floodplain and local increases in slope, markedly enhance channel instability.

River responses to human disturbance vary markedly across the planet, affected by factors

such as environmental setting, population pressure (today and in the past), and level of economic/industrial development. Human-induced changes to the boundary conditions within which rivers operate bring about nonuniform responses to the nature and rate of landscape changes. Timeframes of river adjustment, and the character/extent of human impacts, vary markedly from system to system. Different reaches within a catchment are typically at differing stages of adjustment to differing forms of human and natural disturbance. Individual forms of human disturbance seldom occur in isolation from others. Contemporary river forms and processes record system adjustments to the totality of these disturbance impacts, and their interconnected consequences. It is often very difficult to isolate the consequences of any one form of disturbance from others (e.g., Bravard et al., 1997, 1999; Landon et al., 1998; Liebauldt et al., 1999; Kondolf et al., 2002). This challenge is compounded by the inherent natural variability of any given reach. In many instances, responses may represent a legacy from past events or off-site impacts triggered from elsewhere in the system. These considerations, and associated sets of lags, will fashion future situations.

Disturbance is ongoing. Progressive adjustments to natural events and cumulative human impacts ensure that riverscapes are in a permanent state of adjustment, with marked variability in flow and sediment fluxes over a range of spatial and temporal scales. Understanding the trajectory and rate of change, and the capacity for ongoing and future adjustments, are key considerations for management. Questions to be addressed include:

- will the present trajectory and rate of change continue, and for how long?
- when will available sediment stores be depleted?
- what state is the system moving towards?
- what lagged and off-site impacts are involved?
- is change irreversible?

To resolve such questions, a catchment perspective is required in management endeavors, examining the changing nature of biophysical fluxes and the strength of linkages between different landscape compartments. Understanding of contemporary river forms and processes, tied to interpretations of longer term natural variability derived from studies of river evolution, provides a basis to assess river responses to differing forms of disturbance. From this, predictions of likely future character, behavior, and condition can be made in context of changes to within-catchment linkages of river processes.

7.5 (Ir)reversibility and the river evolution diagram revisited

The extraordinary capacity for human endeavor has brought about a range of dramatic transformations to river forms and processes, many of which are irreversible. Indeed, in many instances, if the river seeks to readjust, other direct measures have been applied to try to make the river behave in a manner that conforms to human/societal interests or expectations. Direct or indirect human impacts may bring about reversible or irreversible changes to rivers. In this book, irreversible geomorphic change is framed in terms of management timeframes of 50–100 years.

Appraisal of the geomorphic consequences of human-induced changes to river courses must be framed in context of natural patterns and rates of adjustment that are characteristically experienced for the given river type and its setting. Of particular importance here is the behavioral regime of the river, as highlighted in Chapter 5. Different types of river have distinct behavioral regimes and associated propensity for adjustment. For example, differentiation can be made among rivers that oscillate around a characteristic state, reaches that fluctuate among multiple states, reaches that systematically adjust to progressive changes in boundary conditions, or sites that respond chaotically to external disturbance events. Human disturbance may transform, negate/accelerate, or induce little change to these behavioral regimes. This level of complexity overlies the natural capacity for adjustment of a river. Inevitably, consequences of these adjustments affect other parts of the system.

These various notions are built into the river evolution diagram in Figure 7.10. On this figure, Zone A represents the natural capacity for adjustment within which a range of river behavior is evident. If direct or indirect human disturbance occurs, the capacity for adjustment can expand or contract depending on whether the range of behav-

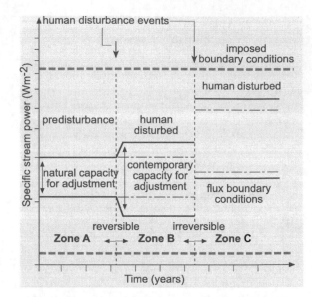

Figure 7.10 Extension of the river evolution diagram to include human disturbance
Human disturbance to river courses often induces an additional layer of complexity to that associated with "natural" disturbance events. This is noted on the river evolution diagram by an expansion in the capacity for adjustment if changes are reversible (Zone B). If changes are irreversible, the capacity for adjustment may be expanded and the position of the inner band may be shifted, such that a different type of river is adopted (Zone C). In some instances, human disturbance may suppress the capacity for adjustment of a river.

ior is accentuated or suppressed. An expanded capacity for adjustment, termed the contemporary capacity for adjustment, is depicted in Zones B and C. If human disturbance expands the range of behavior for that type of river, this marks adjustments away from a natural range of states towards an altered range of states. This may reflect modified rates or abundance of certain forms of adjustment.

Interpretation of the balance between the size of a disturbance event and the ability of a landscape to resist, or accommodate, the impact of the disturbance, is a key consideration in predicting landscape responses to human disturbance. The severity and extent of system response to disturbance determine whether recovery is possible or not. In general, the ability of a river to adjust after disturbance is inverse to the inherent resilience of the system.

In general terms, when rivers are subjected to relatively low levels of impact spread over a considerable period of time, they progressively adjust while maintaining a roughly equivalent state. In systems in which the natural behavioral regime and morphological character of rivers fluctuate among multiple states, gradual and low impact forms of human disturbance may increase the periodicity with which changes among these various states take place and the capacity for adjustment expands. However, these adjustments are unlikely to push the system to a new state that falls outside the contemporary capacity for adjustment. In this instance, although rates of change are modified, adjustments tend to be localized and reversible. These sorts of adjustments tend to occur in relatively resilient systems.

Systems that are resilient to geomorphic change under natural conditions are unlikely to demonstrate significant geomorphic responses, regardless of the nature and extent of human-induced disturbance. In these instances, the capacity for system adjustment is so limited that profound human disturbance may only bring about negligible landscape responses. These reaches absorb the impacts of disturbance, typically through negative feedback mechanisms that enable rapid adjustment after disturbance (Thomas, 2001; Werritty and Leys, 2001). Stability thresholds ensure that a resilient system will only fail under exceptional stress. As a consequence, these reaches tend to experience reversible geomorphic change, with only modest or localized adjustments to their geomorphic configuration.

When reversible geomorphic change occurs, a fundamental shift in the type of river does *not* occur. This is represented by a shift from Zone A to Zone B in Figure 7.10. During these changes, the key defining attributes of the type of river do not change (i.e., the key geomorphic units remain unaltered). However, other structural and functional attributes of the river are considered to be out-of-balance. For example, a sand-bed meandering river with floodplain ridges and swales, instream point bars, pools and riffles, may naturally be expected to have a sinuosity of between 1.7 and 2.0 with occasional cutoff formation occurring every 100 years when operating under a certain set of flux boundary conditions. With human disturbance, the sinuosity of the channel may span a range between 1.5 and 2.2, and cutoffs may occur every 30 years. The key geomorphic structure of this type of river has not been altered, but the rate of adjustment has been accelerated, and the range of adjustment has been widened. Hence, the potential exists for the river to operate outside its natural capacity for adjustment. Ongoing adjustments are reversible.

In contrast, if profound human disturbance is instigated over a short period of time, threshold conditions may be breached, pushing the system outside its long-term range of behavior, and metamorphosis may ensue. This transition may take the form of a relatively simple, one-step transformation such that the system oscillates around a new state, or disturbance may set in train progressive adjustments around multiple states. Regardless of these latter scenarios, changes from the predisturbance condition are likely to be irreversible over management timeframes. These types of responses tend to occur along sensitive reaches that are vulnerable to disturbance and exhibit internal instability (see Chapter 6). Depending on the nature and severity of human disturbance, sensitive reaches can undergo a fundamental and persistent change in their morphology and associated process domain, thereby becoming a different type of river (Werritty and Leys, 2001).

When irreversible change occurs, a wholesale shift in the type of river occurs. This is represented on Figure 7.10 by a shift from Zone A to Zone C if change is induced from a natural state, or a shift from Zone B to Zone C if change is induced by continued/sustained human disturbance. In this case, the key defining attributes of the type of river have changed (i.e., the key geomorphic units have changed) and other structural and functional attributes have been altered, forming a different type of river. Using the same example presented above, if the sand-bed meandering river (represented by Zone B in Figure 7.10) is affected by indirect human disturbance, such that a sediment slug or a headcut passes through the reach, the channel may straighten to a sinuosity of 1.1. Instream geomorphic units may change from bank-attached point bars to midchannel longitudinal bars. The floodplain may change from a lateral accretion system to one that is dominated by vertical accretion of

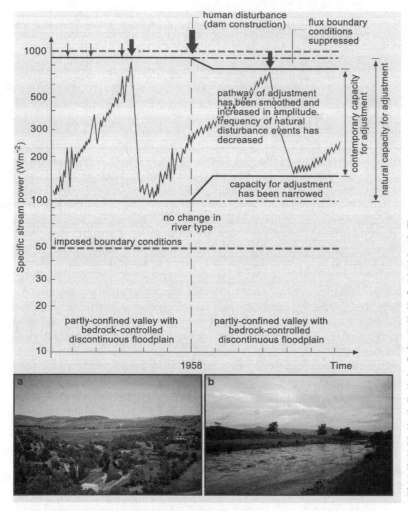

Figure 7.11 Geomorphic responses of the Hunter River, New South Wales, Australia to dam construction
Dam construction along the upper Hunter River did not bring about a change in river type, but the capacity for adjustment of the river was narrowed (see text). This reflects the limited range of possible adjustments that could be experienced by this type of river. Although the flow regime has been markedly altered, the geomorphic structure of the river remains unchanged. Photograph (a) shows the river downstream of the dam, while photograph (b) demonstrates the maintenance of base flow conditions following dam construction. Photographs were kindly provided by Mark Elsley.

sand sheets. A low sinuosity sand-bed river has been formed. The geomorphic character and behavior of this river has been fundamentally and irreversibly altered over management timeframes. Various examples of river evolution diagrams that integrate the impacts of human disturbance are presented in Figures 7.11–7.14.

Figure 7.11 presents changes to the geomorphic character and behavior of the Hunter River, in New South Wales, Australia, following construction of Glenbawn Dam. Prior to 1958, this system operated as a partly-confined river with bedrock-controlled floodplain pockets. The channel comprised an array of gravel point bars, bedrock pools,

and gravel riffles, with discontinuous pockets of floodplain on the insides of bends. Dam construction and the modified flow regime exerted relatively minor changes to the geomorphic structure of this relatively resilient bedrock-controlled river (Erskine, 1985). The river retains the key geomorphic attributes of the same type of river that was evident prior to dam construction (i.e., changes have been reversible in geomorphic terms, as noted on the right-hand side of Figure 7.11). However, the dam has had a range of secondary geomorphic impacts. The dam traps bedload material supplied from the upper catchment. Downstream of the dam, degradation and armor-

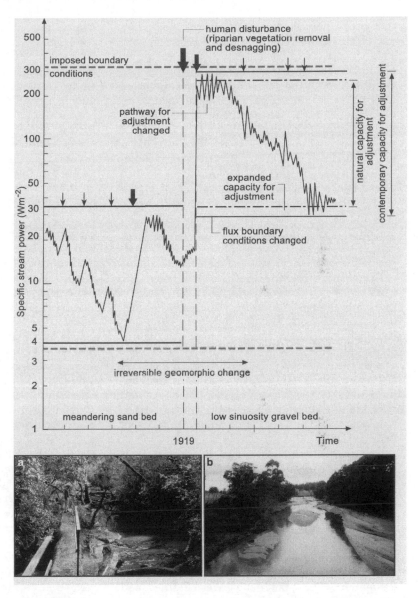

Figure 7.12 Impacts of clearance of riparian vegetation and removal of woody debris on geomorphic changes to the Cann River, Victoria, Australia

Prior to direct forms of human disturbance, Cann River operated as a slowly migrating, slowly accreting river that was subjected to occasional avulsion (see Figure 7.4). Following clearance of riparian vegetation and removal of woody debris, the system became so sensitive to change that the next formative event breached threshold conditions (see text). As the channel incised and expanded, it operated as a high energy system with much greater capacity to transport materials than its predisturbance state. Hence its capacity for adjustment was expanded. Irreversible changes occurred over timeframes of hundreds or thousands of years. The photographs show (a) the adjacent Thurra River which remains in an intact condition, (b) the Cann River as it is today.

ing have occurred. The flow regime has been altered, as water releases from the dam maintain base flow conditions for irrigation purposes. Peak flows have been reduced, and the seasonality of flow has been altered. In many ways the capacity for adjustment of the river has been suppressed, as the range of flux boundary conditions has been reduced. This is represented on the river evolution diagram by the narrowing of the inner band on the right hand side of Figure 7.11. Similarly, the amplitude and frequency of the pathway of adjustment has been reduced, reflecting the lower geomorphic effectiveness of flood events.

Figure 7.12 conveys changes to the Cann River in East Gippsland, Australia following removal of riparian vegetation and desnagging operations in the mid-twentieth century (Brooks et al., 2003; Brooks and Brierley, 2004). Prior to European set-

tlement of the area, the river operated as a low capacity, slowly meandering sand-bed channel with
a high loading of woody debris and rainforest vegetation association on the floodplain (Figure 7.4).
Every few thousand years the river was subjected
to avulsion, as indicated by the natural capacity for
adjustment on the left-hand side of Figure 7.12.
Following human disturbance, and the associated
passage of a sediment slug, river character and behavior were fundamentally altered, as the system
was transformed into a low sinuosity sand-bed
river (depicted on the right-hand side of Figure
7.12). The system has been irreversibly altered
over management timeframes. Channel incision
and lateral expansion have created a low sinuosity
trench that is largely decoupled from the floodplain. Rates of sediment transfer are several orders
of magnitude higher than prior to disturbance. The
capacity for adjustment of the new river system is
far greater than its predecessor. The pathway of adjustment has been altered to reflect the change in
river behavior. As the energy of the system has increased significantly, the new river type sits higher
within the potential range of variability. Based on
sediment supply and transport rates in the contemporary system, it is estimated that it would
take many thousands of years for the system to
recover to its predisturbance state (Brooks and
Brierley, 2004).

Impacts of channelization on the Ishikari River
in Hokkaido, Japan are conveyed in Figure 7.13.
Prior to channelization, this river system was a
low energy, fine-grained meandering river with
a marshland floodplain (left-hand side of Figure
7.13). Large wetlands and cutoffs occurred on the
floodplains. After the Second World War, the city
of Sapporo expanded significantly, and additional
land along the Ishikari River was required for development. The marshlands were drained and
resurfaced with fill, and an extensive channelization scheme was undertaken along the lower
Ishikari River. A canal was dredged and lined with
concrete bricks. Meander bends were cutoff and
plugged, significantly shortening the river, to convey flood flows as efficiently as possible to the sea.
An extensive network of flood control structures
and canals was emplaced, some utilizing the old
channel network. The Ishikari River was irreversibly altered and retains little in the way of its
inherent geomorphic diversity or ecological value.

The energy of the system has likely increased, but
the capacity for adjustment has been severely constrained (right-hand side of Figure 7.13). Water
quantity and sediment supply are stringently controlled through the use of reservoirs and weirs, producing a regularly fluctuating, artificial pathway
of adjustment. Other than localized bank erosion,
little geomorphic adjustment is allowed to occur.

Changes to rivers in Rhone Basin of the French
Pre-Alps in the last 400 years are presented in
Figure 7.14 (based on Bravard et al., 1999; Piégay et
al., 2000). Multiple responses to various forms of
human disturbance have been recorded in these
catchments. Most rivers experienced channel
metamorphosis following deforestation of the
upper catchment and a resulting increase in bedload transport that induced the development of a
braided planform. The capacity for adjustment of
these rivers was high, as indicated on the left-hand
side of Figure 7.14. Wide and shallow channels
transported and stored significant volumes of
gravel, and floodplains were subjected to regular
flooding. In response, channel embankments were
built and gravel extraction became common.
Subsequently, in the mid–late nineteenth century,
afforestation and erosion control management
strategies were applied in the upper catchment, including construction of artificial reservoirs. This
altered the yearly water fluxes, reduced peak discharges, and decreased seasonal flows. As a result
of these management practices, sediment supply
from upstream decreased and incision occurred
downstream. In the case of the Drome River, channel incision averaged around 3 m and extended to
bedrock or an armored layer (Piégay and Schumm,
2003). Channel incision led to the formation of a
single channel within the previous braid plain.
Channel dynamism decreased and artificial levees
were undermined as bends moved and became
cutoff (Piégay et al., 2000). The river had been
transformed to a meandering gravel bed river
(right-hand side of Figure 7.14). Subsequent
encroachment of vegetation into this alluvial
corridor has led to channel constriction and the
formation of inset floodplain surfaces (Bravard et
al., 1999; Piégay and Schumm, 2003). Management
strategies now aim to reinstigate a braided river
system in parts of these catchments through artificial injection of gravel and removal of artificial
sediment storage reservoirs from the upper catch-

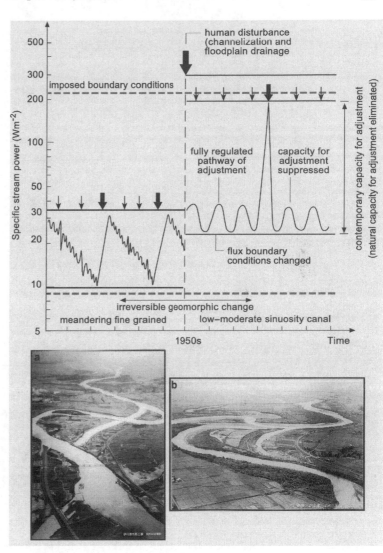

Figure 7.13　Impacts of channelization and floodplain drainage on the Ishikari River, Hokkaido, Japan
Channelization of the Ishikari River in Hokkaido, Japan dramatically altered the geomorphic structure and function of the river. Irreversible geomorphic changes marked the transition from a meandering fine grained system to a local sinuosity canal (see text). Although the capacity for adjustment is shown to have expanded on the river evolution diagram, the river oscillates within a relatively narrow zone most of the time (i.e., the capacity for adjustment has been suppressed and the natural capacity for adjustment has been eliminated). Infrequent high magnitude events flood areas beyond the channel zone. Photographs were kindly provided by Tomomi Marutani.

ment (Bravard et al., 1999). As the energy of the system has decreased over time, the contemporary sinuous single-thread pattern sits at a lower position within the potential range of variability on the river evolution diagram, and has a different pathway of adjustment than the former braided river configuration (Figure 7.14).

The transformation of river courses, whether induced by a thousand cuts, or near-instantaneous changes to boundary conditions, presents an important context with which to interpret likely future pathways and rates of geomorphic adjust-

ments. The contextual information outlined above presents critical insights with which to guide management efforts that strive to work with the variable and dynamic nature of river forms and processes. Appraisal of contemporary river character and behavior in context of former conditions can be used to determine whether human-induced changes to catchment boundary conditions have resulted in irreversible geomorphic changes over management timeframes. Such assessments have major implications for identifying reference conditions against which to assess geomorphic

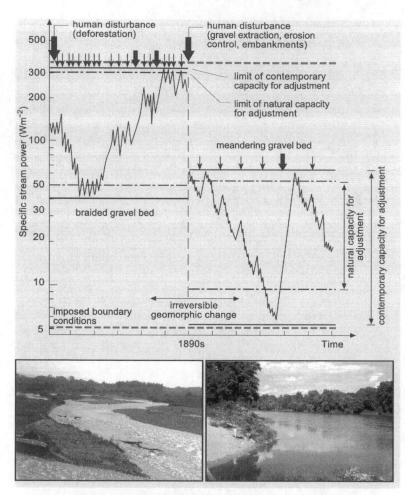

Figure 7.14 The effect of human disturbance on rivers in the French Pre-Alps

Rivers of the French Pre-Alps were in a disturbed state with significant capacity for adjustment in the late nineteenth century (see text). Rural depopulation, along with government reafforestation programs, facilitated the recovery of these systems. Many braided gravel-bed systems returned to a meandering configuration, such that rivers now operate as lower energy system that are subjected to different forms of geomorphic activity (and associated pathways of adjustment).

condition and recovery potential (see Chapters 10 and 11). These considerations determine what character and behavior is considered to provide an appropriate benchmark against which to appraise the contemporary condition of any given reach under prevailing boundary conditions (see Part C).

In the River Styles framework, contemporary attributes of rivers are related to the capacity for change under current conditions, whether that represents an irreversibly altered human-induced set of conditions or otherwise. In making these assessments, the range and rate of contemporary processes along a reach are related to a "natural" condition (i.e., how the reach is expected to look

and behave in the absence of human disturbance). Assessment of whether river response to human disturbance is reversible or permanent is appraised in terms of the assemblage of river forms and processes along the reach (i.e., the assemblage of geomorphic units).

7.6 Synopsis

Regardless of underlying causes, whether natural, purposeful, or unintended/accidental, all river systems are subject to disturbance events that promote adjustments to their behavioral regime. In some instances, change occurs. Most rivers have suffered detrimental effects of human disturbance. These impacts have gathered momentum over time, especially since the nineteenth century. The construction of structures such as dams, levees, and concrete-lined trapezoidal channels, and activities such as sand/gravel extraction, have induced enormous damage to river structure and function. Many channels have been homogenized and effectively separated from their floodplains. Intended modifications have resulted in a range of unintentional consequences, such as changes to flow and sediment transfer regimes, patterns and rates of erosion and sedimentation, hydraulic resistance, and flow velocity. Changes to geomorphic river character and behavior have brought about significant adjustments to habitat availability, species diversity, and aquatic ecosystem functioning.

In many parts of the world, management strategies now strive to work with natural processes and enhance river recovery, aiming to undo the consequences of past actions. In many areas, notable improvements in river condition mark a reversal in the trend of environmental degradation. Learning lessons from past experiences in this age of repair requires detailed documentation, interpretation, and explanation of how past and present human impacts shape the catchment-specific nature and rate of river forms and processes, unraveling these impacts in the light of natural system variability.

Human impacts on river systems vary in type and extent, ranging from site-specific works along a particular reach (e.g., bridge construction or emplacement of a stormwater outlet) to catchment-wide changes in ground cover. Catchment-specific attributes, and variability in the character, extent, history, and rate of human-induced disturbance, ensure that cumulative changes induced by human impacts are system-specific. In many settings it is now impossible for rehabilitation programs to regain some form of predisturbance condition. As such, management efforts must work towards the best-achievable river structure and function given the prevailing boundary conditions under which any given reach operates. The River Styles framework provides a catchment-based physical platform with which to guide management activities that respect the inherent diversity and ever-changing nature of river systems, as outlined in the next part of this book.

PART C

The River Styles framework

(D)evelopment of sustainable management strategies for aquatic ecosystems requires an intimidatingly sophisticated level of knowledge of the spatial context and causal linkages among human actions, watershed processes, channel conditions, and ecosystem response. . . . (W)e should be cautious about our ability to predict ecosystem response based on simplified models of complex systems. . . . (L)andscape management strategies founded upon documented linkages between geomorphological processes and ecological systems should be developed based on sound data and relationships supported by appropriately scaled models, rather than predicated on the predictions of complicated, overparameterized models. . . . (N)o simple cookbooks or manuals . . . can capture the inherent regional complexity or interactions between geomorphic processes, riverine habitat and ecological systems. We can translate understanding based on the general physics that underpins fluvial geomorphology to any region; however, it is much more difficult to generalize how regional differences interact with that physics to structure the manner in which river processes influence ecological systems (and vice-versa). Consequently, we need to pursue regional research programs to develop a sound empirical basis for understanding system behavior and for developing models to usefully extrapolate system behavior into the management arena.

Dave Montgomery, 2001, pp. 247–52

Overview of Part C

The River Styles framework provides a coherent, catchment-wide template for river management activities. Key considerations that underpin the framework include:

• emphasis is placed on linkages between river forms and processes and their capacity to adjust in any given setting. Various attributes of river character are tied directly to interpretations of river behavior. Appreciation of river dynamics lies at the heart of the framework;

• procedures are applied at the catchment scale, focusing on controls on river character and behavior, and their linkages, within any given system;

• appraisals of geomorphic river condition and recovery potential form separate layers of analysis that build on evolutionary trajectories of each reach in the catchment;

• collectively, these insights provide an information base with which foresighting exercises are applied to predict likely river futures, providing a future-focus for management applications.

The first five chapters in this part document the various components of the River Styles framework. In Chapter 8 an overview of the framework is presented, highlighting how the approach breaks down the diversity and changing nature of river forms and processes. Issues considered include an overview of the principles required for an effective river classification scheme, how the River Styles framework addresses these issues, and practical considerations in application of the framework. Chapter 9 documents Stage One of the framework. This entails catchment-scale mapping of river character and behavior, and explanation of downstream patterns of river types. In Chapters 10 and 11 approaches used to analyze geomorphic river condition (Stage Two) and recovery potential (Stage Three) are presented. Human-induced changes to river forms and processes are analyzed in light of natural (ongoing) evolutionary tendencies to frame the (ir)reversibility of river adjustments. From this, the trajectory of likely future geomorphic river condition and/or recovery potential is appraised. Chapter 12 outlines how information on river character and behavior, condition, and recovery potential provides a geomorphic template for river management practice. This forms Stage Four of the River Styles framework. Throughout these chapters, flow diagrams are used to depict the sequence of steps through which information is collated. Various boxes are presented to provide examples of the types of products derived from each stage of the framework. These examples are drawn from the Bega catchment, on the south coast of New South Wales, Australia, where the framework has been applied in its entirety.

CHAPTER 8

Overview of the River Styles framework and practical considerations for its application

There is nothing more basic than categorization to our thought, perception, action, and speech. Every time we see something as a kind of thing, for example, a tree, we are categorizing. Whenever we reason about kinds of things – chairs, nations, illnesses, emotions, any kind of thing at all – we are employing categories. . . . Without the ability to categorize, we could not function at all, either in the physical world or in our social and intellectual lives. An understanding of how we categorize is central to understanding of how we think and how we function, and therefore central to an understanding of what makes us human.

George Lakoff, 1987, pp. 5–6

8.1 Moves towards a more integrative river classification scheme

The inherent complexity of the natural world presents many problems in the development of a workable and comprehensive approach to river classification. Rather than endeavor to create a universal scheme with which to frame management efforts in a prescriptive sense, the approach to breaking down reality adopted in this book provides a learning tool that can be applied to determine the geomorphic components of any given landscape. In no sense, however, do these considerations represent an endpoint for management applications – quite the opposite, in fact!

Part B of this book outlined the conceptual underpinnings of approaches to analyze river character (Chapter 4), behavior (Chapter 5), and change (Chapter 6). Key principles that emerge from these chapters are outlined in Table 8.1. These principles form a series of filters of information upon which the four stages of the River Styles framework are built, as indicated in Figures 8.1 and 8.2. Analyses of geomorphic river character and behavior, viewed from cross-sectional and planform perspectives, provide a platform upon which separate sets of procedures are used to appraise geomorphic river condition (Figure 8.1). Interpretation of river

evolution is applied to determine whether the condition of any given reach of a particular River Style has deteriorated or improved over time. From this catchment-wide appraisal, and using insights into biophysical fluxes and associated linkages, future scenarios are constructed to appraise geomorphic river recovery potential (Figure 8.2). In the examples shown, this is framed in terms of future sediment availability and off-site impacts of change. These types of analysis provide a platform for informed and geoecologically sound river management.

Various practical issues that underpin the approach to geomorphic river classification applied in the River Styles framework are summarized in Table 8.2. Just as important as design attributes of a classification scheme, however, is the way in which it is used. Procedures must be applied in a consistent and nonprescriptive manner. The time-frame of reference and resources used in making the classification must be stated explicitly. Although rigor in application is essential, there are inherent dangers in overly rigid, prescriptive, and inflexible classification schemes. Any classification scheme must be used with caution, guiding observations rather than structuring what is seen (see Miller and Ritter, 1996; Thorne, 1997; Kondolf, 1995; Newson et al., 1998; Goodwin,

Table 8.1 Conceptual underpinnings of an integrative approach to geomorphic classification of rivers.

Approaches to analysis of river character

1. There is significant inherent variability in river morphology in the natural world. Although characteristic forms and assemblages can be identified, site-specific circumstances may result in "unique" attributes of river character and behavior. Adopted procedures used to analyze river character should be applicable across the spectrum of morphological complexity demonstrated by rivers.
2. A nested hierarchical framework provides a meaningful physical basis with which to appraise interactions among various components of river morphology. The valley floor trough provides a key initial guide to river morphology. Notable differences in river morphology are evident for rivers with no floodplains, reaches with discontinuous floodplain pockets, and reaches with continuous floodplains along both banks.
3. All components of a river, including channel shape (bed and bank components), channel planform, and floodplain compartments must be integrated in a meaningful approach to river classification.
4. Procedures used to differentiate among river types should integrate what each river looks like with how it behaves. As geomorphic units characterize the range of erosional and depositional forms demonstrated by rivers, they present a unifying theme for analysis of river character *and* behavior.

Approaches to analysis of river behavior

1. Rivers are never static. They adjust and behave within a set of imposed boundary conditions and in response to a range of biophysical fluxes and disturbance events.
2. Interpretation of form–process associations of geomorphic units and reach-scale evolutionary pathways, as evaluated from assemblages of geomorphic units, must consider both channel and floodplain attributes, framing contemporary adjustments in light of reach history. These analyses must appraise flow stage relations that drive geomorphic activity of any reach.

Approaches to analysis of river change

1. Change, defined as a fundamental alteration to the structure and function of a river, can be induced naturally, or by direct and indirect human disturbance.
2. To identify and understand the causes rather than the symptoms of change, and to predict likely future adjustments, contemporary river character and behavior must be placed in its evolutionary context.
3. Analyses of river evolution must be viewed within a catchment context, recognizing the changing nature of catchment linkages and associated biophysical fluxes over time.

Analysis of controls on river character and behavior

1. Each reach must be placed in context of its landscape setting, assessing the connectivity of biophysical processes between landscape compartments.
2. Effective description is a prerequisite for meaningful explanation and prediction.
3. Procedures used to assess geomorphic river condition and recovery must be appraised as separate layers of information that build on analyses of river character, behavior, and evolution.

1999; Kondolf et al., 2003). Incorporation of these principles to provide a generic river classification scheme is outlined in the following section.

8.2 What is the River Styles framework?

The River Styles framework provides a coherent set of procedures with which to integrate catchment-scale geomorphic understanding of river forms, processes, and linkages. Building on the practical set of objective criteria used in river reach analysis by Kellerhals et al. (1976), and the nested hierarchical framework proposed by Frissell et al.

(1986), the River Styles framework provides a physical basis with which to describe *and* explain the within-catchment distribution of river forms and processes, and predict likely future river behavior.

River Styles record the character and behavior of rivers throughout a catchment, providing a geomorphic appraisal of what a river system looks like, how it behaves, and how it has adjusted over time. This spatially and temporally integrative frame appraises contemporary river morphology and formative processes in light of river change, thereby providing critical insights with which to interpret geomorphic river condition. This forms a

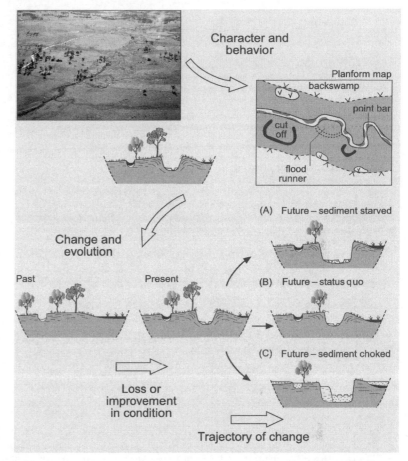

Figure 8.1 Filters of information used in the River Styles framework
The River Styles framework builds upon various filters of information. Analyses of river character and behavior are separated from assessment of geomorphic river condition and recovery potential. In the latter two instances, analyses of river evolution are used to describe how the river adjusted in the past, explain how it presently adjusts, and predict how it is likely to adjust in the future (i.e., whether condition is likely to improve or deteriorate).

basis to predict river futures and the potential for geomorphic river recovery.

The nested hierarchical basis of the framework is structured into five scales: catchments, landscape units, reaches, geomorphic units, and hydraulic units. Catchment-scale conditions dictate the type and configuration of landscape units. Landscape setting, in turn, controls the range of reaches formed along river courses. At the reach scale, River Styles are identified, framed in terms of the valley setting in which a river operates. Distinction is made among confined (no floodplain), partly-confined (discontinuous floodplain), and laterally-unconfined (continuous floodplain) valley settings. River forms and processes are interpreted using a building block, "constructivist" approach to analyze reach-scale assemblages of

geomorphic units. This provides a basis to assess river behavior at differing flow stages, analyzing the capacity of the river to adjust its form. Hydraulic units, which comprise areas of homogeneous substrate and flow type nested within geomorphic units, are used to interpret aquatic habitat patches along river courses.

The River Styles framework does not apply a rigid, prescriptive, top-down approach to pigeonholing reality. Rather, a discrete set of attributes is used to define River Styles and the boundaries between them. Interpretation of form–process associations at the scale of geomorphic units provides a unifying basis with which to analyze the character, behavior, and evolutionary tendencies across the spectrum of morphological diversity demonstrated by rivers. Each River Style comprises a spe-

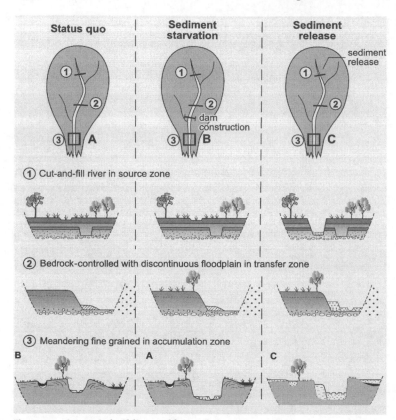

Figure 8.2 Scenario building and foresighting

Appraisal of river evolution, and ongoing adjustments, provides a basis to predict how the river is likely to adjust in the future. Foresighting exercises are used to predict the trajectory of change. In this figure, changes to sedimentary fluxes are used to assess off-site impacts in response to changes elsewhere in the catchment. Maintenance of the status quo is presented in the scenario on the left-hand side. In the middle example, upstream sediment starvation promotes channel enlargement downstream. In the scenario on the right-hand side, release of sediment following incision and channel expansion in headwater reaches has reduced channel capacity in both the midcatchment and lowland reaches, enhancing rates of floodplain sedimentation in the latter instance. In all assessments of river recovery potential, foresighting exercises are undertaken to predict what the most likely future scenarios will be, providing a future focus for river management activities.

cific assemblage of these landforms. Dependent on the valley-setting, channel planform and bed material texture are integrated into river analytical procedures. Building on these notions, attributes used to assess river condition are separated from measures of river character and behavior, ensuring that measures used to assess condition are appropriate for the given river type. From this premise, like is compared with like.

Given system specific conditions, it may be impossible to determine the trajectory and rate of change with accuracy (Shields et al., 2003).

Analysis and interpretation of river condition and recovery potential necessarily entail a degree of subjectivity. Pathways and processes of geomorphic recovery are not necessarily the same or the reverse of those associated with river degradation. For example, channel changes associated with vegetation removal are not the inverse of changes following vegetation reintroduction. Further analysis is required to quantify limiting factors and threshold conditions that inhibit recovery for different types of rivers in different landscape settings. Interpretative procedures must be applied

Table 8.2 Practical considerations in the development of a geomorphic river classification scheme.

The classification scheme must have a clearly defined purpose and target audience

Ideally, schemes are multifunctional and can be used by a range of practitioners. While rigor must be maintained, flexibility is required to incorporate additional layers of information. This may include human-induced variants of river morphology (e.g., urban or regulated rivers), biodiversity value (e.g., fragments, unique attributes, rarity, etc), or assessments of river condition or health.

Procedures should be operationally straightforward and readily and unambiguously communicable, with jargon-free nomenclature

Procedures should be fully and clearly documented, so that consistency in applications can be achieved amongst practitioners. Communicating principles and procedures in a clear, consistent manner is critical. However, a balance must be achieved between overly jargonistic presentation without being simplistic on the one hand, and losing the essence and complexity of the systems being analyzed on the other. A "common language" that is readily accessible to a range of practitioners is required. Limitations of procedures should be stated explicitly. In moving beyond visual assessments or tick-box exercises, appropriate training in fluvial geomorphology is essential.

An appropriate balance between subjectivity and prescriptiveness must be achieved

Classes must be defined using easily measurable parameters that have a diagnostic set of characteristics. Many attributes of rivers that are important in their classification are of a continuous nature, rather than falling readily into discrete classes. As a consequence, river classification, or more specifically their typology, entails components of subjectivity. For example, when new field sites are visited, comparisons are instinctively made with "familiar" sites, categorizing the experience and the site by placing observations into classes based on similarity, uniqueness, or determination of a variant that falls between other situations. However, being overly prescriptive makes redundant the concept of open-endedness, and leads to "black boxing." Boundaries are open along river courses, and a gradation of types may be evident. Inevitably, boundaries change over time, just as the "type" of river may change! A clear statement must be provided of the potential states and behavior that a reach may show whilst still maintaining its character as a particular type. This is a critical criterion in enabling condition assessments to be differentiated from appraisals of character and behavior for differing river types.

The scheme must be generic, open-ended, and flexible

River classification schemes should be applicable in any given landscape setting, allowing the full spectrum of river diversity to be captured, including previously undocumented variants of river. Each field situation must be allowed to speak for itself. If a classification scheme is developed for a subset of rivers and uses parameters that are only relevant to a part of the spectrum of morphological diversity demonstrated by rivers, this should be stated. Geomorphic classification schemes that focus on single attributes of river morphology, whether it be channel shape, channel planform, floodplain type, or valley confinement, fail to appreciate the diversity of river morphology that is evident when these parameters are viewed collectively.

Procedures should be scaleable

There is no inherent size classification for river systems. Reach classifications must be placed in a meaningful spatial (catchment) and temporal (i.e., evolutionary) context. Use of a hierarchical framework enables the interpretation of controls that operate at differing spatial and temporal scales. Applications of river classification schemes at a consistent scale provide a basis to compare and contrast the diversity and patterns of river types at reach, catchment, regional or even national scales. This presents enormous flexibility in the use of the scheme as a tool for management applications.

Must be process-based so that description, explanation, and prediction can occur

Any river classification scheme must be process-based and have evolutionary context. Explanation of river morphology in process terms must be married with an interpretation of system history and configuration. A combination of parameters must be used to identify and interpret key river characteristics and the fluvial processes responsible for generating and maintaining them. As far as practicable, form and process components should be directly linked. This is essential if river character is to be explained, or if the classification scheme is to be used to appraise past and future states. Historical appraisal of channel and floodplain changes are required to place contemporary river forms and processes in an evolutionary context, thereby providing a basis to predict likely pathways of future river adjustment and the potential for river recovery.

Must have practicable, real-world applicability

Classification schemes should guide rehabilitation efforts by indicating the suitability of treatment options for a given river structure of a given condition in a given setting. The scheme should provide a basis to predict system responses to treatments, and associated off-site or lagged responses. The simplest procedure by which to determine a suitable geomorphic structure in rehabilitation programs is to replicate the natural character of "healthy" rivers of the same "type," analyzed in equivalent landscape settings. This provides a guiding image for river rehabilitation applications.

Table 8.3 Key attributes of the River Styles framework as an integrative river classification scheme.

The River Styles framework:

- Works with the ***natural diversity*** of river forms and processes. Due recognition is given to the continuum of river morphology, extending from bedrock-imposed conditions to fully alluvial variants (some of which may comprise unincised valley floors). The River Styles framework can be applied in any environmental setting.

- Is framed in terms of ***generic, open-ended*** procedures that are applied in a catchment-specific manner. Reaches are not "pigeon-holed" into rigid categories; rather, new variants are added to the existing range of River Styles based on a set of discrete attributes (i.e., the valley setting, geomorphic unit assemblage, channel planform, and bed material texture).

- Evaluates ***river behavior***, indicating how a river adjusts within its valley setting. This is achieved through appraisal of the form–process associations of geomorphic units that make up each River Style. Assessment of these building blocks of rivers, in both channel and floodplain zones, guides interpretation of the range of behavior within any reach. As geomorphic units include both erosional and depositional forms, and characterize ALL riverscapes, they provide an inclusive and integrative tool for classification exercises.

- Provides a ***catchment-framed*** baseline survey of river character and behavior throughout a catchment. Application of a nested hierarchical arrangement enables the integrity of site-specific information to be retained in analyses applied at catchment or regional levels. Downstream patterns and connections among reaches are examined, demonstrating how disturbance impacts in one part of a catchment are manifest elsewhere over differing timeframes. Controls on river character and behavior, and downstream patterns of River Styles, are explained in terms of their physical setting and prevailing biophysical fluxes.

- Evaluates recent river changes in context of longer-term landscape ***evolution***, framing river responses to human disturbance in context of the "capacity for adjustment" of each River Style. Identification of reference conditions provides the basis to determine how far from its "natural" condition the contemporary river sits and interpret why the river has changed. Ergodic reasoning is applied to interpret the stage and rate of adjustment of reaches of the same type.

- Provides a meaningful basis to compare type-with-type. From this, the contemporary ***geomorphic condition*** of the river is assessed. Analysis of downstream patterns of River Styles and their changes throughout a catchment, among other considerations, provides key insights with which to determine geomorphic ***river recovery potential***. This assessment, in turn, provides a physical basis to predict likely future river structure and function.

---

carefully to isolate responses of individual causative factors, in light of cumulative adjustments. These considerations, appraised for each type of river and its associated dynamics, and viewed in context of catchment-scale linkages of biophysical processes, must be comprehensively assessed prior to recommending on-the-ground rehabilitation and management actions. So long as classification procedures are used in their intended manner, procedural steps are documented fully, and subjectivity is minimized, they provide a solid basis with which to make reach to reach or broader-scale comparisons. To avoid misinterpretations, limitations, assumptions, and uncertainties must be stated explicitly, and measures of error or risk should be applied.

The River Styles framework does not present a quantitative summary of river character, behavior, condition, or recovery potential for differing river types. Rather, it provides the guiding principles upon which quantitative assessments can be made on a catchment- or (eco)region-specific basis. The

approach is flexible, open-ended, and nonprescriptive. Key attributes of the River Styles framework provide a geomorphic platform for an integrative river classification scheme (Table 8.3).

The River Styles framework comprises four stages (Figure 8.3). Stage One entails identification, interpretation and mapping of River Styles throughout a catchment. This provides a baseline survey of river character *and* behavior. Each River Style is characterized by a distinctive set of attributes, analyzed in terms of channel planform, the geomorphic units that make up a reach, and bed material texture. The identification and interpretation of geomorphic units provide insight into the range of formative processes that reflect the range of behavior of a River Style. River Styles, and their downstream patterns, are then appraised in terms of their landscape setting and the spatial and temporal linkages of geomorphic processes. Analysis of catchment-scale linkages between differing reaches, tributary streams, and the trunk stream, provides guidance into the boundary con-

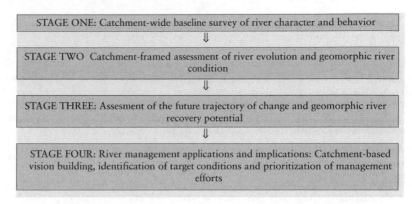

Figure 8.3 Stages of the River Styles framework

ditions that control the nature and rate of river adjustment.

Assessments of river character and behavior must appreciate system dynamic, and the ongoing patterns of adjustment. These are critical underpinnings of the River Styles framework, wherein contemporary river forms and processes (Stage One) are assessed in context of system evolution (Stage Two). This latter component entails determination of the capacity for adjustment of any given type of river, based on analysis of its evolution. This provides the context to determine measures of geomorphic river condition that are directly pertinent to each River Style. Morphodynamic perspectives on the geomorphic make-up of catchments, tied to appraisal of system evolution, provide a predictive context with which to interpret how changes in one part of the catchment have impacted elsewhere, over what timeframe. This provides a basis to predict the pathway of likely future river adjustment. These insights are used to analyze the recovery potential of each reach of each River Style (i.e., assess the likelihood of future improvement or deterioration of river condition; Stage Three). Collectively, these principles provide a geomorphic template with which to frame management applications (Stage Four). As individual catchments comprise unique patterns of River Styles, in which reaches have differing condition and recovery potential, planning for river conservation and rehabilitation is a catchment-specific exercise.

The River Styles framework provides a coherent package of baseline data upon which an array of

additional information can be applied, providing a consistent platform for decision-making for a range of management activities. Insights from application of the River Styles framework are used to identify target conditions for river conservation and/or rehabilitation of each reach, framed in context of a catchment-wide vision (Stage Four). Less impacted sections of a River Style are used to guide the target conditions for river structure in more degraded reaches of river of the same type, replicating the "natural" character of rivers for equivalent landscape settings. A physically based procedure is then applied to prioritize catchment-framed river conservation and rehabilitation strategies.

As the River Styles framework provides an organizational structure for information management, its application enables best use to be made of available information in a precautionary way. This presents a sound basis to collect and compare information, enabling gaps in understanding to be identified and analyzed. Reach or smaller-scale observations are placed in their catchment context, facilitating more detailed analysis of particular areas (i.e., trouble spots). Management applications of the framework are summarized in Table 8.4.

8.3 Scale and resolution in practical application of the River Styles framework

The River Styles framework provides a meaningful basis to classify rivers for comparative purposes while retaining the integrity of scientific input to

Table 8.4 Key management applications of the River Styles framework.

The River Styles framework:

- Provides a basis to **order physical information** in a consistent, coherent, and integrative manner, presenting a systematic and meaningful basis for communication. From this, information gaps, and the need for more detailed assessments of biophysical attributes, can be determined. Catchment-framed assessments provide a template onto which finer scale resolution work can be added, without compromising the integrity of the information base for the catchment as a whole.

- Shows how the physical structure of rivers throughout a catchment provides a **template** to evaluate interactions of biophysical processes. A consistent basis is provided to appraise issues of uniqueness, rarity, naturalness, geodiversity, and representativeness.

- Helps to develop **proactive**, rather than reactive, management strategies that "work with nature," ensuring that site-specific strategies are linked within a reach and catchment-based **"vision."**

- Determines realistic **"target conditions"** for river rehabilitation, focusing management attention on underlying causes of "problems," rather than the symptoms of change. This enables the most appropriate river rehabilitation treatment to be selected (or designed).

- Can be used to more effectively **prioritize** resource allocation to management issues, balancing efforts at river conservation and rehabilitation. This requires differentiation of reaches of high conservation value (in terms of the geodiversity and/or rarity of River Styles) and degraded or stressed rivers. Priorities can be determined within- and between-catchments, presenting an open and transparent physical basis for decision-making.

- Can be used to select representative or reference sites across the range of River Styles in programs to **monitor** river condition and audit the effectiveness of river management strategies. These benchmarking and monitoring procedures can be applied at scales ranging from within-catchment programs through to regional, State or even National river management programs. For example, classification of wild and scenic rivers can be undertaken to determine the "best remaining reaches" of different types of rivers, providing an appraisal of which components of diversity and functioning have been compromised and whether these trends can be reversed.

---

interpret river character and behavior in an open-ended manner. This comprehensive yet adaptable framework allows geomorphologists to collect data on river character and behavior that can be used by other practitioners. Obviously, the use of River Styles reports should reflect the quality of the data that have been recorded. The reliability of a River Styles report will depend on:

- *The skills base of the practitioner and user.* Air photograph and field analytical procedures require training in fluvial geomorphology. The River Styles framework is not a prescriptive black box exercise but a way of reading the landscape. Inevitably, this process is aided by experience and an appropriate skills base. Professional judgment forms part of the chain of reasoning applied in applications of the framework. Various components entail elements of subjectivity that cannot be avoided. Indeed, inherent uncertainties and limitations of knowledge must be addressed in an up-front manner, rather than merely addressing concerns that conform to existing levels of knowledge or understanding. Each practitioner must understand the limitations imposed by the resolution of the work they are completing.

- *The timeframe in which the study is completed.* In general terms, the time available to complete a River Styles analysis dictates the scale at which data are collected and compiled. Consistency of application must be maintained across a catchment, so careful judgment and time management are required. To complete Stage One of the River Styles framework, the office : field ratio is about 2 : 1. Depending on the level of detail required, field analysis can take up to half a day per site. To complete Stages Two and Three, considerable time is spent on evolutionary assessments and analysis of factors that affect geomorphic condition and recovery potential. These are also field intensive exercises, with an office : field ratio of about 1 : 1. Stage Four can be completed entirely from the office, once all information is in-hand.

- *The scale at which data are reported and analyzed.* For many management applications, broad reconnaissance knowledge of the catchment as a whole may suffice. From this information base, specific reaches in which more detailed analyses are required can be identified. Ideally, the scale of analysis should be consistent across the entire catchment, ensuring that identification of River Styles and their boundaries are performed at the same scale. The scale of analysis must be stated explicitly in each assessment. The scale at which data are derived must constrain the way in which the data are used subsequently. Particular care must be taken in comparing site analyses for differing River Styles performed at differing degrees of resolution.

• *Splitting versus clumping.*   The resolution of analysis undertaken in the River Styles framework is dependent on the purpose to which the information is to be utilized. The assessment of "near-uniform" river character and behavior in a reach will vary depending on the scale at which the River Styles framework is applied. There will be no definitive, final statement on variants of River Styles, as no magic number meaningfully summarizes the diversity of river forms and processes. Different end users will prefer a clumped rather than a split approach to the differentiation and labeling of River Styles. Much deliberation will be encountered over whether reaches should be split into individual River Styles, or clumped together as a broader reach of a single River Style in which there is a range or alternating patterns of river character and behavior (sometimes referred to as a segment). Localized features inevitably get buried in broader-scale analyses, but may be very important considerations in finer resolution work (e.g., assessments of geodiversity). Once more, full documentation of analytical procedures is vital.

• *Reach length.*   There are no primary guidelines with which to approach determination of reach length. Obviously this depends on the level of geomorphic diversity that is evident on maps, air photographs, and in the field, and must reflect the purpose of the investigation. In some instances, river morphology may be near-uniform over hundreds of kilometers. Elsewhere, local variability may be induced at-a-point or over hundreds of meters. The scale of catchment, resolution of analysis, and the purpose to which the data are to be utilized are key considerations here. In practical terms, reach differentiation will reflect the amount of time allocated to make a catchment-wide appraisal of river variability and determina-

tion of reach boundaries. Once more, selected procedures must be stated explicitly.

• *Identification of River Styles boundaries.* Inevitably, identification of reach boundaries entails a degree of subjectivity, as the criteria used in their differentiation usually form part of a continuum along which "breaks" may be abrupt, gradual, diffuse, or even alternating. So long as procedures used to define boundaries are stated explicitly, and the type of boundary is recorded, this should not represent an insurmountable obstacle to the practical application of this information. However, practitioners using information in reaches separated by a gradual or diffuse boundary must appreciate that the section of river they are looking at likely has attributes of both upstream and downstream reaches. The key issue here is to interpret what the field situation indicates, rather than use categories outlined on paper!

8.4 Reservations in use of the River Styles framework

The development of the River Styles framework has been an evolutionary process, with countless refinements and reappraisals. Particular problems emerge in putting labels onto River Styles, striving to achieve a balance between consistency, interpretative meaning, and ease of communication. At times this has proved impossible, and borders on the farcical as a dozen or more terms are merged into the label. Putting boxes, boundaries, or labels on nature is NOT the underlying message of the framework. Much remains to be learnt from ongoing debates about how to define a reach, assess how it looks and behaves, interpret how it is likely to change, and apply a label to it. Debate, in itself, is healthy.

The intent of the River Styles framework is to provide a learning tool through which geomorphologists can summarize river character, behavior, condition, and recovery potential and convey these insights to a range of river practitioners. A series of procedural steps is followed to guide geomorphic insights for river management in a meaningful, coherent, and consistent manner. Inevitably, any approach that endeavors to "construct reality" entails a significant degree of simplification. However, as far as practicable, emphasis in applications of the River Styles framework is placed on collection of field evidence that allows the river to "speak for itself," focusing on unique attributes of a reach as much as finding parallels with seemingly equivalent situations. Adoption of this generic, field-based approach does not necessarily equate to, or lead towards, some form of intellectual anarchy, as suggested for equivalent approaches to analysis of river sediments by Walker (1990, p. 779). Rather, it provides a rigorous framework with which to synthesize assemblages of geomorphic units into characteristic "types" of character and behavior for differing River Styles.

River Styles assessments can be used for an array of management and analytical purposes (Brierley et al., 2002). An underlying premise, however, is that the coherency of the framework lies in its use *in its entirety*, thereby providing a solid baseline with which to appraise the past, present, and future geomorphic character, behavior, and condition of rivers. Practical circumstances and time constraints may not permit detailed analyses to be performed. Alternatively, full assessment of evolution, condition, and river recovery may be beyond the needs of some projects. The scope remains for additional layers of information to be added as needs change. In all applications of the River Styles framework, however, definition of River Styles must be undertaken in a rigorous and consistent manner, across a catchment.

In general terms, classification of geomorphic river types (Stage One of the River Style framework) can be achieved with minimal background resources (maps, air photographs, and field access). Dependent upon the experience and skills base of the practitioner, completion of such an exercise may take a significant period of time, if reporting procedures are to be followed in a reliable manner. Frankly, there is little purpose in pursuing such activities unless due diligence is applied, and tasks are completed competently. There is no fundamental reason why these analyses cannot be completed in any environmental setting, regardless of the nature and extent of human modification to river courses. In all settings, a geomorphic context should provide a physical platform for management activities, recognizing concerns for downstream patterns of slope and valley width, the capacity for channel adjustment, bed material caliber and volume, and the connectivity of various processes within the system. It is hard to envisage how effective management can continue independent from such geomorphic information. While the broad structure of the River Styles framework is considered to provide a sound basis with which to undertake these activities, several of the protocols outlined in Chapters 9–12 may need refinement to fit the specific purpose of investigation. However, the principles of the framework are considered to have broad relevance, and have been developed in an open-ended manner so that extensions can readily be added.

In many areas, fundamental insights into river evolution may be lacking, and significant caution must be applied in making inferences from elsewhere. Additional research will doubtless need to be completed to unravel the complexities of evolutionary tendencies and their underlying causality. Hence, significant research may be required before Stages Two and Three of the River Styles framework (assessment of geomorphic condition and recovery potential respectively) can be completed. It is considered to be a far more healthy and constructive response (and investment) to support this research at the outset, rather than endeavor to pigeon-hole reality in a way that may compromise management strategies. Ultimately, the reliability of decisions in natural resources management is contingent upon the information upon which such decisions are based. The environmental outcomes or consequences are also dependent upon the use of reliable information in an effective manner.

Procedures documented in this book outline the principles that underpin the use of the River Styles framework so that practitioners can develop their own classification of River Styles that fits their own catchment. This book does not provide a "how to do it" guide. Rather, the book should be viewed as a "preparation guide," emphasizing how geomorphic insights can be applied in river management practice. Critical questions must be asked about the degree of professional competence required to complete such analyses. Ideally, they would be performed under the auspices of a professionally (and technically) accredited body. In this regard, geomorphology, as a profession, must gain due recognition in its own right.

Specific applications of the framework are intended to be catchment specific and scalar independent. The procedures outlined in this book can be applied with differing degrees of detail within any individual catchment. One of the innate strengths of the River Styles framework is its flexibility. However, in making cross-catchment comparisons, care must be taken to ensure an equivalent level of detail is used from one system to the next.

The reach-based approach to river classification that is developed through application of the River Styles framework in no way precludes more detailed site-specific analyses that may be needed to characterize and interpret river forms and processes. Indeed, as any given reach represents a summary of a range of river character and behavior, a precautionary approach to management applications should always be adopted, such that suggested management treatments are appropriate to the specific problem to be addressed. Such applications must be cognizant of local river dynamics and potential off-site (upstream and/or downstream) considerations. The coherent catchment-framed basis of the River Styles framework is not intended to replace core field-based analytical inquiry. Rather, applications of the framework provide a consistent and meaningful manner to order and organize insights into river character and behavior such that analyses and interpretations of river forms and processes can be translated from one field situation to another, over an array of spatial scales. Hence, site- or reach-specific insights can be framed and evaluated in light of their catchment context.

CHAPTER 9

Stage One of the River Styles framework: Catchment-wide baseline survey of river character and behavior

It is probably the case that the science that attracts us is a reflection of our mentality – those who crave order and certainty become physicists or chemists while those who wonder at variation and complexity become ecologists. River ecologists face one of the stiffest tests of all because of the extreme spatial and temporal complexity of each individual river and the profound variation among rivers.

Colin Townsend, 1996, p. 3

9.1 Introduction

Stage One of the River Styles framework assesses river character and behavior. This is completed within a nested hierarchical approach (Figure 9.1). Top-down assessment of river character and behavior places each reach in context of its boundary conditions at the landscape unit and catchment scales (Figure 9.2). Analysis of controls is performed through examination of downstream patterns of river types along longitudinal profiles. The nature of the landscape unit and valley setting within which any reach is located, along with within-catchment location, determine the process zone functioning experienced by each reach (i.e., whether it operates as a source, transfer, or accumulation zone). Assessment of valley setting provides a measure of valley confinement and the capacity for river adjustment. This provides the entry point into definition of River Styles. A *constructivist approach* to river analysis is applied, building up reach-scale classes of river type based on discrete assemblages of geomorphic units (Brierley, 1996; Figure 9.2). This approach allows field observations of previously unrecorded phenomena to be integrated into a broader classification scheme in a meaningful manner.

This chapter defines the components of Stage One of the River Styles framework. Prior to commencing analysis of River Styles in the field, two related sets of information are collated. First, a review is made of available regional setting resources that pertain to the catchment being investigated. Second, a thorough catchment-wide analysis of air photographs is completed. Alternatively, satellite images may be used. Results of these office-based analyses are presented in a systematic and concise manner. Field investigations are undertaken to extend and validate office-derived insights. Once fieldwork has been completed, an assessment is made of the character and behavior of River Styles across the catchment and controls on their downstream patterns are analyzed. Stage One of the River Styles framework has three steps, as depicted in Figure 9.3.

9.2 Stage One, Step One: Regional and catchment setting analyses

Stage One, Step One sets up the catchment context and collates information that is later required to unravel controls on river character, behavior, and downstream patterns in Stage One, Step Three.

The River Styles framework is designed to be undertaken throughout an entire catchment. However, the specific parameters selected in the derivation of boundary conditions will vary depending on whether the assessment is being undertaken at a regional level (with a number of

Figure 9.1 The River Styles nested hierarchy

The River Styles nested hierarchy is arranged into five key scales at which a range of analyses are undertaken. At the coarsest scale, catchment scale boundary conditions are assessed. Nested within the catchment are areas of relatively homogeneous topography that are called landscape units. The configuration and connectivity of the landscape at this scale provides the key set of imposed boundary conditions within which rivers are formed and operate. They also dictate the flux boundary conditions that drive the character and behavior of rivers by regulating the water and sediment regimes of the catchment and associated vegetation composition. Interpreting controls on river character and behavior occurs at this scale. At finer scales of resolution, hydraulic units are assessed as the basis for habitat assessment. These areas of homogeneous flow and substrate characteristics are nested within geomorphic units which are the building blocks or key landforms that are formed along rivers. Each River Style has a distinct assemblage of geomorphic units that is used to interpret river behavior. A constructivist approach is used to build river morphology and interpret river behavior from its component parts. These two sets of analyses (top-down and bottom-up) are manifest at the reach scale where different River Styles are formed. Modified from Brierley and Fryirs (2000). Reprinted with permission from Springer-Verlag GmbH & Co. KG, 2004.

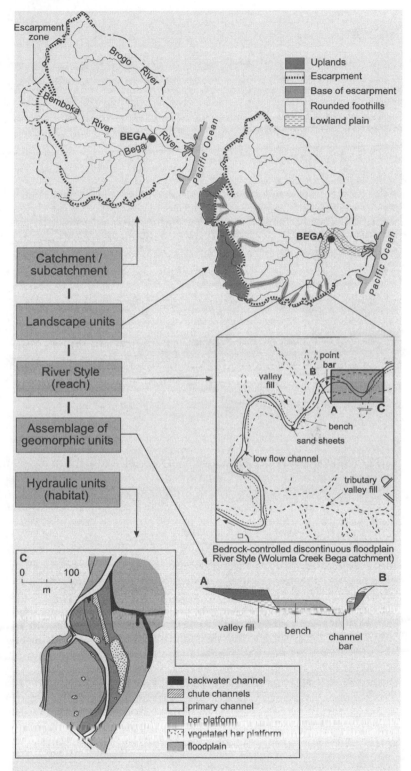

Bedrock-controlled discontinuous floodplain River Style (Wolumla Creek Bega catchment)

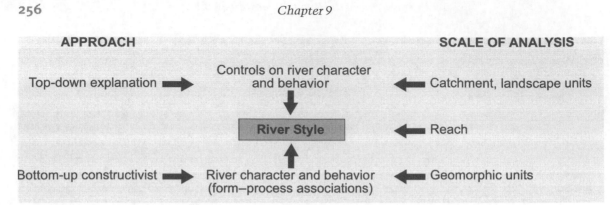

APPROACH   **SCALE OF ANALYSIS**

Top-down explanation ➡ Controls on river character and behavior ⬅ Catchment, landscape units

River Style ⬅ Reach

Bottom-up constructivist ➡ River character and behavior (form–process associations) ⬅ Geomorphic units

Figure 9.2 Approaches and scales of analysis adopted in the River Styles framework
The River Styles framework is arranged in a nested hierarchy of scales extending from catchments to geomorphic units. The structure of this hierarchy allows for top-down explanations of controls on river character and behavior and a bottom-up constructivist approach to interpretation of river character and behavior. These two sets of approaches come together at the reach scale where River Styles are identified and interpreted.

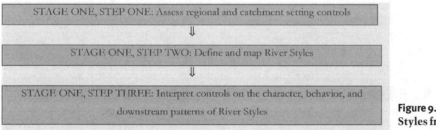
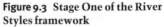

STAGE ONE, STEP ONE: Assess regional and catchment setting controls

⟱

STAGE ONE, STEP TWO: Define and map River Styles

⟱

STAGE ONE, STEP THREE: Interpret controls on the character, behavior, and downstream patterns of River Styles

Figure 9.3 Stage One of the River Styles framework

catchments being assessed), or within an individual catchment. For example, drainage density and catchment shape are not particularly useful for explaining the within-catchment variability of River Styles, but the configuration and relative proportion of landscape units within differing subcatchments will provide useful guidance. In contrast, the interpretation of downstream patterns of River Styles across a region requires a broader focus, in which drainage density, catchment shape, relief factors, regional climate variability, etc., are key factors in the interpretation of the region-wide distribution of River Styles and their downstream patterns. Hence, in the text that follows, a review is provided of the tools that can be molded to suit the level and scale of analysis being undertaken. A flow diagram depicting the procedures undertaken in Stage One, Step One of the River Styles framework is presented in Figure 9.4. Analysis of these various forms of data is used to assess the catch-

ment-scale boundary conditions and controls within which rivers operate.

Skills in Geographic Information Systems (GIS) assist in several components of this assessment, especially relating to derivation of various catchment-scale attributes from manipulation of Digital Elevation Model (DEM) data and associated mapping and graphic skills. Some experience with manipulation of hydrological data aids completion of rainfall-runoff analyses and derivation of associated catchment area–discharge relationships.

These various sources of information are compiled as a Regional Setting chapter within a River Styles report. These analyses should be completed prior to going into the field, as relevant background information can be incorporated into the River Styles analysis, gaps in knowledge are identified, and numerous short cuts may be provided for the field work. Discussions with resource managers

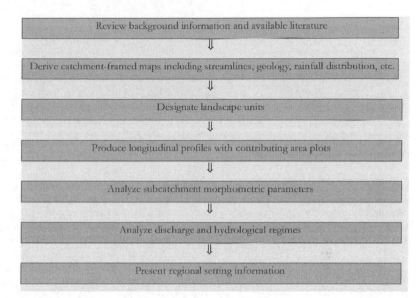

Figure 9.4 Stage One, Step One: Procedures used to produce a regional setting

who have compiled differing components of these data sets, and with technical staff who work in the catchment of concern, will enhance completion of field and analytical tasks.

9.2.1 Background information and review of literature to derive catchment maps

At the outset, background literature and maps of catchment geology, soils, climate, vegetation, land use and settlement history, etc. are compiled. Many maps may already be available in a GIS format from regional offices and local agencies. In regional applications, distinctive catchment characteristics can be identified and compared across the (eco)region. In many catchments, an extensive amount of information can be derived from:
• academic literature including theses, referred publications, books;
• government databases, including commissioned reports;
• consulting reports;
• land systems and topographic map sheets;
• regional resource maps and GIS databases (e.g., soils, geology, land use, etc.);
• local knowledge (historical society, local library, etc.);
• meteorological office data and reports;
• flood history records and gauge station data, etc.

Available GIS data are used to produce a catchment base map showing the river courses under investigation and subcatchment boundaries. Various locational identifiers, such as towns and prominent local landmarks are added to the map, and each primary subcatchment is labeled. If GIS data are not available, 1 : 100,000, 1 : 50,000, or 1 : 25,000 maps (or local equivalents) are used. This provides the catchment template onto which River Styles are subsequently added in Stage One, Step Two of the framework.

9.2.2 Designation of landscape units

In many instances, profound differences in river character and behavior may be evident in differing landscape units. In the River Styles framework, the designation of landscape units builds on the CSIRO Land Systems Unit approach (e.g., Gunn et al., 1969). Various landscape and environmental factors are combined, such as relief, rainfall, elevation, geology, and vegetation coverage. In each land systems unit, environmental factors are sufficiently consistent that a characteristic array of landscape-forming processes occurs, producing distinctive sets of soil, vegetation, and landforms (i.e., ridges, spurs, fans, valley floor, etc.). Hence, landscape units are readily identifiable topographic features with a characteristic pattern

Table 9.1 Parameters used to identify landscape units.

Identifying characteristics	Significance
Physiographic character or landscape morphology	Characteristic pattern of landforms (e.g., shape and size of ridges, shape and smoothness of mountains and hills) can be related to long-term controls on landscape evolution, such as the tectonic setting, structural geology and lithology, rate and extent of escarpment retreat, sea-level adjustments, hydrological and climatic conditions. These in turn dictate the character of the valley setting in which River Styles operate.
Landscape position	Landscape position is important as it dictates the process zone distribution (i.e., whether it acts as a source, transfer or accumulation zone for water and sediment). Characteristic within-catchment patterns of landscape units may be discerned. For example, uplands are commonly found upstream of rounded foothills, which are found upstream of the lowland plain.
Geology	Geological controls on landscape morphology, and hence river character and behavior, are manifest through structural and lithological controls. Structural controls dictate the alignment and configuration of valleys, induced by patterns of folding, faulting, etc., and associated factors that determine the degree and patterns of landscape dissection. Lithological controls determine the availability and caliber of material, dictated in part by the weathering regime. These factors not only influence the structure of a river, but also affect its capacity to adjust (i.e., its sensitivity to change).
Relief	Gross differences in relief provide some indication of the degree to which the landscape is dissected (i.e., a measure of drainage density). This in turn affects the delivery of sediment and water into the river system. As such, the peakedness, geomorphic effectiveness and lagged effects of flow can be determined and inference made about how that may dictate or impact on river morphology.

of landforms. Identification and mapping of landscape units is undertaken on the basis of physiographic character, landscape position, geology, and relief (Table 9.1). Examples of landscape units include: tablelands, uplands, mountains, escarpment, rounded foothills, low-lying hillslopes, and lowland plain. A map showing the distribution of landscape units in the catchment is produced (e.g., Plate 9.1). Elevation, longitudinal valley slope, and valley width are tabulated to characterize each landscape unit (Table 9.2). These descriptors represent fundamental controls on river character and behavior. In many instances, landscape unit boundaries are demarcated by distinct breaks in slope along longitudinal profiles, indicating downstream changes in valley width and elevation that result in a transition in River Style. An example of the summary table of landscape unit attributes for Bega catchment is presented in Table 9.3. These considerations form a basis for assessing controls on the character and behavior of River Styles, as discussed in Stage One, Step Three of the framework, wherein each River Style is viewed in context of its landscape unit setting.

9.2.3 Longitudinal profiles and contributing area

Longitudinal profiles record downstream changes in elevation, and hence slope, along a river. Given that slope is a primary control on river character and behavior, changes in slope along a longitudinal profile often coincide with landscape unit and/or River Styles boundaries. Overlaying longitudinal profiles from different subcatchments can be used to compare downstream changes in slope and assess tributary–trunk relationships. Superimposition of River Styles boundaries onto longitudinal profiles enables analysis and interpretation of controls on the downstream patterns of River Styles in Stage One, Step Three.

In the River Styles framework, longitudinal profiles are constructed using DEM data. Contributing area plots are superimposed onto the longitudinal profiles. This defines the area draining

Table 9.2 Descriptors used to characterize landscape units.

Descriptors of landscape units	Significance
Elevation	Elevation can be used as an explanatory descriptor for landscape position. For example, tablelands in coastal catchments of NSW are generally found in headwater regions at elevations above 1000 m. In contrast, lowland plains are generally observed at elevations below 50 m. Elevation may be a primary control on climate patterns. It must also be noted that elevation can be highly variable for each landscape unit.
Valley slope	Slope is a primary control on the nature and rate of geomorphic processes, whether viewed in terms of the movement of water and sediment on slopes, on the valley floor, or the connection between the two. Breaks in slope along the longitudinal profile often fall at the boundaries of the landscape units. In addition, the slope of the longitudinal profile is one of the key controls on river character and behavior in each landscape unit.
Valley width	Significant changes in valley width commonly define the boundary between landscape units. Valley width is a significant control on the character and behavior of each River Style found in each landscape unit. Although general trends can be discerned, valley widths are often highly variable within landscape units and between subcatchments. Valley width is a key determinant of the valley setting within which a river operates.

into each section of the river (i.e., a surrogate for discharge), providing a visual summary of changes in catchment area along the river. It is often instructive to note (and explain) whether the character and behavior of the trunk stream changes downstream of tributaries. Examples of the products derived from the use of longitudinal profiles are presented in Figure 9.5.

9.2.4 Analysis of catchment morphometric parameters

Geomorphologists have developed a wide range of morphometric parameters with which to characterize landscape morphology. Examples include drainage pattern, drainage density, catchment shape (elongation ratio), stream power, etc. While it is not essential to measure all these parameters in every River Styles assessment, these descriptors may later be used to highlight differences between subcatchments (and assess downstream patterns of River Styles).

9.2.5 Analysis of discharge and hydrological regimes

The timing and frequency of flows influence the capacity of a river to adjust its morphology, while the sequencing of floods affects the geomor-

phic effectiveness of any given event (i.e., its capacity to perform geomorphic work). To assist in the analysis of flow regimes as controls on the range and downstream pattern of River Styles, a series of graphs and tables are produced including:

- regional catchment area-discharge plots for a range of flood recurrence intervals (2, 5, 10, 50, 100 years);
- catchment-specific flood history plots, with an assessment of within catchment variability (if available);
- flow duration curves;
- log Pearson III plots from the most reliable gauge records in the catchment;
- calculation of various flood magnitude indices (e.g., Q_{10}/Q_2).

Hydrological analyses undertaken in applications of the River Styles framework provide an appreciation of what scale of event is the dominant control on river morphology, and how frequently that type of flood occurs. Estimates of stream power are plotted on longitudinal profiles and inundation frequencies are indicated on River Styles cross-sections. From this, triggers for geomorphic changes are assessed at differing positions within the catchment and at differing stages of evolutionary adjustment (Stages Two and Three).

Table 9.3 Parameters used to identify and describe landscape units in Bega catchment.

Parameter/ Landscape unit	Uplands	Escarpment	Base of escarpment	Rounded foothills	Lowland plain
Identifiers					
Physiographic character or landscape morphology	Dissected plateau with relatively deep incised valleys	Steep face incised with deep gorges	Tongue shaped or elongate deep valleys that form downstream from the escarpment	Rounded hills that form ridges dividing each subcatchment. These ridges extend from the base of the escarpment in many cases	Flay, low lying plain with low lying adjacent hillslopes
Landscape position	Atop the escarpment	Between uplands and central catchment	At the base of the escarpment, where valley exists from a gorge	Between the base of the escarpment and the lowlands	Downstream of the rounded foothills where valleys widen significantly. Feeds into the estuary
Geology	Largely granites	Largely granites	Largely granites	Largely granites	Largely granites
Relief	~ up to 400 m	400–600 m	~ 250 m	~ 180 m	~ 15 m
Descriptors					
Elevation (asl)	>600 m reaches a maximum of 1070 m	>200 m	150–400 m	15–200 m	<15 m
Longitudinal valley slope (degrees)	Flat to <3	>15	10–15	3–10	Flat to <3
Valley width	Up to 60 m	<60 m	Up to 300 m	10–150 m	Up to 1500 m

Note:

Five landscape units have been identified in Bega catchment, namely uplands, escarpment, base of escarpment, rounded foothills, and lowland plain. The upland landscape unit is characterized by steep slopes, reflecting dissection of the plateau. It is only prominent in Bemboka, Tantawangalo, and Candelo subcatchments, as the headwaters of other subcatchments lie in the escarpment zone. This differing configuration of landscape units at the upstream end of the subcatchments plays a significant part in determining river morphology in downstream landscape units, especially at the base of the escarpment. Although the base of the escarpment landscape unit is found in all subcatchments, there is pronounced variability in River Style (and response to human disturbance) in this part of the catchment. The differing length of these base of escarpment tongues, and the extent of sediment accumulation in this landscape unit, account for many of the differences in river character in differing subcatchments. In aerial terms, the rounded foothills are the most significant landscape unit in Bega catchment. This landscape unit comprises valley sidewalls of 8–15°, and is dissected by a multitude of lower order channels. The rounded foothills, and the lowland plain, have been almost entirely cleared of vegetation. The lowland plain extends to the Pacific Ocean, although lower Bega River flows through a bedrock-confined reach (Bottleneck Reach) prior to its estuary.

9.2.6 Presentation of the regional setting chapter

Typically, the regional setting chapter of a River Styles report comprises a series of summary tables, plots, and maps. Short paragraphs of text highlight trends and characteristics, but there is little in the way of interpretation. Emphasis is placed on summarizing information that is pertinent to assessment of catchment-scale controls on river character and behavior in Stage One, Step Three. The following format is suggested for this chapter:

- geology;
- soils;

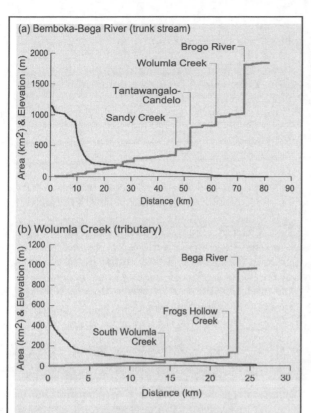

Figure 9.5 Longitudinal profiles and contributing area plots along two contrasting river courses in Bega catchment
(a) All subcatchments that drain directly from the escarpment (e.g., Wolumla Creek) have relatively smooth concave-up forms with occasional bedrock steps downstream of the escarpment zone. These steps act as local base level controls, dictating the slope of valley segments and hence the morphology of river courses. At the base of the escarpment a gentle break in slope is transitional to the rounded foothills landscape unit. These tributary subcatchments tend to be short with relatively small catchment areas. (b) All subcatchments that drain from atop the escarpment are characterized by a distinctly stepped profile in their upper sections where river courses are dissected into the plateau country. This stepped zone is transitional to a concave-up profile downstream of the escarpment zone. However, the break in slope at the base of the escarpment is distinct along these river courses. Again occasional bedrock steps occur along these river courses downstream of the escarpment. These streams tend to be long with significant catchment areas (e.g., Bega River).

- land-use character (including vegetation coverage) and history (including clearance, invasion of exotics, etc.);
- topography (including landscape units, long profiles, catchment morphometric parameters)
- climate;
- hydrological analysis;
- settlement history and population trends.

9.3 Stage One, Step Two: Definition and interpretation of River Styles

9.3.1 Analysis of river character: Parameters used to identify River Styles

River Styles are identified on the basis of a mix of three key parameters: channel planform, the assemblage of geomorphic units that make up a reach (both channel and floodplain components), and bed material texture. The mix varies depending on the degree of insight each parameter provides into river character and behavior in each valley setting. For example, analysis of channel planform for an alluvial river provides the key initial differentiation between River Styles, whereas bed material texture is the key defining characteristic of confined (bedrock-controlled) rivers.

9.3.1.1 Valley setting

The entry point into identification of a River Style is the valley setting. Valley settings are differentiated on the basis of the degree of lateral confinement, expressed by the presence/absence and distribution of floodplains along river courses. The confining medium can be either bedrock valley margin and/or cemented materials preserved in terraces (a form of antecedent control on contemporary river morphology). Valley confinement controls the capacity of the channel to adjust over the valley floor, determining patterns of sediment storage and reworking. Inevitably, the degree to which river morphology reflects an imposed condition extends along a continuum from 100 to 0%. As in all classification schemes, the spectrum of variability must be differentiated into meaningful classes, recognising that some overlap of river character and behavior may occur between classes.

In the River Styles framework, three valley settings are differentiated: confined, partly-confined, and laterally-unconfined, along which floodplains are absent, discontinuous, or continuous respectively. In the latter category, differentiation is made between rivers with continuous channels, and those in which channels are discontinuous or absent. In some instances, continuous floodplains are relatively thin veneers with a bedrock-based channel. Valley settings, in turn, are fashioned in large part by the type of landscape unit within which they lie. For example, confined valley settings are often found in the escarpment landscape unit, partly-confined valleys in the rolling foothills landscape unit, and laterally-unconfined valley settings along lowland plains. Differences in valley cross-sectional morphology, width, and slope directly influence river character and behavior. Within each valley setting a range of River Styles is evident.

While it may be appealing to derive quantitative measures for the three valley setting classes, the following should only be considered as a guide. In the *confined valley setting*, bedrock or terraces are observed along both channel banks. Over 90% of the channel abuts directly against bedrock or terraces. The river course has either no floodplain, or floodplains are restricted to isolated pockets (< 10% of reach length). Channel planform is imposed by valley configuration. For example, if long-term landscape evolution has resulted in a deeply incised and sinuous bedrock valley, the channel must conform to this configuration producing a gorge River Style. Elsewhere, gorges may be straight, as they follow the geologic structure of a region (e.g., along fault lines). In other instances, the channel can be fully contained within terraces or ancient, cemented alluvial deposits that line the valley margin. In almost all cases, bedrock also imposes a vertical control, as bedrock lines the channel bed.

In the *partly-confined valley setting*, between 10 and 90% of the channel abuts directly against bedrock or ancient, cohesive materials. Discrete floodplain pockets occur along the reach, commonly in an alternating or semicontinuous manner. Partly-confined valleys commonly have a sinuous or irregular planform that dictates where floodplains can form (e.g., along the convex banks of bends, or behind bedrock spurs). Along most, but not all rivers found in this valley setting, bedrock also imposes significant base level control, with bedrock outcrops common along the channel bed.

In the *laterally-unconfined valley setting*, less than 10% of the channel margin abuts against bedrock or terrace features. Rivers are laterally unconstrained with continuous floodplains along both channel banks. Banks are deformable, such that the channel is able to mold and rework its boundaries. However, some variants of laterally-unconfined rivers are vertically constrained by bedrock or ancient lag deposits. Rivers found in the laterally-unconfined valley setting are further split on the basis of the continuity of the channel along the valley floor. Reaches with a continuous channel are differentiated from those where the channel is discontinuous or absent.

9.3.1.2 *Geomorphic units*

Analysis of channel and floodplain geomorphic units provides the key tool to interpret reach character and behavior. Given their distinct set of form–process associations, geomorphic units are the key interpretative parameter in the River Styles framework. A bottom-up constructivist approach builds a picture of river character and behavior for any reach, framed in terms of its constituent channel and floodplain components and their interactions. These features can be analyzed across the range of rivers found in different landscape settings, regardless of whether they are confined, partly-confined, or laterally-unconfined variants. While individual types of geomorphic units may be observed in reaches of differing River Styles, a distinct assemblage of geomorphic units occurs along each River Style. For example, pools are evident in many River Styles, although the nature of these pools may be quite variable. Ultimately, it is the assemblage of geomorphic units along a reach, their sedimentological composition, and their mutual association with channel planform and channel geometry, that defines the distinguishing attributes of each River Style.

9.3.1.3 Channel planform

Assessment of channel planform in the River Styles framework builds on conventional notions described in Chapter 4. The following criteria are applied:

1 *The number of channels.* Rivers are identified on the basis of whether they have single channels, up to three channels, more than three channels, or discontinuous/absent channels.

2 *Sinuosity.* The cutoff between low sinuosity and meandering rivers is considered to be 1.3.

3 *Lateral stability.* This describes the degree to which the channel is able to adjust its position across the valley floor, through mechanisms such as lateral migration, thalweg shift, or avulsion. These behavioral attributes are differentiated into differing rates of change, reflecting the likelihood that adjustment will occur. This is dependent, in large part, on bed material texture and available flow energy.

In general terms, direct relationships can be discerned among the attributes used to assess channel planform and the diversity and pattern of geomorphic units along a reach. For example, a laterally migrating meandering channel comprises pool–riffle sequences and point bars, and potentially has a distinct array of geomorphic units on its floodplain (e.g., ridge and swale topography, cutoffs or billabongs, levees, crevasse splays, abandoned channels, etc.).

9.3.1.4 Bed material texture

In the River Styles framework, bed material texture is differentiated on the basis of the dominant caliber of material found along the channel bed. Five key classes are used: bedrock, boulder (b-axis >256 mm), gravel (2–256 mm), sand (0.0625–2 mm), and silt/clay (termed fine-grained) (<0.0625 mm). In general terms, bed material size reflects regional geology (lithology), flow energy, and sediment influx from upstream. Confined-valley settings tend to be dominated by bedrock, with occasional boulder and coarse-textured geomorphic units. Bedrock continues to be a major attribute of rivers in partly-confined valleys. In these settings, bed material size may be locally very variable, with gravels and coarser materials prominent. Floodplain pockets may comprise fine-

grained suspended load deposits. Rivers in the laterally-unconfined valley setting cover the full suite of textures, with bedrock only locally significant. In these various settings, bed material texture, and its relation to bank composition, is a key determinant of river character and behavior.

The relationship between bed material size and the prevailing flow regime not only shapes the assemblage of instream and floodplain geomorphic units, it also influences the capacity of the channel to adjust its form, in terms of both its geometry and planform. Textural differentiation of channel and floodplain compartments, or the bed and banks, may exert a significant influence on the behavioral regime of a river. This is typically reflected in the character and pattern of geomorphic units. For example, a gravel-bed meandering River Style has a different package of geomorphic units to a fine-grained meandering River Style. Fine-grained systems generally have cohesive banks, are laterally stable, and have floodplains dominated by vertical accretion of suspended load materials on the floodplain. In contrast, sand and gravel dominated systems tend to be bedload dominated, have less stable banks, and are prone to lateral accretion and recurrent floodplain reworking. Analysis of bed material texture provides a finer level of resolution in the identification of River Styles. The importance of bed material texture as a defining characteristic varies for differing valley settings.

9.3.2 The River Styles tree

Associations between geomorphic units and channel planform, framed in context of valley setting and differentiated on the basis of bed material texture, are the core criteria used to define River Styles. Because river behavior varies in differing valley settings, different mixes of these parameters are used to classify rivers in different settings. In other words, differentiation of River Styles is not always made on the basis of the same criteria, though geomorphic units are included in ALL interpretations. For example, floodplain considerations provide little insight into the character and behavior of rivers found in a confined valley setting, while various channel planform attributes are irrelevant in nonchannelized settings. Similarly, while the number of channels and their

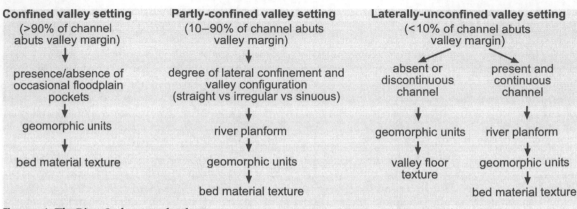

Figure 9.6 The River Styles procedural tree
Each River Style is identified on the basis of its planform, assemblage of geomorphic units, and bed material texture. Depending on the valley setting, different sequences of procedures are applied to identify the River Style. Modified from Brierley et al. (2002). Reproduced with permission from Elsevier, 2003.

stability are dominant attributes of a braided River Style in a laterally-unconfined valley setting, key attributes for other rivers may be, say, bedrock control, or the presence of a discontinuous channel.

Analysis of geomorphic units provides a basis to interpret river character AND behavior. The presence/absence of individual units, and/or packages of these features, provides a meaningful basis to determine boundaries between reaches with differing form–process associations. Placement of these boundaries may range from clear and distinct breaks to gradual and diffuse transitions.

The entry point into the River Styles procedural tree is through identification of the valley setting in which a river is found (i.e., confined, partly-confined, laterally-unconfined; Figure 9.6). Finer levels of differentiation are based on combinations of channel planform, the assemblage of geomorphic units, and bed material texture. Differing procedures are used because differing constraints are imposed on river character and behavior in different valley settings. Hence, distinguishing attributes of River Styles vary depending on the valley setting in which the river occurs. However, a consistent set of procedures is applied to derive a series of generic labels within each valley setting. Differentiation of River Styles becomes progressively more complex as the influence that bedrock exerts on river morphology decreases and the capacity for river adjustment increases (i.e., from confined through partly-confined to laterally-unconfined valley settings).

Each River Style has a discrete set of distinguishing attributes. In many cases, there may be an overlap in the range of attributes. For example, pool–riffle sequences are a common attribute for many River Styles. However, each River Style has unique identifying attributes, or combinations of attributes. Practical application of the River Styles framework and derivation of River Styles trees is a catchment- or region-specific exercise. Examples of a River Styles tree developed for rivers in coastal valleys of New South Wales, and schematic representations of these River Styles, are presented in Figures 9.7 and 9.8.

The first criterion used in analysis of rivers in the confined valley setting is whether reaches are completely bedrock-confined (i.e., no floodplain is evident) or whether occasional floodplain pockets are observed. These rivers are then differentiated on the basis of the type and range of erosional/sculpted geomorphic units found along the channel. Valley slope exerts a primary influence on these features. Channel planform is irrelevant in the differentiation of River Styles in these settings.

All three defining parameters are used to differentiate among River Styles in the partly-confined valley setting. In the first instance, channel planform characterizes the distribution of floodplain pockets based on the relationship between valley

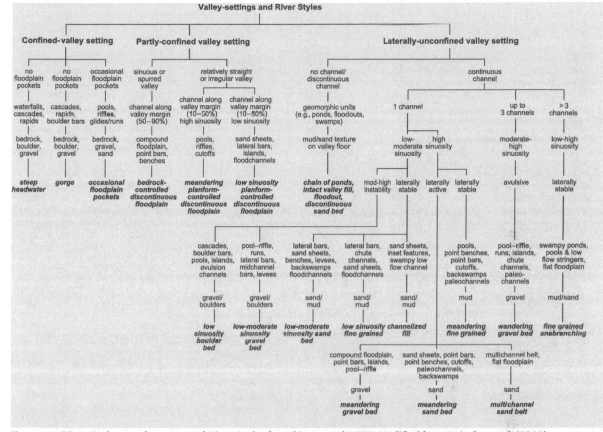

Figure 9.7 River Styles tree for a range of River Styles found in coastal NSW. Modified from Brierley et al. (2002). Reproduced with permission from Elsevier, 2003

configuration and channel alignment. Valley configuration and the ability of the channel to laterally adjust determine channel sinuosity and the degree to which discontinuous or alternating pockets of floodplain occur. This is used to differentiate between bedrock-controlled and planform-controlled discontinuous floodplain sequences. Assemblages of erosional and depositional instream and floodplain geomorphic units, and bed material texture, are assessed to further differentiate among River Styles.

In the laterally-unconfined valley setting, initial identification of River Style is based on the presence or absence of a channel. Reaches with discontinuous channels are analyzed separately based on the assemblage of geomorphic units and the texture of valley floor materials. Differentiation of

laterally-unconfined River Styles with continuous channels is made on the basis of conventional channel planform attributes, namely the number of channels, sinuosity, and lateral stability. In the instream zone, erosional, or depositional forms and bank-attached and midchannel features are differentiated. Bed material texture is then used for finer level differentiation, highlighting differences between boulder, gravel, sand, and fine grained variants. In some instances, laterally unconfined rivers may be bedrock-based.

For highly modified rivers, such as urban streams or regulated rivers (irrigation channels), where river character and behavior are largely imposed, analyses are framed in terms of the "natural" setting of the river. The River Style is then noted as modified (M) and separate layers of

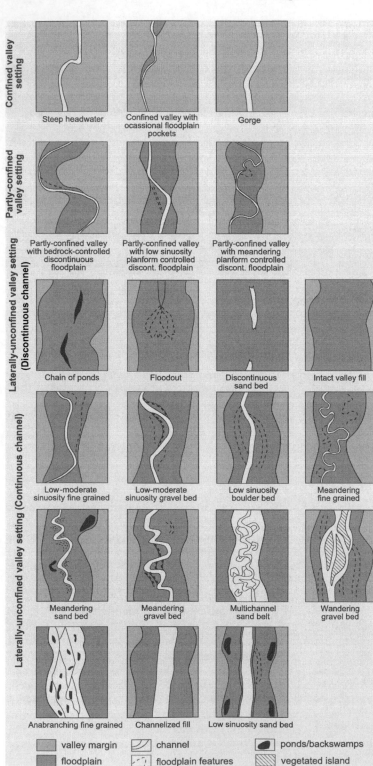

Figure 9.8 Examples of River Styles identified in coastal valleys of NSW. Reproduced from Brierley et al. (2002). Reproduced with permission from Elsevier, 2003

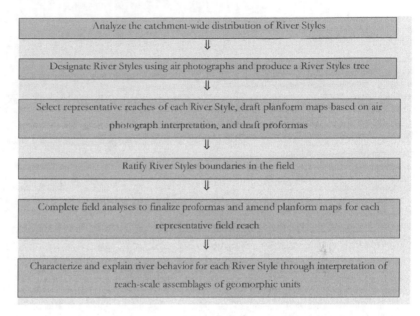

Figure 9.9 consists of a vertical flow chart with the following boxes connected by downward arrows:

Analyze the catchment-wide distribution of River Styles

⇩

Designate River Styles using air photographs and produce a River Styles tree

⇩

Select representative reaches of each River Style, draft planform maps based on air photograph interpretation, and draft proformas

⇩

Ratify River Styles boundaries in the field

⇩

Complete field analyses to finalize proformas and amend planform maps for each representative field reach

⇩

Characterize and explain river behavior for each River Style through interpretation of reach-scale assemblages of geomorphic units

Figure 9.9 **Stage One, Step Two: Definition and interpretation of River Styles**

analysis are undertaken that consider the types of imposed configurations and associated behavioral traits of the river.

9.3.3 Procedures used to identify and interpret River Styles

The definition and interpretation of River Styles is initially undertaken as a desk top exercise. Fieldwork is then undertaken to collect relevant information on river character and behavior for each River Style and ratify boundaries between them. Procedures undertaken in Stage One, Step Two of the River Styles framework are depicted in Figure 9.9.

9.3.3.1 Analysis of the catchment-wide distribution of River Styles

Appraisal of topographic maps in conjunction with the latest set of aerial photographs is undertaken to gain an initial "feel" for the range and distribution of River Styles in the area. The distribution of floodplains along the primary trunk streams in each subcatchment is used to determine the range and pattern of valley settings. Within each valley setting, River Styles are identified using the procedural trees for confined, partly-confined, or

laterally-unconfined valley settings (Figure 9.6). Bed material texture provides a finer level differentiation which is often completed in the field.

9.3.3.2 Designation of River Styles using air photographs and production of a River Styles tree

A catchment- or region-specific River Styles tree is constructed, outlining the distinguishing attributes of each River Style. Each River Style is given a diagnostic name that synthesizes river character and behavior, and a schematic representation is presented. An example of a catchment-specific River Styles tree is presented in Figure 9.10. The distinguishing attributes of River Styles in Bega catchment are presented in Table 9.4.

Boundaries of River Styles are mapped throughout the catchment. The boundaries between River Styles are defined by a change in the diagnostic geomorphic structure of a river. These boundaries can be distinct or gradual. Distinct changes often coincide with tributary–trunk stream confluences, changes in valley gradient (e.g., at bedrock steps) or sudden changes in valley width or morphology associated with lithological or structural changes. Gradual changes are less easily pinpointed. For example, a change from occasional to discontinuous

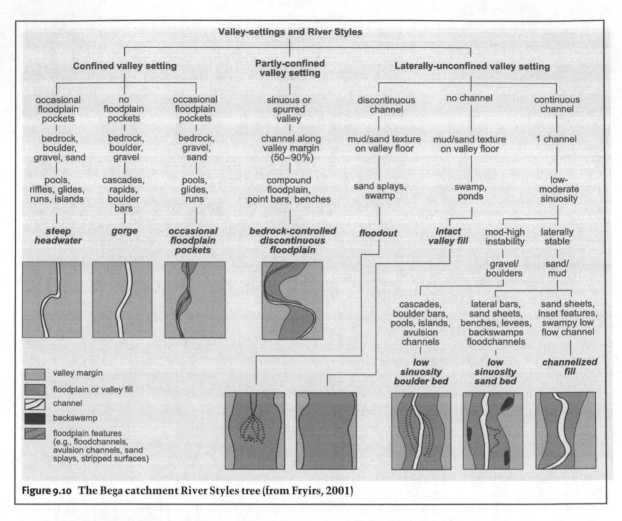

Figure 9.10 The Bega catchment River Styles tree (from Fryirs, 2001)

floodplain pockets may occur over several kilometers. A pragmatic decision needs to be made about the placement of the boundary. In these cases, the boundary is generally placed in the middle of this transition zone and a gradual boundary noted. In all instances, however, whether the boundary is distinct or gradual, the distinguishing attributes should be discernible for River Styles upstream and downstream of each boundary. Things to note while identifying and ratifying boundaries between River Styles are downstream changes in:

• valley morphology (width, slope, and alignment);

• channel alignment on the valley floor and the presence/continuity of floodplain pockets;

• channel planform (i.e., number of channels, sinuosity, indicators of lateral stability);

• floodplain character (flat, leveed, proximal-distal changes, evidence of reworking/scour);

• the presence/absence of geomorphic units and packages of genetically-related assemblages;

• channel geometry (e.g., symmetrical, asymmetrical, compound, irregular);

• bed material texture and associated geomorphic units if visible.

There is no definitive range of River Styles. Using a flexible, open-ended approach to the assessment of river character and behavior retains as much information as possible. It is much easier to clump reaches together in production of the final

Table 9.4 Distinguishing attributes of River Styles in Bega catchment.

River Style	Valley setting/ Landscape unit	River character			River behavior
		Channel planform	Geomorphic units	Bed material texture	
Steep headwater	Confined/ Uplands	Single channel, highly stable channel	Discontinuous floodplain, pools, riffles, glides, runs, vegetated islands	Boulder-bedrock-gravel-sand	Bedrock channel with a heterogeneous assemblage of geomorphic units. Acts to flush sediments through a confined valley. Limited ability for lateral adjustment.
Gorge	Confined/ Escarpment	Single channel, straight, highly stable channel	No floodplain, bedrock steps, pools and riffles, cascades	Boulder-bedrock	Steep, bedrock controlled river with an alternating sequence of bedrock steps and pool–riffle–cascade sequences. Efficiently flushes all available sediments. Channel cannot adjust within the confined valley setting.
Confined valley with occasional floodplain pockets	Confined/ Rounded foothills	Single, straight, channel, highly stable	Discontinuous pockets of floodplain, extensive bedrock outcrops, sand sheets, pools	Bedrock-sand	Found in narrow valleys, these rivers move sediment along the channel bed via downstream propagation of sand sheets. Bedrock induced pools and riffles, and island development occur where sediment availability is limited and the bedrock channel is exposed.
Partly-confined valley with bedrock-controlled discontinuous floodplain	Partly-confined/ Rounded foothills and base of escarpment	Single channel. Sinuous valley alignment, moderately stable	Discontinuous floodplain, point bars, point benches and sand sheets, midchannel bars, pools and riffles, bedrock outcrops	Bedrock-sand	These rivers are found in sinuous valleys. They progressively transfer sediment from point bar to point bar. Sediment accumulation and floodplain formation is confined largely to the insides of bends. Sediment removal occurs along concave banks. Over time sediment inputs and outputs are balanced in these reaches. Floodplains are formed from suspended load deposition behind bedrock spurs.
Low sinuosity boulder bed	Alluvial/ Base of escarpment	Single channel trench consisting of multiple low flow threads around boulder islands, highly stable	Fans extend to valley margins. Channel consists of boulder islands, cascades, runs, pools, bedrock steps	Boulder-bedrock	Lobes of boulder and gravel material have been deposited over the valley floor. The primary incised channel has a heterogeneous assemblage of bedrock and boulder induced geomorphic units that are only reworked in large flood events.

Table 9.4 *Continued.*

River Style	Valley setting/ Landscape unit	River character			River behavior
		Channel planform	Geomorphic units	Bed material texture	
Intact valley fill	Alluvial/ Base of escarpment	No channel	Continuous, intact swamp	Mud-sand	Intact swamps are formed from dissipation of flow and sediment over a wide valley floor as the channel exits from the escarpment zone. Suspended and bedload materials are deposited as sheets or floodout lobes.
Channelized fill	Alluvial/ Base of escarpment	Single, straight channel, unstable	Continuous valley fill, terraces, inset features, sand sheets, sand bars	Sand	Incised channel has cut into the swamp deposits of the intact valley fill River Style. Large volumes of sediment are released and reworked on the channel bed. The channel has a stepped cross-section with a series of inset features and bar forms. These are a function of cut-and-fill processes within the incised channel. Channel infilling, lateral low flow channel movement and subsequent reincision produce the stepped profile.
Floodout	Alluvial/ Rounded foothills	No channel	Continuous intact swamp with floodout	Mud-sand	Formed downstream of an incised channel, this river contains a swamp over which materials supplied from upstream are splayed over the valley floor in a number of lobes.
Low sinuosity sand bed	Alluvial/ Lowland plain	Single macrochannel consisting of an anabranching channel network. Potentially avulsive and unstable	Continuous floodplain with backswamps, levees. Benches, midchannel islands and sand bars	Sand	Found in a broad, low slope valley, the river accumulates sediments in wide, continuous floodplains. These floodplains contain levees and backswamps formed by flow and sediment dispersion over the floodplain. In other sections, floodchannels short circuit floodplain segments at high flow stage. The channel zone is characterized by extensive sand sheets and sand bars. Where these are colonized by vegetation, islands are formed. Where sediments are obliquely accreted against the channel margin, benches are formed.

map, once field analyses are completed, rather than split reaches at a later stage. In the latter in stance, the designation of boundaries is likely to be biased by field observations, which are inevitably constrained by selected field investigations and are not representative of the catchment as a whole. Once the boundaries between River Styles have been identified, a draft catchment-wide map showing the distribution of River Styles is produced (see Plate 9.2). River Styles boundaries are also plotted onto the longitudinal profiles produced in Stage One, Step One.

9.3.3.3 Select representative reaches of each River Style and draft proformas and planform maps using air photograph interpretation

The next stage of analysis involves detailed mapping of the geomorphic structure of representative reaches of each River Style. Pragmatic concerns such as accessibility are considered when choosing representative reaches. Representative field reaches must encompass the full range of river character and behavior for the River Style (i.e., a full range of geomorphic units, and any alternation of patterns that occur within the reach). If the reach is characterized by an alternating pattern, two (or more) representative reaches are required. Hence, the length of reach to be analyzed is dictated by its character and behavior.

Detailed geomorphic base maps are produced for each representative example of a River Style in the catchment. If possible, air photograph mapping of each representative reach is undertaken at a scale of around 1 : 6,000 or smaller. Even at this scale, the resolution of analysis may be too coarse to identify some geomorphic units. In many instances, vegetation cover may impede analyses. Alternatively, differences in vegetation character and cover may aid the identification of some geomorphic units (e.g., wetland vegetation). Geomorphic base maps can be stand-alone products or superimposed on air photographs or topographic contour maps. These maps should show:
- the valley margin;
- the channel(s);
- discernible floodplain pockets and associated geomorphic units;
- discernible instream geomorphic units;

- vegetation cover and character, including any woody debris;
- discernible hillslope sediment sources, such as landslides;
- infrastructure and associated fixed points (bridges, buildings, fencelines, etc.).

Using this analysis, detailed drafts of River Styles proformas are produced. This standard data sheet provides a consistent platform with which to document the character, behavior, and controls on each River Style (Table 9.5). In general, one pro-

Table 9.5 The River Styles proforma.

River Style name
Catchment specific river name and reach name
Defining attributes of River Style (from River Styles tree)
Subcatchments in which River Style is observed:
(catchment specific)
Justification of River Styles boundary

DETAILS OF ANALYSIS	
Representative reach:	
Map sheet(s) air photographs used	
Analysts	
Date	
Upstream grid reference	
Downstream grid reference	
RIVER CHARACTER	
Valley setting	
Channel planform	
Bed material texture	
Channel geometry	
(size and shape)	
Geomorphic units	*Instream – bedrock*
(geometry, sedimentology)	*Instream – alluvial*
	Floodplain
Vegetation associations	*Instream geomorphic units*
	Floodplain geomorphic units
RIVER BEHAVIOR	
Low flow stage	
Bankfull stage	
Overbank stage	
CONTROLS	
Upstream catchment area	
Landscape unit and within-catchment position	
Process zone	
Valley morphology (size and shape)	
Valley slope	
Stream power	

forma is presented for each River Style in the catchment, summarizing the range of data recorded for any given attribute of the River Style. Obviously this range is limited by the site selected for detailed field or air photograph observations. The River Styles proforma comprises four key components:

• *Details of analysis.* This summarizes resources and personnel used to complete the analysis, when the assessment was undertaken, and the location of the reach.

• *River character.* A range of attributes is used to identify and characterize the River Style, including valley setting, channel planform, bed material texture, and the assemblage of geomorphic units. Various characteristics of geomorphic units, such as their geometry and sedimentology, provide the basis to interpret river behavior. Cross-sectional channel geometry is noted and vegetation character is assessed for each geomorphic surface, aiding interpretation of its role in the formation and reworking of geomorphic units, and associated river behavior.

• *River behavior.* Form–process associations of geomorphic units are used to interpret river behavior at low flow, bankfull, and overbank stages.

• *Controls.* The boundary conditions within which the River Style operates are summarized. This includes documentation of the landscape unit and catchment-scale controls within which each reach of the River Style is observed. Many of the details presented in this section are analyzed from resources compiled in Stage One, Step One and Stage One, Step Three of the River Styles framework.

9.3.4 Ratify River Styles boundaries in the field

Field analysis is an integral component of the River Styles framework. It is at this stage that the assemblage of geomorphic units and other geomorphic parameters are assessed at finer resolution along representative reaches of each River Style. Prior to going into the field, the following data should be in-hand:

• pertinent regional setting data compiled in Stage One, Step One;
• landscape units map and longitudinal profiles;

• draft catchment map showing the distribution of River Styles;
• draft versions of planform maps for representative reaches of each River Style, showing geomorphic units wherever discernible;
• draft versions of River Styles proformas for each River Style.

The first task to be undertaken in the field is the ratification of River Styles boundaries. Only conjectural boundaries are visited. This involves locating the boundary on the ground and assessing the distinction between river character and behavior in upstream and downstream reaches to validate the placement of the boundary. Statements justifying the position of the boundaries are noted in the River Styles report and on the River Styles proformas. Once this analysis has been completed, the River Styles map is modified as required.

9.3.5 Complete field analyses to finalize proformas and amend planform maps at each representative field reach

Data on each River Style are recorded in a systematic manner to aid the appraisal and communication of river character and behavior. Proformas are finalized in the field, filling in details on geomorphic units, bed material texture, and river behavior. Annotated photographs, cross-sections, and planform maps accompany these proformas in the final presentation. Dependent on the nature of the exercise, additional data may be required. In all instances, adopted procedures should be clearly articulated and rigorously applied, using standard (accredited) sources whenever possible (e.g., Newbury and Gaboury, 1993; Harrelson et al., 1994; Rosgen, 1996; Thorne, 1999; Rutherfurd et al., 2000).

Depending on the goals of the exercise and the complexity of the reach, field analysis can take up to half a day for each site. A range of attributes is measured (see Table 9.6). Due emphasis is placed on analysis of distinguishing attributes of each River Style, including key characteristics of instream and floodplain geomorphic units. Examples of representative River Styles proformas derived for three types of river in Bega catchment are presented in Tables 9.7–9.9 and Figures 9.11–9.13.

Table 9.6 Procedures used to undertake field analyses in Stage One, Step Two of the River Styles framework.

Task	Procedure
Rectify the planform map	• Walk the reach, amending the geomorphic unit base map as required. • Record the position, assemblage, pattern, and connectivity of instream and floodplain geomorphic units on the map. • Note within reach and downstream changes in the number of channels, lateral stability, sinuosity, channel geometry, bank height and character, sediment storage, channel obstruction, and vegetation structure and character on the map.
Produce a valley wide cross-section and estimate Manning's n for each surface	• Based on the rectified geomorphic unit map, a "representative" site is selected at which a valley-scale cross-section can be completed and various attributes of each geomorphic unit assessed. More than one cross-section may be required to record all geomorphic units and their relative position within the valley. The number of sections analyzed should reflect the complexity (or variability) of geomorphic units in the reach. • Working from left to right (looking downstream), all breaks in slope across the valley are recorded in the survey. The dimensions and geometry of each geomorphic unit are recorded on the proforma. • If the cross-section is to be used as a benchmark for future monitoring or repeat surveys, it should be surveyed to an appropriate datum. • For each geomorphic surface across the cross section, assess the Manning's n roughness coefficient using a visual guide. The valley margin, floodplain, banks, channel bed, and any within channel vegetated features e.g., islands are assessed. This information is used to calculate the spatial distribution of unit stream power for different River Styles in later analyses.
Assess channel geometry and shape	• From the cross-sections calculate the width and the depth of the channel and determine the width : depth ratio. • Characterize the shape of the channel. Is it symmetrical, asymmetrical, compound, or irregular?
Assess the shape, size, and sediment character of each geomorphic unit	• Assess the gross sedimentary character for each instream and floodplain geomorphic unit along the cross-section (i.e., mud, sands, gravels, boulders or bedrock). • Determine whether the sediments within the geomorphic unit are well sorted or poorly sorted, loose or cohesive.
Assess bed material texture	• Across the channel bed, measure the grain size. This level of analysis allows finer resolution differentiation of River Styles, particularly in the laterally-unconfined valley setting. Different procedures are used in gravel-bed versus sand-bed rivers. **For a gravel-bed river:** • Place a 30 m line transect parallel to the channel in the coarsest bar surface. Measure the b-axis of the clast that falls immediately beneath each meter interval (in mm). The b-axis is the diameter of the intermediate axis; i.e., perpendicular to the longest axis. Note 30 clast measurements. **For a sand-bed river:** • Analyze the grain size distribution using a 0.5 Ø particle size analysis card.
Interpret vegetation associations	Vegetation cover and composition, and the loading of woody debris, influence the stability of channel margins, impacting on the capacity of the river to adjust its morphology. Numerous labor-intensive field procedures could be applied to assess the part played by vegetation as a control on river character and behavior. However, in Stage One of the River Styles framework, analyses are restricted to: • Noting the distribution and type of woody debris on the planform map. • Noting the vegetation type and percent cover for instream and floodplain geomorphic units, and the riparian zone.
Photograph the River Style	Representative photographs are taken for each River Style. These are annotated and detailed captions are presented, including the date the photograph was taken and its specific location.

Table 9.7 Proforma for the channelized fill River Style in Bega catchment.

Defining attributes of River Style (from River Styles tree):
This River Style is found in a laterally-unconfined valley setting, where the channel abuts the valley margin < 10% of the time. It has continuous perched, flat topped valley fills along both valley margins. The incised trench has a low sinuosity, is single thread, and is aligned down the center of the valley. When expanding the channel is laterally unstable. The channel comprises an assemblage of sand and/or mud dominated geomorphic units including sand sheets, inset features, bank-attached, and midchannel bars, and a swampy low flow channel.

Subcatchments in which River Style is observed: Greendale, South Wolumla, Wolumla, Reedy, Sandy, Colombo, Pollacks Flat, Numbugga

DETAILS OF ANALYSIS
Representative reach: Wolumla Creek
Map sheet(s) air photographs used: Wolumla 1 : 25,000 topographic sheet; 1994 Bega Run 5 # 40, 41 air photographs
Analysts: Kirstie Fryirs
*Date:*18.06.96
Upstream grid reference: 423187   *Downstream grid reference*: 448224

RIVER CHARACTER

Valley setting	Laterally-unconfined
Channel planform	Single channel aligned down the central axis of the valley. Channel has a low sinuosity and is laterally unstable. Continuous valley flats line both margins of the valley. Valley width is generally less than 300 m wide.
Bed material texture	Continuous, exposed, loose sand sheets composed mainly of 1–0.5 ø sands. Occasional gravels up to 250 mm (*b*-axis). Some mud accumulates along the low flow channel. Occasional bedrock outcrops occur.
Channel geometry (size and shape)	Symmetrical trench-like channel. Channels can be up to 160 m wide and 12 m deep. Compound cross-section consists of numerous terrace, valley fill and benches. Banks are characterized by alternating sequences of mud and sand.
Geomorphic units (geometry, sedimentology)	*Instream – bedrock* • **Bedrock outcrops and steps** – occasionally where channel has incised to bedrock. *Instream – alluvial* • **Sand sheets** – Cover the bed of the incised trench. Surficial gravels can be up to 150 mm (*b*-axis). Are poorly sorted and loose. • **Bank-attached bars and midchannel bars** – Located along the channel margins or on the floor of the trench. Can be several hundreds of meters long and tens of meters wide. Surficial gravels can be up to 150 mm (*b*-axis), but are largely dominated by sands and poorly sorted, loose gravels. • **Inset benches** – Between 2 m and 20 m wide and between 1 m and 6.5 m deep. Highly discontinuous, with up to four levels present in any one cross-section. Comprised of alternating, sand, and thin mud units. Mud units on the surface are up to 200 mm thick and then around 30–50 mm thick down profile. Sand units are up to 200 mm thick and massive, or are thin and planar bedded. • **Swampy low flow channel** – Muds accumulating in the low flow channel zone form a discontinuous channel within the incised trench. *Floodplain* • **Valley fills** – Can be up to 12 m deep and 100 m wide. Continuous along both valley margins. Alternating sand and mud units. Commonly mud units are 3.5–3 ø and up to 2000 mm thick. Sand units are 1–2 ø with up to 3 mm clasts and are up to 2000 mm thick. Some sand units are planar bedded, but most are massive but poorly sorted. Up to 25 bedded units within some exposures. Poor, pasture associations exist. • **Terraces** – Localized, but can be up to 12 m deep and 30 m wide. Interbedded coarse sands and gravels, up to 200 mm (*b*-axis). No mud.

Table 9.7 *Continued.*

Vegetation associations	*Instream geomorphic units* Largely unvegetated, but inset benches and bars can be dominated by tussock or exotic weeds. *Floodplain geomorphic units* Pasture and exotic weed species are the dominant vegetation associations along channel banks. Scattered eucalypts and acacias, as well as exotics in the riparian zone. Localized swamp associations in trapped tributary fills located at valley margins.

RIVER BEHAVIOR

Low flow stage

This River Style is the product of the channelization of intact fills through gully processes. In general, gullies are formed through the headward extension of a headcut. The incision of a channel through a fill alters the hydrological characteristics of the fill, by allowing water to escape the landscape at a greatly increased rate, leaving former swamp surfaces perched above the channel bed.

The within channel assemblage of geomorphic units is not highly varied. Instream geomorphic structure is largely homogeneous, but sediment storage is high. For the majority of the time, a low flow channel will occur through sand sheets and around bank-attached bars. Fine grained muds may be deposited from suspension forming a veneer on the bed of the incised trench. Swampy conditions can result in the formation of a discontinuous low flow channel.

Bankfull stage

All events up to and including the 1 in 100 year event are contained within the incised channel. Along Wolumla Creek the 1 in 100 year event forms a flow depth of around 2 m. This concentration of energy within the incised channel means that high stream powers can be produced and significant volumes of sediment are reworked within the incised trench. In these cut-and-fill landscapes, large flood events lead to significant bed incision (often to bedrock), and subsequent lateral channel expansion via the processes of undercutting and block failure of banks. Sediment movement on the channel bed is high as large volumes of readily available sand material are stored on the bed and in benches. In low–moderate events or the waning stages of large flood events, sediment accumulation on the channel bed and/or the formation of bench features occurs. These features are common in the filling phases and lead to the formation of a stepped channel cross-sectional morphology.

Overbank stage

Deeply incised trench ensures that the valley flats are seldom inundated. Channel is effectively disconnected from its floodplain even in 1 in 100 year events or greater.

CONTROLS

Upstream catchment area	Generally 20 km²
Landscape unit and within-catchment position	Low lying to undulating country at the base of the escarpment.
Process zone	Sediment source zone. In Wolumla catchment over 4 million m³ of material has been released from this River Style following European settlement. Once incised, sediment is efficiently flushed downstream through the incised trench.
Valley morphology (size and shape)	Deep, wide valleys that decrease in width downstream to form a funnel-shaped accommodation space. Valleys generally < 300 m wide.
Valley slope	Ranges from 0.005–0.03 m/m
Unit stream power	On average 1 in 10 year = 440 W/m²; 1 in 100 year = 1140 W/m²

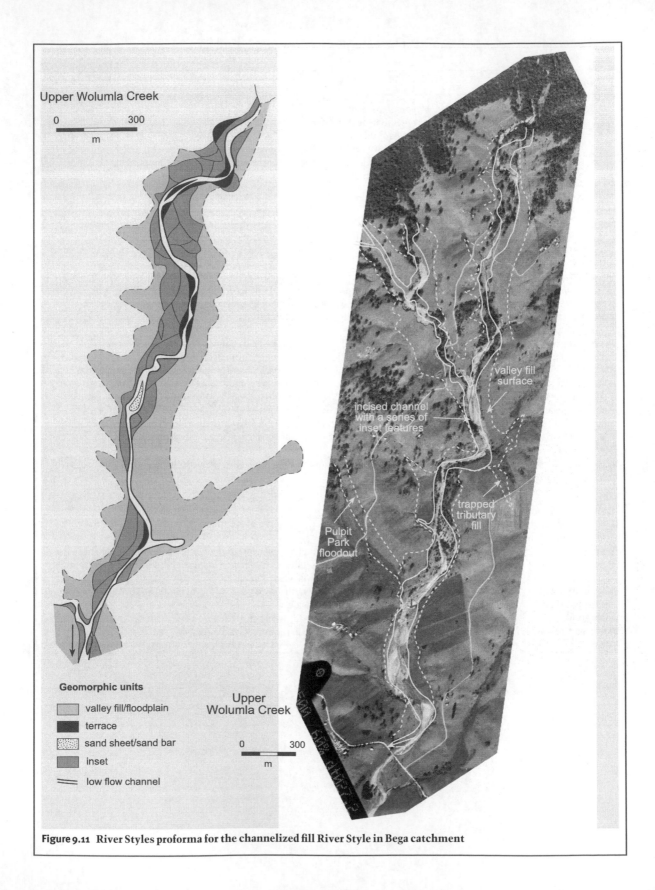

Figure 9.11 River Styles proforma for the channelized fill River Style in Bega catchment

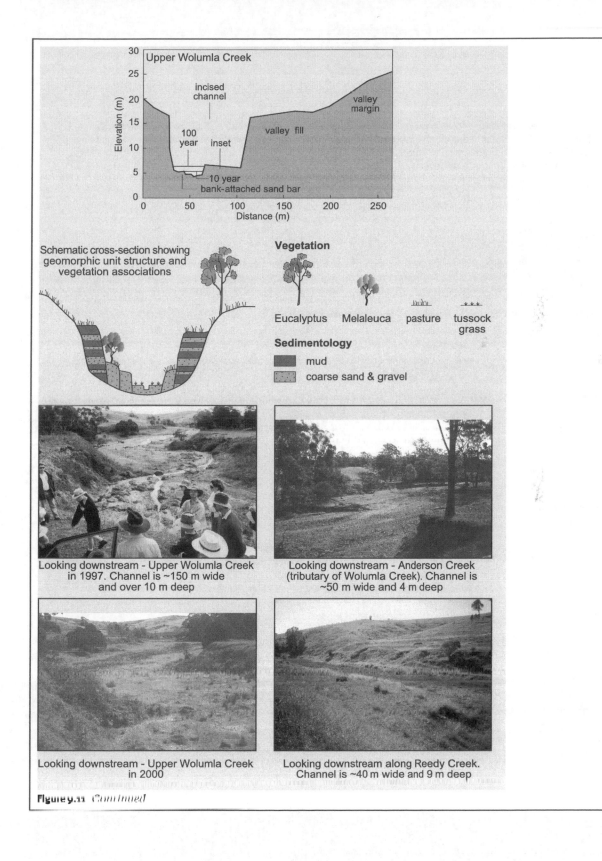

Looking downstream - Upper Wolumla Creek in 1997. Channel is ~150 m wide and over 10 m deep

Looking downstream - Anderson Creek (tributary of Wolumla Creek). Channel is ~50 m wide and 4 m deep

Looking downstream - Upper Wolumla Creek in 2000

Looking downstream along Reedy Creek. Channel is ~40 m wide and 9 m deep

Figure 9.11 Continued

Table 9.8 Proforma for the partly-confined valley with bedrock controlled discontinuous floodplain River Style in Bega catchment.

Defining attributes of River Style (from River Styles tree): Channel abuts valley margin > 50% of the time. Hence, channel morphology and alignment are controlled to a significant degree by the sinuous or irregular valley morphology. Floodplain pockets are discontinuous and occur on the insides of sinuous bends downstream of bedrock spurs, and in irregular valleys where they locally widen behind bedrock spurs. Floodplain pockets are of variable character, but are commonly stepped in response to phases of aggradation and stripping. Hence a series of low terraces, floodplains, and benches can occur.

Subcatchments in which River Style is observed: Greendale, South Wolumla, Wolumla, Reedy, Candelo, Tantawangalo, Sandy, Colombo

DETAILS OF ANALYSIS
Representative reach: Candelo Creek
Map sheet(s) air photographs used: Candelo and Bemboka 1:25,000 topographic sheet, 1994 Bega Run 5 # 36, 37 air photographs
Analysts: Kirstie Fryirs, Rob Ferguson
Date: 18.01.97
Upstream grid reference: 404280   *Downstream Grid Reference:* 414309

RIVER CHARACTER

Valley setting	Partly-confined
Channel planform	Typically single-channeled with low sinuosity, in a valley which is sinuous, producing alternating pockets of floodplain. The channel may locally divide around islands at bends. Despite channel enlargement and floodplain stripping, channel position is generally stable, as it is commonly pinned against the valley margin. Lateral stability is highly variable with significant concave bank erosion occurring on the outsides of bends.
Bed material texture	Bed materials range from sands to gravels with occasional bedrock outcropping; banks are commonly dominated by massive sand units.
Channel geometry (size and shape)	Highly variable shape, ranging from asymmetrical compound channels with multiple floodplain surfaces on insides of bends to symmetrical in some straight sections. Channel is relatively wide and shallow. Channel can be up to 50 m wide and 3 m deep. Frequently bounded by bedrock valley margin on one side, with floodplains on the other.
Geomorphic units (geometry, sedimentology)	*Instream – bedrock*

• **Localized bedrock outcrops and steps.**
Instream – alluvial
• **Pools** – Sand based and located along the concave bank.
• **Compound point bars with chute channels and ridges** – Found along the convex banks of bends. Coarse sands, 0.5–0.0 ø with up to 5 mm clasts.
• **Lateral bank-attached bars** – Found along inflection points between bends. Coarse sands, 0.5–0.0 ø with up to 5 mm clasts.
• **Islands** – localized, found along inflection points between bends.
• **Sand sheets** – Cover channel bed between bars and bedrock outcrops. Coarse sands, 0.5–0.0 ø with up to 5 mm clasts.
• **Benches and point benches** – Primarily occur on convex banks and along inflection points of bends forming a channel marginal step. Interbedded sands with occasional gravels.
Floodplain
• **Floodplain pockets** – can be several hundred meters long and tens of meters wide, generally less than 4 m deep. May be multileveled, comprising a number of stripped floodplain surfaces and valley marginal terraces. Interbedded vertically accreted sequences of medium and coarse sands over basal fine sands.
• **Floodchannels** – generally tens of meters wide and several meters deep. Often run along the valley margin and obliquely across floodplains, often short-circuiting bends.

Table 9.8 *Continued.*

Vegetation associations	*Instream geomorphic units*
	• Largely exposed and unvegetated. Exotic weeds, willows, and pasture cover some bank-attached features.
	Floodplain geomorphic units
	• Dominated by pasture and occasional willows and exotic weeds.

RIVER BEHAVIOR

Low flow stage

In events lower than bankfull, flow is concentrated on the outsides of bends and around midchannel islands, or over a sand sheet. Material is progressively transferred from point bar to point bar producing a balance in sediment inputs and outputs. Although high stream powers can be generated in these partly-confined valley settings, the channel is relatively stable with little capacity for lateral adjustment as it is commonly pinned against bedrock valley margins at concave banks. Under these conditions, the floodplain pockets are protected by bedrock spurs.

Bankfull stage

At bankfull stage, high energy conditions are produced and the channel is prone to widening (where it can locally adjust; i.e., where floodplains occur). Overwidened channels, permit extensive point bar and point bench formation. Sediment movement occurs via episodic erosion of concave banks (via planform controlled erosion processes), and deposition on point bars along the reach. Flow alignment is shifted towards the inside of the bend and erosional surfaces (for which the term "ledge" is preferred), and point bars are formed. These point bar surfaces are often formed and reworked, leading to the formation of compound features comprising discrete assemblages of erosional and depositional features. These can include chute channels and ridges. Pockets of floodplain that are dissected and abandoned, and subsequently colonized by vegetation can form islands. Benches and point benches can reflect channel recovery/contraction after expansion with the "plastering" of sediment along the margins of an overwidened channel. This produces a distinctly stepped feature along the channel margin. Islands that occur in midchannel locations are dissected and pools are scoured.

Overbank stage

Fine-grained sedimentation via vertical accretion occurs on floodplains. Under these conditions, suspended load deposition around vegetation or behind bedrock spurs occurs. Massive sand units result. When high energy flows combine with devegetated conditions, floodplain scour or stripping occurs. Reworking results in the formation of floodchannels as flow short circuits a bend. Stripping occurs when the entire surface of the floodplain is removed. Terrace surfaces are rarely inundated.

CONTROLS

Upstream catchment area	Ranges from 30–200 km² depending on position of reach in the catchment.
Landscape unit and within-catchment position	Found in the rounded foothills landscape unit in middle to upper catchment positions. Valley margins can be relatively shallow.
Process zone	Sediment transfer zone – bedload dominated
Valley morphology (size and shape)	Irregular to sinuous valley which ranges 40–210 m wide. Produces discrete, discontinuous floodplain pockets.
Valley slope	0.005–0.012 m/m depending on location in catchment.
Unit stream power	On average 1 in 10 year = 410 W/m²; 1 in 100 year = 1030 W/m²

9.3.6 Characterize and explain river behavior for each River Style: The assemblage of geomorphic units

Reaches comprise mosaics of geomorphic units that have been constructed and reworked by differing sets of flow events. The contemporary capacity for adjustment and range of river behavior are River Style specific. For any given River Style, the suite of geomorphic units along a reach remains near-consistent over time, reflecting the range of river behavior or system dynamic. Patterns of geomorphic response to events of differing magnitude and recurrence define the range of behavior for each River Style. For example, while a meandering river may laterally adjust its position across a

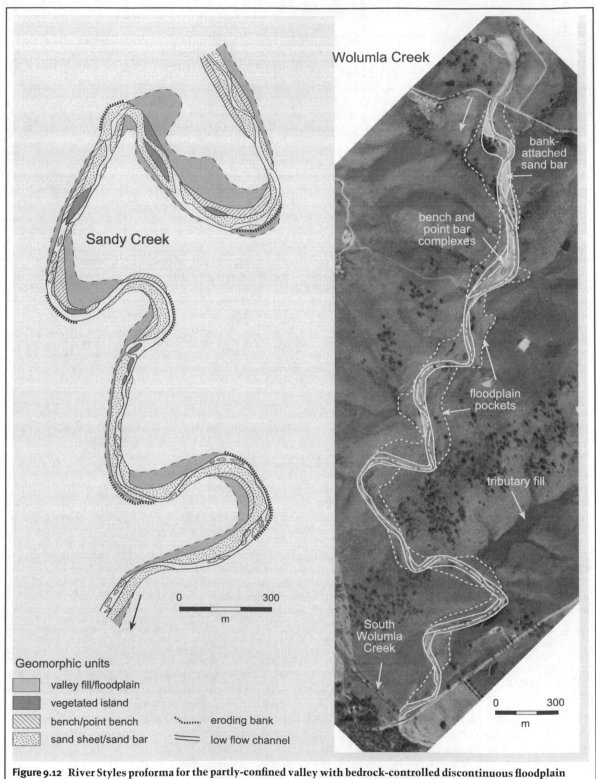

Sandy Creek

Wolumla Creek

bank-
attached
sand bar

bench and
point bar
complexes

floodplain
pockets

tributary fill

South
Wolumla
Creek

0   300
m

0   300
m

Geomorphic units

valley fill/floodplain

vegetated island

bench/point bench   eroding bank

sand sheet/sand bar   low flow channel

Figure 9.12 River Styles proforma for the partly-confined valley with bedrock-controlled discontinuous floodplain
River Style in Bega catchment

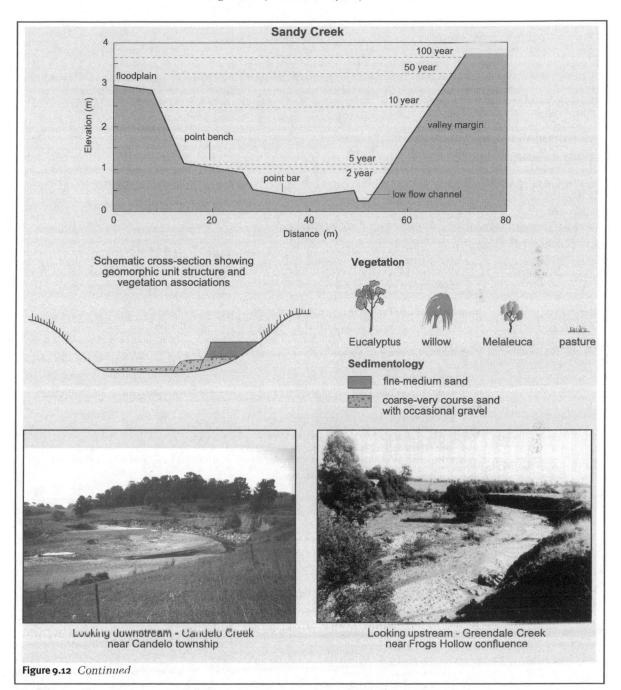

Sandy Creek

floodplain

point bench

point bar

low flow channel

100 year
50 year
10 year
valley margin
5 year
2 year

Schematic cross-section showing
geomorphic unit structure and
vegetation associations

Vegetation

Eucalyptus   willow   Melaleuca   pasture

Sedimentology

fine-medium sand

coarse-very course sand
with occasional gravel

Looking downstream - Candelo Creek
near Candelo township

Looking upstream - Greendale Creek
near Frogs Hollow confluence

Figure 9.12 *Continued*

Table 9.9 Proforma for the low sinuosity sand bed River Style in Bega catchment.

Defining attributes of River Style (from River Styles tree): This River Style is found in an alluvial valley setting, where the channel abuts the valley margin < 10% of the time. It has continuous floodplains along these valley margins. The macrochannel has a low sinuosity, is single thread, and is relatively stable. The channel comprises an assemblage of sand-dominated geomorphic units and the floodplain comprises a levee in proximal locations, extending to backswamps in distal sections. Floodchannels short cut the floodplain. Vegetated bars and sand sheets are dissected at low flow stage, and are set within the macrochannel.

Subcatchments in which River Style is observed: Bega, Brogo

DETAILS OF ANALYSIS
Representative reach: Lower Bega River at Grevillea winery
Map sheet(s) air photographs used: Bega 1:25,000 topographic sheet; 1994 Bega Run 7 # 133, 134 air photographs
Analysts: Kirstie Fryirs
Date: 30.01.97
Upstream grid reference: 508308   *Downstream grid reference:* 527378

RIVER CHARACTER

Valley setting	Laterally-unconfined
Channel planform	Continuous floodplains along both valley margins. Macrochannel is single-thread, laterally stable and of low sinuosity. Low flow channels divide around islands forming an anabranch type network within the macrochannel. Floodchannels short circuit the floodplain.
Bed material texture	Sand sheets line the channel bed. Dominated by 0–0.5 ø sands with gravels up to 15 mm.
Channel geometry (size and shape)	The macrochannel has a symmetrical but irregular shape, given the assemblage of within channel ridges and channel marginal benches. The macrochannel is wide (up to 16 m) and shallow (< 6 m).
Geomorphic units (geometry, sedimentology)	*Instream – bedrock*
	N/a
	Instream – alluvial

- **Runs and shallow pools** – Elongate and shallow. Coarse sands dominated by 0–0.5 ø with gravels up to 15 mm.
- **Midchannel and bank-attached lateral bars** – Elongate, several tens of meters wide and tens of meters wide. Primarily trough cross-bedded sands.
- **Benches** – Average around 5 m deep and 15 m wide. Line the channel banks. Basal planar bedded bar sediments are overlayed with vertical and oblique accretion deposits composed of fine sands and organic silts and muds.
- **Islands** – Elongate. Average around 250 m long, 40 m wide, and 2.5 m deep. Can be up to 500 m long and 75 m wide. Are often elevated as high as the channel banks. Dominantly trough cross-bedded sands interbedded with thin organic layers deposited via trapping of fine materials around vegetation. Basal parts consist of coarse sand and gravels.
- **Sand sheets** – Cover the entire channel bed. Coarse sands dominated by 0–0.5 ø with gravels up to 15 mm.

Floodplain

- **Floodplain** – Can be up to 6 m above the channel bed at the levee crest, or a shallow as 1 m where sand sheets have filled the channel. Interbedded medium sands and fine organic layers. With occasional thick coarse sand units consistent with sand sheet deposits. At depth fine organic rich muds represent the predisturbance floodplain.
- **Levees** – Can be to 6 m above the channel bed. Located at proximal floodplain locations. Interbedded medium sands and fine organic layers. With occasional thick coarse sand units consistent with sand sheet deposits. At depth, fine organic rich muds represent the predisturbance floodplain.
- **Backswamps** – Up to 150 m wide and 300 m long. Located in distal floodplain locations. Upward coarsening, from basal muds to fine sands deposited from overbank flood events.
- **Floodchannels** – Long, thin scour feature, connected to the channel at the head, but not downstream. Cut into a swale on the floodplain. Up to 5 m deep, 15 m wide, and 600 m long. Bed characterized by coarse sands (1–0.5 ø) with occasional gravels up to 25 mm. Exposed and actively eroding bedload materials in high flow.
- **Sand sheets** – Up to 2 m thick, fine towards the valley margin.

Table 9.9 *Continued.*

Vegetation associations	*Instream geomorphic units*
	• Islands and benches are infested with exotic vegetation species including willows, couch, kikuyu, African lovegrass, poplars, privet, and ragweed. Some native species do exist including various Acacia spp. and river oaks. A large number of willow seedlings have also colonized the large sand sheets. Sand sheets that occupy the low flow channels are unvegetated and exposed.
	Floodplain geomorphic units
	• Floodplains cleared for pasture. However, riparian zone is densely vegetated with willows, scattered casuarina, privet, lomandra, lantana, tobacco, kikuyu, and assorted weeds. Backswamps colonized by aquatic vegetation dominated by phragmites and Melaleuca spp.

RIVER BEHAVIOR

Low flow stage

Flow is restricted to shallow runs and pools. In many cases flow is subsurface within extensive sand sheets that cover the channel bed. Hydraulic diversity is limited.

Bankfull stage

At bankfull stage island complexes and midchannel bars are formed and reworked. Large volumes of material are moved through this reach in low-moderate flood events. Sediment is plastered against the channel margins to form obliquely accreted benches. On the falling stage of these events large volumes of sand and organic material are deposited on the islands and benches allowing them to accrete vertically and laterally. The colonization of exotic vegetation in the channel zone aided aggradation and initiated the formation of these large islands and channel marginal benches. Islands within the channel zone are often elevated above the channel, indicating that they are still forming as flows begin to go overbank.

Overbank stage

Shallow, wide, trench-like channel ensures that floodplain inundation occurs in 5–10 year recurrence interval flood events. Floodplains along this River Style are formed largely through vertical accretion processes. In overbank floods, levee–backswamp complexes are formed by distal fining of sediments as flows are spread over the floodplain. In highly sediment charged overbank events, extensive planar sand sheets can be deposited on the floodplain. Floodchannels indicate the potential for significant reworking of floodplain deposits during high flow stage. Alternatively, where riparian vegetation has been disturbed, channel expansion is a common form of floodplain reworking along these river courses.

CONTROLS

Upstream catchment area	Between 500–1840 km^2
Landscape unit and within-catchment position	Found in the most downstream catchment position, along the lowland plain.
Process zone	Accumulation. Bedload dominated.
Valley morphology (size and shape)	Wide, open valley setting typically <1000 m wide. Widen downstream towards the estuary to several kilometers wide.
Valley slope	Range between 0.002–0.0008 m/m and lower close to the tidal limit.
Unit stream power	On average 95 W/m^2 for the 1 in 10 year event; 280 W/m^2 for the 1 in 100 year event.

valley floor, it retains a characteristic pattern of point bars, pools and riffles, and floodplain forms. However, if subjected to considerable disturbance, or in some instances following progressive sets of adjustments, a reach may adopt a "new" set of geomorphic units. This adjustment to river structure and function marks a change to another River Style. These evolutionary notions, and associated procedures with which to appraise geomorphic

river condition, are considered in Stage Two of the River Styles framework.

Differences in river behavior and channel–floodplain linkages vary enormously from River Style to River Style, reflecting their variable capacity for adjustment. Assessment of river behavior is based on the morphology and sedimentology of each geomorphic unit, their position along a reach, and their association with adjacent landforms.

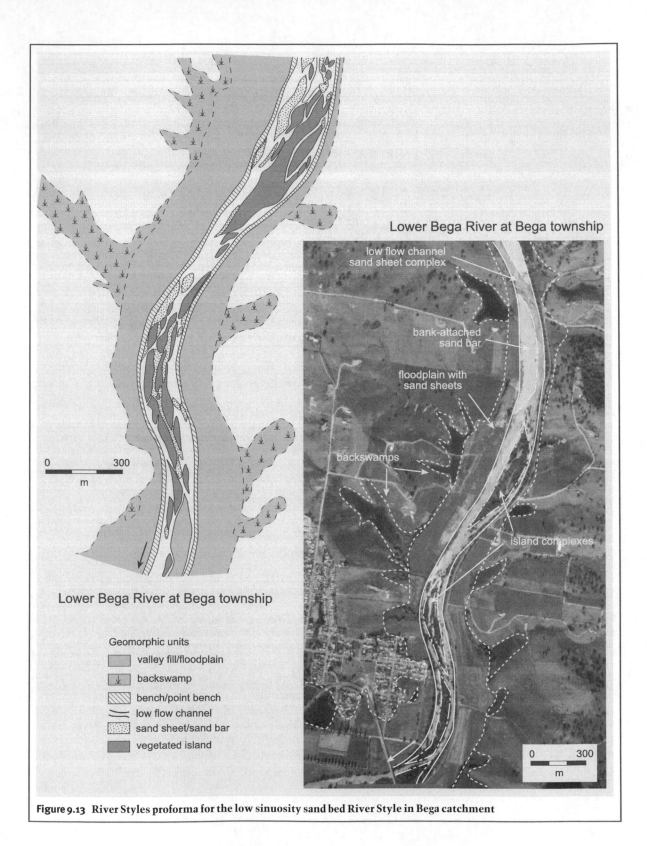

Lower Bega River at Bega township

low flow channel
sand sheet complex

bank-attached
sand bar

floodplain with
sand sheets

backswamps

island complexes

0   300
m

Lower Bega River at Bega township

Geomorphic units

valley fill/floodplain

backswamp

bench/point bench

low flow channel

sand sheet/sand bar

vegetated island

0   300
m

Figure 9.13 River Styles proforma for the low sinuosity sand bed River Style in Bega catchment

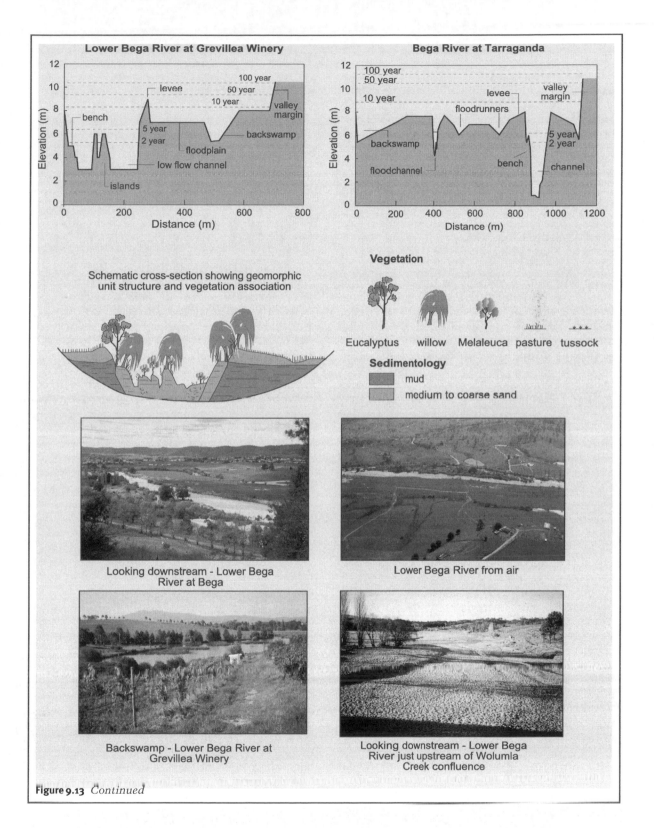

Figure 9.13 *Continued*

Channel and floodplain forms and processes are assessed individually and collectively to provide insight into river behavior at the reach scale. Analysis of the character, distribution, and history of formative processes that generate and rework geomorphic units along a reach provides the key to explanation of river behavior.

The morphology of a geomorphic unit is assessed in terms of its shape and size. In many instances, instream geomorphic units scale relative to the size of the channel, reflecting direct linkage to bankfull discharge. The formation and reworking of each unit reflect the three-dimensional distribution of flow energy at differing flow stages. Geomorphic units respond to floods of differing magnitude in different ways. While a particular landform may be generated during the rising and/or peak stage of a flood event, its resultant morphology reflects the subsequent degree of reworking during waning stages of subsequent, smaller magnitude events. This results in an array of forms of differing scale, geometry, and composition. In some instances, geomorphic units may record the impacts of a high magnitude, low frequency event that occurred in the distant past which recent flows have been unable to modify. Alternatively, all geomorphic units may record the passage of the last major flood. Form–process associations for various channel and floodplain geomorphic units are presented in Chapter 4.

The key to analysis of patterns of geomorphic units lies in unraveling the control exerted by slope, valley width, and flow conditions as determinants of the distribution of flow energy. These considerations influence the distribution of erosional and depositional processes, and the ability of the river to redistribute materials and form geomorphic units. Once established, geomorphic units modify the pattern of flow energy within a reach. Feedback mechanisms result in genetically-linked assemblages of geomorphic units. Any factor that affects flow energy, such as bedrock outcrops, valley alignment, or riparian vegetation and woody debris influence the local pattern of sediment distribution and associated geomorphic units.

Just as individual geomorphic units provide insight into form–process associations for particular landforms, so the assemblage of geomorphic units along a reach guides interpretation of river behav-ior for a particular River Style. Analysis of the assemblage of geomorphic units guides interpretation of the range of erosional, reworking, and depositional processes, and associated landform assemblages, as shaped by the range of flow conditions. The character, pattern, and assemblage of geomorphic units along any reach reflect the totality of adjustments in response to the recent history of flow events. Significant site- or reach-specific adjustment in packages of geomorphic units is to be expected, as each reach has its own history of formative events. Geomorphic skills of "reading the landscape" are vital in interpreting these geomorphic responses to differing sets of flow events.

In the River Styles framework, interpretation of river behavior focuses on the manner and frequency with which materials are deposited and reworked under different sets of energy conditions at low flow, bankfull, and overbank flow stages. Different geomorphic units are formed and reworked at these different flow stages. Collectively, these insights provide an interpretation of river behavior. The character and relative abundance of geomorphic units are appraised in terms of their location and their propensity to be reworked by floods of differing magnitude (i.e., their preservation). The position and connectivity of geomorphic units is noted and used to interpret their configuration and formative processes. Relationships are established for differing flow stages by plotting the inundation frequency onto the surveyed cross-sections for each River Style. Stream power is calculated for these different flow events, providing an indication of the energy conditions under which different geomorphic unit assemblages are formed and reworked, and the frequency with which this takes place. Questions addressed at this stage include: How was the floodplain formed and reworked? How is the broader channel structure formed? How are instream units formed? These insights are summarized to determine the behavior for each River Style, as noted on their proformas.

As the behavioral regimes of rivers vary markedly in differing valley settings, the relevance of these flow stage relationships in the interpretation of river behavior also varies. For example, defining bankfull or overbank stage is meaningless along a confined bedrock channel with no floodplain. In these settings, flows other than the largest events are unlikely to be geomorphically effective. Similarly, bankfull conditions cannot be defined for a laterally-unconfined River Style that does not have a continuous channel. In contrast, in the laterally-unconfined valley setting with a continuous channel, all flow stages provide insight into river structure and function because multiple components of the bed, channel, and floodplain can adjust under different flow conditions. For example, in highly sediment charged situations (e.g., sand-bed braided systems), channel morphology can adjust readily at low flow stage via recurrent and ongoing formation and reworking of instream geomorphic units, while major adjustments via thalweg shift are likely to occur at bankfull and overbank flow stages.

General relationships that characterize geomorphic adjustments of rivers at different flow stages are summarized below.

1 *Low flow stage river behavior – Packages of bedforms on the channel bed.* Bedforms are sparsely vegetated transient features that are readily reworked at low flow stage and are replaced at high flow stage (see Chapter 5). Although their geomorphological significance is often not important in terms of the structural integrity of a reach, flow–sediment interactions are a critical consideration in the assessment of hydraulic and/or habitat diversity (e.g., Thomson et al., 2001).

2 *Bankfull stage river behavior – Packages of geomorphic units that reflect channel structure.* Packages of geomorphic units that reflect the broader channel structure are associated with formative (bankfull) discharge stage. These reflect constraints on the dimension, shape, and alignment of the channel and the distribution of flow within it. Specific combinations of bank-attached and midchannel geomorphic units can be determined. Interpretation of genetic linkages and mutual interactions among instream geomorphic units provides insight into why the channel has adopted its particular geometry and planform (see Chapters 4 and 5).

3 *Overbank stage river behavior – Packages of floodplain geomorphic units.* Packages of floodplain geomorphic units are indicative of channel–floodplain linkages and the manner of channel adjustment on the valley floor (i.e., its planform; see Chapters 4 and 5). At stages beyond bankfull, floodplain formation processes are activated.

Under extreme circumstances, the floodplain may be reworked.

9.4 Stage One, Step Three: Assess controls on the character, behavior, and downstream patterns of River Styles

To develop an understanding of how and why each reach looks and behaves in the manner that it does, a summary assessment of controls on the distribution of River Styles is developed. Critical controls on river behavior may vary from reach to reach. Initial insights are gained by plotting downstream patterns of River Styles onto longitudinal profiles. This provides guidance into slope–discharge relationships and associated ranges of stream power conditions under which differing River Styles are observed. It also presents a basis to examine linkages along river courses, placing each reach in its catchment context. When linked to the nature of the sediment transport regime (i.e., whether the river is bedload, suspended load, or mixed load dominated), process zone functioning is interpreted (i.e., whether the reach operates as a sediment source, transfer, or accumulation zone). These conditions determine the contemporary capacity for adjustment and the pathway of adjustment of a River Style. The synthesis of downstream patterns provides a useful template for management of subcatchments with a relatively similar physical framework.

The critical role of downstream changes in valley confinement is explained in relation to the geological imprint (structure and lithology) and long-term landscape history. This is manifest in the catchment-scale pattern of landscape units. These imposed boundary conditions define the potential range of variability of a River Style. Analyses of these imposed boundary conditions are then tied to flux boundary condition controls.

Isolating controls on river behavior is difficult, and a process of elimination is suggested. As a starting point, determinations can be made as to what has conditioned the position of boundaries between River Styles. Typical questions that are asked at this stage of analysis include:
• what attributes have changed upstream and downstream of the boundary, and why has that change occurred?

• can discernible changes in imposed boundary conditions be isolated in the transitional zones between River Styles? Do transition zones coincide with tributary–trunk stream boundaries as noted by changes in contributing area, breaks in slope, and base level controls along the longitudinal profile, or geological (lithologic and/or structural) changes? Do certain River Styles occur in certain landscape units at certain positions in the catchment?

• can underlying factors that result in changes to flux boundary conditions be expressed in terms of changes to the flow and sediment fluxes? Do cumulative responses reflect a change in stream power? Can these relationships be quantified? What has fashioned patterns of sediment transport and storage along river courses, and their relationships to geomorphic process zones?

Interpretation of controls on river character and behavior may result in assessments that are catchment (or region) specific. As many factors may conspire to create an opportunity for any given type of river to be generated, it is considered to be overly simplistic to quantify controls on the distribution of individual River Styles in a prescriptive manner. River character and behavior do not reflect variability along a single continuum in, say, slope, grain size or valley width. Rather, they reflect a multivariate continuum with an infinite complexity of associations. Any given set of boundary conditions may result in an array of River Styles, and it is unlikely that any particular River Style will only be observed under a "unique" set of conditions such as a specific lithology on a characteristic slope set within a particular range of stream power conditions. Hence, there will be overlap among the circumstances under which any River Style is observed and associated "typical" patterns in space and time.

The best way to assess controls on river character and behavior is to determine the conditions under which all examples of a particular River Style operate. Differences, similarities, and overlaps in controls among River Styles are analyzed and interpreted. Just as differing parameters define the character and behavior of River Styles, so the relative influence of differing controls condition the presence/absence and distribution of River Styles. To interpret the dominant controls on the character and behavior of each River Style, a comparison is made between River Styles to determine which controlling parameters are significantly different. Obviously, depending on the scale of analysis (e.g., regional versus catchment versus subcatchment), and the nature of the environmental setting, other measures may be required, such

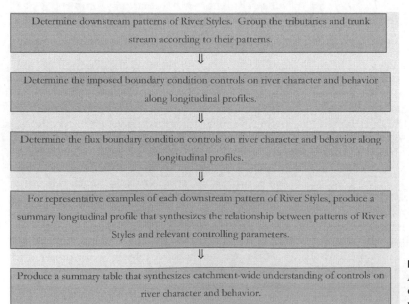

Figure 9.14 Stage One, Step Three: Assessment of controls on the character, behavior and downstream patterns of River Styles

as structural geology, drainage density, stream order, catchment shape, vegetation cover, forms of human disturbance (e.g., urban attributes, irrigation schemes), etc. Beyond this, an assessment is made of how various controls interact to dictate the character and behavior of the River Style. Finally, any anomalies are explained. Procedures used to assess controls on the character, behavior, and downstream patterns of River Styles in a catchment are summarized in Figure 9.14.

patterns, from those that are distinct. Alternating patterns of River Styles can be synthesized in this assessment. This analysis provides a basis to assess the similarity in downstream controls on River Styles and helps explain river character and behavior in different subcatchments. A tree-like diagram is produced showing the tributary patterns of River Styles and how they connect to the trunk stream (Figure 9.15). The pattern is noted on the catchment River Styles map (see Plate 9.2).

9.4.1 Determine downstream patterns of River Styles

Downstream patterns of River Styles are determined for all river courses in the catchment, differentiating those that have similar downstream

9.4.2 Determine imposed boundary condition controls on river character and behavior along longitudinal profiles

River Styles boundaries are placed atop the longitudinal profile-contributing area plots, noting

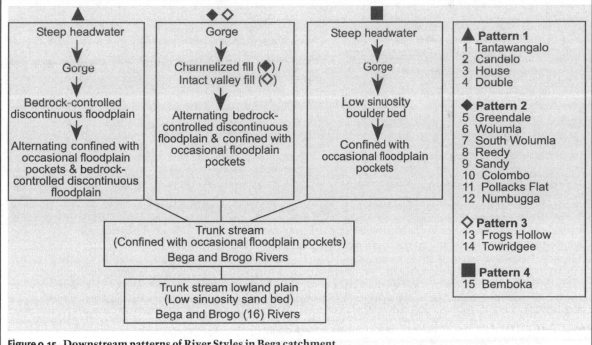

Figure 9.15 Downstream patterns of River Styles in Bega catchment
Four primary downstream patterns of River Styles were identified from 16 subcatchments in Bega catchment:
• those in which headwaters are transitional to long, elongate, bedrock-controlled valleys downstream of the escarpment (Pattern 1);
• those which have large accommodation spaces at the base of the escarpment in which extensive Holocene fills have formed (Pattern 2 rivers have channelized fills, Pattern 3 rivers have intact valley fills);
• those in which boulder fans at the base of the escarpment are transitional to bedrock-controlled valleys and the lowland plain (Pattern 4).
Representative examples of each were chosen to explain the controls on the character, behavior, and downstream pattern of River Styles in the catchment.

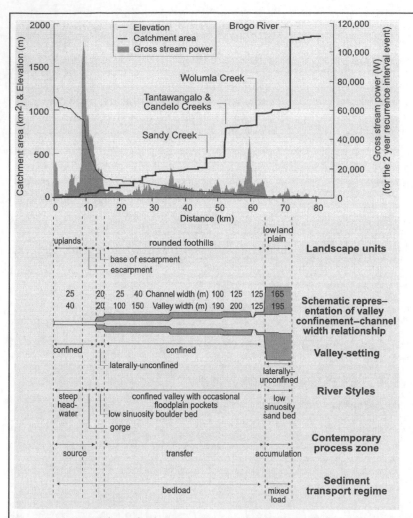

Figure 9.16 Controls on the downstream pattern of River Styles along the Bega River
Significant catchment areas drain from the uplands landscape unit where the steep headwater River Style occurs. The longitudinal profile has a distinct step in the escarpment zone where the gorge River Style is formed. Beyond the escarpment, the longitudinal profile has a relatively smooth, concave-upwards form. Associated with this progressive downstream change in slope, there is progressive downstream widening of both the channel and the valley through the base of the escarpment, the rounded foothills and the lowland plain landscape units. At the base of the escarpment, the low sinuosity boulder bed River Style occurs. This occurs where slopes are high and flow exits from the escarpment zone. Large boulder fans have been deposited. There is a lack of a valley constriction along the downstream margin of the base of escarpment landscape unit. This prevents the accumulation of valley fills at the base of the escarpment along these valleys. Instead bedload materials are transferred through the system, until they reach the lowland plain where they accumulate.

Along the majority of the Bega trunk stream, the confined valley with occasional floodplain pockets River Style occurs. As catchment area increases and discharges increase with the inputs from numerous tributaries, gross stream power along this River Style progressively increases. Peaks occur where bedrock steps occur in the longitudinal profile (e.g., at ~60 km where Kanooka knickpoint is located). It is not until the valley widens and slope decreases significantly around the Wolumla Creek confluence that the transition to a low sinuosity sand bed River Style occurs. This transition coincides with the start of the lowland plain landscape unit and a drop in the gross stream power. No partly-confined valleys are found along this pattern of River Styles in Bega catchment.

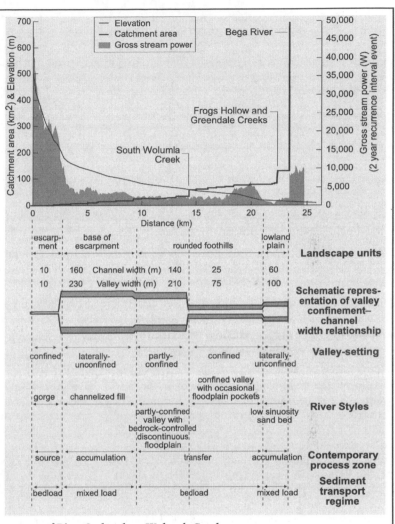

Figure 9.17 Controls on the downstream pattern of River Styles along Wolumla Creek

Most subcatchments in Bega catchment drain directly from the escarpment and have smooth concave-upward longitudinal profiles, with a gentle break in slope at the base of the escarpment. In these tributaries, where the uplands landscape unit is absent, the channelized fill River Style is formed at the base of the escarpment. These laterally-unconfined valley settings are formed under a particular set of catchment boundary conditions. Broad, asymmetrical valleys are formed downstream of a gentle break in slope. The downstream margin of these valleys is characterized by either a significant narrowing of the valley or a bedrock step which gives the valley a funnel shape. Large accommodation spaces store material behind these constrictions. The formation of these laterally-unconfined valley settings at the base of the escarpment is a direct result of interactions between escarpment retreat and valley-sidewall expansion (Fryirs, 2002). When infilling, these valleys are characterized by the intact valley fill River Style. When cutting, these valleys are characterized by the channelized fill River Style. In 8 of the 10 subcatchments that display this downstream pattern of River Styles, only two remain intact (i.e., contain the Intact valley fill River Style at the base of the escarpment).

In the rounded foothills of these subcatchments, the confined valley with occasional floodplain pockets and the partly-confined valley with bedrock-controlled discontinuous floodplain River Styles extend to the trunk stream. Unlike the classical downstream sequence of channel geometries and process zones along long profiles, streams along this pattern of River Styles have large, laterally-unconfined valleys with wide, deep channels at the base of the escarpment. These channels are mixed load in composition, with sands and muds accumulating on the channel bed. These are transitional to narrower, shallower channels in the confined and partly-confined valley settings in the middle and lower sections of the catchment which act to effectively transfer bedload materials through to the lowland plain.

their relation to each valley setting and landscape unit (see Figures 9.16 and 9.17). These imposed boundary condition controls define the potential range of variability of a River Style. For each River Style, the range of valley widths, valley slope, and contributing areas are determined, and presented as part of a "controls" table. The transition zone between River Styles is related to catchment-scale boundary conditions (e.g., geological boundaries), associated landscape units, and the catchment position (and contributing area). By definition, shifts in River Styles are evident whenever changes in valley setting occur. These transitions, in turn, are commonly associated with down-stream changes in landscape units, reflecting land-scape history. Considerable overlap may exist in these imposed boundary conditions, as a range of River Styles can be formed under similar sets of these conditions.

9.4.3 Determine the flux boundary condition controls on river character and behavior along longitudinal profiles

Stream power provides a guide to the energy regime at differing positions along a river. Outputs required to generate stream power include contributing area and a running average of slope. When combined with the catchment area-discharge relationships derived in Stage One, Step One, discharge estimates can be extracted for the 2, 5, 10, 50, and 100 year events and plotted as a continuous data set along the longitudinal profile. When discharge is combined with slope, gross stream power is generated (Reinfelds et al., 2004).

In the River Styles framework, the gross stream power curve is superimposed onto the longitudinal profile-contributing area plot for each subcatchment. If possible, unit stream power estimates are derived for each River Style, recognizing explicitly that the range of estimates may vary markedly for differing reaches (reflecting variability in slope or upstream catchment area, or the geomorphic condition of the reach; see Chapter 10). Generally, only one recurrence interval relationship (e.g., 1 in 2 year event) is depicted, but all values are presented in a summary table of controls on the character, behavior and pattern of River Styles (Table 9.10). The distribution of geomorphic process zones (i.e., source, transfer, and accumula-

tion zones) and associated sediment transport regime (i.e., bedload, mixed load, or suspended load system) are plotted beneath these curves. An estimate is made as to whether each reach is sediment supply- or transport-limited. In general, rivers in confined and partly-confined valley settings act as sediment transfer reaches, as sediments are readily flushed. Rivers in laterally-unconfined valley settings tend to be accumulation zones, unless the reach responds to disturbance by reworking its sediment stores, thereby acting as a sediment source zone.

This analysis provides the basis to interpret process responses of each reach to imposed and flux boundary condition controls, which are major determinants of river character and behavior. A summary representation of the range of controls for all River Styles in Bega catchment is presented in Figure 9.18.

9.5 Overview of Stage One of the River Styles framework

The baseline survey of River Styles integrates catchment-scale controls on rivers with reach-based assessments of river character and behavior through use of a nested hierarchical approach. Classification of River Styles is based initially on valley setting. For differing settings, variable sets of parameters including channel planform, the assemblage of geomorphic units, and bed material texture are used to define River Styles, emphasizing distinguishing attributes in River Styles trees. Proformas are completed for each River Style. Analysis of catchment-scale linkages and boundary conditions aids determination of the controls on river character and behavior for each River Style.

The following products are produced in Stage One of the River Styles framework:
• regional setting chapter;
• River Styles tree that is specific to the catchment;
• catchment-wide map showing the distribution of River Styles;
• River Styles proformas, annotated cross-sections, annotated geomorphic unit planform map and photographs for each River Style in the catchment;

Table 9.10 Controls on river character and behavior in Bega catchment.

River style	Valley slope	Valley width (m)	Catchment area (km²)	Unit stream power (Wm⁻²)					Formative/bankfull recurrence interval (years)
				1 in 2	1 in 5	1 in 10	1 in 50	1 in 100	
Steep headwater	0.02	40	>20	180	415	500	270	390	N/a
Gorge	0.04–0.08	10–40	0–135	685	1730	2305	2530	2770	N/a
Channelized fill	0.005–0.03	<300	<20	100	125	440	1020	1140	>100
Inact valley fill	0.020–0.028	200	<20	3	4	25	70	100	N/a
Low sinuosity boulder bed	0.03	100	>50	70	90	390	900	1190	–
Confined with occasional floodplain pockets (trunk)	0.004–0.006	60–240	100–1000	100	130	390	640	730	>100
Confined valley with occasional floodplain pockets (tributaries)	0.005–0.029	20–80	20–325	165	210	680	1270	1520	2–50
Flood out	0.010	150	<30	3	4	25	70	100	N/a
Partly-confined valley with bedrock-controlled discontinuous floodplain	0.005–0.012	40–210	30–200	95	120	410	820	1030	10–50
Low sinuosity sand bed	0.002–0.0008	100–650	500–1840	30	35	95	220	280	5–10

Note

Gorges are found on high slopes in confined settings that generate high stream powers. This acts to flush materials efficiently through the escarpment zone of the catchment. High stream powers are also generated along the confined valley with occasional floodplain pockets and the partly-confined valley with bedrock-controlled discontinuous floodplain River Styles. While found on lower slopes, their position in the catchment ensures that large discharges are generated in these midcatchment locations. Given their bedrock-controlled character they too act to efficiently flush sediment to the lowland plain. The lowest stream powers in the catchment are generated in the intact valley fill and flood out River Styles which remain unchannelized and effectively dissipate energy over valley fill surfaces. These River Styles act as large sediment sinks.

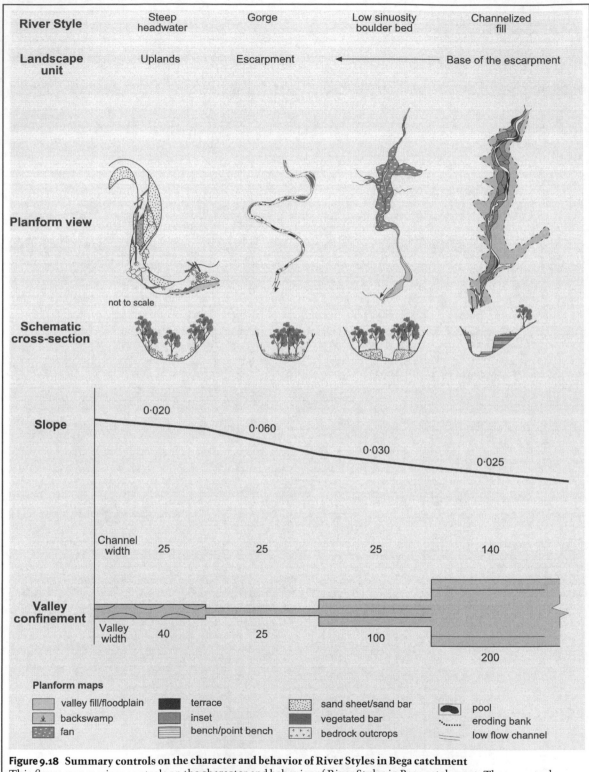

Figure 9.18 Summary controls on the character and behavior of River Styles in Bega catchment
This figure summarizes controls on the character and behavior of River Styles in Bega catchment. These controls include slope and valley confinement. Each reach is placed within its landscape context through analysis of landscape units. Reprinted from Brierley and Fryirs (2000) with permission from Springer-Verlag GmbH & Co. K.G. 2004.

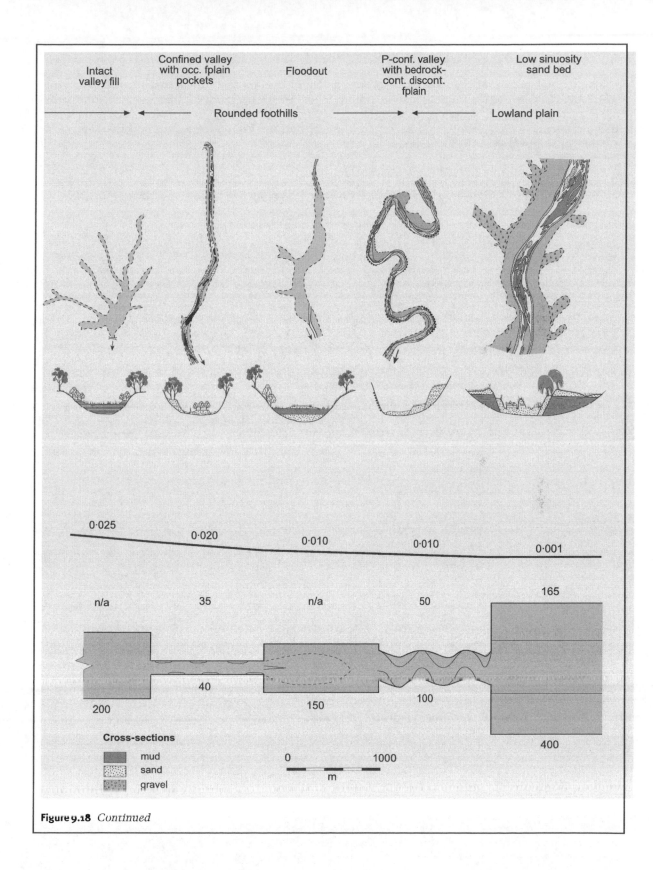

Figure 9.18 *Continued*

• longitudinal profile diagrams with associated assessment of controls for a representative example of each downstream pattern of River Styles in the catchment;
• table of controls for all River Styles in the catchment.

Baseline information derived from analysis of catchment-wide river character and behavior pro-vides a geomorphological baseline for comparing like-with-like, ensuring that reaches of the same River Style are used in analyses of river condition and recovery potential, as outlined in Chapters 10 and 11.

CHAPTER 10

Stage Two of the River Styles framework: Catchment-framed assessment of river evolution and geomorphic condition

Many rivers no longer support valued native species or sustain healthy ecosystems that provide important goods and services.

LeRoy Poff et al., 1997, p. 769

10.1 Introduction

Completion of Stage One of the River Styles framework produces a catchment-wide analysis of differing geomorphic types of river. Inevitably, reaches of any given type do not have a uniform character and behavior. Inherent diversity in river forms and processes may be evident at the local scale, such that a mapped River Style represents a summary sense of character and behavior at the reach scale. In many instances, these differences may reflect variability in the geomorphic condition of differing sections of a reach of a given River Style, as induced by human disturbance (whether direct or indirect, at-a-site or off-site). In this chapter, a set of procedures with which to appraise the geomorphic condition of rivers is documented. This represents Stage Two of the River Styles framework.

Measures of geomorphic river condition record deviations from a natural or expected state in any given reach (i.e., how human disturbance has altered river character and behavior). In this book, river condition is defined as a measure of the capacity of a river to perform functions that are expected for that river within the valley setting that it occupies (Table 10.1). The further a reach sits from its reference condition, the poorer its geomorphic condition. When appraised in terms of the capacity for adjustment that is appropriate for the given boundary conditions (i.e., the reference condition), each reach is placed into a good, moderate, or poor condition category.

To frame the assessment of geomorphic river condition, human-induced changes to river forms and processes must be viewed in context of the inherent evolutionary tendencies of the system. In applying a generic set of procedures in each field situation, elements of subjectivity are encountered. So long as limitations are recognized, they do not present an insurmountable problem. Indeed, much is to be gained by thinking through and discussing these issues. Regardless of the challenges faced, assessments of river condition (or health) constitute an integral part of the river management process, providing a critical platform for environmental decision-making and associated actions.

Given that each River Style records the character and behavior of reaches that operate within an equivalent set of boundary conditions, comparison of reaches of the same River Style provides an ideal basis to assess river condition. Geomorphic river condition is appraised in context of the capacity for adjustment of the River Style and the degree to which contemporary measures of geomorphic structure and function for the reach have moved away from the reference condition. As geomorphic structure and function, and associated adjustments, are predictable for each River Style, a platform is provided with which measures of river condition can be evaluated in a consistent yet flexible manner, ensuring that appropriate criteria are measured to make this determination. River condition is appraised in terms of channel planform, channel geometry, bed character, and the geomorphic unit assemblage along a reach.

Table 10.1 Definition of terms used to describe the geomorphic condition of a reach.

Term	Definition in the River Styles framework
River condition	A measure of the capacity of a river to perform functions that are expected for that river within the valley setting that it occupies. The contemporary geomorphic state of a reach relative to a "natural" or "expected" reference condition of the same type of river. Assessment of river condition requires an understanding of: • the spatial distribution of river types • how those rivers behave • river dynamics (i.e., river evolution), and • forms, extent, and impact of human disturbance, including an appraisal of whether this change has been irreversible.
Capacity for adjustment	Morphological adjustments brought about by the changing nature of biophysical fluxes that do not record a wholesale change in river type.
Degrees of freedom	The ability of differing components of a river system to adjust, measured in terms of bed character, channel attributes, and planform attributes.
Relevant geoindicators	Parameters used to interpret and explain system structure, function, and condition for each degree of freedom. "Relevant geoindicators" provide a reliable and relevant signal about the condition of a reach. The geoindicators measured are River Style specific.
Desirability criteria	Assessment of the appropriateness of each relevant geoindicator for each River Style. A question is posed for each geoindicator to produce a set of desirability criteria to identify a reference condition and assess the geomorphic condition of a reach.
Natural river (Natural reference condition)	A "natural" river is dynamically adjusted so that geomorphic structure and function operate within a capacity for adjustment that is appropriate for that type of river, given the prevailing boundary conditions. A "natural reference condition" is considered to be a river that is operating in the absence of human disturbance. Changes to this "intact" or "predisturbance" condition are considered to be reversible.
Expected reference condition	A prehuman disturbance reference condition is largely irrelevant for many river systems. Hence, expected reference conditions are identified against which the geomorphic condition of a reach is assessed. Three types of expected reference condition differentiate among situations in which the reach has been: • reversibly altered by human disturbance; • irreversibly altered by indirect human disturbance; • irreversibly altered by direct human disturbance.
Irreversible geomorphic change	A wholesale shift in the geomorphic unit structure, planform, and bed material texture, such that the river operates in a fundamentally different manner to its former state with no prospect of return over a 50–100 year timeframe. This transition in the behavioral regime marks the adoption of a different type of river.
Good condition	River character and behavior are appropriate for the River Style given the valley setting and within-catchment position. Geomorphic structures are in the right place and operating as expected for the River Style. These reaches have a near-natural potential for ecological diversity and associated vegetation associations.
Moderate condition	Certain characteristics are out-of-balance or inappropriate for the River Style. Localized degradation of river character and behavior is typically marked by modified patterns of geomorphic units. Key geomorphic structures are in the wrong places. Locally anomalous processes are occurring. In general, these reaches have poor vegetation associations and/or cover.
Poor condition	River character is divergent from the natural reference condition. Abnormal or accelerated geomorphic behavior occurs. Key geomorphic units are located inappropriately along the reach, and processes are out-of-balance or anomalous. These reaches generally have low levels of bank vegetation and/or are weed infested. If fundamental threshold conditions are breached, irreversible geomorphic change would transform the reach to a new River Style.

Differentiation of good, moderate, and poor condition states is based on analysis of the variables that have the capacity to adjust for a particular River Style. If geomorphic adjustment occurs within the contemporary capacity for adjustment of the River Style, reaches are considered to be in good geomorphic condition. For example, breaching of an intrinsic threshold may promote formation of a cutoff along a gravel bed meandering river— a healthy adjustment for this type of river. Alternatively, geomorphic adjustment can occur towards the extremes of the contemporary capacity for adjustment, inducing shifts between moderate and poor condition reaches.

Solid baseline understanding on the diversity of river character and behavior is required to identify appropriate benchmarks against which to assess river condition in a way that meaningfully compares like with like. Understanding of river evolution is required to determine the nature, timing and extent of change to river structure and function, and to assess whether this change is irreversible. Assessments of river evolution span the period in which environmental conditions have remained relatively uniform. This provides a basis with which to appraise the extent of degradation resulting from human disturbance. Within this timeframe, lag effects and threshold breaches (whether natural or human-induced) are identified and interpreted. Evolutionary perspectives also provide a basis to identify reference conditions in the absence of field-based remnants using ergodic reasoning.

Human-induced adjustments away from a reference condition vary for different types of river, because of their inherent capacity for adjustment. Appraisal of the various states of adjustment for differing River Styles provides a core basis to identify appropriate reference conditions for different reaches. Discernible (predictable) patterns of response to human disturbance are identified for different River Styles. Such notions are framed in context of the (ir)reversibility of river responses to human disturbance. From this, stages of adjustment away from the "natural" state are identified. Appraisals of river condition are assessed in light of these different forms of evolutionary pathways for each River Style. Fundamental principles underpinning the assessment of river condition in the River Styles framework are outlined in Table 10.2.

Only certain geoindicators provide reliable and relevant insights into system structure and function, and hence the geomorphic condition of a River Style (cf., Costanza, 1992; Elliott, 1996; Osterkamp and Schumm, 1996; Rogers and Biggs, 1999). Interpretation of these geomorphic indicators must be accompanied by solid process-based understanding of the river type under investigation. Geoindicators are analogous to *vital ecosystem attributes* suggested by Aronson et al. (1995). Geoindicators used to assess geomorphic condition are River Style specific, relating directly to the degrees of freedom that characterize the capacity for adjustment for each type of river (Table 10.1).

Identifying whether irreversible geomorphic change has occurred provides the benchmark to determine which River Style should be used to identify the reference condition. In some instances this reflects an intact (prehuman disturbance) state (Table 10.1). In general, however, reference conditions are framed in terms of an expected state. Differentiation is made between settings in which human disturbance has brought about reversible or irreversible changes to river character and behavior. In the latter instance, the river has become a new River Style, and measures of geomorphic river condition are appraised in terms of the best attainable state for that particular River Style.

The condition of a reach at the time of the assessment records cumulative responses to a range of disturbance events. The time since the last flood and the geomorphic effectiveness of that flood will influence the observed condition. For example, if a reach has been subjected to a recent large flood event, its geomorphic condition may be poorer than another reach of the same River Style that did not experience that event. The role of preconditioning and the geomorphic effectiveness of floods on the geomorphic condition of rivers are specific to the reach and catchment under investigation and should be noted at the time of the assessment.

Assessment of river condition in Stage Two of the River Styles framework is field intensive and requires solid assessment of river evolution. Depending on the nature of the River Style under investigation, this can be undertaken in a cursory manner at the same time as the ratification of the River Styles boundaries in Stage One, Step Two. It

Table 10.2 Principles used to assess geomorphic river condition. Reproduced from Fryirs (2003) with permission from Elsevier, 2003.

Principle	Explanation
1) Compare like-with-like	Assessment of river character, behavior, and distribution throughout a catchment, or between catchments if required, provides a basis for meaningful comparisons to be made across the range of river types.
2) Measure appropriate parameters for each River Style	As different types of river have different capacities to adjust, certain parameters give a reliable and relevant signal about the condition of a reach, while others are irrelevant or give poor signals. Hence, the range of parameters measured are River Style specific.
3) Place reaches within their evolutionary context	Assessments of river evolution provide a sense of river dynamic, allowing underlying causes of change to be identified. This helps explain the present geomorphic condition of the reach, enabling direct assessment of the underlying causes of that condition, and enhances predictive capability.
4) Define irreversible geomorphic change	Once a reach has been subjected to disturbance impacts, and moves away from its predisturbance condition, an assessment is made whether the reach retains the capacity to return to a condition akin to the intact, predisturbance character and behavior, or whether the boundary conditions are now such that the river operates under a revised set of boundary conditions, and has changed its structure and function. In this latter case the reach must be appraised in terms of a different River Style.
5) Select appropriate reference conditions	A reference condition is identified for the contemporary River Style. Determination of reference conditions is framed in terms of whether irreversible geomorphic adjustments have taken place following human disturbance. Four variants of reference condition are differentiated:
	1 Remnant reaches that have been minimally disturbed by humans, such that geomorphic changes to river character and behavior remain reversible.
	2 Reaches where human disturbance has occurred, but geomorphic changes to river character and behavior remain reversible.
	3 Reaches where change has been induced by indirect human disturbance and irreversible change has resulted.
	4 Reaches where change has been induced by direct human disturbance and irreversible geomorphic change has resulted.
	The reference condition must occur at a similar position in the catchment and operate under near-equivalent catchment boundary conditions. The reference condition can occur within the region or be identified from the evolutionary sequence using ergodic reasoning.

is difficult to assess river condition based solely on interpretation of air photographs.

There are three steps in Stage Two of the River Styles framework (Figure 10.1). All reaches of all River Styles are assessed to produce a catchment-wide map of geomorphic river condition.

10.2 Stage Two, Step One: Determine the capacity for adjustment of the River Style

Procedures undertaken in Stage Two, Step One of the River Styles framework are summarized as a

flow chart in Figure 10.2. These various components are discussed in turn.

10.2.1 Assess the ability of each degree of freedom to adjust for each River Style

To determine the capacity for adjustment of each River Style in a catchment, the question is asked: In what ways can this type of river adjust under the prevailing set of flow, sediment, and vegetation characteristics? Three degrees of freedom are appraised to assess the capacity for adjustment in the River Styles framework, namely, bed character,

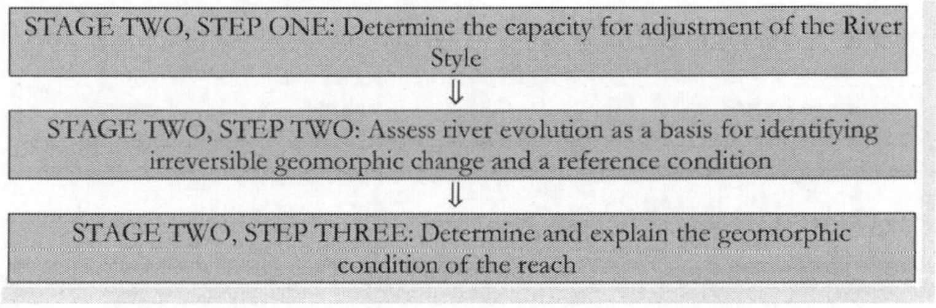

STAGE TWO, STEP ONE: Determine the capacity for adjustment of the River Style

⇓

STAGE TWO, STEP TWO: Assess river evolution as a basis for identifying irreversible geomorphic change and a reference condition

⇓

STAGE TWO, STEP THREE: Determine and explain the geomorphic condition of the reach

Figure 10.1 Steps in Stage Two of the River Styles framework

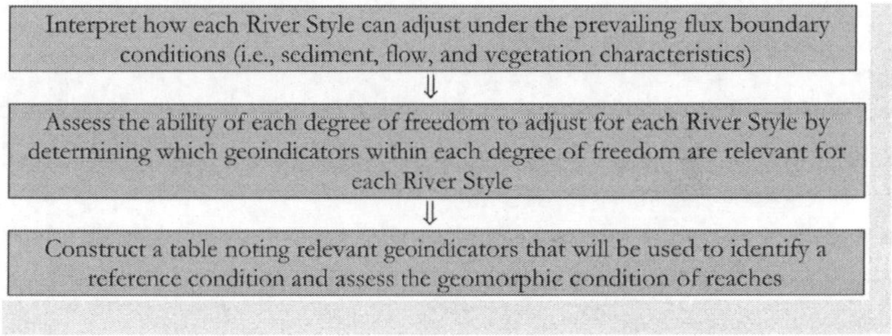

Interpret how each River Style can adjust under the prevailing flux boundary conditions (i.e., sediment, flow, and vegetation characteristics)

⇓

Assess the ability of each degree of freedom to adjust for each River Style by determining which geoindicators within each degree of freedom are relevant for each River Style

⇓

Construct a table noting relevant geoindicators that will be used to identify a reference condition and assess the geomorphic condition of reaches

Figure 10.2 Stage Two, Step One: Determine the capacity for adjustment of the River Style

channel character, and channel planform. These are directly comparable to scales of analysis described in Chapters 4 and 5. The relevance of differing criteria in the determination of the capacity for adjustment of differing River Styles in Bega catchment is presented in Table 10.3. In some cases a reach has significant potential to adjust, whereas in others only localized or minimal adjustment can take place. This type of analysis is used as a simplified measure of the sensitivity of a River Style to change. Reaches with significant capacity to adjust are considered to be sensitive to change, while those with localized adjustment potential are considered to be resilient to change. This provides an initial guide to the types of rivers that are most likely to experience irreversible geomorphic change.

10.2.2 Determine relevant geoindicators for each degree of freedom for each River Style

Within each degree of freedom, a series of geoindicators is assessed to determine the ability of each reach to adjust within its valley setting (Table 10.4). In this framework, bed character is assessed

in terms of grain size and sorting, bed stability, hydraulic diversity, and sediment regime (i.e., whether the reach acts as a sediment source, transfer, or accumulation zone). Channel attributes assessed include the size and shape of a channel, bank morphology, instream vegetation structure, and woody debris loading. Channel planform is assessed in terms of the number, sinuosity, and lateral stability of channels, the assemblage of instream and floodplain geomorphic units, and riparian vegetation structure and composition.

The relevance of each geoindicator in providing insight into the capacity for adjustment of each River Style in a catchment is noted. Only those geoindicators that provide direct insight into how that river adjusts are assessed (i.e., each geoindicator is diagnostic of a certain adjustment process for that River Style). For example, channel geometry is not assessed for rivers in a confined valley setting, whereas it is assessed for a laterally-unconfined river. In general, the range of relevant geoindicators is far greater for alluvial streams where the capacity for river adjustment is at a maximum. Relevant geoindicators for each River Style are subsequently used to identify a reference reach and assess the geomorphic condition of each

Table 10.3 Capacity for adjustment of River Styles in Bega catchment.

River Style	Channel attributes	Channel planform	Bed character	Capacity for adjustment
Confined valley setting				
Steep headwater				Low
Gorge				Low
Occasional floodplain pockets				Low
Partly-confined valley setting				
Bedrock-controlled discontinuous floodplain				Moderate
Laterally-unconfined valley setting				
Intact valley fill				Low
Floodout				Low
Low sinuosity boulder bed				Low
Channelized fill				High
Low sinuosity sand bed				High

Minimal or no adjustment potential

Localized adjustment potential

Significant adjustment potential

Note:
All confined and partly-confined River Styles have limited capacity for adjustment. The intact valley fill and floodout River Styles also have a low capacity for adjustment because of the absence or discontinuity of channels which limits the ability to rework sediments. However, once incised such that the channelized fill River Style develops, these rivers are highly sensitive to change. Because of its coarse boulder texture, the low sinuosity boulder bed River Style is also considered to have a low capacity for adjustment. The laterally-unconfined rivers with continuous channels (i.e., the channelized fill and low sinuosity sand-bed River Styles) have the greatest capacity for adjustment. These rivers are considered to be sensitive to change and have experienced metamorphosis in the period since European settlement.

reach of each River Style in the catchment. To demonstrate this procedure, relevant geoindicators for two laterally-unconfined rivers and a partly-confined river from Bega catchment are presented in Table 10.5.

10.3 Stage Two, Step Two: Interpret river evolution to assess whether irreversible geomorphic change has occurred and identify an appropriate reference condition

Procedures undertaken in Stage Two, Step Two of the River Styles framework are presented as a flow chart in Figure 10.3.

Table 10.4 Measurement procedures for each geoindicator. Reprinted from Fryirs (2003) with permission from Elsevier, 2003.

Degree of freedom / geoindicator	Definition	Examples of geomorphic tools or techniques used to assess each geoindicator
CHANNEL ATTRIBUTES – Channel structure is a function of bed and bank material texture, vegetation cover, bed slope, and discharge		
Size	The width and depth of the channel	• Width : depth ratio and cross-sectional area of the channel relative to the catchment area as it drains
Shape	The cross-sectional form of the channel	• Identification of irregular, compound, symmetrical, or asymmetrical channels
Bank morphology	The shape and character of each bank	• Identification of uniform vertical, uniform graded, faceted, undercut banks
		• Characterization of bank texture using grain size analyses
Instream vegetation structure	The character and density of aquatic and terrestrial vegetation. Linked to the geomorphic structure and flow regime	• Qualitative rating of the composition (native versus exotic) and coverage of vegetation on instream geomorphic surfaces
Woody debris loading	The character and density of woody debris and its relationships to the geomorphic structure and flow regime	• Qualitative rating of the type, alignment, and abundance of woody debris in the channel
CHANNEL PLANFORM – The outline of a river from above is a function of material texture, valley slope, valley setting, and vegetation structure		
Number of channels	Count of the number of channels	• Identification of absent, discontinuous, single, or multichannel variants
Sinuosity of channels	The degree of channel curvature along the length of a river	• The ratio between channel length along the thalweg and valley length along its axis
Lateral stability	The degree to which the channel can move on the valley floor	• Identification of channel expansion, bank erosion, migration, and avulsion processes
Geomorphic unit assemblage	The building blocks of rivers. Each geomorphic unit has a distinct form–process association	• Analysis of form and sedimentology to interpret processes responsible for formation of geomorphic unit
		• Assessment of the juxtaposition and assemblage of units
		• Assessment of channel–floodplain connectivity and unit condition (e.g., signs of reworking, dissection, etc.)
Riparian vegetation	The character and density of vegetation in the riparian zone, linked to the geomorphic structure and flow regime	• Qualitative rating of the composition (native versus exotic), continuity, and structure of vegetation assemblages in the riparian zone
BED CHARACTER – Is a function of flow regime, sediment availability, and the capacity of the reach to transfer materials		
Grain size and sorting	The size, distribution, and arrangement of materials stored and transported on the bed	• Visual estimates of the percent of the bed that comprises different grain size fractions
		• Analysis of sediment distributions on different geomorphic units
Bed stability	Capacity of channel bed to adjust vertically	• Interpretation of vertical bed activity via incision
Hydraulic diversity	The character of flow as it passes over the bed	• Visual water surface flow estimates (see Thomson et al., 2001)
Sediment regime	The storage, transfer, and delivery capacity of the reach. Measures the capacity and/or competence of the reach to transport sediment	• Identifying sediment process zone (i.e., source, transfer, accumulation; Schumm, 1977)
		• Quantitative measure of sediment transport capacity versus sediment availability to interpret supply vs. transport limited reaches

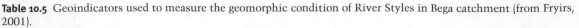

Table 10.5 Geoindicators used to measure the geomorphic condition of River Styles in Bega catchment (from Fryirs, 2001).

Geoindicator / River Style	Partly-confined valley with bedrock-controlled discontinuous floodplain	Channelized fill	Low sinuosity sand bed
Channel attributes			
Size	YES	YES	YES
Shape	YES	YES	YES
Bank morphology	YES	YES	YES
Instream vegetation structure	YES	YES	YES
Woody debris loading	YES	NO	YES
Channel planform			
Number of channels	NO	NO	YES
Sinuosity of channels	NO	NO	YES
Lateral stability	YES	YES	YES
Geomorphic unit assemblage	YES	YES	YES
Riparian vegetation	YES	YES	YES
Bed character			
Grain size and sorting	YES	YES	YES
Bed stability	YES	YES	YES
Hydraulic diversity	YES	YES	YES
Sediment regime	YES	YES	YES

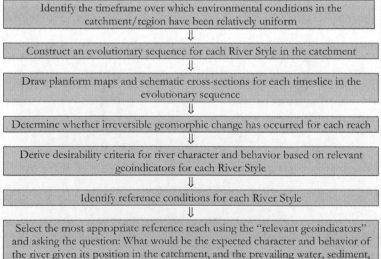

Figure 10.3 Stage Two, Step Two: Interpret river evolution as a basis for identifying irreversible geomorphic change and a reference condition

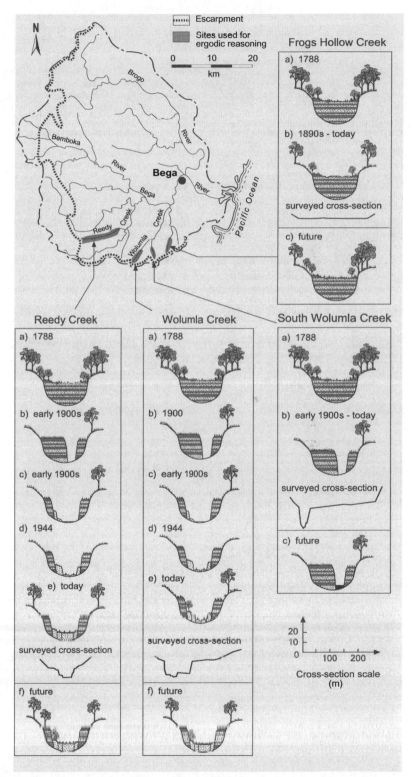

Figure 10.4 Use of ergodic reasoning to determine evolutionary sequences
Using procedures outlined in Chapter 3, evolutionary sequences for the intact valley fill and channelized fill River Styles in Bega catchment were constructed using ergodic reasoning. Each of the four reaches selected for analysis experienced the same sets of geomorphic changes at different times. This analysis has been used to assess how these rivers have adjusted in the past and to predict how reaches that have not experienced the full suite of changes may adjust in the future.

10.3.1 Identify the timeframe over which environmental conditions in the catchment/region have been relatively uniform

The evolutionary sequence of a reach provides an appreciation of how flux boundary conditions have adjusted in response to factors such as human disturbance (and subsequent recovery) over a period where environmental conditions have been relatively uniform. Lag effects and threshold breaches, whether triggered by natural events or human impacts, are identified and interpreted. For example, in coastal valleys of New South Wales, stable environmental conditions extend from the mid–late Holocene to present. Within this period, sea level stabilized and climatic conditions have remained relatively constant. Since 1788, human disturbance associated with colonial settlement has induced significant geomorphic changes along river courses. In other settings, many landscapes continue to be shaped by lagged responses to glaciation/deglaciation or tectonic uplift. Hence, the timeframe over which the evolutionary sequence required to isolate impacts of human disturbance is specific to the setting under investigation.

10.3.2 Construct an evolutionary sequence for each River Style in the catchment

An experienced and skilled fluvial geomorphologist is required to construct an evolutionary sequence for each River Style in the catchment. Field evidence is used to provide detailed knowledge of the temporal sequence of changes along one reach of each River Style. If chronostratigraphic dating is available, time constraints are placed on the

Figure 10.5 Evolution of the intact valley fill and channelized fill River Styles in Bega catchment. Modified from Fryirs (2003). Reprinted with permission from Elsevier, 2003

The stratigraphy of valley fills in Wolumla subcatchment reflects recurrent phases of cutting and filling over the last 6,000 years. However, the present incisional phase is considered to be the largest and most extensive of any that has occurred over this timeframe. Hence in Wolumla subcatchment, and in most other base of escarpment valleys, irreversible geomorphic change has been recorded over the last 200 years.

At the time of European settlement, most base of escarpment valleys in Bega catchment contained the intact valley fill River Style, characterized by unincised swamps with discontinuous drainage lines (a). The valley floor comprised mud and sand with a distinct vegetation pattern dominated by *Melaleuca ericifolia*. Only two analogous intact valley fill River Style reaches remain in Bega catchment (i.e., along Frogs Hollow and Towridgee Creeks (b)).

Stratigraphic and historical portion plans indicate that analogous features occurred along Wolumla Creek in the late 1860s, as "Wolumla Big Flat" is noted on the portion plans. Following anthropogenic disturbance to swamp surfaces, knickpoints retreated through the valley fill of upper Wolumla Creek by 1900 (c). A fundamental shift in the behavioral regime of this river occurred, irreversibly transforming this reach into a channelized fill River Style. Incision was quickly followed by channel expansion, producing a channel that was locally more than 10 m deep and 100 m wide (d). This wide, deep channel comprised continuous sand sheets with occasional bench features, and a poorly defined low flow channel. Riparian vegetation cover was poor. Given low channel roughness, there was limited capacity to retain finer grained materials within the channel. A contemporary version of this phase is evident in upper Numbugga catchment (d).

Air photographs from 1944 indicate that along Wolumla Creek, the incised trench was beginning to infill (e). In some places, over 3 m of material has accumulated on the channel bed while benches have continued to build along channel margins. These act to reduce channel width and depth, and modify channel alignment. Subsequent air photograph runs in 1962, 1971, 1989, and 1994 show little change in geomorphic structure. This reach is now characterized by increased heterogeneity in its geomorphic unit assemblage (f). Channel infilling and narrowing continue to occur, producing a compound channel with a vegetated inset floodplain. Initially, a well-defined low flow channel develops. With time, the low flow channel will become locally swampy and mud will be retained on the channel bed analogous to processes occurring along Reedy Creek (g). There are signs that the channel bed is becoming discontinuous. Eventually this will instigate redevelopment of an intact swamp within the incised trench, inducing greater water retention (increasing base flow) (h). Given the irreversible nature of geomorphic change since European settlement, this latter condition (h) is identified as the expected reference condition under contemporary flux boundary conditions as indicated by the box.

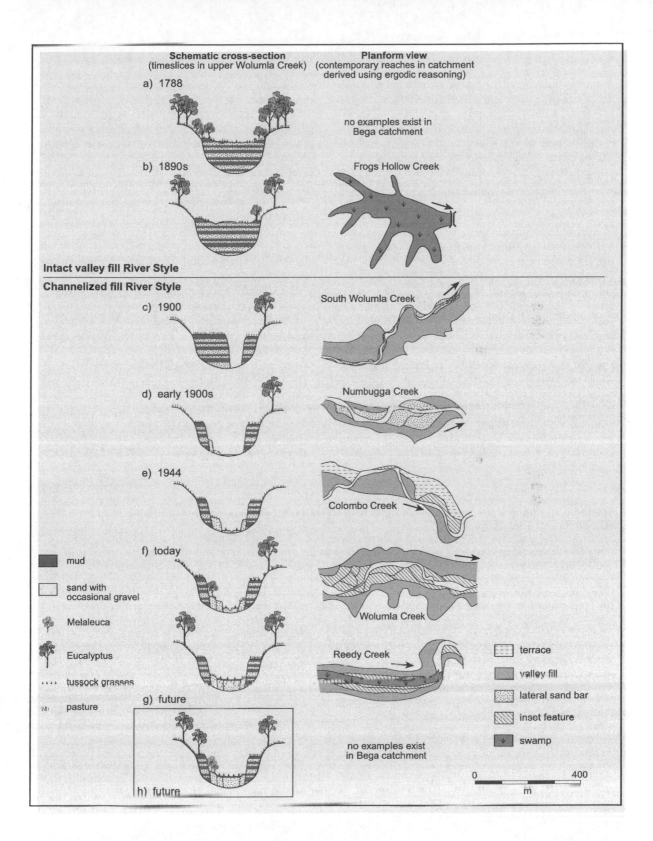

Schematic cross-section
(timeslices in upper Wolumla Creek)

Planform view
(contemporary reaches in catchment
derived using ergodic reasoning)

a) 1788

no examples exist in
Bega catchment

b) 1890s

Frogs Hollow Creek

Intact valley fill River Style

Channelized fill River Style

c) 1900

South Wolumla Creek

d) early 1900s

Numbugga Creek

e) 1944

Colombo Creek

f) today

Wolumla Creek

Reedy Creek

g) future

h) future

no examples exist
in Bega catchment

mud

sand with
occasional gravel

Melaleuca

Eucalyptus

tussock grasses

pasture

terrace

valley fill

lateral sand bar

inset feature

swamp

0   400
m

nature of change, allowing the timeframe and rate of change to be quantified. Field insights are tied to historical information such as portion plans, old photographs, explorers' notes, bridge surveys, sequential sets of air photographs, and historical maps. Where field evidence is poor or has not been preserved, ergodic reasoning is used to fill the gaps in the evolutionary sequence. Selected reference reaches used for ergodic reasoning analysis must be of the same River Style, occupy a similar position in the catchment with near-equivalent channel gradient, and operate under the same set of imposed boundary conditions (e.g., Kondolf and Downs, 1996; Fryirs and Brierley, 2000). An example of how ergodic reasoning has been used to determine the evolutionary sequence for the intact valley fill and channelized fill River Styles in Bega catchment is presented in Figure 10.4.

Timeslices are often constrained by available data. In an Australian context, the following timeslices are helpful: pre-European settlement (field sedimentology, dating techniques, and analysis of portion plans), 1860s–1900s (timing of first photographs), 1940s (military air photograph set), and 1960s–present (subsequent air photograph series, photographs, maps, and contemporary field analy-

ses). Planform and cross-sectional views are constructed for each timeslice. Planform maps are drafted directly from maps and air photographs, while cross-sections present a schematic summary of the geometry, geomorphic unit structure, sedimentology, and vegetation character of the reach. These schematic diagrams summarize the range of river character and behavior along the reach.

Once an evolutionary sequence has been constructed, reaches are placed within the evolutionary context of its River Style. This is used to:
• assess river character and behavior prior to human disturbance and determine changes in the post-human disturbance phase;
• identify how flux boundary conditions have impacted upon the timing and causes of river change;
• determine whether human disturbance has resulted in irreversible geomorphic change over management timeframes;
• identify a reference condition;
• predict future adjustments and the trajectory of change (in Stage 3);
• determine potential creation and restoration conditions for each reach (in Stage 3).

Evolutionary sequences for the channelized fill, the partly-confined valley with bedrock-

---

▶

Figure 10.6 Evolution of the partly-confined valley with bedrock-controlled discontinuous floodplain River Style in Bega catchment (from Fryirs, 2001)

Analysis of the evolution of the partly-confined valley with bedrock-controlled floodplain River Style was conducted along Middle Tantawangalo Creek. Along these river courses the pre-European settlement river was characterized by a narrow, deep channel with significant hydraulic diversity induced by large woody debris and a heterogeneous assemblage of geomorphic units including pools, riffles, bar complexes, islands, etc. (a). Following removal of riparian vegetation, channels widened and the channel bed became increasingly homogeneous (b). As the channel widened, significant volumes of sediment were released. With the additional large inputs of bedload material from upstream, sand sheets covered the channel bed infilling pools and smothering riffles (c). Contemporary versions of this condition are evident along Sandy Creek. The low flow channel is poorly defined and braids atop large sand sheets. No pools are evident. Given the poor cover of riparian vegetation and lack of instream roughness, the reach has limited capacity to retain fine grained sediments. Devegetated banks and erodible bank materials accelerate channel expansion and the rate of lateral channel movement. Convex banks are characterized by point bars and discontinuous pockets of floodplain.

Over time, the low flow channel becomes better defined. A contemporary example of this stage occurs along Middle Tantawangalo Creek (d). Sediment inputs and outputs eventually become roughly balanced, with point bar and point bench storage on the inside of bends and maintenance of sediment throughput on the channel bed. As the channel becomes narrower and deeper, it adopts a more sinuous course. Point bars and point benches store significant volumes of material, as do within-channel ridges which form as a result of vegetation colonization. Pools reemerge as sediment is flushed through the reach. An example of this stage occurs today along Upper Tantawangalo Creek (e). Given that geomorphic change since European settlement has been reversible, but the flux boundary conditions under which this river operates have been altered, the latter condition (e) is identified as the expected reference condition under contemporary flux boundary conditions as indicated by the box.

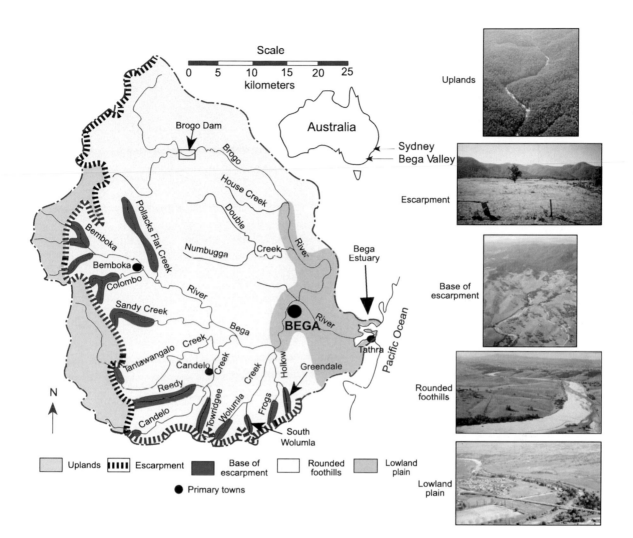

Plate 9.1 Landscape units in Bega catchment (from Fryirs, 2001)

Landscape units are areas of relatively homogeneous topography and morphology that make up a catchment. In Bega catchment, five landscape units were identified. The uplands sit above the escarpment and are characterized by a dissected plateau. The escarpment is a near vertical face that demarcates the uplands from the remainder of the catchment. The base of escarpment landscape unit extends as tongues down the dominant drainage lines. The rounded foothills comprise an undulating topography that extends to the lowland plain which is a relatively flat wide valley that feeds into the estuary.

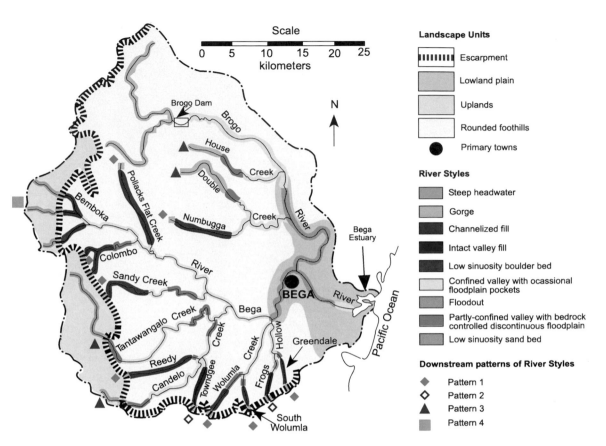

Plate 9.2 The distribution and downstream patterns of River Styles in Bega catchment
Nine River Styles have been identified in Bega catchment. There are three confined variants, one partly-confined River Style, and five laterally-unconfined River Styles. These are arranged in the catchment in four downstream patterns reflecting whether rivers drain from the uplands or the escarpment zone, and the River Style found at the base of the escarpment.

Plate 10.1 The geomorphic condition of rivers in Bega catchment (from Fryirs, 2001)
Considerable variability is evident in the geomorphic condition of rivers across Bega catchment. However, a few notable trends emerge. Intact and good condition reaches are largely confined to the uplands and gorge country. Isolated good condition reaches occur along the resilient confined and partly-confined rivers of the tributaries. Reaches in poor condition are concentrated along laterally-unconfined River Styles (i.e., the channelized fill and low sinuosity sand bed rivers) where pronounced geomorphic changes have occurred. These are the most sensitive River Styles to change. At the base of the escarpment, most reaches of the channelized fill River Style are in moderate or poor condition. This is largely due to the profound changes that have occurred following incision into intact swamps and subsequent channel expansion. Along the trunk stream and lowland plain the cumulative effects of catchment disturbance have been manifest, triggering significant geomorphic deterioration. Hence, the geomorphic condition of these reaches is moderate–poor.

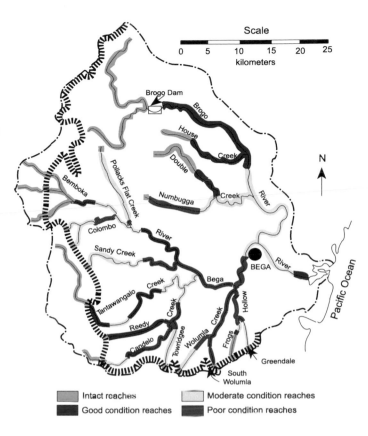

Plate 11.1 The geomorphic recovery potential of reaches in Bega catchment (from Fryirs, 2001)
Given that Bega catchment is largely exhausted of sediment, the potential for river recovery is low (see Fryirs and Brierley, 2001). However, there is significant spatial variability in recovery potential. In general, intact reaches are located in relatively inaccessible headwater sections of the catchment. High recovery potential reaches are largely found along the more resilient confined and partly-confined rivers in the middle sections of the catchment. Reaches with the lowest recovery potential are generally laterally-unconfined, are sensitive to change, and have experienced significant geomorphic change following European settlement. Many of these reaches have been irreversibly altered. Along the lowland plain, the cumulative impacts of catchment disturbance limit the potential for river recovery. It will take thousands of years to ameliorate catchment-wide off-site impacts.

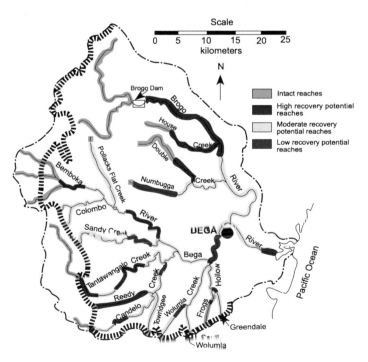

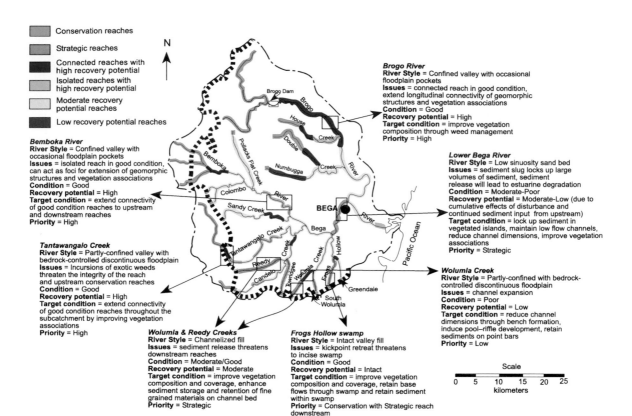

Conservation reaches

Strategic reaches

Connected reaches with high recovery potential

Isolated reaches with high recovery potential

Moderate recovery potential reaches

Low recovery potential reaches

N

Brogo River
River Style = Confined valley with occasional floodplain pockets
Issues = connected reach in good condition, extend longitudinal connectivity of geomorphic structures and vegetation associations
Condition = Good
Recovery potential = High
Target condition = improve vegetation composition through weed management
Priority = High

Bemboka River
River Style = Confined valley with occasional floodplain pockets
Issues = isolated reach in good condition, can act as foci for extension of geomorphic structures and vegetation associations
Condition = Good
Recovery potential = High
Target condition = extend connectivity of good condition reaches to upstream and downstream reaches
Priority = High

Lower Bega River
River Style = Low sinuosity sand bed
Issues = sediment slug locks up large volumes of sediment, sediment release will lead to estuarine degradation
Condition = Moderate-Poor
Recovery potential = Moderate-Low (due to cumulative effects of disturbance and continued sediment input from upstream)
Target condition = lock up sediment in vegetated islands, maintain low flow channels, reduce channel dimensions, improve vegetation associations
Priority = Strategic

Tantawangalo Creek
River Style = Partly-confined valley with bedrock-controlled discontinuous floodplain
Issues = Incursions of exotic weeds threaten the integrity of the reach and upstream conservation reaches
Condition = Good
Recovery potential = High
Target condition = extend connectivity of good condition reaches throughout the subcatchment by improving vegetation associations
Priority = High

Wolumla Creek
River Style = Partly-confined with bedrock-controlled discontinuous floodplain
Issues = channel expansion
Condition = Poor
Recovery potential = Low
Target condition = reduce channel dimensions through bench formation, induce pool-riffle development, retain sediments on point bars
Priority = Low

Wolumla & Reedy Creeks
River Style = Channelized fill
Issues = sediment release threatens downstream reaches
Condition = Moderate/Good
Recovery potential = Moderate
Target condition = improve vegetation composition and coverage, enhance sediment storage and retention of fine grained materials on channel bed
Priority = Strategic

Frogs Hollow swamp
River Style = Intact valley fill
Issues = kickpoint retreat threatens to incise swamp
Condition = Good
Recovery potential = Intact
Target condition = improve vegetation composition and coverage, retain base flows through swamp and retain sediment within swamp
Priority = Conservation with Strategic reach downstream

Scale
0 5 10 15 20 25
kilometers

Brogo Dam, Brogo, House, Creek, Double, Creek, Numbugga, Pollacks Flat Creek, Bemboka, River, Colombo, River, Sandy Creek, BEGA, Bega, River, Pacific Ocean, Tantawangalo Creek, Reedy, Candelo, Towridgee, Wolumla Creek, Frogs Hollow, Greendale, South Wolumla

Plate 12.1 Prioritization of management reaches within the Bega catchment vision (from Fryirs, 2001)
This figure shows the catchment-framed biophysical vision that is being implemented in Bega catchment. It details how this vision can be achieved by implementing the prioritization framework and concentrating river rehabilitation efforts along specific reaches that will enhance the recovery potential of rivers throughout the catchment. In Bega catchment, conservation reaches are largely concentrated in the uplands and escarpment areas within the steep headwater and gorge River Styles. In addition, remnant predisturbance conditions exist along Frogs Hollow Creek in the form of intact valley fill and floodout River Styles. Strategic reaches are located immediately downstream of these swamps. Rehabilitation strategies are being implemented to inhibit knickpoint retreat into these remnant reaches. Strategic reaches are also located along Upper Wolumla and Reedy Creeks. These channelized fill River Styles have supplied extensive volumes of sediment in the period since European settlement (around 10 million m^3). Only now are they accumulating sediments along their channel beds and showing signs of recovery. A strategic value is placed on these reaches to ensure that rehabilitation occurs to enhance sediment storage and prevent reincision and sediment release into downstream reaches. The lowland plain is also given a strategic rating. Although this reach is in poor geomorphic condition and is still experiencing the effects of the sediment slug, and the potential for sediment release into downstream reaches and the estuary is significant. Connected reaches with high recovery potential are concentrated at the base of the escarpment. Many of these reaches have the potential to recover quickly with minimal intervention because of their location downstream of intact or conservation reaches. Isolated reaches with high recovery potential occur along Tantawangalo Creek, Double Creek, and Bemboka River. These reaches are resilient to change and provide foci from which rehabilitation programs can extend into more degraded sections of the catchment. Along Tantawangalo and Double Creeks there is significant potential for the longitudinal connectivity of good condition reaches to be extended along these river courses. The remainder of the catchment is characterized by reaches that have moderate or low recovery potential. These reaches are considered low rehabilitation priorities.

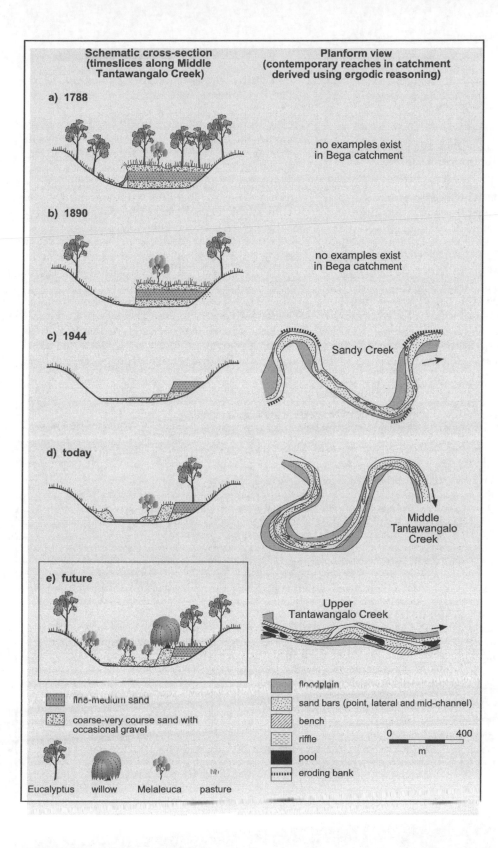

Schematic cross-section
(timeslices along Middle
Tantawangalo Creek)

Planform view
(contemporary reaches in catchment
derived using ergodic reasoning)

a) 1788

no examples exist
in Bega catchment

b) 1890

no examples exist
in Bega catchment

c) 1944

Sandy Creek

d) today

Middle
Tantawangalo
Creek

e) future

Upper
Tantawangalo Creek

fine-medium sand

coarse-very course sand with
occasional gravel

floodplain

sand bars (point, lateral and mid-channel)

bench

riffle

pool

eroding bank

0   400
          m

Eucalyptus   willow   Melaleuca   pasture

controlled discontinuous floodplain pockets, and the low sinuosity sand bed River Styles in Bega catchment are presented in Figures 10.5–10.7.

10.3.3 Has geomorphic change been reversible or irreversible?

Determination of an appropriate reference condition must reflect a realistically attainable river structure and function given the prevailing boundary conditions expressed over management timeframes of 50–100 years. This requires assessment of whether system responses to human disturbance have brought about irreversible changes to river character and behavior. The evolutionary sequence of each reach is used to identify if, how, and when irreversible geomorphic change occurred. Irreversibility is defined as a wholesale shift in the behavioral or process regime of a river that induces a shift to a new River Style. If a shift in River Style

has occurred, the assemblage of geomorphic units, channel planform, and bed character have changed to such a degree that the river operates in a fundamentally different manner to its former state. A return to the predisturbance state will not occur without significant physical intervention or manipulation. The contemporary capacity for adjustment of the River Style requires redefinition for the new River Style and a solid line is used on the evolution diagram to note the change in River Style (e.g., Figures 10.5 and 10.7). Reversible geomorphic change occurs when adjustments occur within the contemporary capacity for adjustment of the River Style under investigation (e.g., Figure 10.6).

In many cases, identification of irreversible change to river character and behavior is a straightforward exercise. For example, regulated flow or urban development result in irreversible alteration to the geomorphic structure of a river. These

---

Figure 10.7 Evolution of the low sinuosity sand-bed River Style in Bega catchment (from Fryirs, 2001)

The evolutionary development of the lowland plain river around Bega township is documented in Brooks and Brierley (1997, 2000). Portion plans dating from the 1850s and paleochannel indicators (i.e., Casuarina lined channel margins) show that the pre-European settlement lowland plain of the Bega and Brogo Rivers was characterized by a deep, narrow channel with a series of pools and riffles (a) (Brooks and Brierley, 1997, 2000). It was a mixed load system with fine grained suspended load material deposited on the floodplain in overbank events. The loading of large woody debris was likely high, and riparian vegetation was dominated by *Casuarina cunninghami* and *Lomandra* spp. The floodplain consisted of an open woodland association and the backswamps were dominated by *Melaleuca* spp. Given the relatively low channel capacity, transfer of water and organic matter to the floodplain was readily maintained. A low sinuosity fine grained River Style occurred along this lower section of the catchment.

The lower course of the Bega River expanded from around 40 m wide to 140 m wide within a few decades of European settlement (c), essentially as a consequence of the removal of riparian vegetation. Photographs from the 1890s show a wide, shallow channel with a homogeneous sand sheet that is free of vegetation. The river has been transformed from a mixed load to a bedload dominated system. Pools have been infilled, and up to 2 m of sand has accumulated on floodplains previously dominated by silt (Brooks and Brierley, 1997). An irreversible change to a low sinuosity sand bed River Style had occurred.

Detailed field investigations indicate that relatively little change to river structure occurred along the lower Bega River between 1920 and 1960 (Brooks and Brierley, 2000) (d). However, since the 1960s, willows and other forms of vegetation (native colonizers and exotics) have choked the channel. This has produced a complex pattern of channel-marginal benches, bars and islands. Channel contraction and the reworking of instream sediments (i.e., the formation of islands) have returned some structural heterogeneity to the channel (f).

Over time the geomorphic function of the lowland plain has changed. In its predisturbance state, this reach acted as a transfer zone, with limited rates of sediment flux. As the channel expanded, significant volumes of sediment were released. Subsequently, however, the lowland plain has stored large volumes of material derived from the upstream sediment slug (Fryirs and Brierley, 2001). As the tail of the sediment slug passes, it is considered likely that the lowland course of Bega River will be characterized by numerous narrow, deep channels in something akin to an anabranching pattern (g). As these channels deepen, sand sheet inundation on the floodplain will be alleviated and the habitat potential and transfer of flow, organics, and fine grained sediment to backswamps will be improved. Given that geomorphic change since European settlement has been irreversible, this latter condition (g) is considered the expected reference condition under contemporary flux boundary conditions (as indicated by the box).

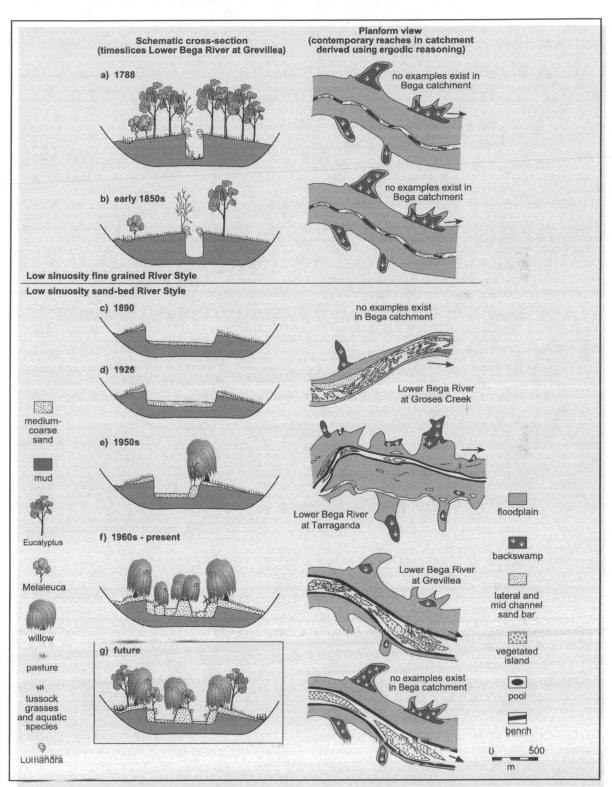

Schematic cross-section
(timeslices Lower Bega River at Grevillea)

Planform view
(contemporary reaches in catchment
derived using ergodic reasoning)

a) 1788

no examples exist in
Bega catchment

b) early 1850s

no examples exist in
Bega catchment

Low sinuosity fine grained River Style

Low sinuosity sand-bed River Style

c) 1890

no examples exist
in Bega catchment

d) 1926

Lower Bega River
at Groses Creek

e) 1950s

Lower Bega River
at Tarraganda

f) 1960s - present

Lower Bega River
at Grevillea

g) future

no examples exist
in Bega catchment

medium-
coarse
sand

mud

Eucalyptus

Melaleuca

willow

pasture

tussock
grasses
and aquatic
species

Lumandra

floodplain

backswamp

lateral and
mid channel
sand bar

vegetated
island

pool

bench

0   500
         m

changes are often accompanied by changes to the sediment and flow regimes. Equivalent changes may result from inadvertent but systematic changes to other flux boundary conditions, such as riparian vegetation structure and cover. In these cases, irreversible change is less easily defined and requires looking into the past to identify shifts in river character and behavior.

Assessing whether irreversible geomorphic change has occurred provides the basis to define an appropriate reference reach. If a reach has experienced irreversible geomorphic change, the condition of the reach must be assessed in light of the newly adopted River Style. In this instance, comparing the contemporary reach with a reference condition of the predisturbance river type is irrelevant in setting realistic management goals. If a river still operates as the River Style that existed in the predisturbance period, such that human-induced adjustments are reversible, reach condition is assessed against a reference condition of this River Style. In these cases, an expected reference condition is designed, reflecting the altered flux boundary conditions under which the river now operates.

The key to this analysis is identifying how disturbance may modify the threshold levels at which irreversible change may occur. For example, in some cases, reduced resistance following the removal of riparian vegetation lowers the threshold levels for geomorphic adjustment via incision and channel expansion. In these cases, small triggers (such as small–moderate flood events) can breach fundamental resistance thresholds, triggering significant geomorphic change. Identification of threshold conditions that trigger fundamental changes to river forms and processes represents a critical consideration for management applications. These insights guide designation of strategic reaches, aiding the identification of reaches where manipulation of river character and behavior will have the most positive impacts, enhancing river recovery potential (Chapter 12).

10.3.4 Derive desirability criteria for river character and behavior based on relevant geoindicators for each River Style

Assessments of relevant contemporary river attributes are used to identify good, moderate, and poor conditions for each River Style. This is framed in terms of the degrees of freedom that are used to assess the capacity for river adjustment for each River Style. Relevant geoindicators identified in Stage Two, Step One are measured to give a reliable and relevant signal about the condition of a reach (cf., Elliott, 1996; Osterkamp and Schumm, 1996). The range of selected geoindicators is River Style specific. Although this procedure entails elements of subjectivity, it is based on fundamental geomorphic principles and process-based understanding of how each River Style works. A table is constructed that includes questions about the "desirability" of each relevant geoindicator for each River Style. One question is asked for each relevant geoindicator (Tables 10.6–10.8).

10.3.5 Identify and select a reference reach for each River Style

Any assessment of river condition must be framed relative to some benchmark or reference reach, thereby providing a determination of the extent to which human-induced changes to river character and behavior fall outside the long-term pattern. In an Australian context, two options seem appropriate:

1 an intact, pristine condition. For example, putting aside the impacts of Aboriginal practices (e.g., use of fire), the initiation of colonization in 1788 provides a suitable reference point. If patterns and rates of river geomorphic change fall outside their "natural" capacity for adjustment, a reach is considered to have deviated from its "intact" condition;

2 an assessment of the "best" condition that can be attained by a river that has been altered by human disturbance, given the prevailing catchment boundary conditions. As noted previously, this "expected" reference condition must separate reaches that have been subjected to reversible and irreversible changes in response to human disturbance.

Although the first option may be preferred from a preservationist perspective, the second option provides a far more practical and realistic perspective with which to appraise geomorphic river condition (cf., Cairns, 1989, 1991; Brookes, 1995; Gore and Shields, 1995; Sparks, 1995). Given the extent of human disturbance to river systems, whether directly induced or an indirect response, comparison with an intact condition may seem little more

Table 10.6 Measures used to assess good condition reaches of the channelized fill River Style in the laterally-unconfined valley setting in Bega catchment. Reproduced from Fryirs (2003) with permission from Elsevier, 2003.

Degrees of freedom and relevant geoindicators	Questions to ask for each reach of the River Style	Questions that must be answered YES
Channel attributes		
• Size	• Is channel size appropriate given the catchment area, the prevailing sediment regime, and the vegetation character? (i.e., is the channel overwidened, overdeepened, or does it have an appropriate width : depth ratio?)	3 out of 4
• Shape	• Is channel shape appropriate along the reach? (i.e., does the channel have a compound shape, with inset surfaces within a symmetrical trench?)	
• Bank morphology	• Are banks eroding in the right places and at the right rate? (signs of deterioration include vertical or undercut banks along the reach)	
• Instream vegetation structure	• Is the instream vegetation structure appropriate? (i.e., is aquatic vegetation colonizing the bed of the incised channel?)	
Channel planform		
• Lateral channel stability	• Is the lateral stability of the channel appropriate given the texture and slope of the reach? (signs of deterioration include channel expansion and low flow channel reworking of bed materials)	2 out of 3
• Assemblage of geomorphic units	• Is the assemblage, pattern, and condition of instream and floodplain geomorphic units appropriate? Are key units present? (i.e., does the reach have a series of insets and a swampy channel bed with no signs of reworking such as dissection, stripping, or undercutting?)	
• Riparian vegetation	• Is the continuity and composition of riparian vegetation near-natural with few exotics?	
Bed character		
• Grain size and sorting	• Is the grain size, sorting, and organization of materials in different geomorphic units appropriate? (i.e., are sands stored in insets, and mud and organic matter stored on the channel bed?)	3 out of 4
• Bed stability	• Is bed stability appropriate? (signs of bed instability or disturbance will include incision into sands or to bedrock)	
• Hydraulic diversity	• Is the sediment storage/transport function of the reach appropriate? (i.e., is it acting as a sediment accumulation zone?)	
• Sediment regime	• Are roughness characteristics and the pattern of hydraulic diversity along the reach appropriate? (i.e., does the reach have a swampy channel bed with a series of inset bench features?)	

than an academic exercise. This is NOT to say that insights from near-intact reaches do not provide fundamental guidelines to the "natural" structure and function of rivers, and associated implications for geodiversity, biodiversity, conservation, aquatic ecosystem functioning, etc. Unfortunately, however, relatively few remnants remain, and these typically form an unattainable target condition for river management practices.

The objective in identifying a reference reach is to determine a morphological configuration that is compatible with the prevailing flux boundary conditions. Viewed in this way, channel attributes, channel planform (including the assemblage

of geomorphic units), and bed character must be appropriate for the River Style under investigation (cf., Hughes et al., 1986; Brookes and Shields, 1996; Rhoads and Herrick, 1996). The approach used to identify a reference condition in the River Styles framework is framed in terms of whether irreversible geomorphic change has occurred following human disturbance. This is determined by the sensitivity of the River Style to change (i.e., its capacity to adjust), and how it responds to alterations in flux boundary conditions (i.e., flow and sediment transfer, and vegetation associations)

The nature and extent of human disturbance are unlikely to be uniform across any particular catch-

Table 10.7 Measures used to assess good condition reaches of the partly-confined valley with bedrock-controlled discontinuous floodplain River Style in Bega catchment (from Fryirs, 2001).

Degrees of freedom and relevant geoindicators	Questions to ask for each reach of the River Style	Questions that must be answered YES
Channel attributes		
• Size	• Is channel size appropriate given the catchment area, the prevailing sediment regime, and the vegetation character? Is the channel functionally connected to floodplain pockets? (i.e., is the channel overwidened, overdeepened, or does it have an appropriate width : depth ratio?)	4 out of 5
• Shape	• Is channel shape appropriate along the reach? (i.e., is it symmetrical or compound at inflection points and asymmetrical at bends?)	
• Bank morphology	• Are banks eroding in the right places and at the right rate? (concave bank erosion is a natural process along these rivers)	
• Instream vegetation structure	• Is there woody debris in the channel and/or potential for woody debris recruitment? (these reaches often have wood induced pools on bends)	
• Woody debris loading	• Is the instream vegetation structure appropriate?	
Channel planform		
• Lateral channel stability	• Is the channel positioned correctly on the valley floor and the lateral stability of the channel appropriate given the texture and slope of the reach? (signs of instability include channel expansion and accelerated rates of concave bank erosion)	2 out of 3
• Assemblage of geomorphic units	• Is the assemblage, pattern and condition of instream and floodplain geomorphic units appropriate? Are key units present? (i.e., are there instream point benches, point bars, pools, riffles and island complexes, and floodplain floodrunners with no signs of deterioration such as floodplain stripping and/ or sand sheet deposition?)	
• Riparian vegetation	• Is the continuity and composition of the riparian vegetation near-natural, with few exotics?	
Bed character		
• Grain size and sorting	• Is the grain size, sorting, and organization of materials in different geomorphic units appropriate? (i.e., is there a mix of sands in point bars and islands, occasional gravels, and organics in the pools and fine sands on the floodplain?)	3 ou of 4
• Bed stability	• Is bed stability appropriate? (signs of bed instability or disturbance will include incision along inflection points between bends and formation of an homogeneous bed morphology)	
• Hydraulic diversity	• Are roughness characteristics and the pattern of hydraulic diversity along the reach appropriate?	
• Sediment regime	• Is the sediment storage/transport function of the reach appropriate for the catchment position? (i.e., is it a sediment transfer zone where inputs and outputs of sediment are balanced over time via movement of material from point bar to point bar?)	

ment. Hence, a range of altered conditions is likely to be evident. Four types of reference condition can be differentiated:

1 *Remnant reaches that have been minimally disturbed by humans, such that geomorphic changes to river character and behavior remain re-* *versible*. These reaches represent "natural" reference conditions and are considered to be "intact". No further condition assessments are carried out for these river reaches.

2 *Reaches where human disturbance has occurred, but geomorphic changes to river character*

Table 10.8 Measures used to assess good condition reaches of the low sinuosity sand-bed River Style in the laterally-confined valley setting in Bega catchment (from Fryirs, 2001).

Degrees of freedom and relevant geoindicators	Questions to ask for each reach of the River Style	Questions that must be answered YES
Channel attributes		
• Size	• Is channel size appropriate given the catchment area, the prevailing sediment regime, and the vegetation character such that the channel and flood plain are functionally connected? (i.e., is the channel overwidened, overdeepened, or does it have an appropriate width : depth ratio?)	4 out of 5
• Shape	• Is channel shape appropriate along the reach? (i.e., compound or irregular)	
• Bank morphology	• Are banks eroding in the right places and at the right rate?	
• Instream vegetation structure	• Is there woody debris in the channel and/or potential for woody debris recruitment? (wood induced pools may be evident)	
• Woody debris loading	• Is the instream vegetation structure appropriate?	
Channel planform		
• Number of channels	• Does the reach have the right number of channels positioned correctly on the valley floor? (signs of change may include the formation of avulsive networks or floodchannels on the floodplain)	4 out of 5
• Sinuosity of channels	• Is the sinuosity of the channel appropriate given the texture and slope of the reach?	
• Lateral channel stability	• Is the lateral stability of the channel appropriate given the texture and slope of the reach? (signs of degradation include channel expansion and low flow channel reworking of bed materials)	
• Assemblage of geomorphic units	• Is the assemblage, pattern and condition of instream and floodplain geomorphic units appropriate? Are key units present (i.e., does the reach have islands, benches, levees, backswamps?)	
• Riparian vegetation	• Is the continuity and composition of the riparian zone near-natural, with few exotics?	
Bed character		
• Grain size and sorting	• Is the grain size, sorting, and organization of materials in different geomorphic units appropriate? (i.e., are there sands and organics in the channel, fines on the floodplain and mud in backswamps?)	3 out of 4
• Bed stability	• Is bed stability appropriate? (signs of bed instability or disturbance may include incision into sand bed materials)	
• Hydraulic diversity	• Are roughness characteristics and the pattern of hydraulic diversity along the reach appropriate?	
• Sediment regime	• Is the sediment storage/transport function of the reach appropriate for the catchment position? (i.e., is it a sediment transfer or accumulation zone?)	

and behavior remain reversible. This may reflect either inbuilt resilience for the River Style under consideration, or relatively low levels of disturbance. The capacity for adjustment has been broadened for the River Style. This scenario is characteristic of rivers where wholesale changes in character and behavior are unlikely, but the river retains its core structure and function. Many examples are found in partly-confined valley set-

tings. In these cases, the river still operates as the River Style that existed in the predisturbance period and human-induced adjustments are reversible. Therefore, reach condition is assessed against an expected reference condition of this River Style, reflecting the altered flux boundary conditions under which the river now operates. For example, a "natural" predisturbance river may have been characterized by a 10 m wide channel, but the "expected" postdisturbance river operates under altered sediment and flow regimes, such that it is expected to have a channel width of 20 m. Hence, the condition of contemporary reaches of that River Style must be assessed relative to the 20 m wide expected reference condition.

3 *Reaches where change has been induced by indirect human disturbance and irreversible change has resulted.* In this case, the contemporary capacity for adjustment must be reframed for the new River Style. Condition assessments are made relative to an "expected" reference condition for the new River Style. This scenario is characteristic of adjustment along highly sensitive rivers (e.g., lowland alluvial rivers).

4 *Reaches where change has been induced by direct human disturbance and irreversible geomorphic change has resulted.* In this case, condition assessments must be made relative to an "expected" reference condition for the new River Style. This scenario is characteristic of adjustments along regulated, urban, and channelized systems. In these cases, different sets of analyses must be undertaken, as the condition assessment outlined in this book is inappropriate. Additional layers of information are required, reflecting variants of human-imposed river condition.

A decision-making tree is applied to identify which reference condition should be used for any given situation (Figure 10.8). This provides a generic tool to assess river responses to disturbance. Reference conditions against which to assess geomorphic river condition are defined for each River Style unless they are intact (reference condition 1) or are highly modified rivers (reference condition 4), in which case the condition assessment is not performed.

Reference conditions are identified for each River Style in the catchment. Options include:

1 use available remnant reaches;

2 where remnant reaches do not exist, identify a reference reach using the evolutionary sequence and ergodic reasoning;

3 derive an "expected" structure and function based on analyses of river character and behavior performed in Stage One of the River Styles framework.

A reference reach of a River Style is identified by asking the question: What would be the expected character and behavior of the river given its posi-

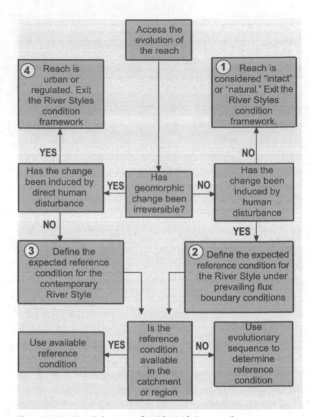

Figure 10.8 **Decision tree for identifying a reference condition. Modified from Fryirs (2003). Reprinted with permission from Elsevier, 2003**

tion in the catchment, and the prevailing water, sediment, and vegetation regimes? The parameters measured are River Style specific and are the same as those used to assess the capacity for adjustment. In general, for a reach to be used as a reference condition, most of the geoindicators used to assess the capacity for adjustment must operate "appropriately" for that River Style (i.e., the reach must satisfy all the requirements outlined in the "desirability" table for the River Style under investigation; Figures 10.5–10.7).

10.4 **Stage Two, Step Three: Interpret and explain the geomorphic condition of the reach**

Assessments of geomorphic river condition, in the River Styles framework, are based on a solid understanding of river character, behavior, and evolution throughout a catchment. Each reach is placed

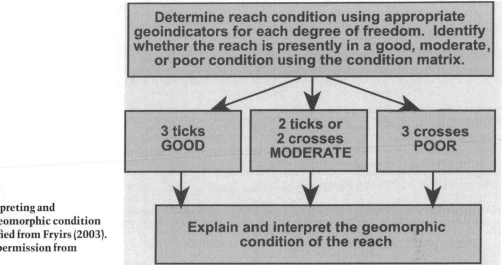

Determine reach condition using appropriate geoindicators for each degree of freedom. Identify whether the reach is presently in a good, moderate, or poor condition using the condition matrix.

3 ticks
GOOD

2 ticks or
2 crosses
MODERATE

3 crosses
POOR

Explain and interpret the geomorphic condition of the reach

Figure 10.9 Interpreting and explaining the geomorphic condition of a reach. Modified from Fryirs (2003). Reprinted with permission from Elsevier, 2003

in context of its evolutionary sequence, identifying whether irreversible geomorphic change has occurred. River condition is framed relative to a defined reference condition to determine how far from "desirable" the geomorphic structure and function of differing reaches of river are. Procedures and decision-making processes used to assess the geomorphic condition of reaches are summarized in Figure 10.9.

10.4.1 Determine reach condition using the degrees of freedom and good, moderate, and poor condition matrix

A matrix is developed to assess the condition of a reach, whereby the practitioner assigns ticks or crosses in answer to the table of "desirability questions" constructed for each River Style. Procedures used to measure each geoindicator are presented in Table 10.4. Each relevant geoindicator must be measured and interpreted to gain a full appreciation of the condition of the reach and to answer "desirability questions" appropriately. Once fieldwork is completed, the results are synthesized across the entire reach, comparing the observed conditions with the reference condition. Depending on the number of Yes/No responses to questions asked about the "desirability" of reach character and behavior, a tick is assigned for each degree of freedom. For example, revisiting Table

Table 10.9 Determining the geomorphic condition of a reach of a River Style. Reproduced from Fryirs (2003) with permission from Elsevier, 2003.

Geomorphic river condition	Channel attributes	Channel planform	Bed character
Good	✓	✓	✓
Moderate	✓	✗	✓
Moderate	✗	✓	✗
Poor	✗	✗	✗

10.6, if a reach of the channelized fill River Style receives 3 out of 4 Yes responses regarding channel attributes, that degree of freedom is assigned a tick. In general, however, for a tick to be assigned for a degree of freedom, most geoindicators must be considered to be "desirable." In many instances, the geoindicators measured within each degree of freedom are interrelated. For example, shape, size, and bank morphology are dynamically adjusted such that a change in one leads to adjustments in the others. The overall result is a measure of the condition for that degree of freedom. This assessment provides an indication of how far from the reference condition the geomorphic structure and function of different reaches of each River Style are. The further a reach sits from the reference condition for a certain parameter, the poorer its condition. The observed structure and function of each

Table 10.10 Explaining the geomorphic condition of reaches of the channelized fill River Style in Bega catchment. Reproduced from Fryirs (2003) with permission from Elsevier, 2003.

Degree of freedom	Good condition	Moderate condition	Poor condition
Channel attributes	Compound, stepped cross-sectional form within a wide, deep incised trench. Bank erosion minimal. Channel bed dominated by aquatic swamp vegetation and tussock grasses. Benches colonized by some hardy vegetation including Melaleucas.	Compound, stepped cross-sectional form within a wide, deep, incised trench. Localized bank erosion and slumping occurs. Bench units colonized by some hardy vegetation. Occasional tussock grasses on sand bars and along the low flow channel.	Large, symmetrical, overwidened incised trench. Near vertical, exposed banks with significant erosion along the entire reach. Within-channel units are unvegetated.
Channel planform	No lateral adjustment of incised trench. Swamps and a poorly-defined or discontinuous channel characterizes the channel bed. Multiple benches line the channel margin. Increased within-channel sedimentation may promote reconnection of channel and floodplain processes. Scattered riparian strip. Valley fill surfaces are dominated by pasture.	Limited planform adjustment of incised trench. Bench features occur along channel banks. Well-defined low flow channels shift over trench floor reworking sand sheets and bars. Floodplain perched above low flow channel, disconnecting channel from floodplain processes. Little or no riparian strip. Valley fill surfaces are dominated by pasture.	Incised trench experiencing accelerated rates of lateral expansion and bed lowering (incision). Multiple low flow stringers atop sand sheet produce an array of midchannel and lateral bars. Floodplain disconnected from the channel given the incised nature of the fill. No riparian strip. Valley fill surfaces are dominated by pasture.
Bed character	Segregated sediment mix, with sands in benches, and mud and organic matter accumulation on the trench floor. Bed stable and aggrading in sediment accumulation reaches.	Moderately segregated sediment mix, with coarse sands in benches, and finer sand in the low flow channel. Low flow channel redistributes and reorganizes sediment locally within the incised trench, improving bed material organization. Moderate bed stability as trench infills. Sand accumulating on the channel bed. Acting as a sediment accumulation or transfer zone.	Bedload dominated with limited capacity to retain finer grained materials. Still releasing sediment from valley fill. Sediment on the channel bed is loose, poorly segregated, and poorly sorted. Poor bed stability. Bed may still be incising. High rates of material reworking and sediment transport. Acting as a sediment source zone.
Photograph	Reedy Creek	Wolumla Creek	Anderson Creek (Wolumla tributary)

Table 10.11 Explaining the geomorphic condition of reaches of the partly-confined valley with bedrock-controlled discontinuous floodplain River Style in Bega catchment (from Fryirs, 2001).

Degree of freedom	Good condition	Moderate condition	Poor condition
Channel attributes	Channel has a relatively low width : depth ratio, with significant local variability induced by bedrock outcrops, vegetation, and woody debris. Natural or low rate of erosion of concave banks. Cross-sectional form is asymmetrical on bends and irregular at inflection points. Islands and bars are vegetated with hardy shrubs and aquatic grasses.	Channel has a high width : depth ratio. Asymmetrical-compound geometry at bend apices; symmetrical at inflection points. Localized erosion of concave banks. The channel has expanded along the reach, including at inflection points. Point bars and benches remain largely unvegetated or dominated by exotics. No woody debris.	Channel has a high width : depth ratio. The channel is overwidened along the reach, including at inflection points. Asymmetrical shape at bend apices, symmetrical shape at inflection points. Accelerated rates of concave bank erosion and channel expansion along the entire reach. No within-channel vegetation or woody debris.
Channel planform	Low sinuosity, single channel within a sinuous valley. Moderate lateral stability. Occasional bedrock outcrops, division of flow around islands and bank-attached bars and sand sheets. Bar–island–riffle complexes are separated by pools. Occasional bank attached bars and sand sheets. Point bars and point benches occur on bends. Discontinuous pockets of floodplain may be scoured around large trees and shrubs. Continuous or scattered riparian corridor consists mainly of natives. Banks are lined with lomandra. Point bars are colonized by tussock. Some hardy shrubs on benches.	Low sinuosity, single channel within a sinuous valley. Laterally unstable on concave banks. Point benches, point bars, concave benches, localized sand sheets, well-defined low flow channel. Occasional bedrock outcrops. Discontinuous pockets of floodplain either scoured or stripped. Poor riparian vegetation cover. Floodplain dominated by pasture.	Low sinuosity, single channel within a sinuous valley. Laterally unstable at concave banks and inflection points reflecting channel expansion. Point bars, sand sheets with localized bedrock outcrops, multistringed low flow channel. Discontinuous pockets of floodplain either stripped, characterized by short cutting floodchannels or extensive sand sheets. No riparian vegetation. Floodplain dominated by pasture.

Table 10.11 *Continued.*

Degree of freedom	Good condition	Moderate condition	Poor condition
Bed character	Well-segregated bedload, with discrete pockets of material of different texture. Some gravel deposits in riffles. Bars and benches comprise sands, with fine-grained floodplain. Organic matter accumulation is high in pools and on the floodplain. Bed is stable with no signs of incision or aggradation. High instream roughness (vegetation and woody debris) promote localized deposition of fine-grained materials and organics. Balance maintained between sediment input and output along the reach. Acts as a sediment transfer zone.	Poorly sorted material distribution. Sand sheets present local homogeneity, reducing roughness and the range of hydraulic diversity. Bed is unstable with significant material movement. A balance is maintained between sediment input and output along the reach. Acts as a sediment transfer zone.	Near homogeneous sand sheet. Sediment stored as homogeneous instream and with a poorly sorted material distribution. Bed is unstable with significant aggradation and/or incision occurring. Hydraulically homogeneous. Reach is sediment transport limited and acts as an accumulation zone.
Photograph	Upper Tantawangalo Creek	Candelo Creek	Wolumla Creek

Table 10.12 Explaining the geomorphic condition of reaches of the low sinuosity sand bed River Style in Bega catchment (from Fryirs, 2001).

Degree of freedom	Good condition	Moderate condition	Poor condition
Channel attributes	Channel has a relatively low width : depth ratio with an irregular shape induced by woody debris and riparian vegetation. *Lomandra* lined channel and numerous aquatic species.	Channel has a compound shape with a relatively high width : depth ratio (including each low flow thread). Within-channel ridges colonized by native shrubs and grasses. Low flow channels clear of vegetation. No woody debris.	Channel has a symmetrical shape and a high width : depth ratio. Channel may be overwidened and have a relatively homogeneous structure. No instream vegetation exists.
Channel planform	Single thread, moderately sinuous aligned down the centre of the valley. Laterally stable channel. Continuous floodplains with backswamps and levees. Localized within-channel bars, pools and riffles, the distribution of which is controlled by woody debris. High channel–floodplain connectivity maintains backswamps. Occasional drapes of sand around vegetation on the floodplain. Densely vegetated riparian zone dominated by *Casuarina cunninghamiana* with open vegetation on the floodplain. Backswamps contain tussock and are lined with Melaleucas.	Anabranching network of ridges and low flow channels. Channel expansion and lateral instability has occurred. Potential for avulsion into floodplain floodchannels. Sand sheets characterize the floodplain and backswamps. Active channel–floodplain processes maintained. Floodchannels are active in high flows. Continuous floodplain, consisting of levees and backswamps. Complex within-channel and channel-marginal assemblage of units, comprising benches, sand sheets, midchannel bars, ridges, etc. Pools are infilled. Good riparian vegetation coverage, but consists mainly of exotic vegetation. Backswamps have aquatic vegetation associations. Floodplain dominated by pasture.	Low sinuosity, single channel with accelerated rates of channel expansion. Multiple low flow channels flow atop sand sheets. Potential for avulsion into floodchannels. Continuous floodplain, consisting of levees, backswamps, and extensive sand sheets. Lateral bars, sand sheets, shallow runs, and midchannel bars. Pools have been infilled. Channel–floodplain connectivity is high, but out-of-balance with extensive sand sheets over the entire floodplain. Backswamps are infilling with bedload materials. Little or no riparian vegetation cover. Floodplain dominated by pasture.

Table 10.12 *Continued.*

Degree of freedom	Good condition	Moderate condition	Poor condition
Bed character	Well-segregated material distribution. Mud and organic matter accumulate in backswamps. Levees and floodplains comprise sand and organic accumulations. Mix of sand and mud instream, with efficient trapping by bankside vegetation and woody debris. Bed is stable and hydraulic diversity is high given the vegetation and woody debris loading. Acts as a sediment transfer zone.	Sand-dominated, with occasional deposition of finer materials around vegetation on midchannel ridges and on the floodplain. Bed is stable, but sediment reorganization is occurring as flow redistributes sediment into well-defined islands and ridges. A series of well-defined low flow channels are formed. Hydraulic diversity is limited. Acts as a sediment accumulation zone.	Bed materials dominated by sands forming a planar homogeneous channel bed. Local sediment redistribution as multiple low flow stringers shift over the sand sheets. Bed stability is low given high rates of sediment accumulation. Channel is sediment transport limited.
Photograph	No examples exist in Bega catchment	 Lower Bega River @ Grevillea	 Lower Bega River @ Wolumla Creek

reach places it in a good, moderate, or poor geomorphic condition (Table 10.1).

Three ticks for a reach places it in a good geomorphic condition category (Table 10.9). Reaches in good geomorphic condition are defined as those in which river character and behavior are appropriate for the River Style in that valley setting and that position in the catchment. Geomorphic structures are in the right place and operating as expected for that River Style. This reach should cross-compare closely with the reference condition. Along some of these reaches, the geomorphic condition may be near-intact.

Three crosses places a reach in the poor condition category (Table 10.9). Poor condition reaches are defined as those in which river character is divergent from the reference condition and abnormal or accelerated geomorphic behavior/change occurs. The breaching of a fundamental threshold could push the reach into a new River Style causing irreversible geomorphic change. Key geomorphic units are located inappropriately along the reach, and processes are out-of-balance with the geomorphic structure of the reach. The geomorphic structure of that reach would be given a poor condition rating compared to the reference reach.

Moderate condition reaches sit between these two extremes (with either two crosses, or two ticks for any of the degrees of freedom; Table 10.9). Certain characteristics of the reach are out-of-balance or inappropriate for that River Style.

10.4.2 Interpret and explain the geomorphic condition of the reach

Once the condition of each reach has been determined, a table is constructed for good, moderate, and poor condition reaches of each River Style in the catchment that details and explains how each degree of freedom has adjusted. When tied to the evolutionary sequences, this allows the causes rather than the symptoms of change to be identi-

fied. Analysis of geomorphic responses to altered flux boundary conditions and the associated sequence of events that result in changes to geomorphic condition are assessed. These tables provide a template for repeat surveys, where the geomorphology of the reach can be monitored to determine if improvement has taken place. They also outline the geomorphic parameters that require manipulation to improve the condition of the reach (Tables 10.10–10.12). When completed, a catchment-wide map showing the condition of each reach of each River Style is constructed (Plate 10.1).

10.5 Products of Stage Two of the River Styles framework

A range of products is produced from Stage Two of the River Styles framework:
- capacity for adjustment table for each River Style in the catchment;
- table of relevant geoindicators used to determine reference conditions and good, moderate, and poor condition reaches for each River Style in the catchment;
- planform and cross-section evolutionary sequences for each River Style noting where irreversible change has occurred;
- tables of "desirability criteria" used to assess the geomorphic condition of each reach of each River Style, framed in terms of relevant geoindicators for each degree of freedom;
- tables noting the "ticks and crosses" for each reach of each River Style in the catchment, framed in terms of the three degrees of freedom;
- tables that outline how each degree of freedom has adjusted along good, moderate, and poor condition reaches of each River Style. Photographs should accompany these tables;
- a catchment-wide map showing the distribution of good, moderate, and poor condition reaches.

CHAPTER 11

Stage Three of the River Styles framework: Prediction of likely future river condition based on analysis of recovery potential

Restoration is an acid test of our ecological understanding because if we do not understand the processes at work in an ecosystem we are unlikely to be able to reconstruct it so that it works.

A.D. Bradshaw, 1996, p. 7

11.1 Introduction

Effective management strategies that "work with nature" must appreciate of the trajectory of change of the river. For example, if the river was left alone, would its condition deteriorate or improve? Notions of geomorphic river recovery encapsulate a sense of how a river has adjusted from its "natural" condition following human disturbance, and what that river is adjusting towards. While changes to river morphology must be considered to be irreversible (in practical terms) in many river systems, some rivers have proven to be remarkably resilient to change, while others have started on a pathway towards recovery. Recovery is a natural process that reflects the self-healing capacity of river systems. In this book, *geomorphic river recovery is defined as the trajectory of change towards an improved condition*. Assessing the pathway of geomorphic river recovery is a predictive process.

Recovery rarely reflects an orderly, progressive, and systematic process. Nonlinear dynamics and threshold-induced responses present considerable challenges in determining the likelihood that any particular trajectory will be followed, and when that will occur. Different components of a system adjust at different rates, such that different reaches undergo transitions between different states at different times. Multiple potential trajectories occur for any given river type dependent on the condition of each reach and its likely responses to future disturbance events, along with prevailing, system-

specific driving factors and timelags. In the River Styles framework, three trajectories of change are determined, namely: degradation, restoration, and creation (Figure 11.1). On these trajectories, five different states of adjustment can be differentiated, namely: intact, degraded, turning point, restored, and created conditions (Table 11.1).

As each River Style operates under a specific set of boundary conditions and has a distinctive character and behavior, natural recovery processes can be identified for each River Style. When framed in terms of its evolution and capacity for adjustment, this enables prediction of likely future adjustments for each River Style. Assessments of underlying causes and mechanisms of change must build on solid understanding of river character, behavior, and downstream patterns, along with appraisal of geomorphic condition, as outlined in Stages One and Two of the River Styles framework.

The route by which a reach has attained its present geomorphic condition has a significant impact on its future pathway of recovery. Understanding of past geomorphic change provides a means to explain the timing, rate, and magnitude of change. In the River Styles framework, reach-based evolutionary timeslices that outline changes to river character and behavior are analyzed to determine the stage of adjustment of each reach of each River Style and predict the likely trajectory of future change. Assessment of geomorphic river recovery entails two steps, focusing on the trajectory of change and analysis of recovery potential (Figure

Table 11.1 States of adjustment used to describe river recovery.

Term	Definition
River recovery	Trajectory of change towards an improved condition. River recovery is not simply the reverse of river degradation.
Trajectory of change	The pathway along which a reach adjusts following disturbance. Three trajectories are identified in the River Styles framework, degradation, restoration, and creation.
Recovery potential	The capacity for improvement in the geomorphic condition of the reach over the next 50–100 years. Restoration (return to predisturbance condition) or creation (development of a new condition) can occur. Determining river recovery potential requires an understanding of the linkages of geomorphic processes, off-site impacts, and limiting factors within a catchment.
Intact	A river that has operated in the absence of human disturbance such that the geomorphic characteristics and behavioral attributes are consistent with the predisturbance state. Reaches are often sufficiently robust to "bounce back" to their intact condition following disturbances.
Degraded	A reach that has moved away significantly from its intact condition, and has not commenced along a recovery pathway. The river continues to adjust to disturbance, and form–process associations are out of balance.
Turning point	A transitional stage, used to describe a bifurcation in the reach's evolution that marks a transition from degradation to recovery. Future river adjustments may push the river in one of three directions: on the continuing path of degradation, onto the restoration pathway, or onto the creation pathway. In many instances, these reaches show initial signs of recovery.
Restoration	Reversible geomorphic change has occurred and recovery towards a predisturbance state follows disturbance. Ultimately, these reaches have the potential to regain a near-intact condition.
Creation	Recovery towards a new alternative state that did not previously exist at the site. The character and behavior of these reaches does not equate to a predisturbance state. Rather, the river is adjusting towards the best attainable state given the prevailing flux boundary conditions. All rivers that have experienced irreversible geomorphic change are recovering along a creation pathway.

11.2). These analyses are a system-specific exercise based on findings from Stages 1 and 2 of the River Styles framework.

A step-by-step guide with which to derive the system-specific knowledge that provides a future focus and predictive capability for river management activities is outlined below.

• Describe and explain the character and behavior of each reach in the catchment, highlighting the potential for each reach to adjust its form and change in response to disturbance events and off-site impacts. Interpret controls on why the reach looks and behaves the way it does, based on an understanding of the boundary conditions under which each reach operates, and interpret the direction and rate of change should any of the boundary conditions be modified.

• Appraise river evolution to determine the history, pathway, and rate of adjustment of each reach. Detailed evolutionary frameworks enable responses to human disturbance to be framed in light of the longer-term pattern and rate of changes that reflect the natural capacity for adjustment for

that type of river. If possible, threshold conditions under which river change occurred should be isolated. From this, the trajectory of change is appraised, predicting whether the reach will continue to deteriorate, operate as a different type of river under a modified set of boundary conditions, or readjust back towards its "intact" state.

• Link all reaches within a catchment and evaluate off-site impacts of disturbance. Assessment of river recovery potential interprets whether a reach will recover along a restoration trajectory or a creation trajectory and the relative timeframe over which this will occur. This is framed in terms of limiting factors that may inhibit river recovery, and catchment specific pressures that will impact on the future state of the system. An assessment must be made of how imposed pressures and limiting factors affect different types of rivers at different positions in a catchment, and how these various responses are linked and interact. This is completed by examination of the patterns and rates of physical fluxes (sediment, flow and vegetation associations), how they have changed over time, and how they are likely to adjust in the future. This must build on an understanding of the linkages of geomorphic processes and past/future changes to their connectivity and coupling that modify the pattern and rate of propagation of disturbance responses through the system. Barriers and buffers that bring about lagged responses must

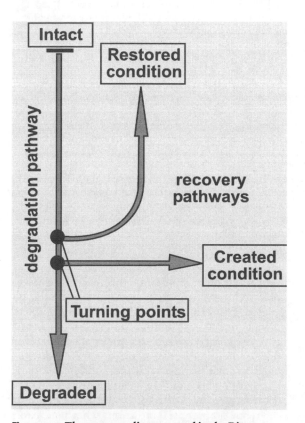

Figure 11.1 The recovery diagram used in the River Styles framework showing the degradation scale on the left and the two recovery pathways on the right. Reprinted from Fryirs and Brierley (2000), with permission from © V.H. Winston & Son, Inc., All rights reserved

be interpreted. Geomorphic responses to any given sequence of natural disturbance events or imposed pressures are likely to be highly variable, as the effectiveness of different geomorphic processes is fashioned by the specific condition of the river at the time of the event(s). The spatial pattern of river types in a catchment, along with inherent variability in their sensitivity to disturbance and the site/reach/catchment specific nature of disturbance events, allow patterns and rates of river change to be identified and potential off-site impacts to be isolated.

• Integration of these various forms of information is used to determine likely future trajectories of change. Predictions of future responses to disturbance events should cover a spectrum of future scenarios based on whether boundary conditions remain the same or are altered by limiting factors and pressures.

The inherent diversity, complexity, and uncertainty of natural systems, and human modifications to these systems, ensure that predictions of future states represent, at best, an approximation of reality. The precautionary principle must always be applied. Measures of risk or likelihood that particular future states will be attained should be provided. In making these "best-guess" predictions, realistic appraisal of contemporary forms, processes, and condition provides a critical starting place. From this, determinations of various scenarios for environmental futures can be assessed using foresighting exercises, including appraisals of "what it won't be like." The complexity and differing degree of connectivity of natural systems ensure that geomorphic responses to interference or modification are highly variable. In some instances, perturbations result in progressive and systematic responses. Elsewhere, complex and/or chaotic responses may be noted. This makes extrapolation exceedingly difficult. Hence, the only reasonable outcome is "Know Your Catchment."

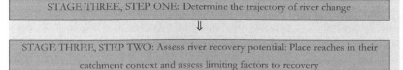

STAGE THREE, STEP ONE: Determine the trajectory of river change

⇓

STAGE THREE, STEP TWO: Assess river recovery potential: Place reaches in their catchment context and assess limiting factors to recovery

Figure 11.2 Steps in Stage Three of the River Styles framework

Table 11.2 Definition of terms used to describe river recovery potential.

Term	Definition
High recovery potential	Reach is in good geomorphic condition and is located in a position where the potential for deleterious impacts is minimal. These reaches are commonly found in upstream parts of catchments.
Moderate recovery potential	Reach is either resilient to change but in moderate of poor geomorphic condition, or is in good condition, but sits downstream of a poor condition reach. The potential for off-site impacts and limiting factors propagating into the reach is high.
Low recovery potential	Reach is in poor geomorphic condition, is sensitive to change or sits at a position in the catchment where pressures and limiting factors are likely to have negative off-site impacts that will impact directly on the future condition of the reach. Often these reaches sit in the most downstream sections of the catchment, where the cumulative effects of disturbance are manifest.

In the River Styles framework, the direction of adjustment (i.e., the trajectory) is indicated by determining the most likely future recovery condition. However, it is far more difficult to provide a quantitative measure of the rate of river recovery. Hence, river recovery is framed in relative terms, in which reaches with low recovery potential will take considerably longer to recover than those with high recovery potential (Table 11.2). Quantification of rates of change is a site- or reach-specific endeavour, as reaches of a given River Style may be at different phases of adjustment to differing forms of disturbance. In addition, specific geomorphic properties are unlikely to be uniform from system to system, as local-scale factors induce differing patterns and rates of adjustment.

The first step in assessment of geomorphic river recovery is to identify stages of adjustment of each reach of each River Style. Each reach is placed on pathways of degradation or recovery and predictions are made about the direction of change. The geomorphic recovery potential of each reach is determined by assessing the connectivity of reaches and interpreting limiting factors and pressures. In Stage Three of the River Styles framework, all reaches of all River Styles in the catchment are assessed, and a map of high, moderate, and low recovery potential reaches is produced.

11.2 Stage Three, Step One: Determine the trajectory of change

Following disturbance, a reach can adjust along a creation, restoration, or degradation pathway (Figure 11.1). The vertical line on the left-hand side of the recovery diagram represents the continuum from an intact to a fully degraded condition. The contemporary character and behavior of the reach can lie at any position along this degradation pathway, depending on its geomorphic condition, its sensitivity to disturbance, the form and extent of disturbance, and time since disturbance.

If a natural system is resilient to disturbance, it operates within its natural capacity for adjustment and remains close to an intact condition. Ongoing processes maintain the predisturbance character and behavior. If human disturbance is severe, such that a threshold condition is breached, or pervasive degradation undermines the integrity of the reach, the river cannot self-adjust, and falls along the degradation pathway. Such reaches have moved outside their natural capacity for adjustment. Reaches in good geomorphic condition sit high on the degradation pathway, while reaches in poor geomorphic condition sit low on the degradation pathway.

Reaches that show initial signs of recovery are considered to be at the turning point on Figure 11.1. The transition to recovery can occur at any stage along the sliding scale of the degradation pathway. The pathways of river adjustment depicted on the right-hand side of Figure 11.1 differentiate between a restoration and a creation trajectory (Table 11.1). Restoration is defined as a return towards the characteristic state that reflected predisturbance conditions (Table 11.1). Creation is defined as recovery towards a new, alternative condition that did not exist previously at the site.

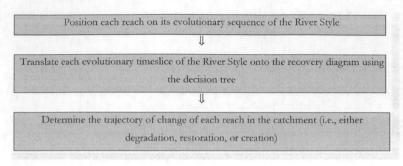

Figure 11.3 Stage Three, Step One: Determining the trajectory of change

Reaches on the creation trajectory are adjusting to altered flux boundary conditions.

The future recovery trajectory is dependent on the history of change (i.e., whether irreversible geomorphic change has occurred), present reach conditions, and prevailing flux boundary conditions. The further down the degradation scale a reach sits, the less likely it is that it will recover along the restoration pathway, and creation is the most likely future scenario. Reaches that have experienced reversible geomorphic change have the potential to recover along the restoration pathway. These reaches are generally in good or moderate geomorphic condition and sit higher on the degradation pathway. However, if flux boundary conditions have been severely altered, creation is underway. These reaches tend to be in poor condition and sit low on the degradation pathway. Ultimately, both the restoration and creation pathways reflect the operation of self-healing processes that improve river condition. By definition, reaches that have experienced irreversible geomorphic change are unable to recover along the restoration pathway. As such, they recover along a creation pathway irrespective of their contemporary geomorphic condition.

Since effective rehabilitation strategies work with the contemporary condition and trajectory of river changes, the position of each reach on the recovery diagram must be determined (Figure 11.1). The procedure used to assess the trajectory of change in the River Styles framework is outlined in Figure 11.3. With the evolutionary framework of river changes in-hand (Stage Two, Step Two), the position of each reach on the recovery diagram is determined using a decision tree. Based on this

analysis, stages of adjustment for each reach and trajectory of change are interpreted.

11.2.1 Position each reach on the evolutionary sequence of the River Style

The first step in determining the trajectory of change is to apply ergodic reasoning and position the reach under investigation on the evolutionary sequence for each River Style constructed in Stage Two, Step Two.

11.2.2 Translate each evolutionary timeslice onto the recovery diagram using the decision tree

The second step towards assessing the trajectory of change involves translating the entire evolutionary sequence of the River Style onto the recovery diagram. The position of each timeslice relates directly to its geomorphic condition and past evolutionary adjustments. Figure 11.4 summarizes the decision-making process that is applied for each reach in the catchment.

11.2.2.1 Is the reach intact?

The first decision involves evaluating whether the reach has developed a range of attributes such that contemporary river condition does not differ from an intact (predisturbance) state. Intact reaches operate within the natural capacity for adjustment of a River Style in the absence of human disturbance. A comparison is made between the character and behavior of the intact version of a River Style and the contemporary reach.

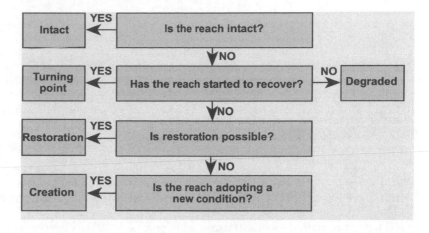

Figure 11.4 Decision tree for determining the trajectory of change of a reach. Modified from Fryirs and Brierley (2000) with permission from © V.H. Winston & Son, Inc., All rights reserved

If the geomorphic character and behavior of the reach are not significantly different from the pre-disturbance condition, the reach is considered to be intact.

11.2.2.2 Has the reach started to recover?

A reach that is recovering has moved to the right-hand side of Figure 11.1. Assessment of whether geomorphic recovery is underway is determined through interpretation of historical information and knowledge about system structure and function derived in Stages One and Two of the River Styles framework. In particular, analysis of changes to the assemblage of geomorphic units within a reach provides a key indicator of whether recovery is underway. For example, the formation of benches that narrow enlarged channels are an initial recovery mechanism. Similarly, the reemergence of pools and the redefinition of a low flow channel following the passage of a sediment slug are indicative of geomorphic recovery. Reaches that show initial signs of recovery are considered to be at the turning point (Figure 11.1). These reaches are generally in moderate geomorphic condition.

If the reach has been altered, and is experiencing progressive deterioration in its geomorphic condition, it is placed in the degraded category. These reaches are generally in poor or moderate geomorphic condition. In degrading reaches, the range of geomorphic units is inappropriate for the environmental setting. For example, sand sheets that have infilled pools or cover the floodplain indicate that there is an oversupply of sediment to that reach. Such reaches are adjusting their character and behavior to this change in sediment regime. Specific indicators of degradation or modified rates of geomorphic change may be detected, without any indication that processes of geomorphic recovery are underway.

11.2.2.3 Is restoration possible?

The final two steps in the decision tree shown in Figure 11.4 determine whether a reach is adjusting along a restoration or creation pathway. The recovery trajectory is dependent on the history of change (i.e., whether irreversible geomorphic change has occurred), the present reach condition, and the flux boundary conditions under which the reach operates (i.e., altered or unaltered). If flux boundary conditions have not been significantly altered along reaches that have experienced reversible geomorphic change, there is potential for restoration to occur. These reaches are generally in good geomorphic condition and sit high on the degradation pathway. Restoration is underway when the geomorphic unit assemblage of a reach approximates the intact, predisturbance character and behavior. For example, the reinstigation of swamps and discontinuous watercourses are indicative of geomorphic restoration along cut-and-fill rivers.

11.2.2.4 Is the reach adopting a new condition?

Reaches that have experienced irreversible geomorphic change recover along the creation pathway irrespective of their contemporary geomorphic condition. These reaches do not have the capacity to return to a predisturbance character and behavior under prevailing boundary conditions, and restoration is no longer a viable option. In this case, a "new" condition is defined, and reaches recover towards the best attainable condition for the contemporary River Style. Reaches that have experienced irreversible geomorphic change were identified in Stage Two, Step Two of the River Styles framework. If irreversible change led to a shift in River Style such that the reach now displays geomorphic structures that were not present at any stage in the evolutionary sequence, a solid line is placed across the degradation pathway of the recovery diagram. This depicts the irreversible nature of change and adjustment along the creation trajectory (Figures 11.5 and 11.7).

Reaches that have experienced reversible geomorphic change can recover along either a restoration or creation pathway, depending on the geomorphic condition of the reach and the degree to which human disturbance has altered the flux boundary conditions under which the reach operates (Figure 11.6). Reaches that are in poor geomorphic condition, and which operate under altered catchment boundary conditions, sit lower on the degradation scale and are less likely to recover along a restoration pathway. Under these conditions, creation is the most likely future scenario. The further down the degradation scale a reach sits, the less likely it is that it will recover along a restoration pathway.

11.2.3 Determine the trajectory of change of each reach in the catchment (i.e., either degradation, restoration, or creation)

Using the recovery diagram for each River Style, a direct assessment is made of the trajectory of change of each reach in the catchment. The position of each reach in the evolutionary sequence for that River Style determines where it sits on the recovery diagram. In turn, the position of the reach on the recovery diagram determines its trajectory of change. If the reach sits on the degradation path-

way it shows signs of continued deterioration. If the reach sits at the turning point, the nature of future flow and sediment regimes will dictate whether the reach continues to deteriorate or moves onto a recovery pathway. If the reach shows signs of recovery, it sits on either a creation or restoration pathway. Examples of the resultant diagrams are presented in Figures 11.5–11.7.

11.3 Stage Three, Step Two: Assess river recovery potential: Place reaches in their catchment context and assess limiting factors to recovery

The effectiveness and extent of river recovery processes reflect a wide range of catchment-specific considerations. Reaches of different River Styles are at different phases of degradation and recovery. The character and mechanisms of recovery differ from system to system conditioned by the type of river, the extent of disturbance, reach position relative to off-site or lagged disturbance responses, and a range of physical issues related to limiting factors operating within the catchment. Assessment of limiting factors and pressures that constrain the recovery of landscapes and their associated ecosystems enables the adoption of management strategies that minimize the impacts of these constraints. Such practices address underlying causes of change, rather than their symptoms. These factors must be assessed on a catchment-by-catchment and reach-by-reach basis. In the River Styles framework, the changing nature of within-catchment linkages of physical processes is examined, relating disturbances in one part of a catchment to off-site impacts and lagged responses elsewhere.

Given that rivers are evolving, adjusting entities, a reach can sit at any position on the degradation pathway and, given favorable conditions, may move onto a recovery pathway at any point. However, even though a reach may sit on a recovery trajectory, the river may not have the "potential" to move along that pathway. *Recovery potential is defined as the capacity for improvement in the geomorphic condition of a reach in the next 50–100 years.* Under the same set of environmental conditions, a reach may potentially adjust along multiple recovery trajectories. Analysis of the pathway of adjustment, and prediction of

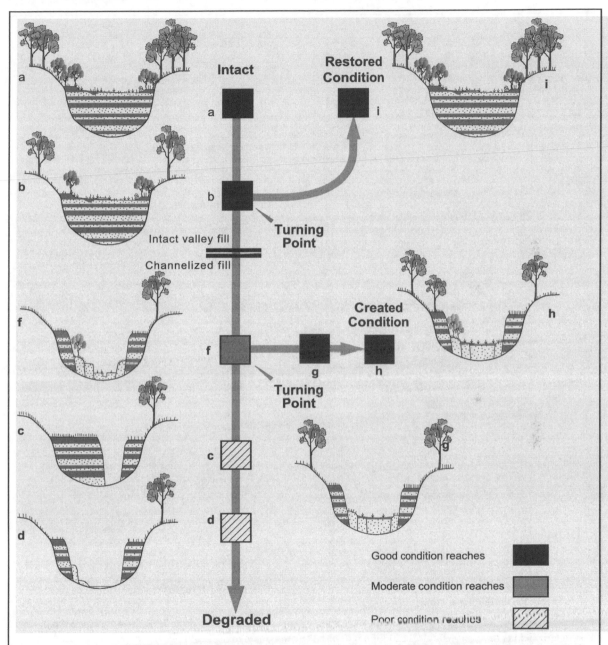

Figure 11.5 Trajectories of change for reaches of the channelized fill River Style in Bega catchment (from Fryirs, 2001). The letters equate to the evolutionary stage in Figure 10.5
In Bega catchment, reaches that are sensitive to change have experienced irreversible change in River Style (represented by the black lines on the degradation pathway). In this case, the river has changed from an intact valley fill River Style to a channelized fill River Style. Reaches of the latter now operate under altered catchment boundary conditions. These reaches are unlikely to recover along a restoration pathway over the next 50–100 years. Hence, *creation* of a new condition is underway along these reaches.

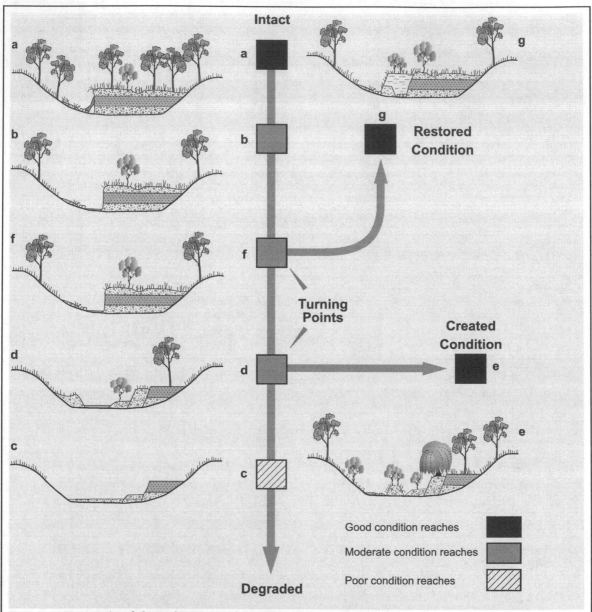

Figure 11.6   Trajectories of change for reaches of the partly-confined valley with bedrock controlled discontinuous floodplain pockets River Style in Bega catchment. Capital letters equate to the evolutionary stage in Figure 10.6. Modified from Brierley et al. (2002). Reproduced with permission from Elsevier, 2003

River Styles that have experienced reversible geomorphic change have the potential to adjust along either a *restoration* or *creation* pathway. Reaches of the partly-confined valley with bedrock controlled discontinuous floodplain pockets River Style fall into this category. The trajectory taken depends largely on the condition of the reach and the degree to which human disturbance has altered the flux boundary conditions. The poorer the condition of the reach, the lower it sits on the degradation pathway of the recovery diagram, and the less likely it is that restoration will occur. In these cases creation is underway.

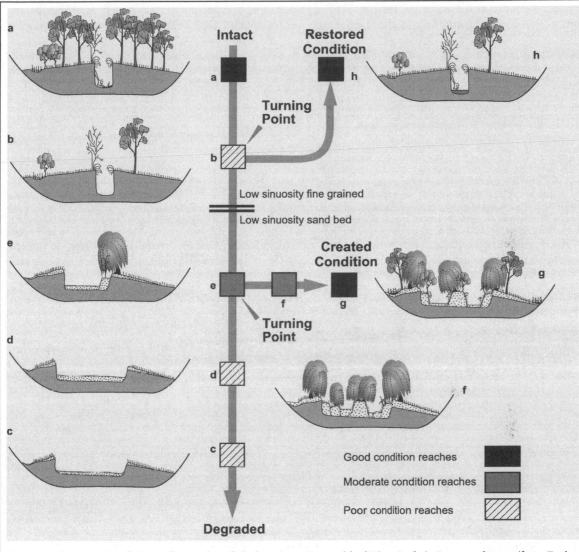

Figure 11.7 **Trajectories of change for reaches of the low sinuosity sand-bed River Style in Bega catchment (from Fryirs, 2001). Capital letters equate to the evolutionary stage in Figure 10.7**
In Bega catchment, reaches that are sensitive to change have experienced irreversible change in River Style (represented by the black lines on the degradation pathway). In this case, the river has changed from a low sinuosity fine-grained River Style to a low sinuosity sand-bed River Style. Reaches of the latter now operate under altered catchment boundary conditions. These reaches are unlikely to recover along a restoration pathway over the next 50–100 years. Hence, this recovery pathway has been eliminated for this River Style. Creation of a new condition is underway along these reaches.

recovery potential, requires analysis of limiting factors to recovery and the connectivity of geomorphic processes, placing each reach within its catchment context. This foresighting process entails analysis of river character and behavior, sensitivity to change, proximity to a threshold, geomorphic condition, position in the catchment, and limiting factors and pressures, building on analyses completed in Stages One and Two of the River Styles framework. Procedures used to assess river recovery potential are shown in Figure 11.8.

11.3.1 Determine reach sensitivity and geomorphic condition

Analyses completed in Stage Two of the River Styles framework are used to determine the sensitivity of a reach to change and its geomorphic con-

dition. Sensitivity to change reflects the capacity for adjustment of each River Style. Reaches with significant adjustment potential are sensitive to change, while those with limited or localized adjustment potential are resilient to change. In Stage Two, Step Three, reaches are differentiated into good, moderate, and poor condition categories. In general, adjustments to sensitive reaches drive propagation of geomorphic change through a system.

11.3.2 Assess limiting factors and pressures in the catchment

The capacity of a reach to realize its full recovery potential is dependent on limiting factors and pressures that operate within a catchment, and their lagged and off-site impacts. In the River Styles

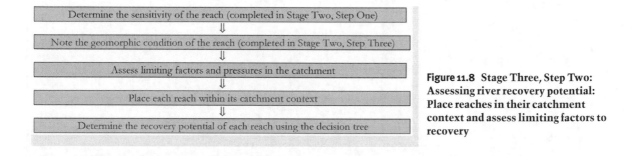

Figure 11.8 Stage Three, Step Two: Assessing river recovery potential: Place reaches in their catchment context and assess limiting factors to recovery

Figure 11.9 Post-European settlement alluvial sediment budget for Bega catchment. Reproduced from Fryirs and Brierley (2001) with permission from Elsevier, 2003

At the time of European settlement, around 55 million m^3 of alluvial material was stored along river courses in the Bega system. Of this, around 23 million m^3 of material has been released since the 1850s. Almost half of the sediment released (around 10 million m^3) has been sourced from incised valley fills. Channel expansion along tributary and trunk streams has yielded an additional 4 million m^3. Just over 6 million m^3 has been restored as readily reworked instream storage units (e.g., sand sheets and channel bars) in the contributing catchment. This material is progressively being transferred through the system as the tail of the sediment slug. Most sediment yielded from the contributing catchment (around 14 million m^3) has been efficiently flushed to the lowland plain through bedrock-controlled midcatchment reaches, with a sediment delivery ratio of around 70%. The contributing catchment is effectively exhausted of sediment.

Along the lowland plain, significant channel expansion immediately following European settlement yielded around 2 million m^3 of material. In subsequent decades, however, the role of this reach was altered. Of the 16 million m^3 of material supplied to the lowland plain, the majority (around 12 million m^3) is stored in within-channel sand sheets, ridges, bars, and benches, and on the floodplain as sand sheets. Most of the instream sediments have now been stabilized by exotic vegetation. The lowland plain acts as a large sediment sink. Low slope and a wide valley combine to ensure that the sediment slug remains in this part of the landscape. Of all the material released from Bega catchment, just over 3.5 million m^3 has been transferred to the estuary, with a catchment sediment delivery ratio of around 16%. This sort of analysis provides the baseline upon which the linkage and cascading of sediment can be assessed, allowing the recovery potential of each reach to be assessed.

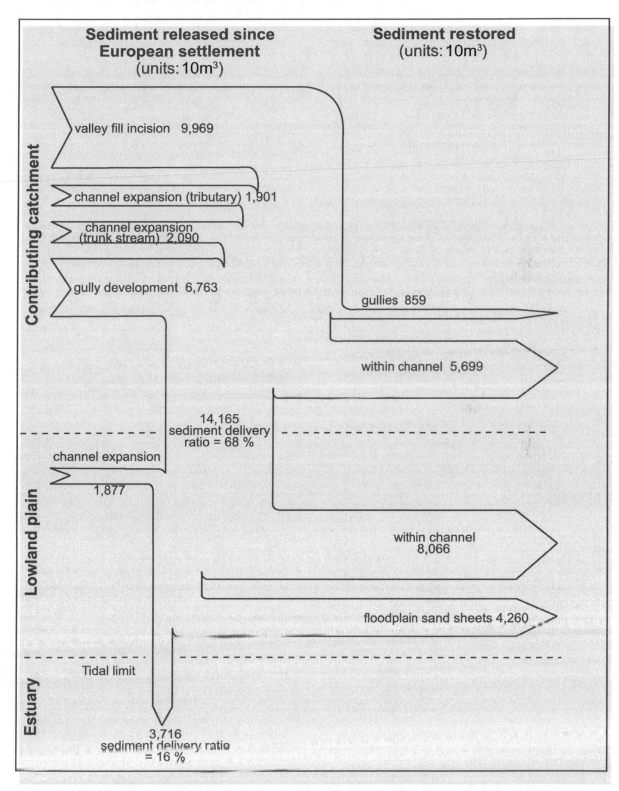

Sediment released since European settlement (units: 10m³)

Sediment restored (units: 10m³)

Contributing catchment

valley fill incision 9,969

channel expansion (tributary) 1,901

channel expansion (trunk stream) 2,090

gully development 6,763

gullies 859

within channel 5,699

14,165 sediment delivery ratio = 68 %

Lowland plain

channel expansion

1,877

within channel 8,066

floodplain sand sheets 4,260

Estuary

Tidal limit

3,716 sediment delivery ratio = 16 %

framework, limiting factors to recovery are defined as catchment-specific physical considerations that constrain the potential of a reach to move along a recovery trajectory. Limiting factors include sediment availability and transport capacities, discharge considerations, and vegetation distribution, character, and composition. Pressures are human-induced practices that can be either internal or external to the catchment. They include land-use change, vegetation, and water management, climate change, social and political perspectives, etc. Their consequences may be either negative or positive in environmental terms. The importance of any limiting factor or pressure differs not only between catchments, but between reaches within a catchment. Because of the changing nature and rate of (bio)physical linkages in a catchment, different reaches are subjected to markedly variable off-site and/or lagged impacts. These factors are also important when deriving catchment-framed visions (Stage Four of the River Styles framework). Examples of various limiting factors and pressures, and procedures used to assess them, are outlined below.

11.3.2.1 Deriving a catchment sediment budget

Given the emphasis on geomorphic structure and function, and the movement of sediment through catchments, sediment availability presents a critical constraint on the river recovery potential. Spatial predictability of disturbance response is influenced primarily by the patterns of sediment stores within a catchment, and controls on the degree to which they are likely to be reworked. Ultimately, success in predicting patterns and rates of sediment flux throughout a catchment, and associated river changes, is constrained by knowledge of the residence time of different sediment storage units. Identification of sites that are most sensitive to change is critical in assessment of the frequency with which perturbations are likely to occur. Such notions determine, for example, whether reaches are likely to be subjected to sediment starvation, sediment slugs will be generated and conveyed, or fluxes will remain roughly constant over time.

In the River Styles framework, particular emphasis is placed on analysis of the bedload component, as this is the primary determinant of river morphology. Specific consideration is given to the relationship between bedload transport capacity and sediment availability in each reach. This determines the cascade and lagged effects of sediment release following disturbance events, shaping the prospects for geomorphic recovery (Fryirs and Brierley, 2001). The downstream pattern of River Styles provides a useful basis for these analyses.

Deriving a catchment-scale sediment budget is a significant exercise in its own right. Products from this sort of work are presented in (Figures 11.9 and 11.10; Fryirs and Brierley, 2001). At the very least, analysis should examine the distribution and blocking effect of buffers, barriers, and blankets within a catchment. This provides a basis for determining from where in the catchment sediments of varying caliber will be derived, and over what timeframe it will be delivered to different parts of the system.

11.3.2.2 Hydrological analyses

Whether brought about by flow regulation or as secondary responses to altered ground cover, changes to flow conditions may modify river morphology. As channel geometry adjusts, the stream power conditions under which rivers operate are modified. If threshold conditions are breached in sensitive reaches, the river becomes more prone to adjustment, possibly compromising the potential for geomorphic river recovery. Determining the stream power conditions under which each type of river operates and the threshold conditions under which change is likely to occur, gives some indication as to whether reaches will be event sensitive or event resistant. This provides a basis to predict where in the catchment change is likely to occur, and where off-site impacts will be manifest.

11.3.2.3 Vegetation analyses

Different River Styles have differing geomorphic responses to vegetation removal, with different consequences for river recovery. In some instances, the removal of resisting factors along the valley floor may enhance erosion, potentially pushing the channel beyond critical threshold

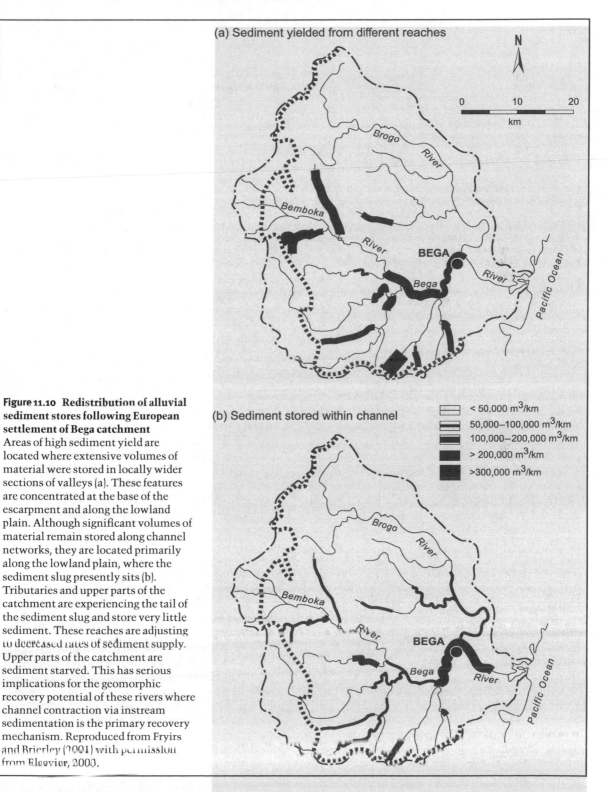

(a) Sediment yielded from different reaches

N

0   10   20
km

Brogo River

Bemboka

River

BEGA

Bega   River

Pacific Ocean

(b) Sediment stored within channel

< 50,000 m³/km
50,000–100,000 m³/km
100,000–200,000 m³/km
> 200,000 m³/km
>300,000 m³/km

Brogo River

Bemboka

River

BEGA

Bega   River

Pacific Ocean

Figure 11.10 Redistribution of alluvial sediment stores following European settlement of Bega catchment
Areas of high sediment yield are located where extensive volumes of material were stored in locally wider sections of valleys (a). These features are concentrated at the base of the escarpment and along the lowland plain. Although significant volumes of material remain stored along channel networks, they are located primarily along the lowland plain, where the sediment slug presently sits (b). Tributaries and upper parts of the catchment are experiencing the tail of the sediment slug and store very little sediment. These reaches are adjusting to decreased rates of sediment supply. Upper parts of the catchment are sediment starved. This has serious implications for the geomorphic recovery potential of these rivers where channel contraction via instream sedimentation is the primary recovery mechanism. Reproduced from Fryirs and Brierley (2001) with permission from Elsevier, 2000.

Table 11.3 Examples of internal and external pressures on river systems.

Internal to the catchment

- Direct alterations to the sediment and water regimes of rivers and catchments, such as water and sediment extraction (e.g., irrigation) and artificial storage
- Water allocation and water reform initiatives
- Land management practices (e.g., intensity of cropping or stock densities)
- Alterations to land cover (either afforestation or deforestation)
- Indirect changes associated with riparian vegetation and woody debris removal or placement

External to the catchment

- Management responses to natural occurrences such as droughts, flood events, salinity
- Management structures and government/political-induced pressures
- Global environmental change
- Population structure change/trends
- Climate change (e.g., greenhouse or El Nino Southern Oscillation)

---

conditions. This may compromise the potential of the system to recover (e.g., Brooks et al., 2003). The expansion of exotic vegetation, and associated patterns of seed dispersion, may also affect recovery potential (e.g., Brooks and Brierley, 2000). Alternatively, if native seed sources remain intact, recovery may be enhanced decades after initial clearance. Mapping of the distribution of vegetation and seed sources provides valuable insight into geomorphic recovery potential.

11.3.2.4 Assessment of pressures

Any assessment of geomorphic river recovery potential must be framed in context of human-induced pressures that will impact on the operation of processes in the future, and their interactions in the physical environment. Table 11.3 outlines some of these pressures. The range of pressures varies from system to system. In the River Styles framework, particular consideration is given to pressures that threaten the conservation of intact reaches. If excess pressures are applied to a system, management responses aim to reduce or negate the influence of these pressures. An assess-

ment is made of whether the nature and distribution of point and nonpoint impacts are likely to change over a given timeframe, and whether the extent of pressures will increase or decrease. In determining the recovery potential of systems and deriving appropriate catchment-based visions, strategic measures must be put in place to address future negative impacts. The key is to identify where social and economic policies can minimize environmental degradation and enhance river recovery.

11.3.3 Place each reach in its catchment context

The linkage of geomorphic processes throughout a catchment is dictated largely by the spatial configuration and downstream pattern of River Styles. This in turn determines the propagation of disturbance responses and off-site impacts experienced elsewhere in a catchment. For example, a reach in good geomorphic condition may sit immediately downstream of a reach in poor condition. The recovery potential of the former will be dictated by the degree of upstream degradation and downstream transmission of degrading processes. Hence, while the downstream reach is in good geomorphic condition, it may only have moderate recovery potential.

Reach position is easily identified by revisiting analyses completed in Stage One, Step Three of the River Styles framework where downstream patterns of River Styles were assessed. To simplify analyses, a "catchment tree" is constructed that depicts the linkage of geomorphic processes (Figure 11.11). The branches of the tree represent examples of each pattern of River Styles found along tributary networks. A number of boxes can be added to represent different segments along the river. In Figure 11.11, this is based primarily on the distribution of valley-settings. These boxes feed into the trunk stream, which is also broken into different segments. Statements on river sensitivity, geomorphic condition, and limiting factors operating on each segment are placed in each box. An analysis is then made of the connectivity and off-site impacts of each segment in both upstream and downstream terms. This forms the basis for determining the recovery potential of each reach in the catchment.

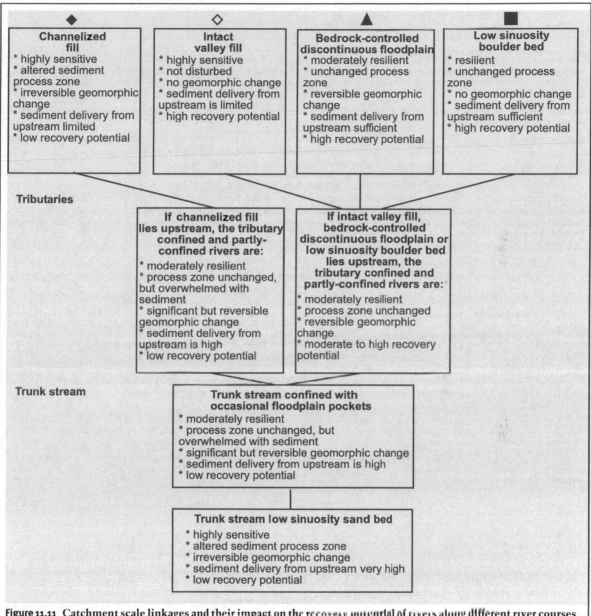

Figure 11.11 Catchment scale linkages and their impact on the recovery potential of rivers along different river courses that display different downstream patterns of River Styles. Symbols refer to the downstream pattern of River Styles in Bega catchment

The recovery potential of rivers in Bega catchment is dependent on the position of the reach in the catchment, its condition and resilience to change, and the upstream availability of sediment that enhances (or diminishes) prospects for recovery. Reproduced from Fryirs and Brierley (2001) with permission from Elsevier, 2003.

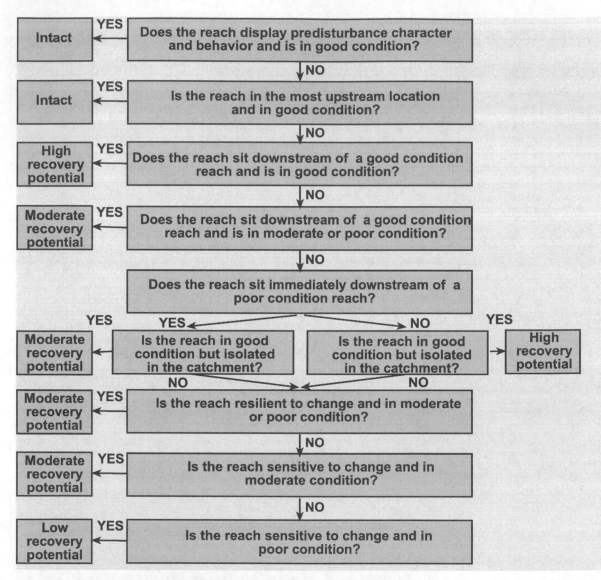

Figure 11.12 Decision-making tree for determining the recovery potential of a reach

11.3.4 Determine the recovery potential of each reach

Using a range of previously compiled information, a decision-making tree is developed to determine the recovery potential of each reach (Figure 11.12). This analysis combines assessments of geomorphic condition, river sensitivity, and position in

catchment. The latter factor summarizes the effect of limiting factors and the connectivity of geomorphic processes throughout the system.

In general terms, reaches in moderate to good condition that sit high in the catchment, close to intact and good condition reaches, are assigned a high recovery potential rating. Impacts of disturbance are likely to be limited, providing an opportunity to recover relatively quickly. Reaches that remain in good condition but are isolated in the catchment are generally resilient to change and will absorb off-site impacts. They are given a high or moderate recovery potential rating. The position of poor condition reaches dictates the recovery potential of remaining reaches. Poor condition reaches are often sensitive to change. Off-site impacts are propagated at differing rates, with differing consequences through the system. Moderate and low recovery potential ratings are assigned according to the sequencing of reaches and their resilience or sensitivity to change. By assessing the recovery potential of each reach of each River Style in the catchment, a map is produced that shows river recovery potential (Plate 11.1).

11.4 Products of Stage Three of the River Styles framework

Four main products are produced in Stage Three of the River Styles framework:

- recovery diagrams for each River Style noting the trajectory of change of each reach;
- tables, maps, or flow diagrams that describe and quantify limiting factors and pressures operating in the catchment;
- a "catchment tree" noting the linkage of geomorphic processes and off-site impacts;
- a catchment map showing the recovery potential of each reach.

These products, together with those produced in Stages One and Two, provide a physical template with which to create a catchment vision, identify target conditions for river rehabilitation, and prioritize strategies for river management in Stage Four of the River Styles framework.

CHAPTER 12

Stage Four of the River Styles framework: Implications for river management

There is a great mismatch between the rate at which human activities may damage a system and the rate at which that system can be restored. . . . Political and managerial "short-termism" is inimical to worthwhile restoration . . . Partnerships between scientists and resource managers are crucial to the successful execution of restoration projects. . . . (T)hese partnerships need to be forged at the planning stages of the project and not after the restoration measures have been initiated.

Sam Lake, 2001a, p. 370

12.1 Introduction: River rehabilitation in the context of river recovery

Ecologically sustainable river management strategies will only be achieved if adopted procedures work with the natural behavior of river systems, balancing efforts at river conservation and rehabilitation. A future-focus is required for such activities, ensuring that river rehabilitation strategies fit the local environmental setting framed within a catchment-scale "vision" that integrates linkages of biophysical processes. In the River Styles framework, river rehabilitation is viewed as a process of recovery enhancement, in which management efforts strive to help the river to adjust naturally, improving river condition over the short–medium term. The degree to which river rehabilitation projects enhance recovery and are self-sustaining is constrained by their compatibility with the dominant fluvial processes that determine river morphology. River morphology and vegetation associations must be appropriately reconstructed before sympathetic rehabilitation of riverine ecology will occur.

Understanding river character, behavior, condition, and recovery potential provides an integrative physical platform for river rehabilitation planning. Stage Four of the River Styles framework uses the information derived in Stages One–Three to guide the derivation of management programs that maximize the geoecological potential of river courses. Insights for river rehabilitation are applied at reach, subcatchment, and catchment scales, based upon predictions of likely future changes and associated insights into geomorphic river recovery potential. Within this framework, outstanding examples of particular River Styles, or rare/unique examples of River Styles, are identified and protected, while reaches with the greatest likelihood of recovery are targeted for rehabilitation. A physical vision is created, based on a realistic sense of what is achievable in river rehabilitation terms throughout a catchment. With this vision in-hand, realistic target conditions for river rehabilitation are identified and prioritized for each reach. This provides a foundation to monitor improvement and determine the success of the river rehabilitation strategy.

Stage Four of the River Styles framework has four steps (Figure 12.1). Terms used in Stage Four of the River Styles framework are presented in Table 12.1.

12.2 Stage Four, Step One: Develop a catchment-framed physical vision

Articulation of a vision provides a basis to assess whether management efforts are successful. Vision statements provide a basis to identify clear, measurable objectives, forming a platform with which to assess management performance

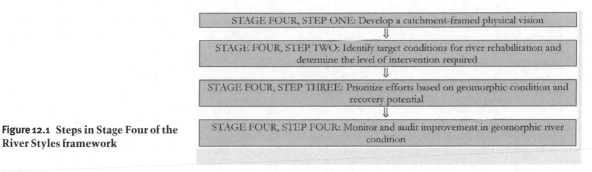

Figure 12.1 Steps in Stage Four of the River Styles framework

STAGE FOUR, STEP ONE: Develop a catchment-framed physical vision

⇩

STAGE FOUR, STEP TWO: Identify target conditions for river rehabilitation and determine the level of intervention required

⇩

STAGE FOUR, STEP THREE: Prioritize efforts based on geomorphic condition and recovery potential

⇩

STAGE FOUR, STEP FOUR: Monitor and audit improvement in geomorphic river condition

Table 12.1 Definition of terms used in Stage Four of the River Styles framework.

Term	Definition in the River Styles framework
River rehabilitation	A process that is undertaken to help a river adapt to a new environment by improving the condition and enhancing the recovery of the system through manipulation of its structure and function.
Catchment-based vision	Set of goals that outline a realistic sense of what is achievable in river rehabilitation terms throughout a catchment. Target conditions are identified, prioritization is undertaken, and monitoring strategies are put in place.
Recovery goal	A defined condition (either restoration or creation) that provides a long-term goal for river rehabilitation.
Target condition	A good or moderate condition state that provides a short–medium term goal for the rehabilitation of reaches in poor geomorphic condition. Target conditions are the stepping stones towards the recovery goal.

through monitoring and auditing programs. Working towards an integrative, catchment-scale vision will improve the likelihood of rehabilitation success in environmental and economic terms. Effective vision statements must be ecosystem based, and related to functional linkages in landscapes. An overarching catchment-framed vision outlines the goals and objectives for river rehabilitation over a 50–100 year timeframe. As noted by Brierley and Fryirs (2001), unless a realistic catchment-scale vision is determined, river rehabilitation activities are unlikely to achieve their intended goals because:

• management programs react to the latest sets of threats or pressures, rather than adopting longer-term, proactive approaches that address underlying causes of problems;

• there is limited transparency in the decision-making process, making it difficult to be strategic in conservation planning and prioritizing management actions;

• catchment linkages of physical processes, and off-site impacts of reach-based programs, are not given due attention;

• monitoring and auditing programs are likely to be ineffective. As such, an underlying sense of achievement may be compromised, impacting on community engagement in the process.

A vision statement envisages an improved state for a system that can be achieved at some stage in the future. The mission, goals, and objectives of environmental projects fit into this overarching vision. The physical vision must be catchment-specific and realistically achievable within a specified timeframe. This helps to effectively "ground" the vision, thereby ensuring that is has biophysical authenticity. A usable vision also needs to be sufficiently flexible and adaptive such that it can be modified in light of changing circumstances, perceptions and understanding, or as targets are achieved. It should also be generative and catalytic, galvanizing, and maintaining action around broadly identified themes (Hillman and Brierley, in press). Key considerations in development of a catchment-based physical vision include:

• ensure that slopes, channels, floodplains, and the estuary/delta are considered as part of the same system;

• environmental measures that take account of landscape patterns, processes, and history, building integrative perspectives that reflect the inherent linkages of the system;

• frame goals in terms of the best that is achievable under prevailing boundary conditions;

• reconnect fragmented ecosystems as far as practicable (and appropriate), building out from remnants. In fragmented and modular ecosystems (and landscapes), connections of physical processes must be retained or enhanced as appropriate. Discontinuity of physical fluxes is a natural part of some river systems. Elsewhere, barriers to transfer processes may be removed;

• manage for naturalness within a "living river" perspective, allowing rivers to "run free" whenever practicable;

• attain a balance in efforts at river conservation and rehabilitation. Conserve and, where appropriate, enhance remaining seminatural ecosystems. Protect "unique" and "rare" attributes, regardless of their present condition, striving to maintain essential ecological processes and preserve genetic diversity. While promoting heterogeneity and diversity as underlying goals, remember that some ecosystems are naturally simple;

• maximize environmental and social benefits for the capital outlay by targeting rehabilitation programs. Apply an open and transparent prioritization program, ensuring that measures are sympathetic to local needs and values;

• apply flow management strategies that enhance habitat diversity and edge effects (ecotones), considering the need for refugia at all flow stages. Aim to maintain, whenever possible, the vagaries of the natural hydrological cycle, including river and wetland functions and processes. Maintain floodplain inundation, as appropriate to the local setting;

• strive to regain, if appropriate, a continuous riparian corridor, with native vegetation dominant. Fence off the riparian zone, and/or apply effective weed management and maintenance programs, as required;

• document the desired sediment transfer role of all reaches, taking into account the need to trap or release sediments as required;

• develop and apply representative monitoring programs, and respond to lessons learnt.

In the River Styles framework, a five-phase approach is applied to derive a catchment-based vision (Figure 12.2). The first phase entails compiling the coherent package of physical information upon which to build a catchment vision from analyses completed in Stages One–Three of the River Styles framework. The second phase entails developing a realistic sense of what can be achieved for each reach, assessed within an overarching catchment-framed vision. The primary task is to identify target conditions for river conservation/rehabilitation within the catchment. This forms Stage Four, Step Two of the River Styles framework. The third phase develops the priorities and sequencing of actions that will lead to the most efficient, cost-effective, and integrative plan of attack with which to work towards the vision. Procedures by which this is achieved form Stage Four, Step Three of the River Styles framework. The fourth phase determines the practical tools required by managers to implement effective on-the-ground conservation and rehabilitation strategies that will achieve the reach-based target conditions and ultimately the catchment-framed vision. Finally, phase five outlines monitoring procedures that help to reframe the vision over time. These latter two phases use established tools and techniques that are not considered here.

12.2.1 Phase 1: Establish an appropriate information base to work from

Vision building will not be successful without a solid understanding of river character, behavior, evolution, process linkages, geomorphic condition, the trajectory of change, and river recovery potential (Figure 12.3). The first layer of information required is knowledge on what is out there and what is left, regardless of its condition. Baseline information on river diversity and representativeness constrains what is physically and financially achievable in efforts at rehabilitation, enabling options to be pursued in an environmentally just manner. By adding layers on river recovery, the geomorphic platform provides a means to predict

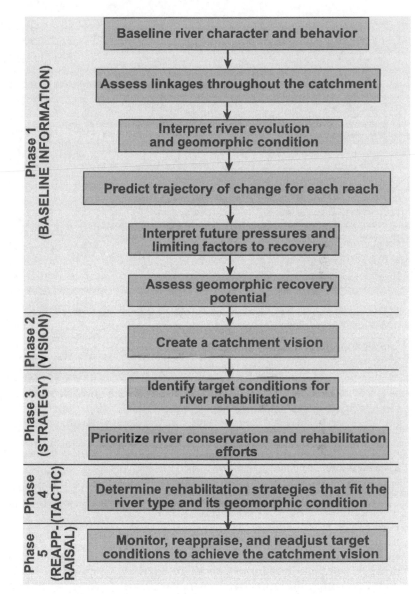

Figure 12.2 Steps towards creating a catchment-framed vision (modified from Brierley and Fryirs, 2001)

future change (i.e., the trajectory a reach will take) and the recovery potential of that reach. For example, if a reach is left alone, will it continue to degrade, has it changed irreversibly, or is the reach recovering towards a predisturbance state? Catchment-wide maps of river character and behavior, condition, and recovery potential provide the geomorphic insight that is required to create a catchment-based physical vision within which

reach scale target conditions for conservation and rehabilitation are identified, and management efforts are prioritized.

12.2.2 Phase 2: Create a catchment vision

Vision statements must extend beyond generalized motherhood statements that have limited operational utility, providing an explicit expression

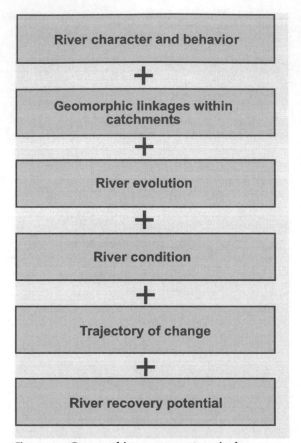

Figure 12.3 Geomorphic components required to create a catchment-framed vision

of catchment-scale goals. Strategic visions are clear, but flexible. They are inspirational to those involved in the project, providing solid foundations and reference points in the rehabilitation process. Carefully crafted visions serve as focal points for community-oriented river management programs, generating considerable goodwill and commitment. To aid this process, it is helpful to pick iconic themes, such as extending the abundance and diversity of fish populations (Mersey Basin Campaign, England), seeing one's toes (Chesapeake Bay, United States), and swimming (Brisbane River, Australia). Achieving these seemingly straightforward goals requires consideration of complex physical responses and interactions.

Catchment-framed physical visions must be based on a solid geomorphic template. They must

be realistically achievable in social, practical, and financial terms over a 50–100 year timeframe. Derivation of a realistic vision must consider ongoing and likely future pressures and limiting factors that will be experienced in the catchment, and associated appraisals of prospective environmental changes. The vision should aim to enhance the recovery of the system. Target conditions and long-term goals for river rehabilitation should be identified (see Stage Four, Step Two) and incorporated into the details of the vision. Questions that need to be asked include:

- what are we trying to achieve?
- what do we want the river to be like?
- what are we managing for?
- how can we achieve our goals?

The answers to these questions must incorporate statements pertaining to the interaction and linkages of geomorphic structure and function as manifest through sediment supply and storage relations. They must also address linkages with other biophysical processes, such as water transfer, storage, and quality, vegetation coverage, composition, and succession, habitat availability, viability and variability, and measures of ecological functioning, such as food web processes, nutrient flux, organic matter processing, etc. Ultimately, a vision entails developing an overarching statement and details of how it will be achieved (e.g., Table 12.2).

12.2.3 Phase 3: Derive a meaningful strategy to achieve the biophysical vision

Effective river management planning entails many small steps in working towards the final vision, requiring a clear view of the proposed final outcome and a coherent set of rehabilitation projects that provide a tangible route to success. In some situations, better long-term outcomes, in environmental and economic terms, may be gained by allowing the system to undergo natural recovery (i.e., the do nothing option). The nature and rate of catchment-scale biophysical linkages are critical determinants of the effectiveness of self-healing processes and tendencies.

Procedures used to decide where in the catchment to start and the associated plan of activities should be logical, testable, and transparent. A series of readily achievable target conditions is

Table 12.2 A biophysical vision for Bega catchment. Modified from Brierley et al. (2002). Reproduced with permission from Elsevier, 2003.

Overall vision: Our vision for Bega catchment is to achieve a healthy and diverse river environment that maintains connections of core ecosystem processes in an environmentally sustainable way, balancing the needs of all users.

What are we trying to achieve? Community and government working together, with nature, to improve the health of riverine ecosystems. The Bega Catchment Integrated River Health Package has set priorities for on-ground works that integrate sediment and water storage and delivery issues, exotic weed eradication and planting of native vegetation, water quality objectives, enhancing ecological recovery potential, cost effectiveness, and "demonstration" value.

What are we managing for? The aim is to return the river system to a sustainable (self-healing) geomorphic and ecological condition, minimizing the need for ongoing (reactive) maintenance. This will be achieved by enhancing river recovery throughout the catchment and working with the character and behavior of streams.

What do we want the river to be like? Implementation of management activities strive to work towards a healthier, catchment-wide river system with a natural sediment regime, improved water quality, native vegetation, and ecological associations.

Issue	Long-term vision	Short-term action
Sediment regime	• Lock up sediment in cut- and-fill River Styles at the base of the escarpment. • Maintain balance between sediment input and output along midcatchment reaches. • Maintain remnant swamps and floodouts along Frogs Hollow Creek and lower order drainage lines that act as sediment sinks. • Lock up sediment along the lowland plain while maintaining natural levels of output to the estuary.	**Upper catchment** • Protect remnant swamps and floodouts from knickpoint retreat. • Cattle exclusion and fencing off. • Revegetate riparian and within-channel geomorphic surfaces to stabilize sediment stores. • Emplace bed control structures to retain sediment in within-channel swamps. **Middle–lower catchment** • Develop riparian revegetation programs to reduce rates of channel expansion, and associated removal of floodplain sediment. • Emplace bank control structures to reduce sediment loss. • Strategically place large woody debris structures stabilize to in-channel sediments and induce pool development. • Establish cattle access points to reduce bank and bed degradation.
Vegetation associations	• Remove willows and reestablish native vegetation associations. • Reinstate a continuous riparian corridor.	• Sustain strategic willow management programs. • Replant native vegetation that suits the riparian environment for each River Style, using species that are indigenous to the region.
Water regime	• Maintain base flow conditions and water storage in remnant swamps and floodouts for drought proofing and ecological refugia. • Reduce time of travel and stream powers by flattening the hydrograph i.e., reduce flood peaks. • Introduce environmental flows within a water management plan that satisfy instream and wetland requirements.	• Conserve and protect swamps and floodouts from knickpoint retreat. • Undertake riparian and within-channel revegetation programs. • Increase channel roughness through woody debris placement and revegetation of instream geomorphic surfaces. • Maintain pools under low flow conditions and enhance floodplain wetland connectivity through water allocation and irrigation licensing.
Ecological associations	• Enhance native terrestrial and aquatic ecological associations. • Reinstigate channel–floodplain connections (e.g., between channel habitat and floodplain wetlands). • Improve water quality and organic matter retention. • Maintain and improve the viability of remnant ecological niches in swamps and floodouts.	• Reduce channel capacities to reinstigate channel–floodplain connectivity. This requires sediment storage and revegetation at appropriate places along each River Style. • Protect remnant swamps and floodouts. • Supply and retain organic matter in the system through native revegetation programs. • Reintroduce woody debris along appropriate streams.

required (Stage Four, Step Two). While all endeavors are made to determine the best overall outcome, some reaches may have to be compromised to achieve this goal. In other words, collective enhancement of rivers in broad-scale terms is unlikely to be achieved without some short-term losses. This requires prioritizing where in the catchment the greatest likelihood of success will be (Stage Four, Step Three). By effectively prioritizing efforts, tangible success can be achieved in a strategic manner rather than spreading efforts so thinly that little is achieved.

Principles from geomorphology, rainforest ecology, and bush regeneration suggest that conservation-oriented visions must emphasize the maintenance of those parts of landscapes that retain a good condition. Building from that platform, effective river management strategies work outwards from conservation sites to the most degraded sites in the catchment. This not only helps reestablish a continuous riparian strip; it also maximizes the protection given to reaches in good condition (or reaches of rare river types). Given the community focus of many river rehabilitation projects, and the underlying emphasis on the return for money spent, working at sites with a high likelihood of success provides a sound management strategy in biophysical, socioeconomic, and environmental terms (e.g., Frissell et al., 1993; Brierley, 1999; Erskine and Webb, 1999; Rutherfurd et al., 1999; 2000; 2001a; Brierley and Fryirs, 2000; Brierley et al., 2002). However, as noted by Collins and Montgomery (2001) and Gellis et al. (2001), projects must also be targeted strategically to conserve threatened or rare remnant reaches (and/or habitats), and proactive (preventative) measures must be applied to inhibit threatening processes (e.g., conveyance of sediment slugs into high value reaches, erosion prevention).

In Stage Four, Step Three of the River Styles framework, a prioritization framework with which to determine a "plan of attack" is outlined. Through application of this framework, interpretations of the spatial and temporal causes of change provide a means to identify where manipulation is needed to enhance recovery. The most cost-effective and efficient reach-based strategies that work towards a catchment vision can then be employed. In this way, the likelihood of rehabilitation strategies in one part of a catchment having a negative effect (or working against those occurring elsewhere) is negated.

12.2.4 Phase 4: Determine tactics to achieve target conditions and the catchment-based vision

Appropriate techniques must be applied to address targeted river rehabilitation problems. The most suitable treatment may vary markedly for different problems and for different river types. Adopted techniques must fit the character, behavior, and condition of the reach. Off-site impacts should be minimized. Appropriate tools for river rehabilitation practice are addressed elsewhere (e.g., Gore, 1985; Newbury and Gaboury, 1993; Rutherfurd et al., 2000).

12.2.5 Phase 5: Monitor, reappraise, and readjust target conditions

Implementation of management strategies will modify the interactions of biophysical processes. Hence, target conditions and priorities for conservation and rehabilitation must be subjected to ongoing reappraisal. The vision/goals may be extended once certain target conditions have been achieved. Alternatively, pressures such as climate change, major flood/drought events, or other unexpected consequences may require perspectives to be reframed. Adaptive management principles tied to effective monitoring and maintenance programs enable flexibility to be maintained in the rehabilitation process. Learning from mistakes, and successes, enables continual progress to be made towards the vision, maintaining momentum in the process. Sampling procedures used for monitoring must be pertinent to the type of river under investigation. Hence, measures used for discontinuous watercourses, swamps, anabranching, or anastomosing rivers, gorges, etc. must reflect the array of landforms, vegetation associations, and hydraulic diversity evident in these settings, ensuring that due emphasis is placed on both channel and floodplain forms. This enables comparison of like with like in a meaningful manner, ensuring that monitoring programs relate to the "natural" range of diversity evident along a river course.

12.3 Stage Four, Step Two: Identify target conditions for river rehabilitation and determine the level of intervention required

Nested within an overarching vision is a suite of short–medium term target conditions. Determining target conditions involves assessing what is realistically achievable for each reach in the catchment over a short–medium timeframe (i.e., over years to decades). These represent stepping-stones towards longer-term goals. Target conditions and long-term goals are defined for particular time periods for each reach of each River Style in the catchment. As target conditions and goals are achieved, the catchment vision is reappraised.

If they are to be sustainable over the long term, target conditions for each River Style must be designed within an integrative, catchment perspective. Given that geomorphic responses to disturbance at the reach scale vary depending on the River Style, its position in the catchment (i.e., the pattern and condition of upstream and downstream reaches), the connectivity of geomorphic processes throughout the catchment, and off-site impacts, target conditions must be assessed in a reach- and catchment-specific manner. Working with the contemporary character and behavior of each reach, strategies are developed to enhance geoecological recovery towards the reference condition. Efforts at river rehabilitation must appreciate the stage and direction of river degradation and/or recovery for each River Style (i.e., whether the geomorphic condition of the river is improving, or continues to deteriorate). Hence, interpretation of the trajectory of river change (Stage Three of the River Styles framework) aids identification of attainable conditions for river rehabilitation.

Due regard must also be given to potential off-site impacts, ensuring that balanced perspectives on sediment transfer are determined. For example, it may be pointless to expend significant effort and resource on "fixing" a downstream reach if a large sediment slug sits immediately upstream. Assessment of river condition and recovery potential completed in Stages Two and Three of the River Styles framework provides the tools needed to identify target conditions and goals for river rehabilitation. Creation and restoration goals equate to the expected or natural condition of a River Style operating in a sustainable manner under the prevailing boundary conditions, providing appropriate target conditions for river rehabilitation. Information on minimally impacted reaches of each River Style guides the identification of target conditions for channel alignment, geomorphic unit assemblage, vegetation character, sediment distribution, and channel–floodplain relationships for reaches in poorer condition. Hence, reaches in good geomorphic condition are used to design the target conditions for river structure and function of reaches in moderate or poor geomorphic condition. Examples of the types of analysis undertaken are presented in Figures 12.4–12.6.

Information on the condition and recovery potential of each reach in the catchment is used to determine the level of intervention and the type of manipulation required to attain a sustainable river structure and function, and the level of risk associated with rehabilitation of each reach. The timeframe of recovery is related to the recovery potential of the reach, and the level of intervention required is related to the condition of the reach. Reaches in good condition with high recovery potential will require minimal intervention, with prospects for visible results relatively quickly. Obviously, as the condition and recovery potential of reaches deteriorates, the required level of intervention increases and the rate of recovery decreases. In poor condition reaches, direct intervention and manipulation may be required. The scientific insight provided through application of the condition and recovery potential frameworks is used to define, for each reach, the parameters that require manipulation to enhance recovery towards the target condition. These analyses are used as a benchmark against which to assess whether improvement has occurred (Stage Four, Step Four).

12.4 Stage Four, Step Three: Prioritize efforts based on geomorphic condition and recovery potential

Among the many challenges facing resource managers is how to prioritize expenditure on river management practices, whether within an individual catchment or between river systems. Such decisions cannot be made in a systematic and rig

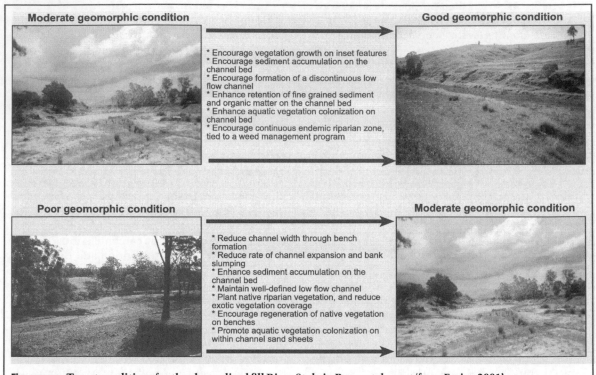

Figure 12.4 Target conditions for the channelized fill River Style in Bega catchment (from Fryirs, 2001)
Reaches in different geomorphic condition are used to determine target conditions for river rehabilitation. Those reaches in good and moderate condition are used to determine the geomorphic structure and vegetation associations required for river rehabilitation along reaches in poor condition. In this example the channelized fill River Style is presented.

orous manner without catchment-framed baseline surveys and a catchment-vision. A prioritization framework determines the sequencing of actions that can be applied to achieve the catchment-based vision. Defining what is achievable in big-picture terms allows prioritization frameworks for river conservation and rehabilitation programs to be effectively applied. These prioritization frameworks ensure that the most cost-effective and efficient reach based strategies work towards the catchment vision.

While economic, cultural, and social values place obvious constraints on application of prioritization frameworks, a physical template forms a critical basis for decision-making. By applying a physical prioritization framework, reaches where the greatest likelihood of success is likely to be attained are identified. In many cases, the rehabilitation of strategic reaches triggers positive responses elsewhere, enhancing prospects for recovery throughout the catchment.

The prioritization procedure applied in the River Styles framework is presented in Figure 12.7. Particular emphasis is placed on "unique" attributes of a catchment (e.g., rare types of river), regardless of their condition, and those rivers that remain intact (i.e., remnants of predisturbance conditions). These reaches form the basis from which to work outwards into more degraded sites in the catchment. Emphasis is then placed on strategic reaches that protect conservation priorities, and reaches that have high recovery potential. Finally, more difficult tasks are contemplated, in the most degraded parts of the catchment.

The philosophical perspective that underpins the prioritization strategy for management efforts in the River Styles framework is as follows (Brierley and Fryirs, 2000):

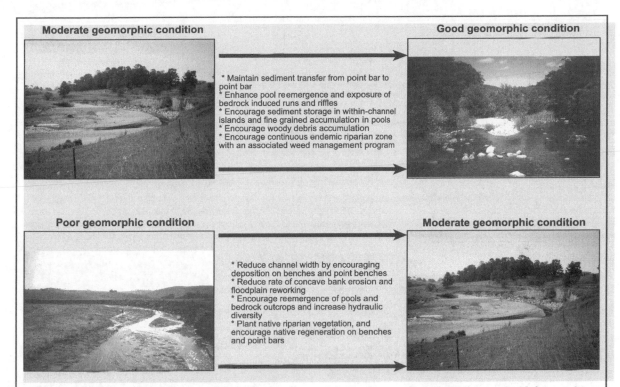

Figure 12.5 Target conditions for the partly-confined valley with bedrock-controlled discontinuous floodplain pockets River Style in Bega catchment (from Fryirs, 2001)

Reaches in different geomorphic condition are used to determine target conditions for river rehabilitation. Those reaches in good condition are used to determine the geomorphic structure and vegetation associations required for river improvement. In this example the partly-confined valley with bedrock-controlled discontinuous floodplain pockets River Style is presented, showing how reaches in moderate geomorphic condition are initially used as a target condition for reaches in poor condition. Once the moderate condition is attained, reaches in good condition provide the ideal target condition for further rehabilitation.

1 *Conservation precedes rehabilitation.* Since habitat conservation is the key to maintaining the biodiversity of aquatic ecosystems, preservation of remaining near-intact fragments of river courses is the first priority. Refugia must be retained under differing flow conditions (e.g., ensuring that pools continue to support water under low flow conditions). Inevitably, the distribution of conservation reaches is fragmented. Most "near-intact" reaches are in national parks in relatively inaccessible parts of the landscape. Given their isolation, these remnants are often unrepresentative of predisturbance conditions in a wider context. However, the systematic, but inadvertent, destruction of former ecological attributes of river courses makes these remnants all the more precious in terms of their

conservation and heritage values. Hence, putting aside the protection of infrastructure and equivalent site-specific requirements, emphasis is placed on conservation of reaches that have unique geomorphic structure, or are remnants of predisturbance conditions.

2 *Strategic reaches* with potentially threatening off-site impacts are the second priority. Particular attention is given to reach or point-impacts that threaten conservation reaches. Irrespective of their geomorphic condition, these reaches must be targeted early in the river rehabilitation process. In many cases, the effects of rehabilitation of strategic reaches will be propagated throughout a catchment, enhancing the natural recovery of adjoining reaches. Identification of strategic reaches

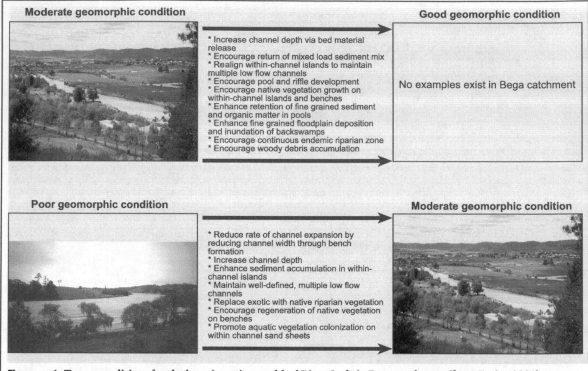

Figure 12.6 Target conditions for the low sinuosity sand-bed River Style in Bega catchment (from Fryirs, 2001)
Reaches in different geomorphic condition are used to determine target conditions for river rehabilitation. Those reaches in good and moderate condition are used as target conditions for river rehabilitation programs that tackle reaches in poor condition. These target conditions provide a basis to identify the geomorphic structure and vegetation associations required for river improvement. In this example the low sinuosity sand bed River Style is presented.

shifts the emphasis of river rehabilitation practice from reactive to proactive decision-making. Preemptive management strategies target reaches that are adjusting "unnaturally" or show signs of accelerated or anomalous behavior yet retain inherent ecosystem values. These reaches have the greatest potential to change, impacting negatively in terms of loss of values and associated off-site impacts. Within this category, infrastructure/asset protection and mitigation of negative societal impacts are also considered as priorities.

3 The next strategy is to work in reaches with *high natural recovery potential*, thereby maximizing the likelihood of management success. Whenever possible, the ideal sites to commence rehabilitation programs are connected to those parts of the catchment in which river character and behavior

are in good condition (e.g., high conservation value reaches), such that longer-term strategies can build on greater lengths of river which have appropriate river structures for their setting. Isolated reaches with high recovery potential may provide important foci or starting points for river rehabilitation. In this way, the longitudinal connectivity of reaches in good geomorphic condition is maximized. Less impacted sections of a River Style are used to assess appropriate target conditions for more degraded river reaches of the same River Style. Strategies that mimic the character and behavior of good condition reaches of the particular River Style will be most cost-effective in the long term, and will require minimal on-going maintenance. As recovery is already underway, a "do-nothing" option may be quite feasible in these

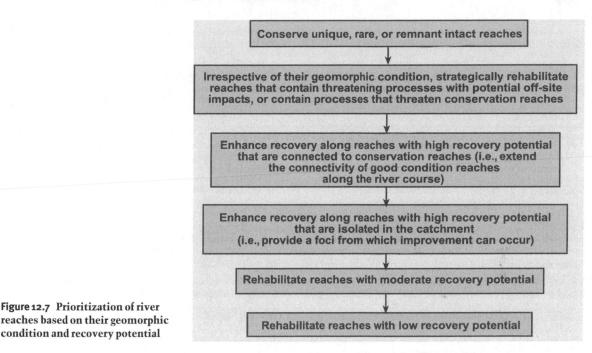

Figure 12.7 Prioritization of river reaches based on their geomorphic condition and recovery potential

reaches. Elsewhere, minimally invasive approaches based on riparian vegetation management may facilitate accelerated recovery.

4 *Consider more difficult tasks.* In degraded reaches that are experiencing sustained adjustment, costly river rehabilitation programs may not yield substantive outcomes, adversely impacting on community confidence in the river management process. Ultimately, this may prolong or compromise the achievement of the vision. Degraded reaches often require invasive rehabilitation techniques with costly on-going maintenance. Although conventional river engineering practices can be employed, the most cost-effective strategy may simply be to wait for these reaches to regain some sort of physical balance before adoption of intervention strategies. In many cases, management success along these reaches will only be attainable at reasonable cost once rehabilitation has been achieved in upstream reaches. Over time, cumulative, off-site improvement may aid the recovery of low recovery potential reaches. Hence, framing goals in context of the catchment vision aids the prospects for rehabilitation success throughout the system.

Application of the River Styles prioritization procedure in working towards the catchment-based vision for Bega catchment is presented in Plate 12.1.

12.5 Stage Four, Step Four: Monitor and audit improvement in geomorphic river condition

Effective monitoring of river rehabilitation programs is a prerequisite for appraisal of successes and failures. Unfortunately, such endeavors have largely been overlooked in the past. This oversight reflects, in part, the short-term nature of many rehabilitation projects and associated funding, a lack of "requirement" to undertake monitoring, limited baseline information against which to monitor change, and poor articulation of objectives of what to monitor and why (Kondolf, 1995; Kondolf and Michcli, 1995). Underlying these factors is a lack of a clearly specified vision and measurable objectives. Procedures for monitoring and auditing must be considered at the outset of any program. Postproject appraisals that endeavor to learn from experience must be tied to preproject data (Downs and Kondolf, 2002). Adoption of adaptive management principles, whereby clearly definable and realistic goals are regularly reappraised and readjusted, optimizes the likely success of monitoring programs, aiding the reinforcement of ideas and learning.

Stratification of landscapes into areas dominated by similar processes aids the design and implementation of effective assessment and monitoring strategies (Montgomery, 1999, 2001). If stratified too finely, local dynamics are hard to interpret because they are driven largely by external processes. If landscape stratification is too coarse, it may be difficult to delineate causal linkages. Hence, the spatial scale at which a monitoring program is applied strongly influences its usefulness for ecosystem management. Data must be framed in context of the disturbance history, evolutionary tendency, and linkages of biophysical processes that operate in the system of concern.

The River Styles framework provides a multi-scalar, process-based procedure by which monitoring programs can be structured in a manner that relates to the "natural" range of diversity evident along a river course, comparing like with like in a meaningful manner (see Boulton, 1999). Catchment-wide River Styles maps can be used to select representative sampling points, reflecting inherent variability in river character and behavior, and measures of geomorphic river condition. Sampling strategies can then target relevant geo-ecological attributes of different types of rivers, reflecting the array of differing landforms, vegetation associations, and hydraulic diversity evident in these different settings.

If issues of monitoring and auditing are to be addressed in a comprehensive manner, an inventory of river diversity and condition at State and/or National levels is required. Without this inventory, the uniqueness or rarity of different types of river character and behavior cannot be determined. Various structural and practical constraints have inhibited the development of truly integrative biomonitoring programs (Armitage and Cannan, 1998).

12.6 Products of Stage Four of the River Styles framework

Products derived from Stage Four of the River Styles framework include:
• tables outlining the catchment-based vision and plan of attack;
• photographs and tables demonstrating target conditions and actions required to move a reach along the recovery trajectory and improve its condition;
• catchment-based prioritization map, with boxes that demonstrate the actions that will be taken to enhance the recovery of representative and strategic reaches and achieve the catchment vision;
• a geomorphic platform upon which representative auditing procedures can be developed and applied.

CHAPTER 13

Putting geomorphic principles into practice

People that see the glass as half full will be the successful agents of change. They will recognize an opportunity to fill the glass. They will be the ones whom uncertainty will stimulate rather than intimidate. They will be the ones who predict the future by creating it.

Henry N. Pollack, 2003, p. 239

13.1 Introduction

In this concluding chapter, the place of geomorphology in river management practice is framed in terms of broader considerations that confront the uptake of scientific notions in environmental management, and emerging issues at the crossover between scientific and community values. Naturally, these perspectives vary markedly from place to place, and from society to society. Indeed, competing agendas are commonly evident both internally within the water resources management arena, and in relation to other forms of natural resource (environmental) management.

Cultural, aesthetic, and psychological associations with rivers fashion the values with which societies appraise what they want rivers to be like. Differing social and cultural perspectives result in starkly differing perceptions of environmental "problems." Such value judgments are far from static. They change as rivers themselves change, with sense of place, or in response to personal experience, background, or stage of life. Some laud river diversity and dynamic adjustments; others fear change and crave the comfort of the stable and the familiar. Some see recreational opportunities where others see potential resource applications or hazards. Depending on the location and associated societal infrastructure, the character, rate, and consequences of river changes may be of significant concern or largely irrelevant to humans. Development of cutoff channels or wholesale avulsion may be lauded from afar as a healthy part of aquatic ecosystem revival when it occurs in intact riparian rainforests. However, inundation of local urban areas built on floodplains, catastrophic erosion of farming land, or siltation of navigation

channels are viewed as "natural" hazards that present significant threats and considerable expense to society. To an ecologist or geomorphologist, "natural" river change is healthy. However, to a design engineer, whose mandated role may be to "stabilize," "train," or "improve" a river reach, undue change may result in questioning of their professional competency. The river management process entails integration of these and many other values. As in many things in life, a sense of loss requires association, connection, and appreciation of present surroundings.

Four primary themes are addressed in this chapter. First, the place of geomorphology in environmental science is considered, emphasizing the need for field-based insights as bases for practical, on-the-ground applications (Section 13.2). Various components of geomorphic inquiry that underpin these applications are discussed in Section 13.3. Community engagement in river management, and uptake of geomorphic considerations in this emerging arena, are addressed in Section 13.4, highlighting the need for adaptive management. Finally, the way in which these various issues are approached in applications of the River Styles framework is summarized in Section 13.5.

13.2 Geomorphology and environmental science

Concern for river health, and broader concern for the state of the world's environment and the sustainability of human lifestyles, has triggered some remarkable improvements in environmental health. The capacity for reform and creative engagement in the face of adversity bodes well for ongoing and future endeavors in conservation, restoration, and rehabilitation. The uptake of scientific principles in these activities bears a strong relation to societal structure, what can be afforded, and the priorities placed on such endeavors. This reflects, in part, relative affluence and the luxury of choice.

In a sense, rivers can be viewed as barometers of landscape condition, or catchment health. Improvements to river condition are contingent on researchers, managers, and the community working together to establish sustainable, long-term management strategies that "work with nature." Ongoing programs target the sustainability and biodiversity of aquatic ecosystems, and the development of more environmentally and socially just river rehabilitation activities. The values of vibrant, living rivers and their associated ecosystems are increasingly recognized. Although considerable successes have been generated in river rehabilitation projects, bigger, more visionary ventures are required to challenge the imagination (Shields et al., 2003). However, this is not a matter of resolution via technofix. Far from it. Sustained outcomes require grass roots campaigns and collective commitment. There is no such thing as a short-term solution. Transient responses must be reinvigorated and expanded if long-term benefits are to be achieved.

Scientists face considerable challenges in ensuring that the best use is made of available knowledge. Inevitably, understanding is incomplete, it has often been developed at inappropriate scales, and cross-disciplinary perspectives are seldom readily integrated in a coherent manner. The reductionist, discipline-bound nature of scientific inquiry has constrained endeavors to embrace more inclusive, transdisciplinary, ecosystem-based approaches. This is more than just "putting the pieces back together." It requires knowledge on how components and scales are linked and interact both spatially and temporally. Striving to restore ecosystem values is much more challenging than single focus engineering projects such as erosion or flood control works. Principles of ecosystem management hypothesize that resource use can be redesigned to continue without jeopardizing the long-term integrity of natural ecosystems (Montgomery, 2001). An ecosystem approach to natural resources management recognizes that key biophysical processes should be the main

guide to intervention in river systems rather than technology or economics.

Lack of coherency has prompted societal crises of confidence in the guidance (or debate) proffered by the scientific community. In part this reflects failures in communication of the uncertainties and complexities that characterize biophysical interactions. Complete understanding will never be attained, but significant insight is available to guide management applications. Black and white perspectives based on linear cause and effect thinking are unlikely to yield realistic outcomes that maintain credibility and confidence in the scientific process. The nonlinear behavior of many natural systems places significant constraints on the reliability of predictions of future scenarios. However, this does not negate the critical leadership role that scientists must adopt in the development and implementation of visionary management programs.

These issues go hand-in-hand with concerns for quality control and the training received by new practitioners in environmental science. Across most of the western world, few would argue that educational standards are improving, whether at High School or Tertiary level (regardless of bureaucratic manipulation of statistics). Ultimately, genuine uptake of science is reflected in meaningful outcomes that mark improvements in environmental health on-the-ground. There will always be the need for pure research, but geomorphologists have a responsibility to engage with communities, communicate their knowledge, and promote practical outcomes that build on their applied research.

In seeking to make geomorphology more accessible, its origins as a field science must be emphasized, presenting a practical basis for on-the-ground applications. Considerable rewards are to be gained through this experience. This is not an exercise in three-dimensional graphic design on a screen, but reading real-world landscapes. It is a battle of hearts and minds and hands. Critically, this battle cannot be won by individuals, however talented or inspiring they may be. Working together is step one on the pathway to success. Commitment to the process of environmental management, through collective ownership of decisions and learning from outcomes are pivotal components of effective practice. In no way does this seek to stifle individual flair and creativity. Such skills are often in limited supply and are avidly sought. It is the cultivation and meaningful integration of talent in constructing future visions and guiding practical steps to achieving them that presents the core challenge. These are matters of societal will, institutional and legal support, and moral/ethical leadership.

The issues raised above may appear to be far removed from geomorphic concerns for river diversity, patterns of physical linkages, and evolutionary trajectories. However, science is not undertaken in a vacuum. There is much more to be gained than the production of a nicely-bound dissertation or thesis, the findings from which may scarcely see the light of day. Perhaps this is a good time for reflection, to reappraise approaches to scientific-inquiry, and consider how greater societal rewards can be engendered from the process.

13.3 Geomorphology and river management: Reading the landscape to develop practices that work with river diversity and dynamism

Biophysical principles underlie the achievement of sustainable ecological practices that maintain (or enhance) the integrity of aquatic ecosystems. All endeavors in environmental management have a landscape context. Landscapes integrate responses to natural and human-induced disturbance events. All activities, whether resource developments or rehabilitation treatments, have off-site impacts that are manifest over a range of spatial and temporal scales. In physical terms, the catchment-scale provides the most appropriate basis to interpret these interactions, thereby providing a critical template with which to frame management programs. Understanding of geomorphic processes, and determination of appropriate river structure at differing positions in catchments, is critical in effective, sustainable rehabilitation of river courses.

Ultimately, however, river health will never be improved if it is framed either in terms of poor science, or poor application of good science. Substandard products that bear little relevance to the particular system or issue of concern present critical limitations in environmental practice. All too often, rehabilitation design has been based on an armory of approaches derived from manuals,

rather than genuinely allowing the river to speak for itself, considering site- or reach-specific issues, and the manner of river adjustment. These themes prompted the development of the River Styles framework and the writing of this book.

So, what do fluvial geomorphologists need to do to see more effective uptake and enhancement of their work? First and foremost, the landscape basis for geoecological inquiry must be reiterated and developed. Habitat availability and viability must be tied to geomorphic structure and function, and the patterns and rates of biophysical fluxes must be related to notions of geomorphic connectivity in different landscape settings. These considerations form the platform for "thinking like an ecosystem," reinforcing the need for catchment-scale planning activities that place due regard on linkages, off-site impacts, and lagged responses.

Second, geoecological notions of ongoing, cumulative responses to disturbance events must underpin interpretations of how river systems work. Concerns for river condition and recovery potential, and the sensitivity of any given reach to change (including notions of vulnerability, susceptibility, and proximity to a threshold), form the platform for determination of a dynamic, integrative physical template that presents a future focus for research and management activities. In striving to improve the geomorphic condition of rivers through rehabilitation activities, principles from geomorphology can be applied to determine what is realistically achievable in any given reach, framing appraisals of recovery potential within a catchment context.

Third, overarching visions at ecosystem, regional, and catchment scales must integrate programs in land management, flow management, vegetation (forest) management, wetland management, management of parks and reserves, soil conservation programs, estuary management, etc., providing a realistic sense of what can be achieved in biophysical terms. Foresighting (scenario-building) exercises may help to appraise catchment or regional priorities and/or limitations that may constrain what can be achieved in management terms. Such insights present a basis to differentiate among potential benefits and costs of "leave it alone," "targeted conservation/rehabilitation options," and "let's spend lots of money" options.

Fourth, appropriate communication strategies that extol the virtues of grounded, real-world knowledge of the natural world must be developed, emphasizing reservations in the use of averaged or inappropriately modeled data. Most river practitioners cherish the fact that they do not live in a world of norms. Although reassurance is gained through familiarity, especially if preconceived notions seem to work, a buzz of excitement beckons in the discovery of something that is "new." Other than reaches that require hard engineering structures to protect infrastructure, reappraisals of thinking have moved beyond single-function "solutions" that aim to impose stability upon a river towards more environmentally sympathetic techniques. A radical and enduring change is underway in river rehabilitation practice (Williams, 2001; Hillman and Brierley, in press), in which fluvial geomorphology has emerged as a core component for river management practice, providing solutions to problems on-the-ground as well as guiding various policy developments and aiding legal reform (Gilvear, 1999; Rhoads et al., 1999; Rutherfurd et al., 2001b; Brierley et al., 2002).

13.4 The river management arena

Good rehabilitation practice moves beyond technical competence and efficiency to embrace a range of social, cultural, political, moral, and aesthetic qualities (Carr, 2002). Inevitably, these values vary from place to place, with differing historical overtures. Major ecological rehabilitation will not be undertaken unless human society approves the goals and objectives, and aspires to maintain the integrity of the rehabilitated ecosystem (Cairns, 1995). Neither technically feasible goals nor scientifically valid goals will be possible in the absence of societal acceptance. Ideally, the community provides the purpose and motivation for the project, guiding what it is hoped will be achieved. Input is also required to implement, maintain, and monitor projects. Response to feedback ensures that outcomes are encapsulated within an adaptive management process. The presentation of such projects has important educational qualities for both landowners and river managers. Partnership approaches to river rehabil-

itation develop awareness, education, and support for achieving mutual goals. The way that river rehabilitation projects are presented to a wider audience and the way in which the audience can become a participant are crucial components of the rehabilitation process (Boon, 1998).

Environmental decision-making is essentially an ethical and political rather than a scientific or technical task (Hillman, 2002). Social attitudes determine the likelihood of success. Will, commitment, and engagement are required to attain sustainable environmental outcomes. A pervasive sense of "duty of care" must underlie this process. Approaches to stakeholder involvement have been variously termed participation, partnership, community involvement, or multistakeholder processes (Hillman and Brierley, in press). Phrases such as "capacity building," "strengthening of communities," and "community engagement" are now an essential part of the vocabulary of environmental management generally. In striving for a fair-go in river rehabilitation practice, a commitment to environmental justice is required (Hillman, subm.). Imposition of noninclusive, nonconsultative "solutions" fails to engage river communities, externalizing concern for river health as someone else's problem. Top-down or bottom-up approaches, in themselves, are unlikely to achieve sustainable, long-term success in environmental management.

Failure to incorporate communities into river management programs has resulted in widespread alienation from the decision-making process, and a failure to tap into local knowledge and resources. To redress this concern, greater emphasis must be placed on efforts that enhance prospects for the emerging "middle-ground" between science and management (see Table 13.1; Carr, 2002). Bringing groups together to generate a shared vision enhances the commitment and focus needed for a successful project. The derivation of a shared vision requires the reconciliation of a range of potentially conflicting interests. The visioning process itself may have large payoffs, through dialogue and recognition of differences. Such engagement is time-consuming and must be adequately resourced. The process starts with listening and clear communication.

Increasingly, community groups no longer expect governments of any ilk to fix problems. An equal disregard is often held for researchers and

Table 13.1 The emerging middle ground in environmental management (based on Carr, 2002, p. 199).

Top-down approach

Potential to:	Danger of:
• Shape local practice in light of national and international forces • Promote efficient utilization and equity in distribution of national/state resources • Develop coherent planning and administrative support among various institutional levels • Provide access to technical and research-based information and associated on-the-ground tools	• Lack of awareness of local needs and conditions • Difficulty in identifying and coordinating local contributions to national programs • Undue emphasis on larger, more visible groups and large-scale projects • Departmental and disciplinary-based barriers to effective communication • Short-term politically expedient actions • Simplistic reductionist framing of environmental problems in purely biophysical terms • Institutional and ideological barriers to local participation • Formula/prescriptive approach to community groups • Challenge of disciplinary chasms and institutional barriers

⇓

Middle-ground approach

- Integrate the benefits and address the dangers of top-down and bottom-up approaches to environmental management through applying good practice through:
- Institutional and legal reform that accommodate the emergence of local organizations and community resource centers (or knowledge networks) across regional or State boundaries
- Engagement with as wide a range of practitioners as possible, striving for representative coverage
- A shared commitment to vision building, built on a common information base and effective communication/facilitation
- Maintaining flexibility through adaptive management, embracing experimentation, and meaningful monitoring
- Due regard for process, rather than purely focusing on outcomes
- Adherence to principles of environmental justice, procedural fairness and intergenerational equity
- Application of a consensus framework, ensuring sufficient time is spent on negotiation, decision-making, planning, action, and monitoring
- Rewarding success and learning from failures, appreciating the historical focus of river rehabilitation activities
- Linking training, education programs, and successional planning arrangements

⇑

Bottom-up approach

Potential to:	Danger of:
• Develop local approaches to catchment planning • Develop and implement monitoring programs that are appropriate to local conditions • Ensure effective utilization and equity in distribution of local resources • Share perspectives and empower local communities through communication and/or negotiation and self-generation activities • Promote local action based on ownership of problems	• Duplication of effort, wasting local resources • Parochial attitudes, not seeing the broader picture • Inappropriate local expectations of achievements • Entrenched leadership not successfully helping the group to progress, and associated challenges presented by burn-out of champions, succession planning, "sharing" of responsibilities, etc. • Lack of group-process skills and an inability to evolve and mature • Uncertainty about whether empowerment truly brings with it "responsibility" and capacity to continue in the light of failure – it may seem too hard, and it's all too easy to walk away • Perception that land users lack skills and education required for environmental management • Challenge of access to information and its coherency/use in decision-making

notional "expert groups." Unless communities are engaged in the process, they will always look to blame someone if things go awry – whether the local management agency, the expert consultants brought in to appraise options, or the local/ state/national government of the day. Recurrent "failures," or even perceptions of failure, may compromise local community goodwill and commitment towards rehabilitation programs, negating the potential for ongoing maintenance. However, if collective ownership of outcomes is achieved, such that lessons are learnt, the rehabilitation process should be considered to be a success. Management efforts will have greatest likelihood of success if there is mutual respect among managers, stakeholders/community representatives, researchers, and others involved in the processes, implementation, and auditing of environmental management. Learning by doing recognizes that each failure is a stepping stone to success. Alternatively, each success enhances the prospects for progressive and sustained reinforcement of ideas and practice. Once gained, momentum must be maintained and enhanced. Mistakes only continue to be a problem if society continues to repeat them (Hobbs, 2003).

A mutual commitment to learning and knowledge transfer, and collective ownership of management plans, is required if long-term programs are to achieve sustainable outcomes. Appropriate communication and environmental education services are fundamental to the process of mutual learning that underpins effective environmental management (Mance et al., 2002). Mechanisms must be set in place for critically based dissemination and use of information. The mind-set within which information is gathered, knowledge is developed, and understanding is communicated present critical constraints on the use of scientific insights. The intent of what is said, and what the target audience actually hears, may be two very different things. This is much more than an issue of word selection and sentence construction. Selective hearing is a part of human nature. To overcome this issue, ownership of information and progressive reappraisal, reinforcement and extension are key components of the adaptive management process (Hillman and Brierley, 2002). To engender trust at the outset, baseline data must integrate scientific and local knowledge through collective dialogue and informed debate. With all information on the table, an open, transparent, and consultative approach is required to prioritize a schedule of on-the-ground works, ensuring that environmentally just strategies attain a balance between conservation and rehabilitation activities. An accompanying commitment to maintenance and auditing must go hand-in-hand with this process.

The push towards greater community involvement in river and catchment management demands that rather than adoption of prescriptive approaches, individual systems must be managed in a flexible manner on the basis of what is actually happening within each river system. Educational tools that assess how catchments work must stress linkages, complexities, and the inherent uncertainties of many environmental outcomes, and place site-specific issues within a total catchment context. Traditionally, management decision-making has typically been framed over short time-frames, with a perception that the river operates as a simple, linear system (Petts, 1984). However, rivers change in episodic and complex ways, dependent on certain thresholds. Practitioners must learn to distance themselves from obvious/visible problems, viewing site-specific issues in their broader (catchment) context. Unfortunately, broadly scoped projects often lack the motivation, planning, support, and funding to be successful. Ultimately, however, everyone is guided by results, and the prospects for long-term success are enhanced by catchment-framed, inclusive, and visionary programs. Research programs must be implemented to accompany these schemes at the outset. For example, design of long-term, catchment-scale projects enables short-term hypotheses of critical ecosystem mechanisms or processes to be investigated (Lake, 2001a).

Adaptive management principles promote concern for process and context, rather than simply emphasizing the short-term outcomes of any given activity. Efforts at river rehabilitation must continue regardless of limitations of knowledge. In many regions, formalized knowledge of river character and behavior is rudimentary, and it is inappropriate to transfer our knowledge from elsewhere in an uncritical manner. In the absence of background understanding, the precautionary principle should be followed.

Natural resource management must continue regardless of limitations imposed by financial and other constraints. Because of its timescale, complexity, and transdisciplinary nature, coping with uncertainty should be a goal of river management, rather than attempting to remove it or using it as an excuse for inaction (Dovers and Handmer, 1995; Clark, 2002). Implicit in setting priorities is the recognition that it is unlikely that everything can be conserved everywhere, so scarce resources must be allocated in ways that can be expected to produce the best outcomes overall (Hobbs and Kristjanson, 2003). Approaches must determine where the greatest benefits will be achieved in a cost-effective manner over a realistic timeframe. As river management entails multiple goals, not all of which are necessarily complementary, open and transparent procedures must be used to ensure accountability is maintained in the prioritization process. Is it more appropriate to spend huge amounts on saving the last remaining individuals of a species on the brink of extinction or to invest in protecting habitat that is used by many other species? Alternatively, is it better to invest in purchasing and managing small patches of good quality habitat or in rehabilitating larger tracts of currently degraded habitat? All too often, there is a preference for dealing with urgent care for charismatic species rather than implementing longer-term preventative measures. For example, if a reach downstream is subject to rehabilitation initiatives, while upstream areas lie on the brink of releasing large stores of sediment, socially constructed priorities may ultimately be unsuccessful due to impacts from outside the reach. Priority areas are likely to account for only a small percentage of the total, meaning that large areas will not be a priority (Hobbs and Kristjanson, 2003). However, the local community in a nonpriority area is likely to think otherwise and see its local surroundings as a priority! Prioritization of rehabilitation projects with a preservation first approach has proven to be the most effective approach to allocation of resources (Boon, 1998). Catchment framed, biophysically informed management visions are crucial to prioritizing reaches. Ultimately, single-interest rehabilitation projects that tackle a particular part or function of an ecosystem are unsustainable and inequitable.

Increased awareness or activity do NOT necessarily equate to success in bringing about substantive change. Ultimately, efforts at restoration or rehabilitation must demonstrate tangible achievements or more effective outcomes than the "do nothing" option, whereby natural processes enable self-sustaining, cost free, improvement of its own accord (Bradshaw, 1996). Jackson et al. (1995) note that success of rehabilitation programs should be demonstrable within 10–50 years. This timeframe is verifiable: if rehabilitation will result in an improvement in ecosystem health in 50 years or less, the evidence for this should be visible in 1–10 years. As this timeframe falls within one human lifetime, it is possible to hold those who inflicted the damage accountable for repairing it.

Society must be aware of the real cost to fix things if appropriate investment is to be made in land repair practices. In some instances, the costs of repair may be less than the costs for prevention! Maximizing the opportunity to "get it right" at the outset will potentially save considerable sums of money. Inefficiency in the execution of the project is avoided by doing things in the right order. If, during this process, practitioners become overwhelmed by the complexity or enormity of the task, their efforts are likely to be compromised. A clear strategy articulates small but progressive steps along the way.

The range of biophysical scales at which stream rehabilitation must operate is seldom matched by equivalent institutional structures, as most institutional arrangements are sociopolitical rather than spatial in origin (Rogers, 1998; Dovers, 2001; Tippett, 2001). Institutional structures need to be flexible and adaptive, employing a holistic approach to management of river systems that incorporates knowledge generation and commitment to a process of learning. Agencies must have the mission, mandate, resources, authority, and skills to effectively manage rivers. Policy, planning, legal, and institutional arrangements must ensure that programs are developed and applied in a socially and environmentally just manner, with a genuine and practical sense of "best management practice." Leadership of the management process must be sustained through succession planning, recognizing the ongoing requirement for training as understanding improves and staff change.

Interdisciplinary learning and systems training must be incorporated into management practices. All "revolutions in perspective" require recurrent inputs that foster the processes that drive change, with appropriate doses of patience, persistence, and resilience.

Whether research and/or management institutions are ready to address this challenge is a matter of conjecture. The notion of integrative science strongly implies a break with reductionist, single-discipline research and management, a tradition underpinned by institutional structures within academic and government agencies. However, this is something of a redundant issue; the challenge is already upon us. In many parts of the world, community groups await collective engagement and mutual guidance in the design, implementation, and maintenance of river rehabilitation projects. Fluvial geomorphologists, among numerous disciplinary specialists, have a moral and social responsibility to engage in the management process. Hopefully, future generations will view the intellectual guidance proffered by contemporary fluvial geomorphologists not only in terms of the communication of knowledge, but also in terms of what has actually been achieved through use of that knowledge.

13.5 Use of the River Styles framework in geomorphology and river management

Appropriate information frameworks present a basis for inclusive, informed debate in river management, providing guidance on the inherent complexity and uncertainty of river systems. This enables gaps in knowledge to be identified, and limitations of understanding to be recognized. The River Styles framework provides a structured set of procedures with which to collect, synthesize, manage, and communicate catchment-specific information. Interpretation of controls on geomorphic river character, behavior and evolution is used to explain contemporary river condition and recovery potential. This promotes the adoption of proactive strategies that work towards a clearly articulated and realistic vision. Although developed in an Australian context, the approach is generic and open-ended, enabling procedures to be applied in any situation. Doubtless extensions and modifi-

cations to the procedure will be required as additional issues arise.

Specialist geomorphological training and stringent quality control procedures are required to ensure that technical standards and protocols are applied in river rehabilitation practice (e.g. Thorne, 1997; Raven et al., 1998; Montgomery, 2001). As noted by Schumm (1991, p. 58), an investigator's experience and perspective may be crucial in solving a problem, while an investigator's bias may prevent a solution. In striving to maintain professionalism and quality assurance, application of the River Styles framework has been developed using short courses and an accreditation procedure.

It is recognized implicitly that the River Styles framework is scientifically based, while river management decision-making is a consultative processes, driven by multiple stakeholders with differing sets of agendas. However, the availability and delivery of coherent information must provide a foundation premise for effective decision-making.

Several core themes in this book warrant final comment:
• ecosystem thinking requires a landscape context;
• rivers are critical linking elements of landscapes and should be viewed in their catchment context;
• remarkable diversity of river structure and function presents a wide range of aquatic habitat in different settings;
• the connected nature of river systems ensures that impacts in one area may have considerable consequences elsewhere, over widely ranging spatial and temporal scales;
• the natural disturbance regimes under which rivers operate ensure that "equilibrium" behavior should not be expected and does not provide an appropriate basis for management practice;
• the memory of any given landscape, and ongoing adjustments to cumulative disturbance events, makes it difficult to discern specific cause-and-effect relationships and predict future trajectories of change.

Geomorphological perspectives that emanate from management applications of the River Styles framework include:
• respect diversity, striving to work with "natural" form–process interactions at the reach scale,

their ongoing adjustments, and responses to off-site, catchment-scale disturbance;

• work with system dynamics, recognizing that many geomorphological systems demonstrate nonlinear, nonequilibrium behavior. Separation of behavior from change provides a useful layer in analysis of evolutionary tendencies;

• use nested hierarchical procedures to break down rivers into meaningful components for analysis and communication. However, ensure that these components fit together in management applications, maintaining the integrity of ecosystems and associated linkages at different spatial and temporal scales;

• a catchment-framed geomorphic template provides a basis to assess biophysical processes along river courses, unraveling causality in assessment of controls and responses to disturbance;

• focus attention on the underlying causes of problems associated with river changes, rather than their symptoms;

• use evolutionary insights of river adjustment and change to describe how a river has adjusted in the past, explain how it is adjusting presently, and predict its likely future trajectory of change;

• use appraisals of river character and behavior as a basis to interpret river condition and recovery potential, comparing like with like in a meaningful way.

The River Styles framework provides a research and management tool with which to develop appropriate catchment-specific understanding. The ultimate success of this framework should be measured through its use as a learning tool and its application as a guide for planning on-the-ground river management activities. And finally, "Don't underestimate the challenge." Be realistic in framing goals, working from a premise that strives to "underpromise and overdeliver." Ultimately, no-one is better off if the ecological integrity of the river is compromised.

References

Aadland, L.P. (1993) Stream habitat types: their fish assemblages and relationship to flow. *North American Journal of Fisheries Management* **13**, 790–806.

Abbe, T.B. and Montgomery, D.R. (1996) Large woody debris jams, channel hydraulics and formation in large rivers. *Regulated Rivers: Research and Management* **12**, 201–221.

Abbe, T.B., Montgomery, D.R. and Petroff, C. (1997) Design of stable in-stream wood debris structures for bank protection and habitat restoration: An example from the Cowlitz River, WA. In: Wang, S.S.Y., Langendoen, E.J. and Shields, F.D. Jr. (eds.) *Proceedings of the Conference on Management of Landscapes Disturbed by Channel Incision*, University of Mississippi, pp. 809–816.

Abernethy and Rutherfurd, I. (1998) Where along a river's length will vegetation most effectively stabilise stream banks? *Geomorphology* **23**, 55–75.

Abrahams, A.D., Li, G. and Atkinson, J.F. (1995) Step-pool streams: Adjustment to maximum flow resistance. *Water Resources Research* **31**, 2593–2602.

Adams, W.M. (1996) Future Nature. A Vision for Conservation. Earthscan, London.

Allen, J.R.L. (1965) A review of the origin and characteristics of recent alluvial sediments. *Sedimentology* **5**, 89–191.

Allen, J.R.L. (1970) Studies in fluviatile sedimentation: A comparison of fining-upwards cyclothems, with special reference to coarse-member composition and interpretation. *Journal of Sedimentary Petrology* **40**, 298–323.

Amoros, C. and Bornette, G. (2002) Connectivity and biocomplexity in waterbodies of riverine floodplains. *Freshwater Biology* **47**, 761–776.

Amoros, C., Roux, A.L. and Reygrobellet, J.L. (1987) A method for applied ecological studies of fluvial hydrosystems. *Regulated Rivers: Research and Management* **1**, 17–36.

Andrews, E.D. (1982) Bank stability and channel width adjustment, East Fork River, Wyoming. *Water Resources Research* **18**(4), 1184–1192.

Andrews, E.D. (1984) Bed-material entrainment and hydraulic geometry of gravel-bed rivers in Colorado. *Geological Society of America Bulletin* **95**, 371–378.

Andrews, E.D. (1986) Downstream effects of Flaming Gorge Reservoir on the Green River, Colorado and Utah. *Geological Society of America Bulletin*, **97**, 1012–1023.

Andrews, E.D. (1996) Downstream effects of Flaming Gorge Reservoir on the Green River, Colorado and Utah. *Geological Society of America Bulletin* **97**, 1012–1023.

Angermeier, P.L. and Karr, J.R. (1994) Biological integrity versus biological diversity as policy directives. *BioScience* **44**, 690–697.

Archer, D. and Newson, M. (2002) The use of indices of low variability in assessing the hydrological and in-stream habitat impacts of upland afforestation and drainage. *Journal of Hydrology* **268**, 244–258.

Armitage, P.D. and Cannan, C.E. (1998) Nested multi-scale surveys in lotic systems – tools for management. In: Bretschko, G. and Helesic, J. (eds.) *Advances in River Bottom Ecology*. Backhuys Publishers, Leiden, The Netherlands, pp. 293–314.

Aronson, Lj., Dhillion, S. and Le Floch'h, E. (1995) On the need to select an ecosystem of reference, however imperfect: A reply to Pickett and Parker. *Restoration Ecology* **3**, 1–3.

Bail, M. (1998) *Eucalyptus*. Text Publishing, Melbourne.

Bailey, P.B. and Li, H.W. (1992) Riverine fishes. In: Calow, P. and Petts, G.E. (eds.) *The Rivers Handbook: Hydrological and Ecological Principles*. Blackwell, Oxford, UK, pp. 251–281.

Baker, V.R. (1977) Stream channel responses to floods with examples from central Texas. *Bulletin of the Geological Society of America* **88**, 1057–1071.

Baker, V.R. (1978) Adjustment of fluvial systems to climate and source terrain in tropical and subtropical environments. *Fluvial Sedimentology. Canadian Society of Petroleum Geologists. Memoir* **5**, 211–230.

Baker, V.R. and Costa, J.E. (1987) Flood power. In: Mayer, L. and Nash, D. (eds.) *Catastrophic Flooding*. Allen and Unwin, Boston, pp. 1–21.

Baker, V.R. and Pickup, G. (1987) Flood geomorphology of the Katherine Gorge, Northern Territory, Australia. *Geological Society of America Bulletin* **98**, 635–646.

Barinaga, M. (1996) A recipe for river recovery? *Science* **273**, 648–1650.

Bathurst, J.C. (1993) Flow resistance through the channel network. In: Beven, K. and Kirkby, M.J. (eds.) *Channel Network Hydrology*. Wiley, Chichester, pp. 69–98.

Bathurst, J.C. (1997) Environmental river flow hydraulics. In: Thorne, C.R., Hey, R.D. and Newson, M.D. (eds.) *Applied Fluvial Geomorphology for River Engineering and Management*. Wiley, Chichester, pp. 69–93.

Beechie, T.J., Collins, B.D. and Pess, G.R. (2001) Holocene and recent geomorphic processes, land use, and salmonid habitat in two North Puget Sound river basins. In: Dorava, J.M., Montgomery, D.R., Palcsak, B.B. and Fitzpatrick, F.A. (eds.) *Geomorphic Processes and Riverine Habitat*. American Geophysical Union, Washington, DC, pp. 37–54.

Beeson, C.E. and Doyle, P.E. (1995) Comparison of bank erosion at vegetated and non-vegetated channel bends. *Water Resources Bulletin* **31**(6), 983–990.

Begin, Z.B. and Schumm, S.A. (1984) Gradational thresholds and landform singularity: significance for Quaternary studies. *Quaternary Research* **21**, 267–274.

Benda, L. and Dunne, T. (1997a) Stochastic forcing of sediment supply to channel networks from landsliding and debris flow. *Water Resources Research* **33**, 2849–2863.

Benda, L. and Dunne, T. (1997b) Stochastic forcing of sediment routing and storage in channel networks. *Water Resources Research* **33**, 2865–2880.

Beven, K. (1981) The effect of ordering on the geomorphic effectiveness of hydrologic events. In: Davies, T.R.H. and Pearce, A.J. (eds.) *Erosion and Sediment Transport in Pacific Rim Steeplands*. IAHS-AISH Publication No.132, pp. 510–26.

Beven, K. and Carling, P. (eds.) (1989) *Floods: Hydrological, Sedimentological and Geomorphological Implications*. John Wiley and Sons, Chichester, UK. 290pp.

Beven, K. and Kirkby, M.J. (eds.) (1993) *Channel Network Hydrology*. Wiley, Chichester, UK. 319pp.

Billi, P., Hey, R.D., Thorne, C.R. and Tacconi, P. (eds.) (1992) *Dynamics of Gravel-Bed Rivers*. Wiley, Chichester, U.K. 673pp.

Bird, J.F. (1982) Channel incision along Eaglehawk Creek, Gippsland, Vistoria. *Proceedings of the Royal Society of Victoria* **94**, 11–22.

Bird, J.F. (1985) Review of channel changes along creeks in the northern part of the Latrobe River basin, Gippsland, Victoria, Australia. *Zeitschrift fur Geomorphologie. Suppl Bnd.* **5**, 97–111.

Bisson, P.A. and Montgomery, D.R. (1996) Valley segments, stream reaches and channel units. In: Hauer, F.R. and Lamberti, G.A. (eds.) *Methods in Stream Ecology*. Academic Press, San Diego, California, pp. 23–52.

Bledsoe, B.P. and Watson, C.C. (2001) Logistic analysis of channel pattern thresholds: Meandering, braided and incising. *Geomorphology* **38**, 281–300.

Bluck, B.J. (1971) Sedimentation in the meandering River Endrick. *Scottish Journal of Geology* **7**(2), 92–138.

Bluck, B.J. (1976) Sedimentation in some Scottish rivers of low sinuosity. *Transactions of the Royal Society of Edinburgh* **69**, 425–456.

Bluck, B.J. (1979) Structure of coarse grained braided stream alluvium. *Transactions of the Royal Society of Edinburgh* **70**, 181–221.

Blum, M.D. and Salvatore, V., Jr. (1989) Response of the Pedernales River of central Texas to Late Holocene climatic change. *Annals of the Association of American Geographers* **79**(3), 435–456.

Blum, M.D. and Salvatore, V., Jr. (1994) Late Quaternary sedimentation, lower Colorado River, Gulf coastal plain of Texas. *Geological Society of America Bulletin* **106**, 1002–1016.

Boon, P.J. (1992) Essential elements in the case for river restoration. In: Boon, P.J., Calow, P. and Petts, G.E. (eds.) *River Conservation and Management*. Wiley, Chichester, pp. 10–33.

Boon, P.J. (1998) River restoration in five dimensions. *Aquatic Conservation: Marine and Freshwater Ecosystems* **8**, 257–264.

Bosch, J.M. and Smith, R.E. (1989) The effects of afforestation of indigenous scrub forest with eucalyptus on streamflow from a small catchment in the Transvaal, South Africa. *South African Forestry Journal* **150**, 7–17.

Boulton, A.J. (1999) An overview of river health assessment: Philosophies, practice, problems and prognosis. *Freshwater Biology* **41**, 469–479.

Bradshaw, A.D. (1996) Underlying principles of restoration. *Canadian Journal of Fish and Aquatic Science* **53**(Suppl. 1), 3–9.

Brakenridge, G.R. (1984) Alluvial stratigraphy and radiocarbon dating along the Duck River, Tennessee: Implications regarding floodplain origin. *Geological Society of America Bulletin* **95**, 9–25.

Brakenridge, G.R. (1985) Rate estimates for lateral bedrock erosion based on radiocarbon ages, Duck River, Tennessee. *Geology* **13**, 111–114.

Brandt, S.A. (2000) Classification of geomorphological effects downstream of dams. *Catena* **40**, 375–401.

Bravard, J.P., Landon, N., Peiry, J.L. and Piégay, H. (1999) Principles of engineering geomorphology for managing channel erosion and bedload transport, examples from French rivers. *Geomorphology* **31**, 291–311.

Bravard, J., Amoros, C., Pautou, G., Bornette, G., Bournaud, M., des Châtelliers, M., Gibert, J., Peiry, J.L., Perrin, J. and Tachet, H. (1997) River incision in south-east France: Morphological phenomena and ecological effects. *Regulated Rivers: Research and Management* **13**, 75–90.

Brayshaw, A.C. (1984) Characteristics and origin of cluster bedfroms in coarse-grained alluvial channels. *Memoir – Canadian Society of Petroleum Geologists* **10**, 77–85.

Brice, J.C. (1964) *Channel patterns and terraces of the Loup rivers in Nebraska*. United States Geological Survey Professional Paper, P 0422-D, pp. D1–D41.

Brice, J.C. (1983) Planform properties of meandering rivers. In: Elliot, C.M. (ed.) *River Meandering: Proceedings of the Conference on Rivers '83*. American Society of Civil Engineers, New York, pp. 1–15.

Bridge, J.S. (1984) Large-scale facies sequences in alluvial overbank environments. *Journal of Sedimentary Petrology* **54**(2), 583–588.

Bridge, J.S. (1985) Paleochannel patterns inferred from alluvial deposits: A critical evaluation. *Journal of Sedimentary Petrology* **55**, 579–589.

Bridge, J.S. (2003) *Rivers and Floodplains: Forms, Processes and Sedimentary Record*. Blackwell Publishing, Oxford, U.K.

Bridge, J.S. and Gabel, S.L. (1992) Flow and sediment dynamics in a low sinuosity, braided river: Calamus River, Nebraska Sandhills. *Sedimentology* **39**, 125–142.

Bridge, J.S., Smith, N.D., Trent, F., Gabel, S.L. and Bernstein, P. (1986) Sedimentology and morphology of a low-sinuosity river: Calamus River, Nebraska Sand Hills. *Sedimentology* **33**, 851–870.

Brierley, G.J. (1991) Bar sedimentology of the Squamish River, British Columbia: Definition and application of morphostratigraphic units. *Journal of Sedimentary Petrology* **61**, 211–225.

Brierley, G.J. (1996) Channel morphology and element assemblages: A constructivist approach to facies modelling. In: Carling, P. and Dawson, M. (eds.) *Advances in Fluvial Dynamics and Stratigraphy*. Wiley Interscience, Chichester, pp. 263–298.

Brierley, G.J. (1999) River Styles: An integrative bio-physical template for river management. In: Rutherfurd, I. and Bartley, R. (eds.) *Proceedings of the 2nd Stream Management Conference*. Cooperative Research Centre for Catchment Hydrology, Melbourne, Adelaide, pp. 93–99.

Brierley, G.J. and Fitchett, K. (2000) Channel planform adjustments along the Waiau River, 1946–1992: Assessment of the impacts of flow regulation. In: Brizga, S. and Finlayson, B. (eds.) *River Management: The Australasian Experience*. John Wiley and Sons, Chichester, pp. 51–71.

Brierley, G.J. and Fryirs, K. (1998) A fluvial sediment budget for upper Wolumla catchment, South Coast, N.S.W. *Australian Geographer* **29**, 107–124.

Brierley, G.J. and Fryirs, K. (1999) Tributary-trunk stream relations in a cut-and-fill landscape: a case study from Wolumla catchment, N.S.W., Australia. *Geomorphology* **28**, 61–73.

Brierley, G.J. and Fryirs, K. (2000) River Styles, a geomorphic approach to catchment characterisation: Implications for river rehabilitation in Bega Catchment, NSW, Australia. *Environmental Management* **25**(6), 661–679.

Brierley, G.J. and Fryirs, K. (2001) Creating a catchment-framed biophysical vision for river rehabilitation programs. In: Rutherfurd, I., Sheldon, F., Brierley, G. and Kenyon, C. (eds.) *Third Australian Stream Management Conference Proceedings: The Value of Healthy Rivers, 27–29 August, 2001, Brisbane*. Cooperative Research Centre for Catchment Hydrology, Melbourne, pp. 59–65.

Brierley, G.J. and Hickin, E.J. (1991) Channel planform as a non-controlling factor in fluvial sedimentology: the case of the Squamish River floodplain, British Colombia. *Sedimentology* **38**, 735–750.

Brierley, G.J. and Hickin, E.J. (1992) Floodplain development based on selective preservation of sediments, Squamish River, British Columbia. *Geomorphology* **4**, 381–391.

Brierley, G.J., Ferguson, R.J. and Woolfe, K.J. (1997) What is a fluvial levee? *Sedimentary Geology* **114**, 1–9.

Brierley, G.J., Cohen, T.J., Fryirs, K. and Brooks, A.P. (1999) Post-European changes to the fluvial geomorphology of Bega catchment, Australia: implications for river ecology. *Freshwater Biology* **41**, 1–10.

Brierley, G.J., Brooks, A.P., Fryirs, K. and Taylor, M.P. (in press) Did humid-temperate rivers in the Old and New Worlds respond differently to clearance of riparian vegetation and removal of woody debris. *Progress in Physical Geography*.

Brierley, G.J., Fryirs, K., Outhet, D. and Massey, C. (2002) Application of the River Styles framework as a basis for river management in New South Wales, Australia. *Applied Geography* **22**, 91–122.

Brizga, S.O. and Finlayson, B.L. (1990) Channel avulsion and river metamorphosis: The case of the Thompson River, Victoria, Australia. *Earth Surface Processes and Landforms* **15**, 391–404.

Brookes, A. (1985) River channelization; traditional engineering methods, physical consequences and alternative practices. *Progress in Physical Geography* 9(1), 44–73.

Brookes, A. (1987) Recovery and adjustment of aquatic vegetation within channelisation works in England and Wales. *Journal of Environmental Management* 24, 365–382.

Brookes, A. (1988) *Channelized rivers; perspectives for environmental management*. John Wiley & Sons, Chichester.

Brookes, A. (1989) Alternative channelisation procedures. In: Gore, J.A. and Petts, G.E. (eds.) *Alternatives in regulated river management*. CRC Press, USA, pp. 139–162.

Brookes, A. (1994) River channel change. In: Calow, P. and Petts, G.E. (eds.) *The River Handbook; Hydrological and Ecological Principles*. Blackwell Science, Oxford, UK, pp. 55–75.

Brookes, A. (1995) River channel restoration: Theory and practice. In: Gurnell, A. and Petts, G.E. (eds.) *Changing River Channels*. John Wiley and Sons, Chichester, pp. 369–388.

Brookes, A. and Shields, F.D. (1996) Perspectives on river channel restoration. In: Brookes, A. and Shields, F.D. (eds.) *River Channel Restoration: Guiding Principles for Sustainable Projects*. John Wiley and Sons, Chichester, pp. 1–19.

Brooks, A.P. and Brierley, G.J. (1997) Geomorphic responses of lower Bega River to catchment disturbance, 1851–1926. *Geomorphology* 18, 291–304.

Brooks, A.P. and Brierley, G.J. (2000) The role of European disturbance in the metamorphosis of the Lower Bega River. In: Brizga, S. and Finlayson, B. (eds.) *River Management: The Australasian Experience*. John Wiley & Sons, Chichester, pp. 221–246.

Brooks, A.P., Abbe, T.B., Jansen, J.D., Taylor, M. and Gippel, C.J. (2001) Putting the wood back into rivers: An experiment in river rehabilitation. In: Rutherfurd, I., Sheldon, F., Brierley, G. and Kenyon, C. (eds.) *Third Australian Stream Management Conference Proceedings: The Value of Healthy Rivers, 27–29 August, 2001, Brisbane*. Cooperative Research Centre for Catchment Hydrology, Melbourne, pp. 73–80.

Brooks, A.P. and Brierley, G.J. (2002) Mediated equilibrium: The influence of riparian vegetation and wood on the long-term evolution and behaviour of a near pristine river. *Earth Surface Processes and Landforms* 27(4), 343–367.

Brooks, A.P., Brierley, G.J. and Millar, R.G. (2003) The long-term control of vegetation and woody debris on channel and floodplain evolution: Insights from a paired catchment study in southeastern Australia. *Geomorphology* 51, 7–29.

Brooks, A.P. and Brierley, C.J. (2004) Framing realistic

river rehabilitation programs in light of altered sediment transfer relationships: Lessons from East Gippsland, Australia. *Geomorphology* 58, 107–123.

Brooks, A.P., Gehrke, P.C., Jansen, J.D. and Abbe, T.B. (2004) Experimental reintroduction of woody debris on the Williams River, NSW: Geomorphic and ecological responses. *River Research and Applications* 20, 513–536.

Brown, A.G. (1997) *Alluvial Geoarchaeology: Floodplain Archaeology and Environmental Change*. Cambridge University Press, Cambridge, UK.

Brunsden, D. (1980) Applicable models of long term landform evolution. *Zeitschrift für Geomorphologie, Suppl. Bnd.* 36, 16–26.

Brunsden, D. (1993) Barriers to geomorphological change. In: Thomas, D.S.G. and Allison, R.J. (eds.) *Landscape Sensitivity*. John Wiley and Sons, Chichester, pp. 7–12.

Brunsden, D. (1996) Geomorphological events and landform change. *Zeitshrift fur Geomorphologie, Suppl. Bnd.* 40(3), 273–288.

Brunsden, D. and Thornes, J.B. (1979) Landscape sensitivity and change. *Transactions of the Institute of British Geographers* NS4, 463–484.

Bull, W.B. (1991) *Geomorphic Responses to Climatic Change*. Oxford University Press, New York.

Bunn, S.E. and Arthington, A.H. (2002) Basic principles and ecological consequences of altered flow regimes for aquatic biodiversity. *Environmental Management* 30, 492–507.

Burch, G.J., Bath, R.K., Moore, I.D. and O'Loughlin, E.M. (1987) Comparative hydrological behaviour of forests and cleared catchments in South Eastern Australia. *Journal of Hydrology* 90, 19–42.

Burkham, D.E. (1972) *Channel changes of the Gila River in Safford valley, Arizona, 1864–1970*. United States Geological Survey Professional Paper, P 0655-G, pp. G1–G24.

Cairns, J.J. (1988) Increasing diversity by restoring damaged ecosystems. In: Wilson, E.O. (ed.) *Biodiversity*. National Academy Press, Washington D.C., pp. 333–343.

Cairns, J.J. (1989) Restoring damaged ecosystems: Is predisturbance condition a viable option? *The Environmental Professional* 11, 152–159.

Cairns, J.J. (1991) The status of the theoretical and applied science of restoration ecology. *The Environmental Professional* 13, 186–294.

Cairns, J.J. (1995) Ecological integrity of aquatic systems. *Regulated Rivers: Research and Management* 11, 313–324.

Cairns, J.J. (2000) Setting ecological restoration goals for technical feasibility and scientific validity. *Ecological Engineering* 15, 171–180.

Cant, D.J. and Walker, R.G. (1978) Fluvial processes

and facies sequences in the sandy braided South Saskatchewan River, Canada. *Sedimentology* **25**, 625–648.

Carr, A. (2002) *Grass Roots and Green Tape. Principles and Practices of Environmental Stewardship.* The Federation Press, Leichardt, NSW.

Carson, M.A. (1984) The meandering-braided river threshold: A reappraisal. *Journal of Hydrology* **73**, 315–334.

Chappell, J. (1983) Thresholds and lags in geomorphologic changes. *Australian Geographer* **15**(3), 357–366.

Charlton, F.G., Brown, P.M. and Benson, R.W. (1978) *The Hydraulic Geometry of Some Gravel Rivers in Britain.* Hydraulic Research Station, Wallingford.

Chin, A. (1989) Step pools in stream channels. *Progress in Physical Geography* **13**, 391–407.

Chin, A. (1999) The morphologic structure of step-pools in mountain streams. *Geomorphology* **27**(3–4), 191–204.

Chorley, R.J. (1969) The drainage basin as the fundamental geomorphic unit. In: Chorley, R.J. (ed.) *Water, Earth, and Man.* Methuen and Co. Ltd., Canada.

Chorley, R.J., Schumm, S.A. and Sugden, D.E. (1984) *Geomorphology.* Methuen and Co. Ltd., New York.

Church, M. (1972) Baffin Island sandurs: A study in Arctic fluvial processes. *Geological Survey of Canada Bulletin*, 216, 208pp.

Church, M. (1983) Patterns of instability in a wandering gravel bed channel. In: Collinson, J.D. and Lewin, J. (eds.) *Modern and Ancient Fluvial Systems.* Special Publication of the International Association of Sedimentologists, pp. 169–180.

Church, M. (1992) Channel morphology and typology. In: Calow, P. and Petts, G.E. (eds.) *The Rivers Handbook.* Blackwell, Oxford, pp. 126–143.

Church, M. (2002) Geomorphic thresholds in riverine landscapes. *Freshwater Biology* **47**, 541–557.

Church, M. and Jones, D. (1982) Channel bars in gravel-bed rivers. In: Hey, R.D., Bathurst, J.C. and Thorne, C.R. (eds.) *Gravel-bed Rivers: Fluvial Processes, Engineering and Management.* John Wiley & Sons, Chichester, pp. 291–338.

Church, M. and Miles, M.J. (1982) Discussion of processes and mechanisms of bank erosion. In: Hey, R.D., Bathurst, J.C. and Thorne, C.R. (eds.) *Gravel-Bed Rivers.* Wiley, Chichester, pp. 259–268.

Church, M. and Ryder, J.M. (1972) Paraglacial sedimentation: a consideration of fluvial processes conditioned by glaciation. *Geological Society of America Bulletin* **83**, 3059–3072.

Church, M. and Slaymaker, O. (1989) Disequilibrium of Holocene sediment yield in glaciated British Columbia. *Nature* **337**, 452–454.

Clark, M.J. (2002) Dealing with uncertainty: Adaptive approaches to sustainable river management. *Aquatic*

Conservation: Marine and Freshwater Ecosystems **12**, 347–363.

Cohen, T.J. and Brierley, G.J. (2000) Channel instability in a forested catchment, a case study from Jones Creek, East Gippsland, Australia. *Geomorphology*, **32**, 109–128.

Collins, B.D. and Montgomery, D.R. (2001) Importance of archival and process studies to characterizing pre-settlement riverine geomorphic processes and habitat in the Puget Lowland. In: Dorava, J.M., Montgomery, D.R., Palcsak, B.B. and Fitzpatrick, F.A. (eds.) *Geomorphic Processes and Riverine Habitat.* American Geophysical Union, Washington, DC., pp. 227–243.

Cooke, R.U. and Reeves, R.W. (1976) *Arroyos and Environmental Change in the American South-West.* Clarendon, Oxford.

Cooper, S.D., Diehl, S., Kratz, K. and Sarnelle, O. (1998) Implications of scale for patterns and process in stream ecology. *Australian Journal of Ecology* **23**, 27–40.

Corning, R.V. (1975) Channelisation: Shortcut to nowhere. *Virginia Wildlife* February, 6–8.

Cosgrove, D. and Petts, G.E. (eds.) (1990) *Water, Engineering and Landscape.* Belhaven Press, London, U.K.

Costa, J.E. (1974) Response and recovery of a Piedmont watershed from tropical storm Agnes, June 1972. *Water Resources Research* **10**, 106–112.

Costa, J.E. (1975) Effects of agriculture on erosion and sedimentation in the piedmont province, Maryland. *Geological Society of America Bulletin* **86**, 1281–1286.

Costa, J.E. and O'Connor, J.E. (1995) Geomorphically effective floods. In: Costa, J.E., Miller, A.J., Potter, K.W. and Wilcock, P.R. (eds.) *Natural and Anthropogenic Influences in Fluvial Geomorphology, Geophysical Monograph 89.* American Geophysical Union, Washington D.C, pp. 45–56.

Costanza, R. (1992) Towards and operational definition of ecosystem health. In: Costanza, R., Norton, B.G. and Haskell, B.D. (eds.) *Ecosystem Health: New Goals for Environmental Management.* Island Press, Washington, D.C., pp. 239–256.

Crosby, A.W. (1986) *Ecological Imperialism: The biological expansion of Europe, 900–1900.* Cambridge University Press, UK.

Crozier, M.J. (1999) The frequency and magnitude of geomorphic processes and landform behaviour. *Zeitschrift für Geomorphologie, Suppl. Bnd.* **115**, 35–50.

Davies-Colley, R.J. (1997) Stream channels are narrower in pasture than in forest. *New Zealand Journal of Marine and Freshwater Resources* **31**, 599–608.

Davis, R.J. and Gregory, K.J. (1994) A new distinct

mechanism of river bank erosion in a forest catchment. *Journal of Hydrology* **158**, 1–11.

Desloges, J.R. and Church, M. (1987) Channel and flood-plain facies in a wandering gravel-bed river. In: Ethridge, F.G., Flores, R.M. and Harvey, M.D. (eds.) *Recent Developments in Fluvial Sedimentology, Special Publication Number 39.* Society of Economic Paleontologists and Mineralogists, Tulsa, Oklahoma, pp. 99–109.

Diamond, J. (1997) *Guns, germs and steel: a short history of everybody for the last 13,000 years.* Jonathan Cape.

Dollar, E.S.J. (2000) Fluvial geomorphology. *Progress in Physical Geography* **24**(3), 385–407.

Dorava, J.M., Montgomery, D.R., Palcsak, B.B. and Fitzpatrick, F.A. (2001) Understanding geomorphic processes and riverine habitat. In: Dorava, J.M., Montgomery, D.R., Palcsak, B.B. and Fitzpatrick, F.A. (eds.) *Geomorphic Processes and Riverine Habitat.* American Geophysical Union, Washington, DC., pp. 3–4.

Dovers, S. (2001) *Institutions for Sustainability.* Australian Conservation Foundation.

Dovers, S.R. and Handmer, J.W. (1995) Ignorance, the precautionary principle and sustainability. *Ambio* **24**, 92–97.

Downs, P.W. (1995) River channel classification for channel management purposes. In: Gurnell, A. and Petts, G. (eds.) *Changing River Channels.* John Wiley & Sons, Chichester, UK, pp. 347–365.

Downs, P.W. and Kondolf, G.M. (2002) Post-project appraisals in adaptive management of river channel restoration. *Environmental Management* **29**(4), 477–496.

Duncan, D.J. (2001) *My Story as Told by Water.* Sierra Club, San Francisco.

Dunne, T. (1979) Sediment yield and land use in tropical catchments. *Journal of Hydrology* **42**, 281–300.

Dury, G.H. (1964) *Principles of Underfit Streams.* U.S. Geological Survey Professional Paper, 452-A.

Ebersole, J.L., Liss, W.J. and Frissell, C.A. (1997) Restoration of stream habitats in the Western United States: Restoration as re-expression of habitat capacity. *Environmental Management* **21**(1), 1–14.

Elliott, D.C. (1996) A conceptual framework for geoenvironmental indicators. In: Berger, A.R. and Iams, W.J. (eds.) *Assessing Rapid Environmental Geoindicators; Changes in Earth Systems.* A.A. Balkema, Rotterdam, The Netherlands, pp. 337–349.

Erskine, W.D. (1985) Downstream geomorphic impacts of large dams: the case of Glenbawn Dam, NSW. *Applied Geography* **5**, 195–210.

Erskine, W.D. and Melville, M.D. (1983) Sedimentary properties and processes in a sandstone valley: Fernances Creek, Hunter Valley, New South Wales. In: Young, R.M. and Nanson, G.C. (eds.) *Aspects of*

Australian Sandstone Landscapes. Australian and New Zealand Geomorphology Group, Special Publication No.1, pp. 94–105.

Erskine, W.D. and Webb, A.A. (1999) A protocol for river rehabilitation. In: Rutherfurd, I. and Bartley, R. (eds.) *Proceedings of the 2nd Stream Management Conference, Adelaide.* Cooperative Research Centre for Catchment Hydrology, Melbourne, pp. 237–243.

Erskine, W., McFadden, C. and Bishop, P. (1992) Alluvial cutoffs as indicators of former channel conditions. *Earth Surface Processes and Landforms* **17**, 23–37.

Erskine, W.D., Terrazolo, N. and Warner, R.F. (1999) River rehabilitation from the hydrogeomorphic impacts of a large hydro-electric power project: Snowy River, Australia. *Regulated Rivers: Research and Management* **15**, 3–24.

Everard, M. and Powell, A. (2002) Rivers as living systems. *Aquatic Conservation: Marine and Freshwater Ecosystems* **12**, 329–337.

Eyles, R.J. (1977) Changes in drainage networks since 1820, southern tablelands, N.S.W. *Australian Geographer* **13**, 377–387.

Fairweather, P.G. (1999) State of environment indicators of river health: Exploring the metaphor. *Freshwater Biology* **41**, 211–220.

Farrell, K.J. (1987) Sedimentology and facies architecture of overbank deposits of the Mississippi River, False River region, Louisiana. In: Etheridge, F.G., Flores, R.M. and Harvey, M.D. (eds.) *Recent Developments in Fluvial Sedimentology.* Special Publication of the Society for Economic Palaentology and Minerology, Vol 39, pp. 111–121.

Farrell, K.M. (2001) Geomorphology, facies architecture, and high resolution, non-marine sequence stratigraphy in avulsion deposits, Cumberland Marshes, Saskatchewan. *Sedimentary Geology* **139**, 93–150.

Ferguson, R.I. (1987) Hydraulic and sedimentary controls of channel planform. In: Richards, K. (ed.) *River Channels: Environment and Process.* Blackwell, Oxford, pp. 129–158.

Ferguson, R.J. and Brierley, G.J. (1999a) Downstream changes in valley confinement as a control on floodplain morphology, lower Tuross River, New South Wales: A constructivist approach to floodplain analysis. In: Miller, A.J. and Gupta, A. (eds.) *Varieties of Fluvial Form.* John Wiley & Sons, Chichester, pp. 377–407.

Ferguson, R.J. and Brierley, G.J. (1999b) Levee morphology and sedimentology along lower Tuross River, south-eastern Australia. *Sedimentology* **46**, 627–648.

Fisher, S.G. (1997) Creativity, idea generation, and the functional morphology of streams. *Journal of the North American Benthological Society* **16**, 305–318.

Fisher, S.G., Grimm, N.B. and Comes, D. (1990) Hierarchy, spatial configuration and nutrient cycling in a desert stream. *Australian Journal of Ecology* **23**, 41–52.

Fisk, H.N. (1944) *Geological investigation of the alluvial valley of the lower Mississippi River.* Mississippi River Commisssion Waterways Experiment Station, Vicksburg, Mississippi.

Fisk, H.N. (1947) *Fine-grained alluvial deposits and their effects on Mississippi River activity.* Mississippi River Commission Waterways Experiment Station, Vicksburg, Mississippi.

Flannery, T. (1994) *The Future Eaters: An Ecological History of the Australasian Lands and People.* Reed Books, Chatswood, NSW.

Fletcher, A. (2001) *The Art of Looking Sideways.* Phaidon Press Ltd, London.

Fonstad, M.A. (2003) Spatial variation in the stream power of mountain streams in the Sangre de Cristo Mountains, New Mexico. *Geomorphology* **55**, 75–96.

Friedman, J.M., Osterkamp, W.R. and Lewis, W.M. (1996) The role of vegetation and bed-level fluctuations in the process of channel narrowing. *Geomorphology* **14**, 341–351.

Friedman, J.M., Osterkamp, W.R., Scott, M.L. and Auble, G.T. (1998) Downstream effects of dams on channel geometry and bottomland vegetation: Regional patterns in the Great Plains. *Wetlands* **18**, 619–633.

Frissell, C.A. and Nawa, R.K. (1992) Incidence and causes of physical failure of artificial habitat structures in streams of western Oregon and Washington. *North American Journal of Fisheries Management* **12**, 182–197.

Frissell, C.A., Liss, W.J. and Bayles, D. (1993) An integrated, biophysical strategy for ecological restoration of large watersheds. In: Spangenbery, N.E. and Potts, D.F. (eds.) *Changing Roles in Water Resources Management and Policy.* American Water Resources Association, Herdon, VA, pp. 449–456.

Frissell, C.A., Liss, W.J., Warren, C.E. and Hurley, M.D. (1986) A hierarchical framework for stream habitat classification: Viewing streams in a watershed context. *Environmental Management* **10**, 199–214.

Frothingham, K.M., Rhoads, B.L. and Herricks, E.E. (2001) Stream geomorphology and fish community structure in channelized and meandering reaches of an agricultural stream. In: Dorava, J.M., Montgomery, D.R., Palcsak, B.B. and Fitzpatrick, F.A. (eds.) *Geomorphic Processes and Riverine Habitat.* American Geophysical Union, Washington, DC., pp. 105–117.

Fryirs, K. (2001) *A geomorphic approach for assessing the condition and recovery potential of rivers: Application in Bega Catchment, South Coast, New South Wales, Australia.* PhD Thesis, Macquarie University, Sydney, 147pp.

Fryirs, K. (2002) Antecedent controls on river character, behaviour and evolution at the base of escarpment in Bega catchment, South Coast, New South Wales, Australia. *Zeitschrift fur Geomorphologie* **46**(4), 475–504.

Fryirs, K. (2003) Guiding principles for assessing geomorphic river condition: Application of a framework in the Bega catchment, South Coast, New South Wales, Australia. *Catena* **53**, 17–52.

Fryirs, K. and Brierley, G.J. (1998) The character and age structure of valley fills in Upper Wolumla Creek catchment, South Coast, New South Wales, Australia. *Earth Surface Processes and Landforms* **23**, 271–287.

Fryirs, K. and Brierley, G.J. (1999) Slope-channel decoupling in Wolumla catchment, New South Wales, Australia: The changing nature of sediment sources following European settlement. *Catena* **35**, 41–63.

Fryirs, K. and Brierley, G.J. (2000) A geomorphic approach for the identification of river recovery potential. *Physical Geography* **21**(3), 244–277.

Fryirs, K. and Brierley, G.J. (2001) Variability in sediment delivery and storage along river courses in Bega catchment, NSW, Australia: implications for geomorphic river recovery. *Geomorphology* **38**, 237–265.

Galay, V.J. (1983) Causes of river bed degradation. *Water Resources Research* **19**(5), 1057–1090.

Gardiner, G.L. (1988) Environmentally sensitive river engineering: Examples from the Thames catchment. *Regulated Rivers: Research and Management* **24**, 365–382.

Gellis, A.C., Cheama, A. and Lalio, S.M. (2001) Developing a geomorphic approach for ranking watersheds for rehabilitation, Zuni Indian Reservation, New Mexico. *Geomorphology* **37**, 105–134.

Gilvear, D.J. (1999) Fluvial geomorphology and river engineering: Future roles utilising a fluvial hydrosystem framework. *Geomorphology* **31**, 229–245.

Gippel, C.J. (1995) Environmental hydraulics of large woody debris in streams and rivers. *Journal of Enviromental Engineering* May, 388–395.

Goldrick, G., Brierley, G., J. and Fryirs, K. (1999) *River Styles in the Richmond Catchment.* A report to the NSW Department of Land and Water Conservation. Macquarie Research Limited, Sydney, pp. 113.

Gomez, B., Phillips, J.D., Magilligan, F.J. and James, L.A. (1997) Floodplain sedimentation and sensitivity: Summer 1993 flood, Upper Mississippi River Valley. *Earth Surface Processes and Landforms* **22**, 923–936.

Goodwin, C.N. (1999) Fluvial classification: Neanderthal necessity or needless normalcy. *Wildland Hydrology* June/July, 229–236.

Gore, D., Brierley, G.J., Pickard, J. and Jansen, J. (2000) Anatomy of a floodout in semi-arid eastern Australia. *Zeitschrift für Geomorphologie, Suppl. Bnd.* **122**, 113–139.

Gore, J.A. (ed.) (1985) *The Restoration of Rivers and*

Streams: Theories and Experience. Butterworth Publishers, Boston, USA.

Gore, J.A. (2001) Models of habitat use and availability to evaluate anthropogenic changes in channel geometry. In: Dorava, J.M., Montgomery, D.R., Palcsak, B.B. and Fitzpatrick, F.A. (eds.) *Geomorphic Processes and Riverine Habitat*. American Geophysical Union, Washington, DC., pp. 27–36.

Gore, J.A. and Shields Jr., F.D. (1995) Can large rivers be restored? *BioScience* **45**(3), 142–152.

Graf, W.L. (1977) The rate law in fluvial geomorphology. *American Journal of Science* **277**, 178–191.

Graf, W.L. (1979) Rapids in canyon rivers. *Journal of Geology* **87**, 533–551.

Graf, W.L. (1982) Spatial variations of fluvial processes in semi-arid lands. In: Thorn, C.E. (ed.) *Space and Time in Geomorphology*. Allen and Unwin, London, UK, pp. 193–217.

Graf, W.L. (1983a) The arroyo problem – palaeohydrology and palaeohydraulics in the short term. In: Gregory, K.J. (ed.) *Background to Palaeohydrology: A Perspective*. John Wiley and Sons, Chichester, pp. 279–302.

Graf, W.L. (1983b) Downstream changes in stream power in the Henry Mountains, Utah. *Annals of the Association American Geographers* **73**(3), 373–87.

Graf, W.L. (1988) Definition of flood plains along arid-region rivers. In: Baker, V.R., Kochel, R.C. and Patton, P.C. (eds.) *Flood Geomorphology*. John Wiley and Sons, New York, pp. 231–242.

Graf, W.L. (1999) Dam nation: A geographic census of American dams and their hydrologic impacts. *Water Resources Research* **35**, 1305–1311.

Grant, G.E. and Swanson, F.J. (1995) Morphology and processes of valley floors in mountain streams, Western Cascades, Oregon. In: Costa, J.E., Miller, A.J., Potter, K.W. and Wilcock, P.R. (eds.) *Natural and Anthropogenic Influences in Fluvial Geomorphology, Geophysical Monograph 89*. American Geophysical Union, Florida, pp. 83–101.

Grant, G.E., Swanson, F.J. and Wolman, M.G. (1990) Pattern and origin of stepped-bed morphology in high-gradient streams, Western Cascades, Oregon. *Geological Society of America Bulletin* **102**, 340–352.

Gregory, K.J. (1995) Human activity and palaeohydrology. In: Gregory, K.J., Starkel, L. and Baker, V.R. (eds.) *Global Continental Palaeohydrology*. John Wiley & Sons, Chichester, pp. 151–172.

Gregory, K.J., Davis, R.J. and Tooth, S. (1993) Spatial distribution of coarse woody debris dams in the Lymington Basin, Hampshire, UK. *Geomorphology* **6**, 207–224.

Gregory, S.V., Swanson, F.J., McKee, W.A. and Cummins, K.W. (1991) An ecosystem perspective of riparian zones. *BioScience* **11**, 540–551.

Grissinger, E.H. (1982) Bank erosion of cohesive materials. In: Hey, R.D., Bathurst, J.C. and Thorne, C.R. (eds.) *Gravel-Bed Rivers*. Wiley, Chichester, pp. 273–287.

Gunn, R.H., Story, R., Galloway, R.W., Duffy, P.J.B., Yapp, G.A. and McAlpine, J.R. (1969) *Lands of the Queanbeyan-Shoalhaven Area, ACT and NSW*. Land Research Series No. 24. Commonwealth Scientific and Industrial Research Organisation (CSIRO), Melbourne, Australia.

Gupta, A. and Fox, H. (1974) Effects of high magnitude floods on channel form: a case study in Maryland Piedmont. *Water Resources Research* **10**, 499–509.

Gurnell, A.M., Piégay, H., Gregory, S.V. and Swanson, F.J. (2002) Large wood and fluvial processes. *Freshwater Biology* **47**, 601–619.

Gustavson, T.C. (1978) Bed forms and modern stratification types of modern gravel meander lobes, Nueces River, Texas. *Sedimentology* **25**, 401–426.

Hack, J.T. (1960) Interpretation of erosional topography in humid temperate regions. *American Journal of Science* **258-A**, 80–97.

Hagerty, D.J. (1991) Piping/sapping erosion. I. Basic considerations. *Journal of Hydraulic Engineering* **117**, 991–1008.

Halwas, K.L. and Church, M. (2002) Channel units in small, high gradient streams on Vancouver Island, British Columbia. *Geomorphology* **43**, 243–256.

Harper, D.M. and Everard, M. (1998) Why should the habitat level approach underpin holistic river survey and management? *Aquatic Conservation: Marine and Freshwater Ecosystems* **8**, 395–413.

Harrelson, C.C., Rawlins, C.L. and Potyondy, J.P. (1994) *Stream channel reference sites: An illustrated guide to field technique*. USDA Forest Service. General Technical Report, RM-245, Fort Collins, CO, U.S.

Harrison, S. (2001) On reductionism and emergence in geomorphology. *Transactions of the Institute of British Geographers*, **26**, 327–339

Harvey, A.M. (2001) Coupling between hillslopes and channels in upland fluvial systems: Implications for landscape sensitivity illustrated from the Howgill Fells, northwest England. *Catena* **42**, 225–250.

Harvey, A.M. (2002) Effective timescales of coupling within fluvial systems. *Geomorphology* **44**, 175–201.

Hawkins, C.P., Kershner, J.L., Bisson, P.A., Bryant, M.D., Decker, L.M., Gregory, S.V., McCullough, D.A., Overton, C.K., Reeves, G.H., Steedman, R.J. and Young, M.K. (1993) A hierarchical approach to classifying stream habitat features. *Fisheries* **18**, 3–12.

Heede, B.H. and Rinne, J.N. (1990) Hydrodynamic and fluvial morphologic processes: Implications for fisheries management and research. *North American Journal of Fisheries Management* **10**, 249–268.

Henderson, F.M. (1961) Stability of alluvial channels.

Journal of the Hydraulics Division, Proceedings of the American Society of Civil Engineers **87**, 109–38.

Henry, C.P. and Amoros, C. (1995) Restoration ecology of riverine wetlands: I. A scientific base. *Environmental Management* **19**, 891–902.

Herget, J. (2000) Holocene development of the River Lippe Valley, Germany; a case study of anthropogenic influence. *Earth Surface Processes and Landforms* **25**(3), 293–305.

Heritage, G.L., Charlton, M.E. and O'Regan, S. (2001) Morphological classification of fluvial environments; an investigation of the continuum of channel types. *Journal of Geology* **109**(1), 21–33.

Heritage, G.L., van Niekerk, A.W. and Moon, B.P. (1997) A comprehensive hierarchical river classification system. *GeookoPlus* **4**, 75–84.

Heritage, G.L., van Niekerk, A.W. and Moon, B.P. (1999) Geomorphology of the Sabie River, South Africa: an incised bedrock-influenced channel. In: Miller, A.J. and Gupta, A. (eds.) *Varieties of Fluvial Form*. John Wiley and Sons, Chichester, pp. 53–79.

Hey, R.D. (1982) Gravel bed rivers: Form and process. In: Hey, R.D., Bathurst, J.C. and Thorne, C.R. (eds.) *Gravel Bed Rivers: Fluvial Processes, Engineering and Management*. John Wiley and Sons, Chichester, pp. 5–13.

Hey, R.D. (1994) Environmentally sensitive river engineering. In: Calow, P. and Petts, G.E. (eds.) *The Rivers Handbook, Vol 2*. Blackwell, Oxford, UK, pp. 337–362.

Hey, R.D. (1997) Stable river morphology. In: Thorne, C.R., Hey, R.D. and Newson, M.D. (eds.) *Applied Fluvial Geomorphology for River Engineering and Management*. Wiley, Chichester, U.K., pp. 223–236.

Hey, R.D. and Thorne, C.R. (1983) Hydraulic geometry of mobile gravel-bed rivers *Proceedings of the Second International Symposium on River Sedimentation*. Water Resource and Electric Power Press, Nanjing, China, pp. 713–723.

Hey, R.D. and Thorne, C.R. (1986) Stable channels with mobile gravel beds. *Journal of Hydraulic Engineering* **112**(8), 671–689.

Hickin, E.J. (1969) A newly identified process of point bar formation in natural streams. *American Journal of Science* **267**, 999–1010.

Hickin, E.J. (1974) The development of meanders in natural river-channels. *American Journal of Science* **274**, 414–442.

Hickin, E.J. (1983) River channel changes: Retrospect and prospect. *International Association of Sedimentologists Special Publication* **6**, 61–83.

Hickin, E.J. (1984) Vegetation and river channel dynamics. *Canadian Geographer* **28**(2), 111–126.

Hickin, E.J. (1986) Concave-bank benches in the floodplains of Muskwa and Fort Nelson Rivers, British Columbia. *The Canadian Geographer* **30**(2), 111–122.

Hickin, E.J. and Nanson, G.C. (1975) The character of channel migration on the Beaton River, Northeast British Columbia, Canada. *Geological Society of America Bulletin* **86**, 487–494.

Hickin, E.J. and Nanson, G.C. (1984) Lateral migration rates of river bends. *Journal of Hydraulic Engineering* **110**(11), 1157–1167.

Higgs, E. (2003) *Nature by Design: People, Natural Process and Ecological Restoration*. The MIT Press, Cambridge, Massachusetts, US.

Hillman, M. (2002) Environmental justice: A crucial link between environmentalism and community development? *Community Development Journal* **37**(4), 349–360.

Hillman, M. (in press) The importance of environmental justice in stream rehabilitation. *Ethics, Place and Environment*.

Hillman, M. and Brierley, G.J. (2002) Information needs for environmental flow allocation: A case study from the Lachlan River, New South Wales, Australia. *Annals of the Association of American Geographers* **92**, 617–630.

Hillman, M. and Brierley, G.J. (in press) A critical review of catchment-scale stream rehabilitation programs. *Progress in Physical Geography*.

Hobbs, R.J. (1994) Landscape ecology and conservation: Moving from description to application. *Pacific Conservation Biology* **1**, 170–176.

Hobbs, R.J. (2003) Ecological management and restoration: Assessment, setting goals and measuring success. *Ecological Management and Restoration* **4**, 2–3.

Hobbs, R.J. and Kristjanson, L.J. (2003) Triage: How do we prioritize health care for landscapes? *Ecological Management and Restoration* **4**, 39–45.

Hobbs, R.J. and Norton, D.A. (1996) Towards a conceptual framework for restoration ecology. *Restoration Ecology* **4**(2), 93–110.

Hogan, D.L., Bird, S.A. and Hassan, M.A. (1998) Spatial and temporal evolution of small coastal gravel-bed streams: Influence of forest management on channel morphology and fish habitats. In: Klingeman, P.C., Beschta, R.L., Komar, P.D. and Bradley, J.B. (eds.) *Gravel-Bed Rivers in the Environment*. Water Resources Publications, Highlands Branch, Colorado, pp. 365–392.

Hollis, G.E. (1975) The effect of urbanization on floods of different recurrence interval. *Water Resources Research* **11**(3), 431–435.

Hooke, J.M. (1979) An analysis of the processes of river bank erosion. *Journal of Hydrology* **42**, 39–62.

Hooke, J.M. (1980) Magnitude and distribution of rates of river bank erosion. *Earth Surface Processes* **5**, 143–157.

Hooke, J.M. (1995) River channel adjustment to meander cutoffs on the River Bollin and River Dane, northwest England. *Geomorphology* **14**, 235–253.

Hooke, J.M. (1999) Decades of change: Contributions of geomorphology to fluvial and coastal engineering and management. *Geomorphology* **31**, 373–389.

Hooke, J.M. (2003) Coarse sediment connectivity in river channel systems: A conceptual framework and methodology. *Geomorphology* **56**, 79–94.

Hooke, R. (2000) On the history of humans as geomorphic agents. *Geology (Boulder)* **28**(9), 843–846.

Howard, A.D. (1967) Drainage analysis in geologic interpretation: a summation. *The American Association of Petroleum Geologists Bulletin* **51**(11), 2246–2259.

Howard, A.D. (1980) Thresholds in river regimes. In: Coates, D.R. and Vitek, J.D. (eds.) *Thresholds in Geomorphology*. George Allen and Unwin, London, U.K., pp. 227–258.

Howard, A.D. (1987) Modelling fluvial systems: Rock-, gravel- and sand-bed channels. In: Richards, K. (ed.) *River channels: Environment and Process*. Blackwell, Oxford, UK, pp. 69–94.

Howard, A.D. and Dolan, R. (1981) Geomorphology of the Colorado River in the Grand Canyon. *Journal of Geology*, **89**, 269–298.

Huckleberry, G. (1994) Contrasting channel responses to floods on the middle Gila River, Arizona. *Geology* **22**, 1083–1086.

Hughes, F.M.R. (1997) Floodplain biogeomorphology. *Progress in Physical Geography* **21**, 501–529.

Hughes, R.M., Larson, D.P. and Omernik, J.M. (1986) Regional reference sits: A method for assessing stream potential. *Environmental Management* **10**(5), 629–635.

Hupp, C.R. (1992) Riparian vegetation recovery patterns following stream channelisation: A geomorphic perspective. *Ecology* **73**(4), 1209–1226.

Hupp, C.R. and Osterkamp, W.R. (1996) Riparian vegetation and fluvial geomorphic processes. *Geomorphology* **14**, 277–295.

Hupp, C.R. and Simon, A. (1991) Bank accretion and the development of vegetated depositional surfaces along modified alluvial channels. *Geomorphology* **4**, 111–124.

Ikeya, H. (1981) A method for designation for areas in danger of debris flow. In: Davies, T.R.H. and Pearce, A.J. (eds.) *Erosion and Sediment Transport in Pacific Rim Steeplands*. International Association of Hydrological Sciences Publication, No.132, pp. 576–588.

Jackson, L.L., Lopoukhine, N. and Hillyard, D. (1995) Ecological restoration: A definition and comments. *Restoration Ecology* **3**, 71–75.

Jackson, R.G. (1976) Depositional mode of point bars in the Lower Wabash River. *Journal of Sedimentary Petrology* **46**, 579–594.

Jacobson, R.B., Laustrup, M.S. and Chapman, M.D. (2001) Fluvial processes and passive rehabilitation of the Lisbon Bottom side-channel chute, Lower Missouri River. In: Dorava, J.M., Montgomery, D.R., Palcsak, B.B. and Fitzpatrick, F.A. (eds.) *Geomorphic Processes and Riverine Habitat*. American Geophysical Union, Washington, DC.

Jaeggi, M.N.R. (1989) Channel engineering and erosion control. In: Gore, J.A. and Petts, G.E. (eds.) *Alternatives in Regulated Rivers Management*. CRC Press, Boca Raton, USA, pp. 163–183.

Jain, V. and Sinha, R. (2004) Fluvial dynamics of an anabranching river system in Himalayan foreland basin, Baghmati river, north Bihar plains, India. *Geomorphology* **60**, 147–170.

James, L.A. (1991) Incision and morphologic evolution of an alluvial channel recovering from hydraulic sediment mining. *Geological Society of America Bulletin* **103**, 723–736.

James, L.A. (1993) Sustained reworking of hydraulic mining sediment in California: G.K. Gilbert's sediment wave model reconsidered. *Zeitschrift fur Geomorphologie. Suppl Bnd.* **88**, 49–66.

James, L.A. (1999) Time and the persistence of alluvium: River engineering, fluvial geomorphology, and mining sediment in California. *Geomorphology* **31**, 265–290.

Jansen, J.D. and Nanson, G.C. (2004) Anabranching and maximum flow efficiency in Magela Creek, northern Australia. *Water Resources Research* **40**(4), W04503, doi:10.1029/2003 WR002408.

Johnson, A.C., Swanston, D.N. and McGee, K.E. (1990) Landslide initiation, runout and deposition within clearcut and old growth forests of Alaska. *Journal of the American Water Resources Association* **36**, 17–30.

Johnston, P. and Brierley, G.J. (subm.) Late Quaternary evolution of floodplain pockets along Mulloon Creek, New South Wales, Australia. *The Holocene*.

Jungwirth, M., Muhar, S. and Schmutz, S. (2002) Re-establishing and assessing ecological integrity in riverine landscapes. *Freshwater Biology* **47**, 867–887.

Junk, W.J., Bayley, P.B. and Sparks, R.E. (1989) The flood pulse concept in river-floodplain systems. *Special Publication of the Canadian Journal of Fisheries and Aquatic Science* **106**, 110–127.

Kasai, M., Marutani, T. and Brierley, G.J. (2003) Patterns of sediment slug translation and dispersion following typhoon-induced disturbance, Oyabu Creek, Kyushu, Japan. *Earth Surface Processes and Landforms* **29**, 59–76.

Kasai, M., Marutani, T. and Brierley, G.J. (in press) Channel bed adjustments following major aggradation in a steep headwter setting: Findings from Oyabu Creek, Kyushu, Japan. *Geomorphology*.

Kashef, A.A.I. (1981) Technical and ecological impacts of the High Aswan Dam. *Journal of Hydrology* **53**(1–2), 73–84.

Keller, E.A. (1971) Areal sorting of bed-load material: The hypothesis of velocity reversal. *Geological Society of America Bulletin* **82**, 753–756.

Keller, E.A. (1972) Development of alluvial stream channels: A five stage model. *Geological Society of America Bulletin* **83**, 1531–1536.

Keller, E.A. and Brookes, A. (1984) Consideration of meandering in channelization projects; selected observations and judgements. In: Elliott, C.M. (ed.) *River Meandering.* American Society of Civil Engineering, New York, pp. 384–397.

Keller, E.A. and Melhorn, W.N. (1973) Bedforms and fluvial processes in alluvial stream channels: selected observations. In: Morisawa, M. (ed.) *Fluvial Geomorphology, 'Binghampton' Symposia in Geomorphology, International Series No. 4.* Allen and Unwin, London, pp. 253–283.

Keller, E.A. and Melhorn, W.N. (1978) Rhythmic spacing and origin of pools and riffles. *Geological Society of America Bulletin* **89**, 723–730.

Kellerhals, R. and Church, M. (1989) The morphology of large rivers: Characterisation and management. In: Dodge, D.P. (ed.) *Proceedings of the International Large River Symposium.* Canadian Special Publication of Fisheries and Aquatic Science, 106, pp. 31–48.

Kellerhals, R., Church, M. and Bray, D.I. (1976) Classification and analysis of river processes. *Journal of the Hydraulics Division, Proceedings of the American Society of Civil Engineers* **102**(HY7), 813–829.

Johnson, A.C., Swanston, D.N. and McGee, K.E. (2000) Landslide initiation, runout, and depositions within clearcuts and old-growth forests of Alaska. *Journal of the American Water Resources Association,* **36**, 17–30.

Kelly, J.R. and Harwell, M.A. (1990) Indicators of ecosystem recovery. *Environmental Management* **14**(5), 527–545.

Kemp, J.L., Harper, D.M. and Giuseppe, A.C. (2000) The habitat-scale ecohydraulics of rivers. *Ecological Engineering* **16**, 17–29.

Kern, K. (1992) Restoration of lowland rivers: The German experience. In: Carling, P.A. and Petts, G.E. (eds.) *Lowland Floodplain Rivers: Geomorphological Perspectives.* John Wiley and Sons, Chichester, pp. 279–297.

Kershner, J.L., Snider, W.M., Turner, D.M. and Moyle, P.B. (1992) Distribution and sequencing of mesohabitats: Are there differences at the reach scale? *Rivers* **3**, 179–190.

Kirkby, M.J. (1999) Towards an understanding of varities of fluvial form. In: Miller, A.J. and Gupta, A. (eds.) *Varieties of Fluvial Form.* John Wiley & Sons, Chichester, pp. 507–514.

Klimek, K. (1974) The retreat of alluvial river banks in the Wisloka Valley (South Poland). *Geographia Polonica* **28**, 59–75.

Knighton, A.D. (1991) Channel bed adjustment along mine-affected rivers of northeast Tasmania. *Geomorphology* **1**, 221–237.

Knighton, A.D. (1998) *Fluvial Forms and Processes: A New Perspective.* Arnold, London.

Knighton, A.D. (1999) Downstream variation in stream power. *Geomorphology* **29**, 293–306.

Knighton, A.D. and Nanson, G.C. (1993) Anastomosis and the continuum of channel pattern. *Earth Surface Processes and Landforms* **18**, 613–625.

Knox, J.C. (1972) Valley alluviation in southwestern Wisconsin. *Annals of the Association of American Geographers* **62**(3), 401–410.

Knox, J.C. (1977) Human impacts on Wisconsin stream channels. *Annals of the Association of American Geographers* **67**(3), 323–342.

Knox, J.C. (1987) Historical valley floor sedimentation in the Upper Mississippi Valley. *Annals of the Association of American Geographers* **77**, 224–244.

Knox, J.C. (1989) Long- and short-term episodic storage and removal of sediment in watersheds of southwestern Wisconsin and northwestern Illinois. In: Hadley, R.F. and Ongley, E.D. (eds.) *Sediment and the Environment, Proceedings of the Baltimore Symposium.* IAHS Publication No. 184, pp. 157–164.

Knox, J.C. (1993) Large increases in flood magnitude in response to modest changes in climate. *Nature* **361**, 430–432.

Knox, J.C. (1995) Fluvial systems since 20,000 years BP. In: Gregory, K.J., Starkel, L. and Baker, V.R. (eds.) *Global Continental Palaeohydrology.* Wiley, New York, pp. 87–108.

Kochel, R.C. (1988) Geomorphic impact of large floods: Review and new perspectives in magnitude and frequency. In: Baker, V.R., Kochel, R.G. and Patton, P.C. (eds.) *Flood Geomorphology.* Wiley, New York, pp. 169–187.

Koehn, J.D., Brierley, G.J., Cant, B.L. and Lucas, A.M. (2001) *River Restoration Framework.* Land and Water Australia Occasional Paper 01/01, Canberra., pp. 130.

Komura, S. and Simons, D.B. (1967) River-bed degradation below dams. *Journal of the Hydraulics Division, Proceedings of the American Society of Civil Engineers* **HY4**, 1–14.

Kondolf, G.M. (1994) Geomorphic and environmental effects of instream gravel mining. *Landscape and Urban Planning* **28**, 225–243.

Kondolf, G.M. (1995) Geomorphological stream channel classification in aquatic habitat restoration: Uses and limitations. *Aquatic Conservation: Marine and Freshwater Ecosystems* **5**, 127–141.

Kondolf, G.M. (1997) Hungry water: Effects of dams and

gravel mining on river channels. *Environmental Management* **21**(4), 533–551.

Kondolf, G.M. (1998a) Environmental effects of aggregate extraction from river channels and floodplains. In: Bobrowsky, P.T. (ed.) *Aggregate Resources; A Global Perspective*. A.A. Balkema, Rotterdam, Netherlands, pp. 113–129.

Kondolf, G.M. (1998b) Large-scale extraction of alluvial deposits from rivers in California; geomorphic effects and regulatory strategies. In: Klingeman, P.C., Beschta, R.L., Komar, P.D. and Bradley, J.B. (eds.) *Gravel-Bed Rivers in the Environment*. Water Resources Publications, Highlands Branch, Colorado, pp. 455–470.

Kondolf, G.M. (1998c) Lessons learned from river restoration projects in California. *Aquatic Conservation: Marine and Freshwater Ecosystems* **8**, 39–52.

Kondolf, G.M. and Downs, P.W. (1996) Catchment approach to planning channel restoration. In: Brookes, A. and Shields Jr., F.D. (eds.) *River Channel Restoration: Guiding Principles for Sustainable Projects*. John Wiley and Sons, Chichester, pp. 129–148.

Kondolf, G.M. and Larson, M. (1995) Historical channel analysis and its application to riparian and aquatic habitat restoration. *Aquatic Conservation: Marine and Freshwater Ecosystems* **5**, 109–126.

Kondolf, G.M. and Micheli, E.R. (1995) Evaluating stream restoration projects. *Environmental Management* **19**(1), 1–15.

Kondolf, G.M. and Wilcock, P.R. (1996) The flushing flow problem: Defining and evaluating objectives. *Water Resources Research* **32**, 2589–2599.

Kondolf, G.M., Piégay, H. and Landon, N. (2002) Channel response to increased and decreased bedload supply from land use change: Contrasts between two catchments. *Geomorphology* **45**, 35–51.

Kondolf, G.M., Montgomery, D.R., Piégay, H. and Schmitt, L. (2003) Geomorphic classification of rivers and streams. In: Kondolf, G.M. and Piégay, H. (eds.) *Tools in Fluvial Geomorphology*. John Wiley and Sons, Chichester, UK, pp. 171–204.

Koster, E.H. (1978) Tranverse ribs; their characteristics, origin and palaeohydraulic significance. *Memoir – Canadian Society of Petroleum Geologists* **5**, 161–186.

Kouwen, N. and Li, R.M. (1979) Biomechanics of vegetated channel linings. *Proceedings of the American Society of Civil Engineers, Journal of the Hydraulic Division* **106**(HY6), 1085–1103.

Kozarski, S. and Rotnicki, K. (1977) Valley floors and changes of river channel patterns in the north Polish PLain during the late Wurm and Holocene. *Quaestiones Geographicae* **4**, 51–93.

Krzemień, K. (1999) *River Channels – Pattern, Structure and Dynamics. Prace Geograficzne*. Fascicle 104.

Institute of Geography of the Jagiellonian University, Cracow, Poland.

Lakc, P.S. (2001a) Restoring streams: Re-building and re-connecting. In: Rutherfurd, I., Sheldon, F., Brierley, G. and Kenyon, C. (eds.) *Third Australian Stream Management Conference Proceedings: The Value of Healthy Rivers, 27–29 August, 2001, Brisbane*. Cooperative Research Centre for Catchment Hydrology, Melbourne, pp. 369–371.

Lake, P.S. (2001b) On the maturing of restoration: Linking ecological research and restoration. *Ecological Management and Restoration* **2**, 110–115.

Lakoff, G. (1987) *Women, fire and dangerous things: what categories reveal about the mind*. University of Chicago Press, Chicago.

Lamb, H.H. (1977) *Climatic history and the future*. Methuen, London.

Lancaster, S.T., Hayes, S.K. and Grant, G.E. (2001) Modeling sediment and wood storage and dynamics in small mountainous watersheds. In: Dorava, J.M., Montgomery, D.R., Palcsak, B.B. and Fitzpatrick, F.A. (eds.) *Geomorphic Processes and Riverine Habitat*. American Geophysical Union, Washington, DC., pp. 85–102.

Landon, N., Piégay, H. and Bravard, J.P. (1998) The Drome river incision (France): From assessment to management. *Landscape and Urban Planning* **43**, 119–131.

Lane, E.W. (1954) *The Importance of Fluvial Morphology in Hydraulic Engineering*. Hydraulic laboratory report no. 372, United States Department of the Interior, Bureau of Reclamation, Engineering Laboratories, Commissioner's Office. Denver, Colorado.

Lane, E.W. (1957) A study of the shape of channels formed by natural streams flowing in erodible material *Missouri River Division Sediments Series, No. 9*. United States Army Engineer Division, Missouri River, Corps of Engineers, Omaha, Nebraksa, pp. 106.

Lane, S.N. and Richards, K.S. (1997) Linking river channel form and process: Time, space and causality revisited. *Earth Surface Processes and Landforms* **22**, 249–260.

Langbein, W.B. and Schumm, S.A. (1958) Yield of sediment in relation to mean annual precipitation. *Transactions of the American Geophysical Union* **39**(6), 1076–1084.

Lawler, D.M. (1988) Environmental limits of needle ice: A global survey. *Arctic and Alpine Research* **20**, 137–159.

Lawler, D.M. (1992) Process dominance in bank erosion systems. In: Carling, P.A. and Petts, G.E. (eds.) *Lowland Floodplain Rivers: Geomorphological Perspectives*. Wiley, Chichester, pp. 117–143.

Lawler, D.M. (1993) The measurement of river bank

erosion and lateral channel change: A review. *Earth Surface Processes and Landforms* **18**, 777–821.

Lawler, D.M., Thorne, C.R. and Hooke, J.M. (1997) Bank erosion and instability. In: Thorne, C.R., Hey, R.D. and Newson, M.D. (eds.) *Applied Fluvial Geomorphology for River Engineering and Management*. Wiley, Chichester, pp. 137–172.

Leece, S.A. (1997) Non-linear downstream changes in stream power on Wisconsin's Blue River. *Annals of the Association of American Geographers* **87**, 471–486.

Leedy, J.O., Ashworth, P.J. and Best, J.L. (1993) Mechanisms of anabranch avulsion within gravel-bed braided rivers: observations from a scaled physical model. In: Best, J.L. and Bristow, C.S. (eds.) *Braided Rivers*. Geological Society Special Publication, No. 75, London, pp. 119–127.

Leeks, G.J. (1992) Impacts of plantation forestry on sediment transport processes. In: Billi, P., Hey, R.D., Thorne, C.R. and Tacconi, P. (eds.) *Dynamics of Gravel-Bed Rivers*. Wiley, Chichester, U.K., pp. 651–670.

Leeks, G.J., Lewin, J. and Newson, M.D. (1988) Channel change, fluvial geomorphology and river engineering: the case of the Afon Trannon, Mid-Wales. *Earth Surface Processes and Landforms* **13**, 207–223.

Leopold, L.B. and Bull, W.B. (1979) Base level, aggradation and grade. *Proceedings of the American Philosophical Society* **123**(3), 168–196.

Leopold, L.B. and Maddock Jr., T. (1953) *The hydraulic geometry of stream channels and some physiographic implications*. United States Geological Survey Professional Paper, P 0252, pp. 57.

Leopold, L.B. and Wolman, M.G. (1957) *River channel patterns: Braided, meandering and straight*. United States Geological Survey Professional Paper, P 0282-B, pp. 39–85.

Leopold, L.B., Wolman, M.G. and Miller, J.P. (1964) *Fluvial Processes in Geomorphology*. W.H. Freeman and Co., San Francisco.

Leuven, R.S.E.W. and Poudevigne, I. (2002) Riverine landscape dynamics and ecological risk assessment. *Freshwater Biology* **47**, 845–865.

Lewin, J. (1978) Floodplain geomorphology. *Progress in Physical Geography* **2**, 408–437.

Lewin, J. (1987) *Fans, floodplains and deltas*, International geomorphology 1986. Proceedings of the First international conference on geomorphology. Part I. John Wiley & Sons, Chichester, pp. 973–975.

Lewin, J. and Brindle, B.J. (1977) Confined meanders. In: Gregory, K.J. (ed.) *River Channel Changes*. John Wiley and Sons, Chichester, pp. 221–233.

Lewin, J. and Macklin, M.G. (1987) *Metal mining and floodplain sedimentation in Britain*, International geomorphology 1986. Proceedings of the First inter-national conference on geomorphology. Part I. John Wiley & Sons, Chichester, pp. 1009–1027.

Lewin, J., Bradley, S.B. and Macklin, M.G. (1983) Historical valley alluviation in mid-Wales. *Geographical Journal* **18**, 331–350.

Liebault, F. and Piégay, H. (2002) Causes of 20th century channel narrowing in mountain and piedmont rivers of southeastern France. *Earth Surface Processes and Landforms* **27**(4), 425–444.

Liebauldt, F., Clement, P., Piégay, H. and Landon, N. (1999) Assessment of bedload delivery from tri-butaries: the Drone River case, France. *Arctic, Antarctic, and Alpine Research* **31**(1), 108–117.

Ligon, F.K., Dietrich, W.E. and Trush, W.J. (1995) Downstream ecological effects of dams. *BioScience* **45**, 183–192.

Lines, W.J. (1991) *Taming the Great South Land: A history of the conquest of nature in Australia*. Allen and Unwin, Australia.

Lisle, T.E. (1979) A sorting mechanism for a riffle-pool sequence: Summary. *Geological Society of America Bulletin* **90**, 616–617.

Lisle, T.E., Nelson, J.M., Pitlick, J., Madel, J.A. and Barkett, B.L. (2000) Variability of bed mobility in natu-ral, gravel-bed channels and adjustments to sediment load at local and reach scales. *Water Resources Research* **33**, 1971–1981.

Lisle, T.E., Pizzuto, J.E., Ikeda, H., Iseya, F. and Kodama, Y. (1997) Evolution of a sediment wave in an experi-mental channel. *Water Resources Research* **33**, 1971–1981.

Mackel, R. (1974) Dambos: A study of morphodynamic activity on the plateau regions of Zambia. *Catena* **1**, 327–365.

Mackenzie, J.A., van Coller, A.L. and Rogers, K.H. (1999) *Rule based modelling for management of riparian sys-tems*. Centre for Water in the Environment, University of Witwatersrand, WRC Report No 813/1/99, South Africa, pp. 156.

Mackin, J.H. (1948) Concept of a graded river. *Geological Society of America Bulletin* **5**, 463–511.

Macklin, M.G. and Lewin, J. (1989) Sediment transfer and transformation of an alluvial valley floor: the River South Tyne, Northumbria, U.K. *Earth Surface Processes and Landforms* **14**, 233–246.

Macklin, M.G. and Lewin, J. (1997) Channel, floodplain and drainage basin response to environmental change. In: Thorne, C.R., Hey, R.D. and Newson, M.D. (eds.) *Applied Fluvial Geomorphology for River Engineering and Management*. Wiley, Chichester, pp. 15–45.

Macklin, M.G., Rumsby, B.T. and Heap, T. (1992) Flood alluviation and entrenchment: Holocene valley-floor development and transformation in the British up-lands. *Geological Society America Bulletin* **104**, 631–643.

Maddock, I. (1999) The importance of physical habitat assessment for evaluating river health. *Freshwater Biology* **41**, 373–391.

Madej, M.A. (1995) *Changes in channel-stored sediment, Redwood Creek, northwestern California, 1947 to 1980.* United States Geological Survey Professional Paper, P 1454, pp. O1–O27.

Madej, M.A. and Ozaki, V. (1996) Channel response to sediment wave propagation and movement, Redwood Creek, California, USA. *Earth Surface Processes and Landforms* **21**, 911–927.

Magilligan, F.J. (1992) Thresholds and the spatial variability of flood power during extreme floods. *Geomorphology* **5**, 373–390.

Magilligan, F.J., Phillips, J.D., James, L.A. and Gomez, B. (1998) Geomorphic and sedimentological controls on the effectiveness of an extreme flood. *Journal of Geology* **106**, 87–95.

Malanson, G.P. (1993) *Riparian Landscapes.* Cambridge University Press, Cambridge, UK.

Mance, G., Raven, P.J. and Bramley, M.E. (2002) Integrated river basin management in England and Wales: a policy perspective. *Aquatic Conservation: Marine and Freshwater Ecosystems* **12**, 339–346.

Maroulis, J.C. and Nanson, G.C. (1996) Bedload transport of aggregated muddy alluvium from Cooper Creek, central Australia: A flume study. *Sedimentology* **43**, 771–790.

Marsh, G.P. (1864) *Man and Nature.* Scribner, New York.

Massong, T.M. and Montgomery, D.R. (2000) Influence of sediment supply, lithology, and wood debris on the distribution of bedrock and alluvial channels. *Geological Society of America Bulletin* **112**, 591–599.

McCully, P. (1996) *Silenced Rivers: The Ecology and Politics of Large Dams.* Zed Books, New Jersey, U.S.

McDowell, P.F. (2001) Spatial variations in channel morphology at segment and reach scales, Middle Fork John Day River, Northeastern Oregon. In: Dorava, J.M., Montgomery, D.R., Palcsak, B.B. and Fitzpatrick, F.A. (eds.) *Geomorphic Processes and Riverine Habitat.* American Geophysical Union, Washington, DC., pp. 159–172.

McGowen, J.H. and Garner, L.E. (1970) Physiographic features and stratification types of coarse-grained point bars: Modern and ancient examples. *Sedimentology* **14**, 77–111.

McKenney, R. (2001) Channel changes and habitat diversity in a warm-water gravel-bed stream. In: Dorava, J.M., Montgomery, D.R., Palcsak, B.B. and Fitzpatrick, F.A. (eds.) *Geomorphic Processes and Riverine Habitat.* American Geophysical Union, Washington, DC., pp. 57–71.

McMahon, T.A., Finlayson, B.L. Haines, A.T. and Srikanthan, R. (1991) *Global Runoff – Continental*

Comparisons of Annual Flows and Peak Discharges. Catena Paperback, Cremlingen, 166pp.

Melville, M.D. and Erskine, W.D. (1986) Sediment remobilization and storage by discontinuous gullying in humid southeastern Australia *Proceedings of the International Association of Hydrological Sciences Symposium.* IAHS. Publication No. 159, Albuquerque, New Mexico, pp. 227–286.

Miall, A.D. (1985) Architectural-element analysis: A new method of facies analysis applied to fluvial deposits. *Earth Science Reviews* **22**, 261–308.

Middleton, B. (ed.) (1999) *Wetland Restoration, Flood Pulsing and Disturbance Dynamics.* John Wiley & Sons, Chichester, UK.

Millar, R.G. (2000) Influence of bank vegetation on alluvial channel patterns. *Water Resources Research* **36**(4), 1109–1118.

Millar, R.G. and Quick, M.C. (1993) Effect of bank stability on geometry of gravel rivers. *Journal of Hydraulic Engineering* **119**(12), 1343–1363.

Miller, A.J. (1995) Valley morphology and boundary conditions influencing spatial patterns of flood flow. In: Costa, J.E., Miller, A.J., Potter, K.W. and Wilcock, P.R. (eds.) *Natural and Anthropogenic Influences in Fluvial Geomorphology, Geophysical Monograph 89.* American Geophysical Union, Washington DC., pp. 57–81.

Miller, A.J. and Gupta, A. (eds.) (1999) *Varieties of Fluvial Form.* John Wiley and Sons, Chichester, UK.

Miller, J.R. (1991) Development of anastomosing channels in south-central Indiana. *Geomorphology* **4**, 221–229.

Miller, J.R. and Ritter, J.B. (1996) An examination of the Rosgen classification of natural rivers. *Catena* **27**, 295–299.

Miller, S.D., Ritter, D.F., Kochel, R.C. and Miller, J.R. (1993) Fluvial responses to land use changes and climatic variations within the Drury Creek watershed, southern Illinois. *Geomorphology* **6**, 309–329.

Milliman, J.D., Qin, Y., Ren, M. and Saito, Y. (1987) Man's influence on the erosion and transport of sediment by Asian rivers; the Yellow River (Huanghe) example. *Journal of Geology* **95**(6), 751–762.

Milner, A.M. (1994) System Recovery. In: Calow, P. and Petts, G.E. (eds.) *The Rivers Handbook.* Blackwell, Oxford, pp. 76–97

Montgomery, D.R. (1999) Process domains and the river continuum. *Journal of the American Water Resources Association* **35**(2), 397–410.

Montgomery, D.R. (2001) Geomorphology, river ecology, and ecosystem management. In: Dorava, J.M., Montgomery, D.R., Palcsak, B.B. and Fitzpatrick, F.A. (eds.) *Geomorphic Processes and Riverine Habitat.* American Geophysical Union, Washington, DC., pp. 247–253.

Montgomery, D.R. and Buffington, J.M. (1997) Channel-reach morphology in mountain drainage basins. *Geological Society of America Bulletin* **109**(5), 596–611.

Montgomery, D.R. and Buffington, J.M. (1998) Channel processes, classification and response. In: Naiman, R.J. and Bilby, R.E. (eds.) *River Ecology and Management: Lessons from the Pacific Coastal Ecoregion.* Springer-Verlag, New York, pp. 13–42.

Montgomery, D.R. and Piégay, H. (2003) Wood in rivers: interactions with channel morphology and processes. *Geomorphology* **51**(1–3), 1–5.

Montgomery, D.R., Buffington, J.M., Peterson, N.P., Schuett-Hames, D. and Quinn, T.P. (1995a) Stream-bed scour, egg burial depths, and the influence of salmonid spawning on bed surface mobility and embryo survival. *Canadian Journal of Fisheries and Aquatic Science* **53**, 1061–1070.

Montgomery, D.R., Buffington, J.M., Smith, R.D., Schmidt, K.M. and Pess, G. (1995b) Pool spacing in forest channels. *Water Resources Research* **31**, 1097–1105.

Montgomery, D.R., Abbe, T.B., Buffington, J.M., Peterson, N.P., Schmidt, K.M. and Stock, J.D. (1996) Distribution of bedrock and alluvial channels in forested mountain drainage basins. *Nature* **381**, 597–589.

Mosley, M.P. (1987) The classification and characterisation of rivers. In: Richards, K. (ed.) *River Channels: Environment and Process.* Blackwell, Oxford, pp. 295–320.

Murray, A.B. (2003) Contrasting the goals, strategies, and predictions associated with simplified numerical models and detailed simulations. In: Wilcock, P.R. and Iverson, R.M. (eds.) *Prediction in Geomorphology.* American Geophysical Union. Geophysical Monograph 135. Washington, DC., pp. 151–165.

Naiman, R.J. and Bilby, R.E. (eds.) (1998) *River Ecology and Management: Lessons from the Pacific Coastal Ecoregion.* Springer-Verlag, New York.

Naiman, R.J. and Decamps, H. (1997) The ecology of interfaces; riparian zones. *Annual Review of Ecology and Systematics* **28**, 621–658.

Naiman, R.J., Lonzarich, D.G., Beechie, T.J. and Ralph, S.C. (1992) General principles of classification and the assessment of conservation potential in rivers. In: Boon, P.J., Calow, P. and Petts, G.E. (eds.) *River Conservation and Management.* Wiley, Chichester, pp. 93–123.

Nanson, G.C. (1980) Point bar and floodplain formation of the meandering Beatton River, northeastern British Colombia, Canada. *Sedimentology* **27**, 3–29.

Nanson, G.C. (1981) New evidence of scroll bar formation on the Beatton River. *Sedimentology* **28**, 889–891.

Nanson, G.C. (1986) Episodes of vertical accretion and catastrophic stripping: A model of disequilibrium flood-plain development. *Geological Society of America Bulletin* **97**(12), 1467–1475.

Nanson, G.C. and Croke, J.C. (1992) A genetic classification of floodplains. *Geomorphology* **4**, 459–486.

Nanson, G.C. and Hickin, E.J. (1986) A statistical analysis of bank erosion and channel migration in western Canada. *Geological Society of America Bulletin* **97**, 497–504.

Nanson, G.C. and Knighton, A.D. (1996) Anabranching rivers: their cause, character and classification. *Earth Surface Processes and Landforms* **21**, 217–239.

Nanson, G.C. and Young, R.W. (1981) Overbank deposition and floodplain formation on small coastal streams of New South Wales. *Zeitschrift fur Geomorphologie* **25**(3), 332–347.

Nanson, G.C., Barbetti, M. and Taylor, G. (1995) River stabilisation due to changing climate and vegetation during the Late Quaternary in western Tasmania, Australia. *Geomorphology* **13**, 145–158.

Nanson, G.C., Cohen, T.J., Doyle, C.J. and Price, D.M. (2003) Alluvial evidence of major late-Quaternary climate and flow-regime changes on the coastal rivers of New South Wales, Australia. In: Gregory, K.J. and Benito, G. (eds.) *Palaeohydrology: Understanding Global Change.* John Wiley and Sons, Chichester, pp. 233–258.

Nanson, G.C., Young, R.W., Price, D.M. and Rust, B.R. (1988) Stratigraphy, sedimentology and late-Quaternary chronology of the Channel Country of western Queensland. In: Warner, R.F. (ed.) *Fluvial Geomorphology of Australia.* Academic Press, Sydney, pp. 151–175.

Neimi, G.J., DeVore, P., Detenbeck, N., Lima, A., Pastor, J., Yount, J.D. and Naiman, R.J. (1990) Overview of case studies on recovery of aquatic systems from disturbance. *Environmental Management* **14**(5), 571–587.

Newbold, J.D. (1992) Cycles and spirals of nutrients. In: Calow, P. and Petts, G.E. (eds.) *The Rivers Handbook.* Blackwell Scientific Publications, Oxford, pp. 379–408.

Newbury, R.W. and Gaboury, M.N. (1993) *Stream Analysis and Fish Habitat Design: A Field Manual.* Newbury Hydraulics Ltd., British Columbia, Canada.

Newson, M.D. (1992a) *Land, Water and Development: River Basin Systems and their Sustainable Management.* Routledge, Chapman and Hall, London.

Newson, M.D. (1992b) Geomorphic thresholds in gravel-bed rivers: Refinement for an era of environmental change. In: Billi, P., Hey, R.D., Thorne, C.R. and Tacconi, P. (eds.) *Dynamics of Gravel Bed Rivers.* John Wiley and Sons, New York, pp. 3–15.

Newson, M.D. (1997) *Land, Water and Development: Sustainable Management of River Basin Systems.* Routledge, London.

Newson, M.D. and Newson, C.L. (2000) Geomorphology,

ecology and river channel habitat: Mesoscale approaches to basin-scale challenges. *Progress in Physical Geography* 24(2), 195–217.

Newson, M.D., Clark, M.J., Sear, D.A. and Brookes, A. (1998a) The geomorphological basis for classifying rivers. *Aquatic Conservation: Marine and Freshwater Ecosystems* 8, 415–430.

Newson, M.D., Harper, D.M., Padmore, C.L., Kemp, J.L. and Vogel, B. (1998b) A cost-effective approach for linking habitats, flow types and species requirements. *Aquatic Conservation: Marine and Freshwater Ecosystems* 8, 431–446.

Nicholas, A.P., Ashworth, P.J., Kirkby, M.J., Macklin, M.G. and Murray, T. (1995) Sediment slugs: Large-scale fluctuations in fluvial sediment transport rates and storage volumes. *Progress in Physical Geography* 19(4), 500–519.

Nolan, K.M. and Marron, D.C. (1995) *History, causes and significance of changes in the channel geometry of Redwood Creek, northwestern California, 1936 to 1982*. United States Geological Survey Professional Paper, P 1454, pp. N1–N22.

Nordseth, K. (1973) Floodplain construction on a braided river: The islands of Koppangoyene on the River Glomma. *Norsk Geografisk Tidsskrift* 27, 109–126.

Nott, J., Price, D. and Nanson, G.C. (2002) Stream response to Quaternary climate change: Evidence from the Shoalhaven River catchment, southeastern highlands, temperate Australia. *Quaternary Science Reviews* 21, 965–974.

Nott, J., Young, R. and McDougall, I. (1996) Wearing down, wearing back, and gorge extension in the long-term denudation of a highland mass: Quantitative evidence from the Shoalhaven catchment, Southeastern Australia. *Journal of Geology* 104, 224–232.

O'Connor, J.E., Webb, R.H. and Baker, V.R. (1986) Paleohydrology of pool-and-riffle pattern development: Boulder Creek, Utah. *Geological Society of America Bulletin* 97, 410–420.

O'Keeffe, J.H., King, J. and Eekhout, S. (1994) *The characteristics and purpose of river classification*. In: Uys, M.C. (ed.), Classification of rivers, and environmental health indicators. Proceedings of a joint South African/Australian workshop. Feb 7–14 1994. Cape Town, South Africa. Water Research Commission Report No. TT 63/94.

Osman, A.M. and Thorne, C.R. (1988) Riverbank stability analysis. I. Theory. *Proceedings of the American Society of Civil Engineers, Journal of Hydraulic Engineering* 114, 134–150.

Osterkamp, W.R. (1978) Gradient, discharge and particle size relations of alluvial channels in Kansas, with observations on braiding. *American Journal of Science* 278, 1253–1268.

Osterkamp, W.R. and Hedman, E.R. (1982) Perennial

streamflow characteristics related to channel geometry and sediment in Missouri River basin. United States Geological Survey Professional Paper, P 1242, pp. 37.

Osterkamp, W.R. and Schumm, S.A. (1996) Geoindicators for river-valley monitoring. In: Berger, A.R. and Iams, W.J. (eds.) *Assessing Rapid Environmental Geoindicators: Changes in Earth Systems*. A.A. Balkema, Rotterdam, The Netherlands, pp. 97–114.

Page, K.J. and Nanson, G.C. (1996) Stratigraphic architecture resulting from Late Quaternary evolution of the Riverine Plain, south-eastern Australia. *Sedimentology* 43, 927–945.

Paine, D.M. (1985) 'Ergodic' reasoning in geomorphology: Time for a review of the term? *Progress in Physical Geography* 9(1), 1–15.

Palmer, M.A., Ambrose, R.F. and Poff, N.L. (1997) Ecological theory and community restoration ecology. *Restoration Ecology* 5, 291–300.

Park, C.C. (1977) Man-induced changes in stream channel capacity. In: Gregory, K.J. (ed.) *River Channel Changes*. Wiley, Chichester, UK, pp. 121–144.

Park, C.C. (1981) Man, river systems and environmental impacts. *Progress in Physical Geography* 5, 1–31.

Parker, G. (1979) Hydraulic geometry of active gravel rivers. *Journal of the Hydraulics Division, Proceedings of the American Society of Civil Engineers* 105, 1185–1201.

Parsons, H. and Gilvear, D.J. (2002) Valley floor landscape change following almost 100 years of flood embankment abandonment on a wandering gravel-bed river. *River Research and Applications* 18(5), 461–479.

Parsons, M., Thoms, M.C. and Norris, R. (2003) Scales of macroinvertebrate distribution in relation to the hierarchical organization of river systems. *Journal of the North American Benthological Society* 22, 105–122.

Parsons, M., Thoms, M.C. and Norris, R. (2004) Using hierarchy scales to select scales of measurement in multiscale studies of stream macroinvertebrate assemblages. *Journal of the North American Benthological Society* 23(2), 157–170.

Patton, P.C. and Schumm, S.A. (1975) Gully erosion, northwestern Colorado. *Geology* 3, 88–90.

Peiry, J.L. (1987) Channel degradation in the Middle Arve River, France. *Regulated Rivers* 1, 100–188.

Petts, G.E. (1979) Complex response of river channel morphology subsequent to reservoir construction. *Progress in Physical Geography* 3, 329–362.

Petts, G.E. (1984) *Impounded Rivers: Perspectives for Ecological Management*. Wiley, Chichester.

Petts, G.E. (1989) Perspectives for ecological management of regulated rivers. In: Gore, J.A. and Petts, G.E. (eds.) *Alternatives in Regulated River Management*. CRC Press, USA, pp. 3–24.

Petts, G.E. (1996) Sustaining the ecological integrity of large floodplain rivers. In: Anderson, M.G., Walling, D.E. and Bates, P.D. (eds.) *Floodplain Processes*. John Wiley & Sons Ltd, Chichester, UK, pp. 535–551.

Petts, G.E. and Amoros, C. (eds.) (1996) *Fluvial Hydrosystems*. Chapman & Hall, London.

Phillips, J.D. (1992) Nonlinear dynamical systems in geomorphology: Revolution or evolution? *Geomorphology* **5**, 219–229.

Phillips, J.D. (2003) Sources of nonlinearity and complexity in geomorphic systems. *Progress in Physical Geography* **27**(1), 1–23.

Pickett, S.T.A. and White, P.S. (eds.) (1985) *The Ecology of Natural Disturbance and Patch Dynamics*. Academic Press, San Diego.

Pickup, G. (1984) Geomorphology of tropical rivers; I, Landforms, hydrology and sedimentation in the Fly and Lower Purari, Papua New Guinea. *Catena Supplement* **5**, 1–17.

Piégay, H. and Schumm, S.A. (2003) Systems approaches in fluvial geomorphology. In: Kondolf, G.M. and Piégay, H. (eds.) *Tools in Fluvial Geomorphology*. John Wiley and Sons, Ltd, Chichester, UK, pp. 103–132.

Piégay, H., Bornette, G., Citterio, A., Herouin, E. and Moulin, B. (2000) Channel instability as a control on silting dynamics and vegetation patterns within the perifluvial aquatic zones. *Hydrological Processes* **14**(16–17), 3011–3029.

Pitlick, J. and Wilcock, P. (2001) Relations between streamflow, sediment transport, and aquatic habitat in regulated rivers. In: Dorava, J.M., Montgomery, D.R., Palcsak, B.B. and Fitzpatrick, F.A. (eds.) *Geomorphic Processes and Riverine Habitat*. American Geophysical Union, Washington, DC., pp. 185–198.

Pitlick, J. and van Streeter, M.M. (1998) Geomorphology and endangered fish habitats of the Upper Colorado River 2: Linking sediment transport to habitat maintenance. *Water Resources Research* **34**, 303–316.

Pizzuto, J.E. (1984) Equilibrium bank geometry and the width of shallow sand bed streams. *Earth Surface Processes and Landforms* **9**, 199–207.

Pizzuto, J.E. (1987) Sediment diffusion during overbank flows. *Sedimentology* **34**, 301–317.

Pizzuto, J.E. and Meckelnburg, T.S. (1989) Evaluation of a linear bank erosion equation. *Water Resources Research* **25**, 1005–1013.

Poff, N.L. and Allan, J.D. (1995) Functional organization of stream fish assemblages ain relation to hydrological variability. *Ecology* **76**, 606–627.

Poff, N.L. and Ward, J.V. (1989) Implications of streamflow variability and predictability for lotic community structure: a regional analysis of streamflow patterns. *Canadian Journal of Fisheries and Aquatic Science* **46**, 1805–1818.

Poff, N.L., Allan, J.D., Bain, M.B., Karr, J.R., Prestegaard, K.L., Richter, B.D., Sparks, R.E. and Stromberg, J.C. (1997) The natural flow regime: A paradigm for river conservation and management. *BioScience*, **47**(11), 769–784.

Pollack, H.N. (2003) *Uncertain Science . . . Uncertain World*. Cambridge University Press, UK.

Poole, G.C. (2002) Fluvial landscape ecology: addressing uniqueness within the river discontinuum. *Freshwater Biology* **47**, 641–660.

Postel, S. and Richter, B.D. (2003) *Rivers For Life: Managing Water for People and Nature*. Island Press, Washington DC.

Potter, K.W. and Gaffield, S.J. (2001) Watershed assessment with synoptic base-flow surveys. In: Dorava, J.M., Montgomery, D.R., Palcsak, B.B. and Fitzpatrick, F.A. (eds.) *Geomorphic Processes and Riverine Habitat*. American Geophysical Union, Washington, DC., pp. 19–25.

Power, M.E., Sun, A., Parker, G., Dietrich, W.E. and Wootton, J.T. (1995) Hydraulic food-chain models. *BioScience* **45**, 159–167.

Prosser, I.P. and Winchester, S.J. (1996) History and process of gully initiation and development in eastern Australia. *Zeitschrift fur Geomorphologie. Suppl Bnd.* **105**, 91–109.

Prosser, I.P., Chappell, J. and Gillespie, R. (1994) Holocene valley aggradation and gully erosion in headwater catchments, south-eastern highlands of Australia. *Earth Surface Processes and Landforms* **19**, 465–480.

Puckridge, J.T., Sheldon, F., Walker, K.F. and Boulton, A.J. (1998) Flow variability and the ecology of large rivers. *Marine and Freshwater Research* **49**, 55–72.

Rabeni, C.F. and Jacobson, R.B. (1993) The importance of fluvial hydraulics to fish habitat restoration in low gradient streams. *Freshwater Biology* **29**, 211–220.

Raven, P.J., Boon, P.J., Dawson, F.H. and Ferguson, A.J.D. (1998) Towards an integrated approach to classifying and evaluating rivers in the UK. *Aquatic Conservation: Marine and Freshwater Ecosystems* **8**, 383–393.

Reid, I., Frostick, L.E. and Brayshaw, A.C. (1992) Microform roughness elements and the selective entrainment and entrapment of particles in gravel-bed rivers. In: Billi, P., Hey, R.D., Thorne, C.R. and Tacconi, P. (eds.) *Dynamics of Gravel-bed Rivers*. Wiley, Chichester, pp. 251–272.

Reinfelds, I.V. and Nanson, G.C. (1993) Formation of braided river floodplains, Waimakariri River, New Zealand. *Sedimentology* **40**(6), 1113–1127.

Reinfelds, I.V., Cohen, T.J., Batten, P. and Brierley, G.J. (2004) Assessment of downstream trends in channel gradient, total and specific stream power: A GIS approach. *Geomorphology* **60**(3–4), 403–416.

Reisner, M. (1986) *Cadillac Desert*. Penguin Books, New York, U.S.

Renwick, R.H. (1992) Equilibrium, disequilibrium and nonequilibrium landforms in the landscape. *Geomorphology*, **5**, 265–276.

Rhoads, B.L. and Herrick, E.E. (1996) Naturalization of headwater streams in Illinois: Challenges and Possibilities. In: Brookes, A. and Shields Jr., F.D. (eds.) *River Channel Restoration: Guiding Principles for Sustainable Projects*. John Wiley and Sons, Chichester, pp. 331–367.

Rhoads, B.L., Wilson, D., Urban, M. and Herricks, E.E. (1999) Interaction between scientists and nonscientists in community-based watershed management: Emergence of the concept of stream naturalisation. *Environmental Management* **24**(3), 297–308.

Rice, S. and Church, M. (1996) Bed material texture in low order streams on the Queen Charlotte Islands, British Columbia. *Earth Surface Processes and Landforms* **21**(1), 1–18.

Rice, S.P., Greenwood, M.T. and Joyce, C.B. (2001) Tributaries, sediment sources and the longitudinal organization of macroinvertebrate fauna along river systems. *Canadian Journal of Fisheries and Aquatic Sciences* **58**, 824–840.

Richards, K.S. (1982) *Rivers: Form and Process in Alluvial Channels*. Methuen, London.

Richards, K.S. (1999) The magnitude-frequency concept in fluvial geomorphology: A component of a degenerating research programme? *Zeitshrift fur Geomorphologie, Suppl. Bnd.* **115**, 1–18.

Richards, K.S. and Clifford, N.J. (1991) Fluvial geomorphology – structured beds in gravelly rivers. *Progress in Physical Geography* **15**(4), 407–422.

Richards, K.S. and Lane, S.N. (1997) Prediction of morphological changes in unstable channels. In: Thorne, C.R., Hey, R.D. and Newson, M.D. (eds.) *Applied Fluvial Geomorphology for River Engineering and Management*. Wiley, Chichester, pp. 269–292.

Richards, K., Brasington, J. and Hughes, F. (2002) Geomorphic dynamics of floodplains: ecological implications and a potential modeling strategy. *Freshwater Biology* **47**, 559–579.

Richter, B.D., Baumgartner, J.V., Wigington, R. and Braun, D.P. (1997) How much water does a river need? *Freshwater Biology* **37**, 231–249.

Roberts, C.R. (1989) Flood frequency and urban-induced channel change; some British examples. In: Beven, K. and Carling, P. (eds.) *Floods; Hydrological, Sedimentological and Geomorphological Implications*. John Wiley and Sons, Chichester, pp. 57–82.

Rogers, K. (1998) Managing science/management partnerships; a challenge of adaptive management.

Conservation Ecology **2**(2), URL: http://www.consecol.org/vol2/iss2/resp1.

Rogers, K. (2002) *Adopting workable ecological and operational approaches to biodiversity management: The Kruger National Park Rivers example*, Enviro Flows. 2002. Proceedings of the International Conference on Environmental Flows for River Systems, incorporating the 4th International Ecohydraulics Symposium. Unpublished proceedings, Cape Town.

Rogers, K.H. and Bestbier, R. (1997) *Development of a Protocol for the Definition of the Desired State of Riverine Systems in South Africa*. Department of Environmental Affairs and Tourism, Pretoria, South Africa.

Rogers, K.H. and Biggs, H. (1999) Integrating indicators, endpoints and value systems in strategic management of rivers of the Kruger National Park. *Freshwater Biology* **41**, 439–451.

Rosgen, D.L. (1994) A classification of natural rivers. *Catena* **22**, 169–199.

Rosgen, D.L. (1996) *Applied River Morphology*. Wildland Hydrology, Pagosa Springs, Colorado.

Rutherfurd, I.D. (2000) Some human impacts on Australian stream channel morphology. In: Brizga, S. and Finlayson, B. (eds.) *River management: The Australasian Experience*. John Wiley & Sons, Chicester, pp. 11–49.

Rutherfurd, I. and Bartley, R. (eds.) (1999) *Proceedings of the 2nd Stream Management Conference, Adelaide*, Volumes 1 and 2. Cooperative Research Centre for Catchment Hydrology, Melbourne.

Rutherfurd, I. and Walker, M. (eds.) (1996) *Proceedings of the First National Conference on Stream Management in Australia, Stream Management '96, Merrijig*. Cooperative Research Centre for Catchment Hydrology, Melbourne.

Rutherfurd, I.D., Jerie, K. and Marsh, N. (2000) *A Rehabilitation Manual for Australian Streams*, Volumes 1 and 2. Cooperative Research Centre for Catchment Hydrology, and the Land and Water Resources Research and Development Corporation, Canberra.

Rutherfurd, I., Jerie, K. and Marsh, N. (2001a) Planning for stream rehabilitation: Some help in turning the tide. *Water; Journal of the Australian Water and Wastewater Association* **28**(1), 25–27.

Rutherfurd, I., Jerie, K., Walker, M. and Marsh, N. (1999) Don't raise the titanic: How to set priorities for stream rehabilitation. In: Rutherfurd, I. and Bartley, R. (eds.) *Proceedings of the 2nd Stream Management Conference*. Cooperative Research Centre for Catchment Hydrology, Melbourne, Adelaide, pp. 527–532.

Rutherfurd, I., Sheldon, F., Brierley, G. and Kenyon, C.

(eds.) (2001b) *Third Australian Stream Management Conference Proceedings: the value of Healthy Rivers, 27–29 August, 2001, Brisbane*, Volumes 1 & 2. Cooperative Research Centre for Catchment Hydrology, Melbourne.

Sahin, V. and Hall, M.J. (1996) The effects of afforestation and deforestation on water yields. *Journal of Hydrology* **178**(1), 293–309.

Schumm, S.A. (1960) *The shape of alluvial channels in relation to sediment type*. U.S. Geological Survey Professional Paper, 352-B, pp. 17–30.

Schumm, S.A. (1968) River adjustment to altered hydrologic regimen – Murrumbidgee River and paleochannels, Australia. *United States Geolgical Survey Professional Paper* **598**, 65pp.

Schumm, S.A. (1969) River metamorphosis. *Journal of the Hydraulics Division, Proceedings of the American Society of Civil Engineers* **95**, 255–273.

Schumm, S.A. (1977) *The Fluvial System*. John Wiley and Sons, New York.

Schumm, S.A. (1979) Geomorphic thresholds: the concept and its applications. *Transactions of the Institute of British Geographers* **N54**, 485–515.

Schumm, S.A. (1985) Patterns of alluvial rivers. *Annual Review of Earth and Planetary Sciences* **13**, 5–27.

Schumm, S.A. (1991) *To Interpret the Earth: Ten Ways to Be Wrong*. Cambridge University Press, Cambridge.

Schumm, S.A. and Hadley, R.F. (1957) Arroyos and the semiarid cycle of erosion. *American Journal of Science* **255**, 161–174.

Schumm, S.A. and Khan, H.R. (1972) Experimental study of channel patterns. *Geological Society of America Bulletin* **83**, 1755–1770.

Schumm, S.A. and Lichty, R.W. (1963) *Channel widening and flood-plain construction along Cimarron River in southwestern Kansas*. United States Geological Survey Professional Paper, P 352-D, pp. 71–88.

Schumm, S.A. and Lichty, R.W. (1965) Time, space and causality in geomorphology. *American Journal of Science* **263**, 110–119.

Schumm, S.A. and Thorne, C.R. (1989) Geologic and geomorphic controls on bank erosion. In: Ports, M.A. (ed.) *Hydraulic Engineering, Proceedings of the National Conference on Hydraulic Engineering*. ASCE, New Orleans, pp. 106–111.

Schumm, S.A., Harvey, M.D. and Watson, C.C. (1984) *Incised Channels: Morphology, Dynamics and Control*. Water Resources Publication, Littleton, Colorado.

Schumm, S.A., Mosley, M.P. and Weaver, W.E. (1987) *Experimental Fluvial Geomorphology*. Wiley, New York, U.S.

Sear, D.A. (1996) Sediment transport processes in pool-riffle sequences. *Earth Surface Processes and Landforms* **21**, 241–262.

Sear, D.A. and Newson, M.D. (1993) *Sediment and Gravel Transportation in Rivers, Including the Use of Gravel Traps. Final Report C5/384/2*. National Rivers Authority, Bristol, pp. 50.

Sear, D.A., Newson, M.D. and Brookes, A. (1995) Sediment-related river maintenance: The role of fluvial geomorphology. *Earth Surface Processes and Landforms* **20**, 629–647.

Sedell, J.R., Richey, J.E. and Swanson, F.J. (1989) The river continuum concept: a basis for the expected ecosystem behaviour of very large rivers? *Canadian Special Publication of Fisheries and Aquatic Sciences* **106**, 49–55.

Seidl, M.A., Weissel, J.K. and Pratson, L.F. (1996) The kinematics and pattern of escarpment retreat across the rifted continental margin of SE Australia. *Basin Research* **12**, 301–316.

SERI (2002) *Society for Ecological Restoration International web site*. www.ser.org.

Shields, F.D., Jr. and Abt, S.R. (1989) Sediment deposition in cutoff meander bends and implications for effective management. *Regulated Rivers* **4**(4), 381–396.

Shields, F.D., Jr., Copeland, R.R., Klingeman, P.C., Doyle, M.W. and Simon, A. (2003) Design for stream restoration. *Journal of Hydraulic Engineering* **129**(8), 575–584.

Simon, A. (1989a) A model of channel response in disturbed alluvial channels. *Earth Surface Processes and Landforms* **14**, 11–26.

Simon, A. (1989b) The discharge of sediment in channelized alluvial streams. *Water Resources Bulletin* **25**(6), 1177–1188.

Simon, A. (1992) Energy, time and channel evolution in catastrophically disturbed fluvial systems. *Geomorphology* **5**, 345–372.

Simon, A. (1994) *Gradation processes and channel evolution in modified West Tennessee streams: Process, response and form*. United States Geological Survey Professional Paper, 1470.

Simon, A. (1995) Adjustment and recovery of unstable alluvial channels: Identification and approaches for engineering management. *Earth Surface Processes and Landforms* **20**, 611–628.

Simon, A. and Darby, S.E. (2002) Effectiveness of grade-control in reducing erosion along incised river channels: The case of Hotophia Creek, Mississippi. *Geomorphology* **42**, 229–254.

Simon, A. and Thorne, C.R. (1996) Channel adjustment of an unstable coarse-grained stream: opposing trends of boundary and critical shear stress, and the applicability of extremal hypotheses. *Earth Surface Processes and Landforms* **21**, 155–180.

Simons, D.B. and Richardson, E.V. (1966) *Resistance to Flow in Alluvial Channels*. United States Geological Survey Professional Paper, P 0442-J, pp. J1–J61.

Smith, D.G. (1976) Effect of vegetation on lateral migration of anastomosed channels of a glacial meltwater river. *Geological Society of America Bulletin* **87**, 857–860.

Smith, D.G. (1983) Anastomosed fluvial deposits: Modern examples from Western Canada. *Special Publications of the Association of Sedimentologists* **6**, 155–168.

Smith, D.G. (1986) Anastomosing river deposits, sedimentation rates and basin subsidence, Magdalena River, northwestern Colombia, South America. *Sedimentary Geology* **46**, 177–196.

Smith, D.G. and Smith, N.D. (1980) Sedimentation in anastomosed river systems: examples from alluvial valleys near Banff, Alberta. *Journal of Sedimentary Petrology* **50**(1), 157–64.

Smith, D.I. (1998) *Water in Australia: Resources and Management*. Oxford University Press, Melbourne.

Smith, N.D. (1970) The braided stream depositional environment: Comparison of the Platte River with some Silurian clastic rocks, north central Appalachians. *Geological Society of America Bulletin* **81**, 2993–3014.

Smith, N.D. (1974) Sedimentology and bar formation in the Upper Kicking Horse River, a braided outwash stream. *Journal of Geology* **82**, 205–223.

Smith, N.D. (1978) Some comments on terminology for bars in shallow rivers. In: Miall, A.D. (ed.). *Fluvial Sedimentology*. Canadian Society of Petroleum Geologists, Memoir 5, Calgary, pp. 85–88.

Smith, N.D., Cross, T.A., Dufficy, J.P. and Clough, S.R. (1989) Anatomy of an avulsion. *Sedimentology* **36**, 1–23.

Southwood, T.R.E. (1977) Habitat, the templet for ecological strategies. *Journal of Animal Ecology* **46**, 337–365.

Sparks, R.E. (1995) Need for ecosystem management of large rivers and their floodplains. *BioScience* **45**(3), 168–182.

Stanford, J.A. and Ward, J.V. (1993) An ecosystem perspective of alluvial rivers: connectivity and the hyporheic corridor. *Journal of the North American Benthological Society* **12**, 48–60.

Stanford, J.A. and Ward, J.V. (1996) Management of aquatic resources in large catchments: Recognising interactions between ecosystem connectivity and environmental disturbance. In: Naiman, R.J. (ed.) *Watershed Management, Balancing Sustainability and Environmental Change*. Springer-Verlag, New York, pp. 91–124.

Stanford, J.A., Hauer, F.R. and Ward, J.V. (1988) Serial discontinuity in a large river system. *Verhandlungen der Internationalen Vereinigung fur Theoretische und Angewandte Limnologie* **23**, 1114–1118.

Stanford, J.A., Ward, J.V., Liss, W.J., Frissell, C.A., Williams, R.N. Lichatowich, J.A. and Coutant, C.C.

(1996) A general protocol for restoration of regulated rivers. *Regulated Rivers: Research and Management* **12**, 391–413.

Stanford, J.A. and Ward, J.V. (1988) The hyporheic habitat of river ecosystems. *Nature* **335**, 64–66.

Starkel, L. (1987) Man as a cause of seidmentologic changes in the Holocene. *Striae* **26**, 5–12.

Starkel, L. (1991a) The Vistula River valley: A case study for Central Europe. In: Starkel, L., Gregory, K.J. and Thornes, J.B. (eds.) *Temperate Palaeohydrology: Fluvial Processes in the Temperate Zone During the Last 15 000 Years*. John Wiley and Sons, Chichester, pp. 171–188.

Starkel, L. (1991b) Long-distance correlation of fluvial events in the temperate zone. In: Starkel, L., Gregory, K.J. and Thornes, J.B. (eds.) *Temperate Palaeohydrology: Fluvial Processes in the Temperate Zone During the Last 15 000 Years*. John Wiley and Sons, Chichester, pp. 473–495.

Steiger, J., Gurnell, A.M. and Petts, G.E. (2001) Sediment deposition along the channel margins of a reach of the middle River Severn, U.K. *Regulated Rivers: Research and Management* **17**, 443–460.

Stevens, M.A., Simons, D.B. and Richardson, E.V. (1975) Nonequilibrium river form. *Journal of the Hydraulics Division, Proceedings of the American Society of Civil Engineers* **101**(HY5), 557–566.

Stewart, J.H. and LaMarche, V.C. (1967) Erosion and deposition produced by floods of December 1964 on Coffee Creek, Trinity County, California. United States Geological Survey Professional Paper, P 0422-K, pp. K1–K22.

Sundborg, A. (1956) The river Klaralven: A study of fluvial processes. *Geografiska Annaler* **xxxviii**, 128–314.

Takahashi, T., Ashida, K. and Sawai, K. (1981) *Delineation of debris flow hazard areas*. In: Davies, T.R.H. and Pearce, A.J. (eds.), Erosion and Sediment Transport in Pacific Rim Steeplands. International Association of Hydrological Sciences Publication, No. 132, pp. 589–603.

Thomas, M.F. (2001) Landscape sensitivity in time and space: An introduction. *Catena* **42**, 83–98.

Thomas, W.L., Jr. (1956) *Man's Role in Changing the Face of the Earth*, Vol 1. University of Chicago Press, Chicago, US.

Thomson, J., Taylor, M.P. and Brierley, G.J. (2004) Are River Styles ecologically meaningful? A test of the ecological significance of a geomorphic river characterization scheme. *Aquatic Conservation: Marine and Freshwater Ecosystems* **14**, 25–48.

Thomson, J., Taylor, M.P., Fryirs, K.A. and Brierley, G.J. (2001) A geomorphological framework for river characterisation and habitat assessment. *Aquatic Conservation: Marine and Freshwater Ecosystems* **11**, 373–389.

Thorne, C.R. (1982) Processes and mechanisms of river bank erosion. In: Hey, R.D., Bathurst, J.C. and Thorne, C.R. (eds.) *Gravel-bed Rivers: Fluvial Processes, Engineering and Management.* John Wiley & Sons, Chichester, pp. 227–271.

Thorne, C.R. (1997) Channel types and morphological classification. In: Thorne, C.R., Hey, R.D. and Newson, M.D. (eds.) *Applied Fluvial Geomorphology for River Engineering and Management.* Wiley, Chichester, UK., pp. 175–222.

Thorne, C.R. (1999) *Stream Reconnaissance Handbook: Geomorphological Investigation and Analysis of River Channels.* John Wiley and Sons, Chichester.

Thorne, C.R. and Lewin, J. (1979) Bank processes, bed material movement and planform development in a meandering river. In: Rhodes, D.D. and Williams, G.P. (eds.) *Adjustments of the Fluvial System.* Kendall/Hunt, Dubuque, Iowa, pp. 117–137.

Thorne, C.R. and Osman, A.M. (1988) Riverbank stability analysis. II. Applications. *Proceedings of the American Society of Civil Engineers, Journal of Hydraulic Engineering* **114**, 151–171.

Thorne, C.R. and Tovey, N.K. (1981) Stability of composite river banks. *Earth Surface Processes and Landforms* **6**, 469–484.

Tinkler, K.J. and Wohl, E.E. (eds.) (1998) *Rivers over rock: Fluvial processes in bedrock channels.* Geophysical Monograph Series 107. American Geophysical Union, Washington DC.

Tippett, J. (2001) *Integrated Catchment Management and Planning for Sustainability – The Case of the Mersey Basin Campaign,* University of Manchester, Manchester.

Tockner, K., Malard, F. and Ward, J.V. (2000) An extension of the Flood Pulse Concept. *Hydrological Processes* **14**, 2861–2883.

Tockner, K., Schiemer, F. and Ward, J.V. (1998) Conservation by restoration: the management concept for river floodplain system on the Danube River in Austria. *Aquatic Conservation: Marine and Freshwater Ecosystems* **8**, 71–86.

Tockner, K., Ward, J.V., Edwards, P.J. and Kollmann, J. (2002) Riverine landscapes: an introduction. *Freshwater Biology* **47**, 497–500.

Tockner, K., Schiemer, F., Baumgartner, C., Kum, G., Weigand, E., Zweimuller, I. and Ware, J.V. (1999) The Danube restoration project: species diversity patterns across connectivity gradients in the floodplain system. *Regulated Rivers: Research and Management* **15**, 245–258.

Tooth, S. (1999) Floodouts in Central Australia. In: Miller, A. and Gupta, A. (eds.) *Varieties of Fluvial Form.* John Wiley and Sons, Chichester, UK., pp. 219–247.

Tooth, S., McCarthy, T.S., Brandt, D., Hancox, P.J. and Morris, R. (2002) Geological controls on the formation of alluvial meanders and floodplain wetlands; the example of the Klip River, eastern Free State, South Africa. *Earth Surface Processes and Landforms* **27**(8), 797–815.

Toth, L.A. (1996) Restoring the hydrogeomorphology of the channelised Kissimmee River. In: Brookes, A. and Shields Jr., F.D. (eds.) *River Channel Restoration: Guiding Principles for Sustainable Projects.* John Wiley and Sons, Chichester, UK., pp. 369–383.

Townsend, C.R. (1996) Concepts in river ecology: pattern and process in the catchment hierarchy. *Archiv fur Hydrobiologie. Suppl.* **113**, 3–21.

Trimble, S.W. (1974) *Man-induced soil erosion on the southern Piedmont, 1700–1970.* Soil Conservation Society of America, Ankeny, Iowa, pp. 69–126.

Trimble, S.W. (1977) The fallacy of stream equilibrium in contemporary denudation studies. *American Journal of Science* **277**, 876–887.

Trimble, S.W. (1983) A sediment budget for Coon Creek basin in the Driftless Area, Wisconsin, 1853–1977. *American Journal of Science* **283**(5), 454–474.

Triska, F.J. (1984) Role of wood debris in modifying channel morphology and riparian areas of a large lowland river under pristine conditions: a historical case study. *Verhandlungen der Internationalen Vereingung fur Theoretische und Angewandte Limnologie* **22**, 1876–1892.

Trofimov, A.M. and Phillips, J.D. (1992) Theoretical and methodological premises of geomorphological forecasting. *Geomorphology* **5**, 203–211.

Twidale, C.R. (1964) Erosion of an alluvial bank at Birdwood, South Australia. *Zeitschrift fur Geomorphologie* **8**, 189–211.

Uys, M.C. and O'Keeffe, J.H. (1997) Simple words and fuzzy zones: Early directions for temporary river research in South Africa. *Environmental Management* **2**, 517–531.

van Niekerk, A.W., Heritage, G.L., Broadhurst, L.W. and Moon, B.P. (1999) Bedrock anastomosing channel systems: Morphology and dynamics of the Sabie River, Mpumulanga Province, South Africa. In: Miller, A.J. and Gupta, A. (eds.) *Varieties of Fluvial Form.* John Wiley and Sons, Chichester, pp. 33–51.

van Streeter, M.M. and Pitlick, J. (1998) Geomorphology and endangered fish habitats of the Upper Colorado River 1: Historic changes in streamflow, sediment load and channel morphology. *Water Resources Research* **34**, 287–302.

Vannote, R.L., Minshall, G.W., Cummins, K.W., Sedell, J.R. and Cushing, C.E. (1980) The river continuum concept. *Canadian Journal of Fisheries and Aquatic Sciences* **37**, 130–137.

Vita-Finzi, C. (1969) *The Mediterranean Valleys.* Cambridge University Press, Cambridge.

Wadeson, R.A. and Rowntree, K.M. (1998) Application of the hydraulic biotope concept to the classification of instream habitats. *Aquatic Ecosystem Health and Management* **1**, 143–157.

Walker, M. and Rutherfurd, I. (1999) An approach to predicting rates of bend migration in meandering alluvial streams. In: Rutherfurd, I. and Bartley, R. (eds.) *Proceedings of the Second Australian Stream Management Conference, Adelaide.* Cooperative Research Centre for Catchment Hydrology, Melbourne, pp. 659–665.

Walker, R.G. (1990) Facies modelling and sequence stratigraphy. *Journal of Sedimentary Petrology* **60**, 777–780.

Ward, J.V. (1989) The four-dimensional nature of lotic ecosystems. *Journal of the North American Benthological Society* **8**, 2–8.

Ward, J.V. (1998) Riverine landscapes: biodiversity patterns, disturbance regimes and aquatic conservation. *Biological Conservation* **83**, 269–227.

Ward, J.V. and Stanford, J.A. (1983) Serial discontinuity concept of lotic ecosystems. In: Fontaine, T.D. and Bartell, S.M. (eds.) *Dynamics of Lotic Systems.* Ann Arbor Science, Ann Arbor, pp. 29–42.

Ward, J.V. and Stanford, J.A. (1995a) The serial discontinuity concept: extending the model to floodplain rivers. *Regulated Rivers: Research and Management* **10**, 159–168.

Ward, J.V. and Stanford, J.A. (1995b) Ecological connectivity in alluvial river ecosystems and its disruption by flow regulation. *Regulated Rivers: Research and Management* **11**, 105–119.

Ward, J.V., Tockner, K. and Schiemer, F. (1999) Biodiversity of floodplain river ecosystems: ecotones and connectivity. *Regulated Rivers: Research and Management* **15**, 125–139.

Ward, J.V., Tockner, K., Arscott, D.B. and Claret, C. (2002) Riverine landscape diversity. *Freshwater Biology* **47**, 517–539.

Ward, J.V., Tockner, K., Uehlinger, U. and Malard, F. (2001) Understanding natural patterns and processes in river corridors as the basis for effective river restoration. *Regulated Rivers: Research and Management* **17**, 311–323.

Warner, R.F. (1992) Floodplain evolution in a New South Wales coastal valley, Australia: Spatial process variations. *Geomorphology* **4**, 447–458.

Warner, R.F. (1997) Floodplain stripping: another form of adjustment to secular hydrologic regime in Southeast Australia. *Catena* **30**, 263–282.

Wende, R. (1999) Boulder bedforms in jointed-bedrock channels. In: Miller, A.J. and Gupta, A. (eds.) *Varieties of Fluvial Form.* Wiley, Chichester, U.K., pp. 189–210.

Wende, R. and Nanson, G.C. (1998) Anabranching rivers.

Ridge-form alluvial channels in tropical northern Australia. *Geomorphology* **22**, 205–224.

Werner, B.T. (2003) Modeling landforms as self-organized, hierarchical dynamical systems. In: Wilcock, P.R. and Iverson, R.M. (eds.) *Prediction in Geomorphology.* Geophysical monograph 135. American Geophysical Union, Washington, DC., pp. 133–150.

Werritty, A. and Leys, K.F. (2001) The sensitivity of Scottish rivers and upland valley floors to recent environmental change. *Catena* **42**, 251–273.

Wesche, T.A. (1985) Stream channel modifications and reclamation structures to enhance fish habitat. In: Gore, J.A. (ed.) *The Restoration of Rivers and Streams.* Butterworth, Boston, pp. 103–163.

Whiting, P.J. (2002) Streamflow necessary for environmental maintenance. *Annual Reviews in Earth and Planetary Sciences* **30**, 181–206.

Whittaker, J.G. (1987) Sediment transport in step-pool streams. In: Thorne, C.R., Bathurst, J.C. and Hey, R.D. (eds). *Sediment Transport in Gravel-Bed Rivers.* Wiley, Chichester, pp. 545–579.

Whittaker, J.G. and Jaeggi, M.N.R. (1982) Origin of step-pool systems in mountain streams. *Journal of the Hydraulics Division American Society of Civil Engineers* **198**, 758–773.

Wiens, J.A. (2002) Riverine landscapes: taking landscape ecology into the water. *Freshwater Biology* **47**, 501–515.

Wilcock, P. (1997) Friction between science and practice: The case of river restoration. *Eos* **78**(41), 154.

Wilcock, P.R. and Iverson, R.M. (eds.) (2003) *Prediction in Geomorphology.* Geophysical monograph 135. American Geophysical Union, Washington, DC., 256pp.

Williams, G.P. and Wolman, M.G. (1984) *Downstream Effects of Dams on Alluvial Rivers.* United States Geological Survey Professional Paper, P 1286, pp. 83.

Williams, M. (2000) Dark ages and dark areas: Global deforestation in the deep past. *Journal of Historical Geography* **26**, 28–46.

Williams, P.B. (2001) *River Engineering Versus River Restoration,* ASCE Wetlands Engineering & River Restoration Conference, Reno, Nevada.

Williams, P.F. and Rust, B.R. (1969) The sedimentology of a braided river. *Journal of Sedimentary Petrology* **39**, 649–679.

Winkley, B.R. (1982) Response of the lower Mississippi to river training and realignment. In: Hey, R.D., Bathurst, J.C. and Thorne, C.R. (eds.) *Gravel-bed Rivers: Fluvial Processes, Engineering and Management.* John Wiley & Sons, Chichester, pp. 659–681.

Wittfogel, K.A. (1956) The hydraulic civilisations. In: Thomas Jr., W.L. (ed.) *Man's Role in Changing the Face of the Earth.* University of Chicago Press, Chicago, pp. 152–164.

Wohl, E.E. (1992) Bedrock benches and boulder bars: Floods in the Burdekin Gorge of Australia. *Geological Society of America Bulletin* **104**, 770–778.

Wohl, E.E. (1998) Bedrock channel morphology in relation to erosional processes. In: Tinkler, K.J. and Wohl, E.E. (eds.) *Rivers Over Rock: Fluvial Processes in Bedrock Channels. Geophysical Monograph Series 107*. American Gephysical Union, Washington, pp. 133–152.

Wohl, E.E. (2000) *Mountain Rivers*. American Geophysical Union Water Resources Monograph 14, Washington, D.C.

Wolman, M.G. (1959) Factors influencing the erosion of cohesive river banks. *American Journal of Science* **257**, 204–216.

Wolman, M.G. (1967) A cycle of sedimentation and erosion in urban river channels. *Geografiska Annaler* **49A**, 385–395.

Wolman, M.G. and Gerson, R. (1978) Relative scales of time and effectiveness of climate in watershed geomorphology. *Earth Surface Processes and Landforms* **3**, 189–208.

Wolman, M.G. and Miller, J.P. (1960) Magnitude-frequency of forces in geomorphic processes. *Journal of Geology* **68**, 54–74.

Wolman, M.G. and Schick, A.P. (1967) Effects of construction on fluvial sediment, urban and suburban areas of Maryland. *Water Resources Research* **3**(2), 451–464.

Woltemade, C.J. and Potter, K.W. (1994) A watershed modelling analysis of fluvial geomorphic influences on flood peak attenuation. *Water Resources Research* **30**, 1933–1942.

Woodroffe, C.D., Mulrennan, M.E. and Chappell, J. (1993) Estuarine infill and coastal progradation, southern van Diemen Gulf, northern Australia. *Sedimentary Geology* **83**, 257–275.

Woodroffe, C.D., Chappell, J., Thom, B.G. and Wallensky, E. (1989) Depositional mode of macrotidal estuary and floodplain, South Alligator River, Northern Australia. *Sedimentology* **36**, 737–756.

Woodward, G. and Hildrew, A.G. (2002) Food web structure in riverine landscapes. *Freshwater Biology* **47**, 777–798.

Wooten, J.T., Parker, M.S. and Power, M.E. (1996) Effects of disturbance on river food webs. *Science* **273**, 1558–1561.

Wu, J. and Loucks, O.L. (1995) From balance of nature to hierarchical patch dynamics: A paradigm shift in ecology. *The Quarterly Review of Biology* **70**(4), 439–466.

Wyzga, B. (1993) River response to channel regulation; case study of the Raba River, Carpathians, Poland. *Earth Surface Processes and Landforms* **18**(6), 541–556.

Wyzga, B. (1996) Changes in the magnitude and transformation of flood waves subsequent to the channelization of the Raba River, Polish Carpathians. *Earth Surface Processes and Landforms* **21**(8), 749–763.

Xu, J. (1990) Complex response in adjustment of the Weihe River channel to the construction of the Sanmenxia Reservoir. *Zeitschrift für Geomorphologie, N. F.* **34**(2), 233–245.

Xu, J. (1996) Channel pattern change downstream from a reservoir: An example of wandering braided rivers. *Geomorphology* **15**, 147–158.

Young, A.R.M. (1986) Quaternary sedimentation on the Woronora Plateau and its implications for climate change. *Australian Geographer* **17**, 1–5.

Zielinski, T. (2003) Catastrophic flood effects in alpine/foothill fluvial system (a case study from the Sudetes Mts, SW Poland). *Geomorphology* **54**, 293–306.

Zimmerman, A. and Church, M. (2001) Channel morphology, gradient profiles and bed stress during flood in a step-pool channel. *Geomorphology* **40**, 311–327.

Zimmerman, R.C., Goodlet, J.C. and Comer, G.H. (1967) The influence of vegetation on channel form of small streams. *Symposium on River Morphology*. IAHS/AISH, Publication No. 75, Bern, pp. 255–275.

Index

Printed and bound in the UK by
CPI Antony Rowe, Eastbourne